I0031597

RAILWAY PLANNING, MANAGEMENT, AND ENGINEERING

In a rapidly changing world, with increasing competition in all sectors of transportation, railways are currently restructuring their planning, management, and technology. As commercial and pricing policies change and new methods of organization are introduced, a more entrepreneurial spirit is required. At the same time, new high-speed tracks are being constructed and old tracks are being renewed, magnetic levitation trains are in operation, hyperloop systems are being planned, high-comfort rolling stock vehicles are being introduced, logistics and combined transport are being developed. Awareness of environmental issues and the search for greater safety attribute a new role to the railways within the transportation system. Meanwhile, methods of analysis have evolved significantly, principally due to computer applications, the internet revolution, satellite technologies, and artificial intelligence, all of which offer new ways of thinking about and addressing old problems.

Railway Planning, Management, and Engineering aims to fulfill the need for a new scientific approach for railways. It is intended to be of use to railway planners, managers, economists, engineers and students in engineering, transportation, economics, and management. The book is divided into three parts, which deal successively with planning, management, track, rolling stock, safety, and the environment.

V.A. Profillidis is Professor of Transportation Engineering at Democritus Thrace University, Xanthi, Greece. He has acted as a consultant on many railway projects, with 30 years of research, teaching, and professional experience in Railway Planning, Management, and Engineering. He has published 10 books and over 200 papers in scientific journals and conference proceedings. He has served as advisor to the Greek Ministry of Transport, the Greek Railways, and the Greek Ministry of Public Works. He has also been a research associate in the Research Department of the International Union of Railways (UIC) and in the French Railways (SNCF). He holds a Diploma in Civil Engineering, a Diploma in Law, both from Aristotle University of Thessaloniki, a D.E.A. and a Ph.D. from Ecole Nationale des Ponts et Chaussées of Paris, France.

To the memory of my father Aristide

RAILWAY PLANNING, MANAGEMENT, AND ENGINEERING

Fifth Edition

V.A. Profillidis

Section of Transportation,
Democritus Thrace University, Greece

Routledge
Taylor & Francis Group
www.routledge.com

First published 2022
by Routledge
4 Park Square, Milton Park, Abingdon, Oxon OX14 4RN

and by Routledge
605 Third Avenue, New York, NY 10158

Routledge is an imprint of the Taylor & Francis Group, an informa business

© 2022 Vassilios Profillidis

The right of Vassilios Profillidis to be identified as author of this work has been asserted in accordance with sections 77 and 78 of the Copyright, Designs and Patents Act 1988.

All rights reserved. No part of this book may be reprinted or reproduced or utilised in any form or by any electronic, mechanical, or other means, now known or hereafter invented, including photocopying and recording, or in any information storage or retrieval system, without permission in writing from the publishers.

Trademark notice: Product or corporate names may be trademarks or registered trademarks, and are used only for identification and explanation without intent to infringe.

British Library Cataloguing-in-Publication Data
A catalogue record for this book is available from the British Library

Library of Congress Cataloging-in-Publication Data
A catalog record has been requested for this book

ISBN: 9780367350116 (hbk)
ISBN: 9781032341699 (pbk)
ISBN: 9780429329302 (ebk)

DOI: 10.4324/9780429329302

Contents

Preface

Railways emerged as a revolutionary transport technology two centuries ago. Until that time, the fastest way of land transport was achieved with the help of carriages driven by horses, which provided average speeds not exceeding 10 km/h and traveling distances of no more than 50 km (without stopping for rest of horses). In the beginning of the 19th century, the first industrial revolution permitted the massive production of steel, coal, and steam engines at reasonable costs. Pioneer engineers transformed these spectacular technological achievements into an impressive new transport mode, the railway: a steam engine hauls a number of vehicles and the train that is formed moves on two steel rails which are positioned above sleepers placed on the ground. Steam railways increased after 1830 average speeds that were achieved until that time by horse carriages from less than 10 km/h to more than 20 km/h, with maximum speeds of 100 km/h, and allowed to travel hundreds of kilometers without the physical or technical need to stop at regular intervals.

Construction of tracks became in the 19th century a priority for citizens, governments, and companies; thus, steam railways expanded rapidly and transformed the shape of urban, economic, social, and even military activities. People could travel many hundreds of kilometers and goods could reach markets thousands of kilometers away from their production site. Railways satisfied the human need for communication, personal, family, and social contacts, economic development and growth, and for more than one century (1830s ÷ 1950s) they monopolized the market of land transport, a period often characterized as the golden era of railways.

Owing to the massive production of cars during the 1930s in the USA and the 1950s ÷ 1960s in Europe, railways started facing a new reality, the era of competition, and began losing short- and medium-distance traffic, which was gradually shifting to road transport. The competitive environment became even harsher after 1960, with competition for medium- and long-distance routes coming also from the emergence of air transport.

Thus, since the 1960s it has become apparent that railways, though being a technology-oriented activity, ought to adopt modern methods of efficiency and of increase of productivity in order to survive in a competitive environment, an evolution that resulted in introducing planning and management as essential components of railway science.

Railways exploited, adopted, and adapted all technological achievements during the last two centuries: electrification (since 1900), diesel traction (since 1930), light signaling (since 1950) (to increase safety and capacity of track), the internet revolution (since 2000) (electronic tickets and information), satellite technologies (since 2010) (for the monitoring of the movement of any rail vehicle), and artificial intelligence (since 2020) (for comparing and deciding over complex situations). In addition, since the 1990s human societies and most governments have started to realize the need to combat climate change, reduce CO_2 emissions, and ensure sustainability for future generations.

In doing so, railways have been struggling for six decades (1960÷2020) to restructure their technology and management in a rapidly changing world, so as to confront competition, both with other transport modes but also among many rail operators running on the same track. New high-speed tracks are constructed and old tracks are upgraded; high-comfort rolling stock vehicles are being introduced; logistics and combined transport are being modernized. At the same time, new methods of organization and management are introduced; commercial and tariff policies change radically, a more entrepreneurial spirit characterizes all components and levels of railway activity. Parallel to technological and organizational innovations, awareness of environmental issues, daily highway and airport congestion, and search for greater safety give railways a new role within the transport system. Perhaps the only common characteristic of the railway of the 2020s with the railway of the 1830s is that both run on two rails, but with totally different technological and managerial characteristics.

During the last four decades, methods of analysis have significantly evolved, principally due to applications of computer science, new technological achievements in materials, advanced technology and design of the track and the rolling stock, massive use of daily applications of internet and satellite communications, new ways of thinking and approaching old problems, possibilities provided by artificial intelligence. It is estimated that after their expansion (from the 1830s to the 1950s÷1960s) and the crisis (from the 1950s to the 2000s÷2020s) periods, railways are entering a new era of development.

Thus, it has become necessary to come up with a new scientific approach to tackle management and engineering aspects of railways, to have an in-depth understanding of causes and consequences of the various situations and phenomena, and to suggest the appropriate methods and solutions to solve the arising problems.

The fifth edition of this book, though based on previous ones, is enriched with the incorporation of new knowledge attained and aims to address the need for a new scientific approach for railways. It is intended to be of use to railway planners, managers, economists, and engineers, consulting economists and engineers, and students of schools of engineering, transportation, and management.

This wide range of intended readership has led me to divide the book in three distinct parts.

The first six chapters deal with the planning and management of railways and more particularly with issues related to the position and prospects of railways in the transport sector, new technological achievements such as high-speed trains, magnetic levitation, and hyperloop, policy and legislation for railways, methods of forecast of rail demand, costs and economics of railways, methods of pricing, planning and management of railways, and the separation of infrastructure from operation.

The next eleven chapters deal with the track and more particularly with issues related to the mechanical behavior and design of the track system and of its various components (rails, fastenings, sleepers, ballast, subgrade), track layout, transverse effects and derailment, switches and crossings, marshaling yards and railway stations, laying and maintenance of track, and slab track.

The last six chapters deal with rolling stock, signaling, safety, and environmental topics and more particularly with issues related to train dynamics, rail tunnels, design and operation of rolling stock, diesel and electric traction, hydrogen trains, signaling and safety, interoperability, rail traffic management system, level crossings, and the environmental effects of railways.

Each chapter of the book contains the necessary theoretical analysis of the topics under study, recommended solutions, suggested computer software, applications, charts and design of the specific railway component. In this way, the requirement for a theoretical analysis is met and the needs of the railway planner, manager, and engineer for tables, nomographs, regulations, computer applications, and software are satisfied as well.

Railways in Europe have separated the activities related to infrastructure from those related to operation. In other parts of the world, however, railways remain unified. The book addresses both situations (separated and unified railways).

Railways present great differences in the technologies they employ. Some assessments may be valid for one such technology, but not for another. To overcome this problem, standards, specifications, and regulations of the

International Union of Railways (UIC) and of the European Commission and the Agency for Railways of the European Union have been used to the greatest extent possible. Whenever a specific technology or method is presented, the limits of its application are clearly emphasized.

I have tried to take into account the most recent scientific and statistical data, available as of autumn 2021. But in the era of the internet with the possibility for immediate information for everything changing in the world, the readers of the book are afforded with all sources of information, so that they can update and adapt the content of the book to their needs.

I would like to express my thanks to Mr. N.G. Botzoris for his technical assistance.

The writing of a book demands a lot of time, which is usually taken from family activities. I would like to thank my wife and son for their understanding and patience.

Authors aim, in vain, to create a perfect book. However, in science nothing is permanent and everything is evolving rapidly. Thus, I will welcome any views and comments.

V.A. Profillidis
March 2022

List of Abbreviations

AC–DC	Alternating – Direct current
Amtrak	American Train Company
ANN	Artificial Neural Networks
ASCE	American Society of Civil Engineers
ATC	Automatic Train Control
ATO	Automatic Train Operation
ATP	Automatic Train Protection
AVE	Spanish high-speed train
b	Non-compensated centrifugal acceleration or Distance between tracks
BR	(former) British Railways
CBR	California Bearing Ratio
COTIF	Convention for the international carriage by rail
cwr	Continuous welded rail
dB	Decibel
DB	German railways
E	Modulus of elasticity
ECMT	European Conference of Ministers of Transport
ERA	European Union Agency for Railways
ERTMS	European Rail Traffic Management System
EU	European Union
FEM	Finite Element Method
GDP	Gross Domestic Product

ghg	Greenhouse gases
GNSS	Global Navigation Satellite System
GPS	Global Positioning System
h	Cant
h_d / h_e	Cant deficiency / Cant excess
HD	Horizontal defect
HST	High-Speed Train
ICE	German high-speed train
IM	Infrastructure Manager
IRR	Internal Rate of Return
ITF	International Transport Forum
JR	Japanese Railways
LC	Level crossing
LD	Longitudinal defect
LEU	Line Electronic Unit
M	Bending moment
Maglev	Magnetic levitation train
N	Loading cycles
NPV	Net Present Value
NS	Railways of Netherlands
OECD	Organization for Economic Cooperation and Development
OPEC	Organization of the Petroleum Exporting Countries
ORE	Organisme des Recherches et d' Essais (named also ERRI: European Rail Research Institute)
P	Axle load
p-km	Passenger-kilometer
PPP	Public-Private Partnership
PSO	Public Service Obligations
Q	Wheel load
R	Radius of curvature (horizontal) or Running resistance
r	Specific resistance

R^2	Coefficient of determination
RAMS	Reliability, Availability, Maintainability, Safety
RENFE	Spanish railways
Ro-Ro	Roll on-Roll off
R_v	Radius of curvature (vertical)
S_1	Subgrade of poor quality
S_2	Subgrade of medium quality
S_3	Subgrade of good quality
sd	Standard deviation
SNCF	French railways
T	Traffic load
TD	Transverse defect
TGV	French high-speed train
t-km	Tonne-kilometer
TOCs	Train Operating Companies
TSIs	Technical Specifications for Interoperability
UIC	International Union of Railways
UK	United Kingdom
V	Speed or Volt
Y	Transverse force
Z	Traction force
γ	Centrifugal acceleration
μ	Adhesion coefficient
ν	Poisson's ratio

Numerical ranges in this book are indicated with the '÷' symbol to avoid confusion with the minus '-' symbol.

1 Railways and Transport

1.1. Invention and evolution of railways

1.1.1. Historical outline

Since the dawn of human activity to this day, quick and safe transportation of people and goods has been a constant goal of every organized society. It is generally acknowledged that the fundamental innovations in the development of transportation include the invention of the *wheel* (about 3000 B.C.), *navigation* (about 3000 B.C. in Niles river in Egypt, about 2000 B.C. in the sea by Phoenicians), the *railway*, the *automobile* (car, bus), and the *airplane*. Railways, in their present form, made their appearance at the beginning of the 19th century in British mines. Their main characteristic is the guided movement of the wheel by the track through a *metal-to-metal* contact.

However, the forerunners of the railways of our time appeared much earlier than the 19th century. Movement of carriages or wagons on metal guides is illustrated in a 1550 gravure found in Basel, Switzerland, depicting transportation methods employed in the mines of Alsace. The guided movement of carriages in general was already known in Roman times, as witnessed by grooves carved on the stone pavement to facilitate and speed up the movement of carriages.

On Mount Penteli near Athens, from where the white marble of the Parthenon and other classical monuments originated, deep grooves in the rocky ground still bear testimony to the methods employed by ancient Greeks to move marble slabs to the construction sites. Furthermore, the guided movement of carriages was applied in Greek antiquity by laying wooden channels on dirt roads to guide carts. Two channels were adequate for the needs of the day to accommodate one carriage. When two carriages came face to face, the younger driver had to make way for the older one. It is believed that in such an encounter, Oedipus refused to make way and killed the older cart driver coming from the opposite direction, being unaware that it was his father Laïus.

1.1.2. The golden age of railways and successive technical innovations

The development of railways was decisively influenced by the first industrial revolution in the 19th century, the introduction of steam power, and the extensive exploitation of coal and iron mines. The first railway lines began operating in most European countries around 1830 and railway networks attained maximum density at the beginning of the 20th century. A factor contributing to the massive growth of the railways was high speed (by the standards of the time), which enabled fast connections. Steam-powered engines had already achieved (in test runs) impressive performances: 125 km/h in 1850 in Great Britain, 145 km/h in 1895 in France, 210 km/h in 1903 in Germany. Although maximum operating speeds were lower, they contributed to the rapid growth of rail transportation.

The adoption of electric traction in the early 20th century permitted a further development of railways, while the application of signaling and automatic train control in the 1950s facilitated the operation and increased carrying capacity of railways. Major technological innovations during the last six decades drastically changed railway services. These innovations include, among others, high-speed trains, applications of satellite positioning systems (such as GPS, GNSS) and intelligent techniques, technical innovations for the reduction of costs, interoperability techniques to tackle incompatibilities between the various railway technologies, and the use of hydrogen as fuel origin to produce batteries for the movement of trains.

Parallel to advances in technology, innovations in softer forms, such as organization, management, costs, information of passengers, supply of tickets, and additional services during travel, have permitted the railways to improve their competitive position in the transport market. Most of these advances were a result of an extensive use of the internet, but also of applications of artificial intelligence and big data, (see section 1.17).

1.1.3. Railways and their competitors

Times have changed, however, and what was impressive in the early 20th century, soon became less and less satisfactory. Airplanes, passenger cars, buses, and trucks were already offering transportation alternatives at every scale. Given the pressure of competition, railways had to modernize and improve, especially as regards speed, reduction of costs, better organization, and improvement of the services offered. Hence, we come to the era of high-speed trains (see chapter 2) operating at 250÷350 km/h (a speed of 574.8 km/h was attained by French high-speed trains in 2007 in test runs), combined transport (see section 1.10.5), high-volume transport for both passengers (commuter services) and freight (bulk loads), (see sections 1.2.1, 1.10.3, 1.10.4), (9), (13), (15)[*].

---

[*] Figures between parentheses denote references, the list of which is at the end of the book.

Nevertheless, in parallel with conventional railways (based on metal-to-metal contact), experimental research has proceeded since the mid-1970s with techniques, which, although using guided[*] transport (like railways), avoid any contact between the moving vehicle and the bearing infrastructure. These are the aerotrain and the magnetic levitation systems (also known as maglevs), which in test runs have attained speeds of 430 km/h for the aero-train in 1974 and 603 km/h for the maglev in 2015. Since 2004, magnetic levitation systems have been applied and operated at a speed of 431 km/h, (see also sections 2.6, 2.7).

In the early 2020s a new revolutionary technology under test, known as hyperloop, consists in traveling in a tube under low pressure conditions, while using magnetic levitation. Hyperloop systems eliminate rolling resistance and in addition reduce greatly air resistance. They can achieve speeds up to 1,200 km/h, (see section 2.8).

1.1.4. Railways in the era of monopoly and competition

Railways played a catalytic role in the first industrial revolution after 1850. In most cases they have been developed by private companies, which built (and owned) the railway infrastructure they operated, while at the same time providing the appropriate rolling stock and personnel. However returns in railway investments were lower than expected and important deficits soon appeared. As railways played a central role in the economy and security of each country, many governments have nationalized their railways since 1935. Thus, railways became a state monopoly, which had as a positive effect the integrated railway services at the state level and as negative effects the inflexibility and poor adaptation to the evolving requirements of the economy and society.

In some parts of the world (particularly in Europe and the USA) state-owned railways have had, after 1950, a declining share in the transport market, (see sections 1.6, 1.7). As a measure to stop and reverse this situation, the introduction of intra-modal competition has been considered, namely, the operation of many railway companies on the same route. In some countries, like the USA, a railway company kept on owning infrastructure, while at the same time another railway company had the right to run on its infrastructure by paying appropriate charges. In European countries, however, the introduction of intra-modal competition was related to the so-called separation of infrastructure from operation, so as to ensure fair and impartial conditions

---

[*] As guided transport can be defined any method of transport where the direction of travel of the vehicle is constrained by physical means and thus traveling vehicles follow a determined trajectory for their journey. Guided transport includes railways, metro and tram systems, magnetic levitation, and hyperloop systems.

among the many rail operators eventually competing on the same route. Some countries (among them the USA, the United Kingdom, Japan, Canada, etc.) have privatized parts or the whole of their railway activities, (see also sections 3.4, 6.10), (15), (20), (29).

1.2. Characteristics of rail transport

1.2.1. Ability to transport high volumes

The main characteristic of rail transport[*] involves its capability to join several units into trains. The heaviest trains in the world are freight trains transporting bulk commodities such as coal, iron, cereals, etc. Indeed, freight trains of 17,200[**] tonnes[***] with multiple couplings are used daily in the USA, while in Australia trains transporting mineral products exceed 40,000 tonnes, in China 20,000 tonnes, in Canada 20,700 tonnes. With regard to passengers, railways are capable of transporting a great number of people. High-speed trains of the Japanese railways have transported 520,000 passengers between Tokyo and Osaka in one day and regularly about 466,000 people between these two cities (a distance of 552 km[****]), (30).

Another characteristic of rail transport is its one degree of freedom, in comparison to road transport, which has two degrees of freedom. The one degree of freedom makes door-to-door transportation impossible for rail, but favors large-scale use of automatic controls, computers, and electronics. As a result, unit transportation capacity of railways is high, e.g. commuter trains can transport 60,000 passengers per hour and per direction, against only 6,000 for bus systems, (28).

High transport capacity of railways (both for passenger and freight), combined with efficient connectivity to the road network and appropriate transfer facilities, can lead to the achievement of *seamless* door-to-door transport, which is one of the tools to reduce environmental effects of transport systems, (see section 1.2.3 and chapter 23).

---

[*]   The terms transport and transportation are used in the book interchangeable. The same is with the terms mode and means.

[**]   Units concerning thousands and decimals are presented in this book according to the American system. Thus, comma (,) denotes thousands and point (.) denotes decimals.

[***] A tonne (or a metric tonne) equals 1,000 kg (or 9,806.65 Nt). It should be distinguished from the short tonne, a unit used in the United States (1 short tonne = 2,000 pounds = 0.907 metric tonnes) and from the long tonne, a traditional British unit (1 long tonne = 2,240 pounds = 1.016 metric tonnes).

[****] The unit used for distance is kilometer (km) or mile (in the United States, the United Kingdom, and elsewhere), 1 mile = 1.609344 km. The nautical mile is very seldom used, 1 nautical mile = 1.8520 km.

1.2.2. Energy consumption

Rail transport is characterized by the guided movement of wheels on tracks through the metal-to-metal contact, which considerably reduces rolling resistance to less than 3 kg per tonne carried. Accordingly, for the same propulsion force, rail vehicles carry a much larger load than road vehicles. As a result, high-speed rail transport consumes $1/4 \div 1/5$ as much energy as road transport for the same traffic. The comparison becomes more definitive with airplanes, which consume for the same traffic $6 \div 7$ times more energy than railways.

The interests of private companies and groups have not permitted to take into account up until now the factor of energy consumption in transport policies. However, oil reserves all over the world can satisfy needs for a maximum of two generations from now onwards, (Fig. 1.1), and they have been stimulated by low oil prices for two decades $(1983 \div 2003)$, as compared to previous years, (Fig. 1.2). In any case, the remaining years for which oil reserves can satisfy human needs are calculated on today's rates of consumption, without being able to forecast accurately the coming needs in the future of emerging economies, like China, India, Brazil, and others, as well as the effects of electrification of road vehicles. World oil demand was in 2019 101.1 million barrels per day, it collapsed in 2020 and 2021 due to the world sanitary crisis, and is estimated to reach around 110 million barrels per day in 2035, (9), (11).

1.2.3. Environmental performance and safety

Another advantage of rail transport is its much lower environmental pollution. Electric trains produce no emissions, while diesel-powered trains generate much less pollution than cars and buses for the same traffic. Concerning CO_2

Reserves-to-production ratio (RPR, in years) as in 2019
RPR = (amount of known resource) / (amount used per year)

Geographical region	1980	2019
North America	18.1%	14.1%
Central & South America	3.9%	18.7%
Europe & Eurasia	12.2%	9.3%
Middle East	53.0%	48.1%
Africa	7.8%	7.2%
Rest Asia and Pacific	5.0%	2.6%

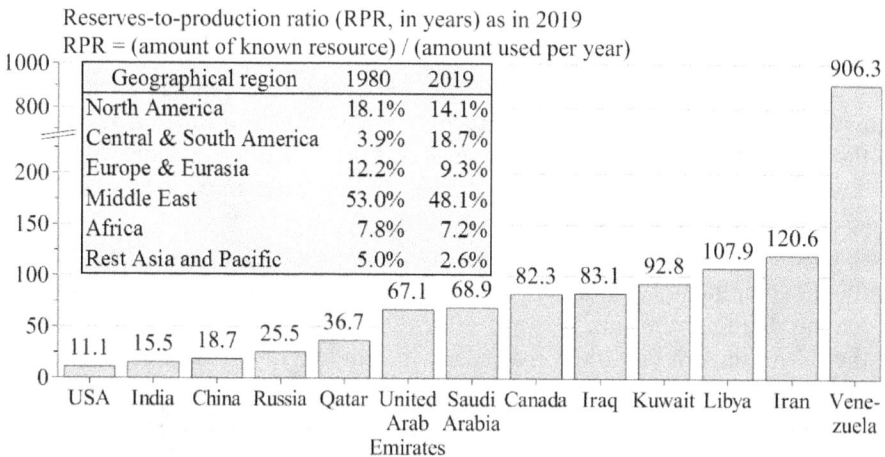

Fig. 1.1. Oil reserves all over the world, (31)

5

Cost of a crude oil barrel (in US$, constant 2020 prices)

Fig. 1.2. Evolution of prices of oil in US dollars from 1861 to 2019, (35)

emissions, high-speed rail passenger transport causes for the same traffic 1/8 CO_2 emissions compared to road passenger transport and 1/11 compared to air transport, (38). Emissions of CO_2 of rail freight are 1/4.5 compared to road and 1/4 compared to inland waterways, (15), (16), (38).

People all over the world have become more sensitive about transport safety. For the same traffic, the risk of a fatality is ten times higher for buses / coaches than for trains, (see section 22.4.6). Railway performance is genuinely impressive.

Finally, land occupation per passenger-kilometer or tonne-kilometer is much less for rail transport than for other transport modes and specifically 2÷3 times less than for road transport. For the purposes of comparison with airplanes, it is noteworthy to mention that the high-speed Paris-Lyon line (a distance of 427 km) occupies as much space as the Paris airport at Roissy.

1.3. Railways and the economy

1.3.1. Economic cycles and railways

The steam railway, the electrified railway, and the high-speed railway have been developed parallel to the three economic cycles of the last two centuries, known also as the three industrial or technological revolutions, (Fig. 1.3).

- the first industrial revolution (curve A in Fig. 1.3) is associated with the development of steam and water power and an extensive mechanization of production,
- the second industrial revolution (curve B in Fig. 1.3) is associated with the development of electricity and the massive production of goods, among them private cars,
- the third industrial revolution (curve C in Fig. 1.3) is associated with the development of electronics, computers and automations, since 1990 of the internet, since 2005÷2010 of an extensive digitalization of almost all human activities, and since 2020 of artificial intelligence.

A. Steal,	B. Oil,	C. Globalization,
Coal,	Electricity,	Computers,
Steam railway,	Electrified	Internet, Digitalization
Inorganic	railway,	High-speed railway,
chemistry	Private car	Airplane

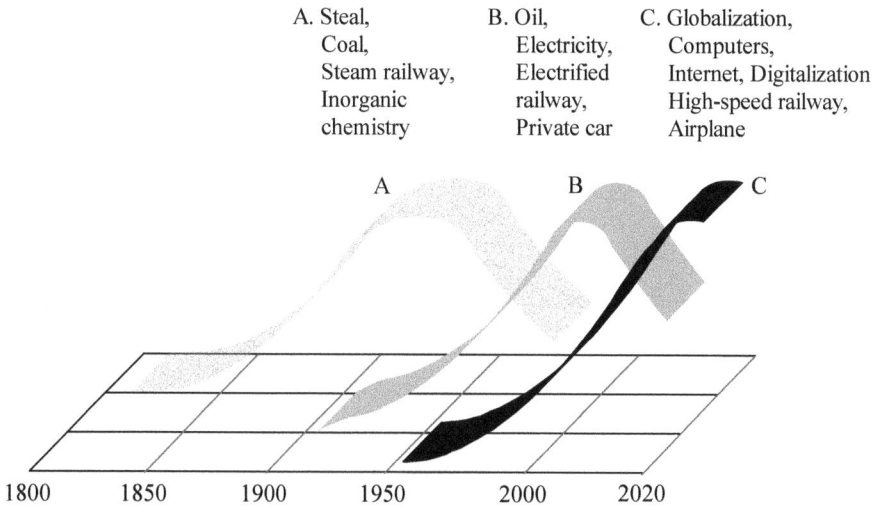

Fig. 1.3. Economic cycles and transport technologies, (26)

However, since 2005 we have witnessed an acceleration of applications of automations, cybernetics, and governance by computers and robots of many human activities. This stage of economic development since 2005 is described by some researchers as the fourth industrial revolution and is characterized by digitalization, possibility of monitoring and surveillance by machines of every human activity, artificial intelligence, (see section 1.17), robotics, the internet, nanotechnology and neurotechnology, biotechnology, and driverless vehicles. Others consider the fourth industrial revolution as a part or even an extension of the third one. Whatever its name, the third decade of the 21st century will be driven by the convergence of digital, biological, and physical innovations. Changes are constantly taking place in an exponential way and a number of traditional aspects are overturned. Some wonder if humans realize the big risk of being governed by machines and technology and not govern and rule them.

1.3.2. Economic growth and railways

Transport activity (both passenger and freight) is strongly affected by the rates of growth of economic activity. For a number of countries and until 1990÷2000 growth rates of passenger and freight transport were almost parallel to the growth rates of Gross Domestic Product (GDP). This situation is referred to as *coupling* between transport and economic activity. However, since 1990÷2000 a break is observed in the link between transport activity and GDP, depicting a situation where the rates of change of GDP are not reflected to the rates of change of transport activity. This situation is referred to as *decoupling*, (9), (21).

Figure 1.4 illustrates for the EU-28 countries the evolution since 1990 of passenger and freight transport and GDP. Freight transport has decoupled from GDP since 1990, whereas passenger transport has decoupled from GDP around 1998. Within the transport sector, growth rates of air transport were much higher than GDP (almost double), whereas growth rates of rail transport were lower than GDP.

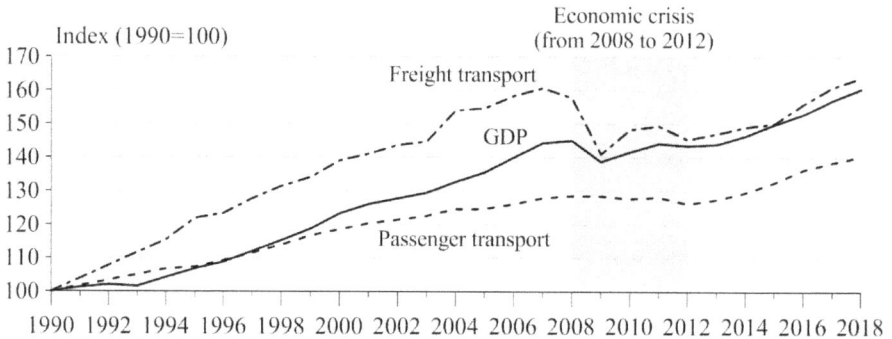

Fig. 1.4. Trends in the passenger traffic (passenger-km) and the freight traffic (tonne-km) in relation to the Gross Domestic Product (GDP) in the EU-28 countries, (2)

1.3.3. Part of revenue spent for transport and contribution of railways to the economy

Expenses for transport represent an essential part of the household budget, around 10%÷14% for the EU-28, 9.6% for the USA, 11.7% for Japan, 12% for Russia, 14% for India, 13.7% for the UK, 14.0% for France, 12.9% for Italy, 12.7% for Spain, (2), (6). The added value of transport to the economy as a percentage of GDP amounts to 6.6 % for the EU-28, 6.3% for the USA, 6.1% for Japan, 4.9% for China, 5.5% for India, 7.8% for Germany, 7.4% for the UK and France. The contribution of railways and of other sectors of transport to the GDP is proportional to the share they have in the national transport market, (16).

1.4. Mobility, sustainability and railways

1.4.1. Mobility and transport evolution

Transport systems (and railways in particular) exist for the satisfaction of human needs for mobility. Without transport, our world would be limited to a few dozen kilometers around the area of our settlement. It is with transport that mankind succeeded to broaden its horizons and the scale of its activities and cover the whole planet, (9).

For centuries and until the early 1800s, mobility was assured by walking or horse power and the average traveling speed worldwide was at the range of 5÷10 km/h. The appearance of railways raised the average traveling speed to 25 km/h in 1900 and later the extensive use of private cars and airplanes to 40 km/h in 2000 and 50 km/h in 2020. It seems that humans are reaching their physical limits of mobility. Thanks to the increase of speed, people in Europe augment since 1800 the amount of kilometers they travel per year by 3%÷3.5%, which leads to the doubling of the traveled distances every 20 years, (9).

While railways from the 19th century and private cars and airplanes in the 20th century were rapidly increasing traveling speeds, the amount of time that individuals were eager to spend for transport has remained almost unchanged in all historical periods and is estimated at 0.8÷1.2 hours per person and per day, with an average value of 1.1, (9). In addition, the more the travel time is reduced, the better use of that time is expected by travelers, (16).

1.4.2. Mobility and sustainability

The climate change in progress and the need to undertake efficient measures to control it, while ensuring mobility and economic growth, have brought up *sustainability* as a new concept for all human activities. Sustainability is specialized in the transport sector in sustainable mobility and sustainable transport, referring to any of the means of transport with low impact on the environment and balancing current and future needs. The most efficient way to achieve sustainable transport is to consider mobility as a service.

1.4.3. Mobility as a service and railways

Mobility as a service is the integration of various forms of transport services into a single mobility service, accessible on demand. It entails a more extensive and rational use of less polluting transport means, such as railways, as well as of walking and cycling. Furthermore, it encourages a collective use of the most polluting transport means such as the private car and leads to schemes of carpooling and car sharing, (see below section 1.5.3).

Mobility as a service describes a shift from personally owned transport means (private car) towards mobility solutions that are consumed as a service and are based on travel needs and not on the possession of a private car. It is estimated that mobility as a service will have positive effects towards a wider use of railway services and existing railway tracks and facilities. Railways will become under this concept one element of a complex mobility system, providing integrated, connected, efficient, user friendly, seamless, end-to-end journeys that combine all available modes of transport. Mobility as a service constitutes a shift from transport device ownership to mobility access usage.

1.5. Railways and the private car

1.5.1. The explosion of private cars

The world has experienced an unprecedented increase in the number of private cars, but this has taken place during different time periods in various countries: after 1930 in the USA, between 1950÷1990 in Western Europe and Japan, after 1990 in Eastern Europe and Russia, after 2000 in China and India. In contrast to the railway which is a massive transport means, new social attitudes and lifestyle opted for the private car as an element of independence and individualism. The periods of explosion of the number of private cars coincided with economic growth and increase in population. Railways did not succeed to convince the public that they are an efficient alternative to the private car and their share in the transport market has been greatly reduced.

It is worth mentioning that the average private car ownership index, that is the average number of inhabitants per 1 private car, is considered to be one among the economic indices for the standard of living of a country. The average value for this index for the EU-15 countries was 1 private car per 5.21 inhabitants in 1970, whereas in 2010 it reached the value of 1 private car per 1.98 inhabitants. The private car ownership index is directly related to per capital GDP of a country, but not proportionally, since it is influenced by the degree of development of the various transport means for each country, its geographical position, etc.

Figure 1.5 illustrates the increase of the private car ownership index between 1990 and 2018 in the world, the EU-28, and the USA. It can be remarked that there is a highly increasing tendency to own a car in the world, a moderately increasing tendency in the EU-28, and an almost stagnation in

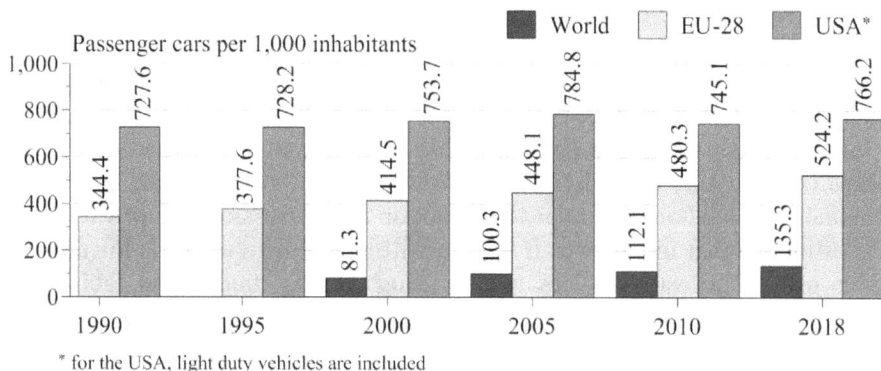

* for the USA, light duty vehicles are included

Fig. 1.5. Average private car ownership index in the EU-28 countries, the USA, and the world, (2)

the USA. The overwhelming preoccupation with the ownership of a private car in the transport market is a result of its inherent characteristics such as door-to-door transport, higher comfort, flexibility, but also lifestyle and owner's image in society.

1.5.2. The electric car

The need to reduce the emissions of gases (such as CO_2) which contribute to the increase of temperature on our planet has led to the rejuvenation of an old invention (since the 1880s), the electric car. An electric car can be fully electric (if it moves based entirely on its batteries) or hybrid (it can use either rechargeable batteries or internal combustion engine). The cost of a lithium-ion battery has dropped since 2015 and the autonomy for higher-cost electric cars already surpasses 500 km, while for medium-cost electric cars is in 2021 around 230÷350 km, (9). It is expected that the cost of electric cars will compete the cost of conventional cars in the late 2020s. Forecasts for the year 2040 estimate annual sales of electric cars worldwide at 40 million, their share in the total car market at around 25%, and their share in the new car market at around 35% or more, (9).

In spite of their environmental performances, electric cars will always be cars, that don't have the capacity to provide massive transport of passengers and freight, as is the case with railways.

1.5.3. Carpooling and car sharing

A private car remains immobile for 95.5 % of its operational life. This fact has led to the consideration to use a private car only when somebody needs it, no matter whether he or she possesses one. Within this consideration, two organizational models have been developed, carpooling and car sharing. In carpooling, car owners share their car with other people for specific purposes. In car sharing, several users share a car fleet owned by a service-providing company, by paying a monthly fee, usually based on both time and mileage.

For urban and medium-distance trips, car sharing or carpooling can become under specific conditions of costs, travel times, and quality of service a competitor for railways.

1.5.4. Driverless (autonomous) vehicles

Driverless (or autonomous) vehicles (cars, buses) are one of the symptoms of the 4th industrial revolution and of automatization in everyday life. Autonomous vehicles can move with little or no human effort, as they can sense and understand their environment with the help of a system of sensors and computer software, which interprets information and identifies the appropriate navigation paths. Extended application of autonomous vehicles is not expected before 2035 and should overcome a number of obstacles such as legal (liability in case of an accident), social (inertia and low acceptance by the society), and technical (how the road infrastructure will be shared by autonomous and conventional vehicles) ones.

Among the scenarios to combat the greenhouse effect, it has been suggested to make the cost of use of public infrastructure by conventional private cars much higher compared to the cost of autonomous vehicles. Within this scenario, mobility for short distances will be satisfied by autonomous shared vehicles (available on demand, like taxis today), that will lead passengers to the terminal of a mass transport system (railway, metro, bus), the last one becoming the basic element of transport and shared mobility, (Fig. 1.6).

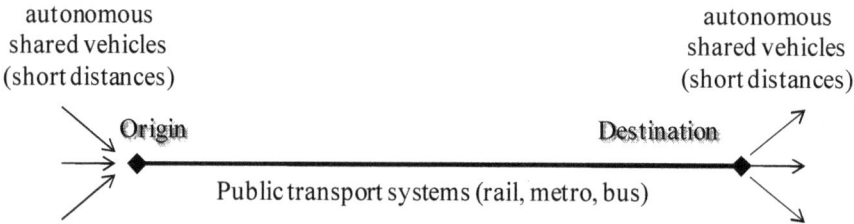

autonomous
shared vehicles
(short distances)

autonomous
shared vehicles
(short distances)

Origin   Destination

Public transport systems (rail, metro, bus)

Fig. 1.6. Autonomous vehicles, railways and shared mobility

1.6. A panorama of passenger traffic of railways around the world

1.6.1.Evolution of passenger traffic and of traveled distance performed by railways around the world

The usual metrics to record and assess transport activity is the number of passenger-kilometers (p-km) for passenger transport and the number of tonne-kilometers (t-km) for freight transport. A passenger-kilometer is a unit to describe a passenger who travels 1 km and a tonne-kilometer is a unit to describe the transport of 1 tonne for 1 kilometer. The terms p-km and t-km take into account both the number of passengers or tonnes transported and the distance traveled.

Figure 1.7 illustrates the evolution of passenger traffic carried by railways worldwide during the last four decades. During that period, railways lost some markets (local and regional transport), due principally to competition from the private car, but gained other markets, due principally to the development of high speeds. Figure 1.7 also illustrates for railways the average traveled distance per capita, which has increasing tendency worldwide; however, this is principally due to long rail trips in China and India. In other parts of the world (particularly in Europe and Japan) the average traveled

12

distance is decreasing, as railways are losing very long and long distance trips (to air transport). The average distance traveled by railway passengers was in 2019 326 km for China, 319 km for the USA, 137 km for India, 78 km for the EU (28 countries), and 29 km for Japan. EU countries and Japan have low traveled distances due to the fact that a great part of railway passengers travel in urban or suburban areas, (1), (6).

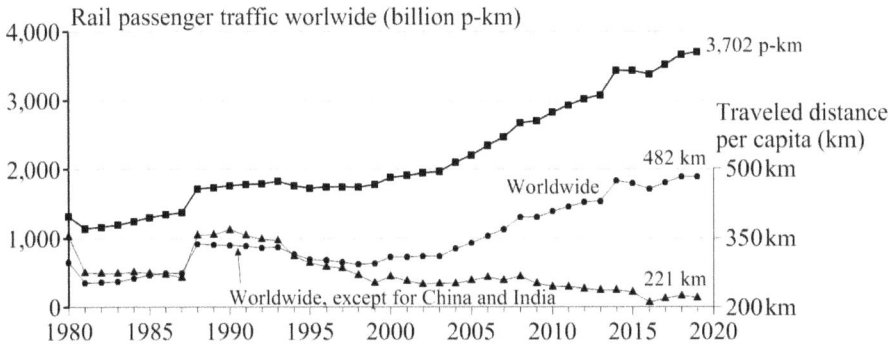

Fig. 1.7. Evolution of rail passenger traffic worldwide and of average traveled distance per capita, (1), (2), (6), (9)

1.6.2 Passenger traffic of railways for some countries of the world and evolution over time

Figure 1.8 illustrates the evolution of passenger traffic of railways for some countries of the world during the last four decades. Railway traffic has been expanding in India and China since 1990, but it has been stagnating in many other parts of the world. In 2019, 90% of world railway passenger traffic was realized in China, India, EU-28, Japan, and Russia.

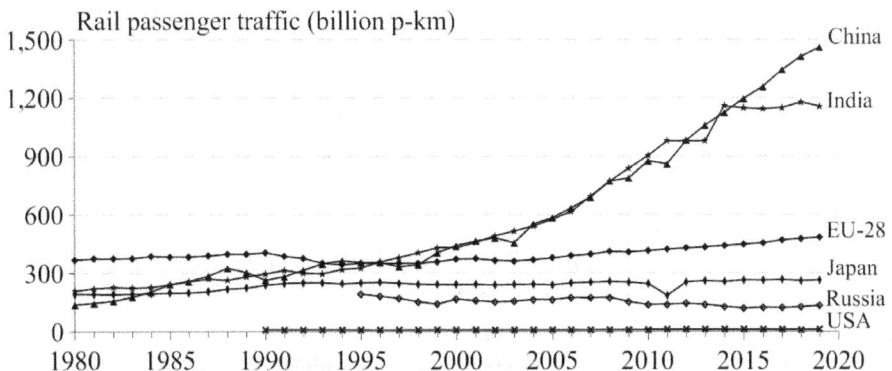

Fig. 1.8. Evolution of passenger traffic of railways for some countries of the world, (1), (2), (6)

Figure 1.9 illustrates passenger traffic[*] of railways for the year 2018 for a number of countries around the world.

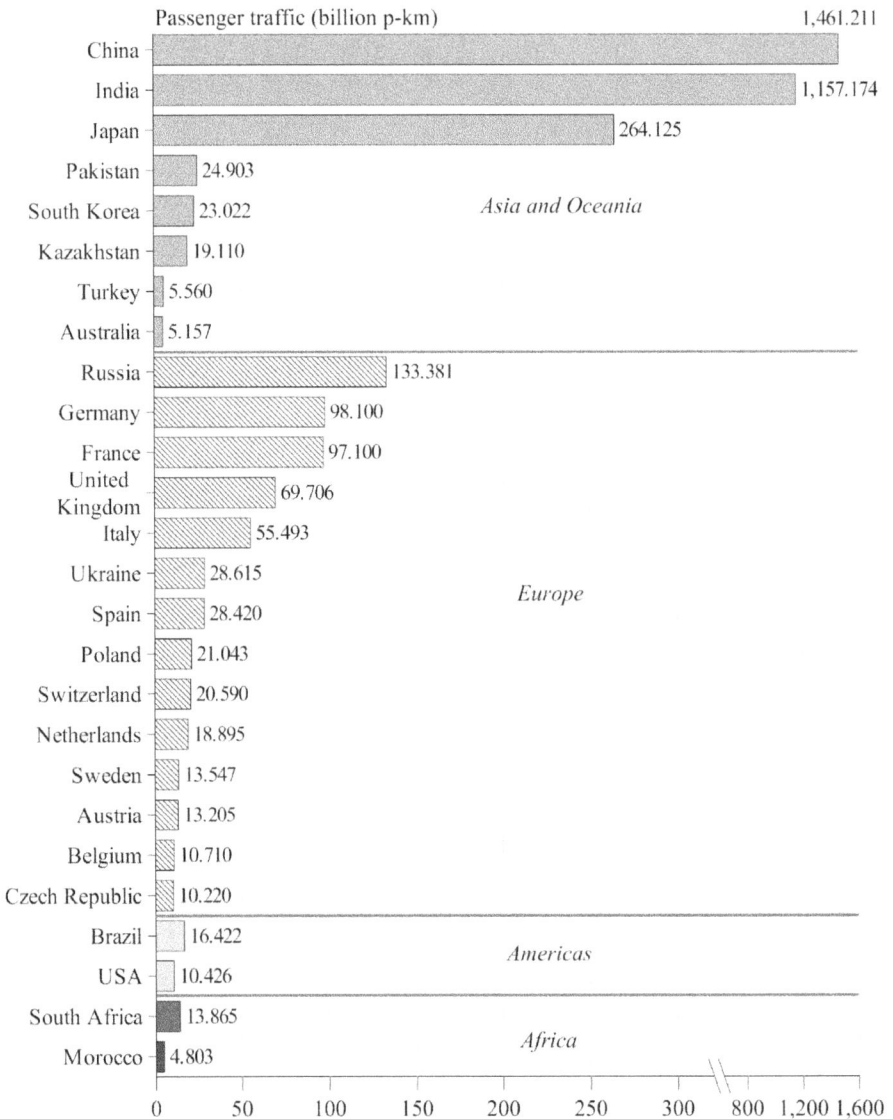

Passenger traffic (billion p-km)

China	1,461.211
India	1,157.174
Japan	264.125
Pakistan	24.903
South Korea	23.022
Kazakhstan	19.110
Turkey	5.560
Australia	5.157

Asia and Oceania

Russia	133.381
Germany	98.100
France	97.100
United Kingdom	69.706
Italy	55.493
Ukraine	28.615
Spain	28.420
Poland	21.043
Switzerland	20.590
Netherlands	18.895
Sweden	13.547
Austria	13.205
Belgium	10.710
Czech Republic	10.220

Europe

Brazil	16.422
USA	10.426

Americas

South Africa	13.865
Morocco	4.803

Africa

0   50   100   150   200   250   300   800   1,200 1,600

Fig. 1.9. Passenger traffic of railways in various geographical areas and countries of the world (2018), (1), (2), (6)

---

[*] Data presented in this book are the most recent ones in spring 2021. The reader, however, can update these data by visiting the EU and UIC internet sites, which are:
- European Union: https://ec.europa.eu/eurostat
- International Union of Railways: https://uic.org/support-activities/statistics/

14

1.6.3. Comparative evolution of passenger traffic for railways and other transport modes

Figure 1.10 illustrates a comparative evolution between 1995 and 2018 of passenger traffic of railways and other transport modes for the EU-28 countries. During that period, the rate of increase of railway traffic was higher compared to traffic of passenger cars, far higher compared to buses, but far lower compared to air transport.

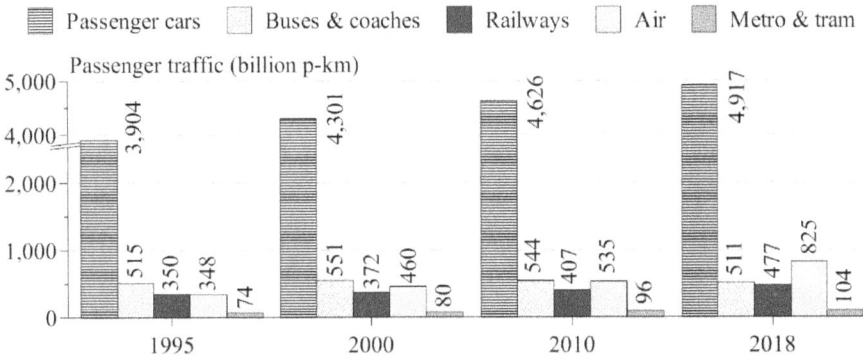

Fig. 1.10. Evolution of passenger traffic for various transport modes for the EU-28 countries, (2)

1.6.4. Share of railways in the national passenger transport market

Share of railways in the national passenger transport market for each country depends mainly on the degree on which the railways meet the requirements of the market and the society, as well as on the degree and orientation of state intervention and policy concerning competition, tariffs, and subsidies. Share of railways in their national land passenger transport market was in 2018 as follows, (Fig.1.11): 52.8% for China (against 69.6% in 1970), 27.9% for Russia (against 50.4% in 1970), 7.7% for the 15 countries of the EU (against 10.4% in 1970), 10.5% for France, 8.8% for the UK, 8.9% for Germany, 6.9% for Spain, 6.2% for Italy, 16% for Switzerland, 0.7% for the USA (against 0.6% in 1970), (2), (6), (13).

Figure 1.12 illustrates the evolution of share of railways in the passenger market for the EU-28 countries. Indeed, railway share in the EU passenger market was around 50% in 1950, but dropped to 10.4% in 1970 and to 7.0% in 2018.

Figure 1.13 illustrates the drop of rail passenger share in the USA from around 15% during the early 1950s to 0.6% during the 1990s and 0.7% in 2018, (1), (6).

Share (%) in the national land passenger transport market (p-km)

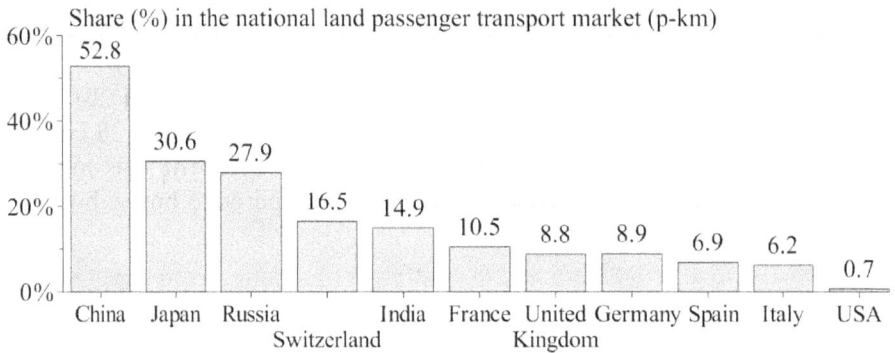

Fig. 1.11. Share of railways in the national land passenger transport market for several countries of the world (data for the year 2018, for China, India, Japan and Russia for the year 2017), (2), (6)

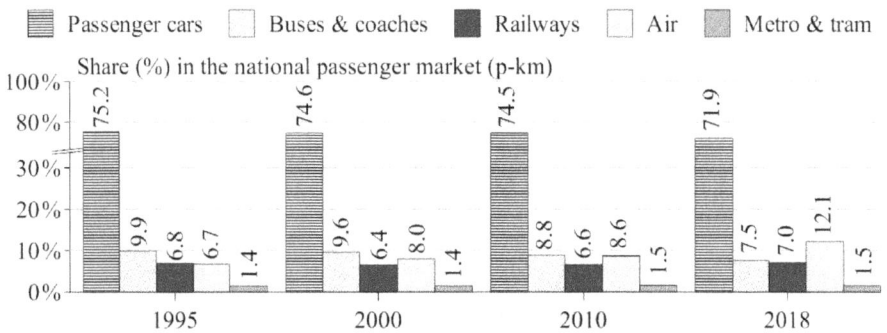

Fig. 1.12. Evolution of share of various transport modes in the passenger transport market of the EU-28 countries, (2)

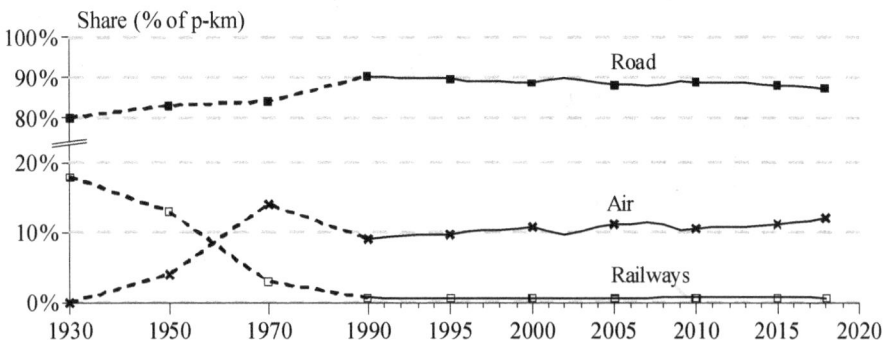

Fig. 1.13. Evolution of share of various transport modes in the passenger transport market of the USA, (1), (6)

16

1.6.5. Growth rates of passenger traffic of railways

Figure 1.14 illustrates growth rates of passenger traffic of railways for the periods 1970÷2007 and 2007÷2016. Effects of the financial-economic crisis of 2008÷2012 on railway traffic are more visible in Russia, Japan, and the EU-28, while China increased growth rates, India kept almost the same growth rates for both of the above periods, and the USA (with however a marginal railway passenger traffic) had surprisingly high growth rates.

1.6.6. Distances with a comparative advantage for rail passenger traffic

Share of railways in the passenger transport market is increased for medium distances (150÷500 km), for which railways have a strong competitive advantage (in relation to other transport modes), as they enter city centers and are not affected by congestion, (Fig. 1.15), (2).

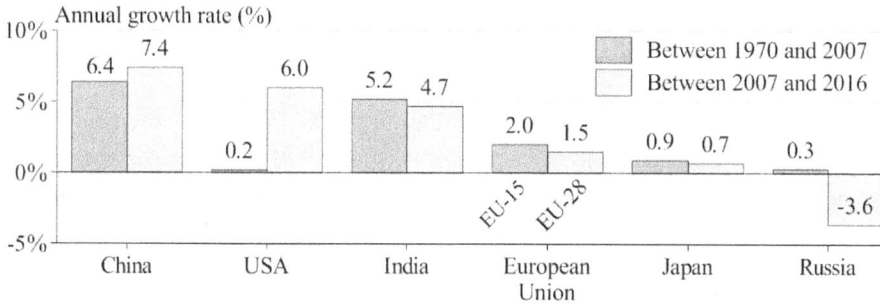

Fig. 1.14. Growth rates (1970÷2016) of rail passenger traffic for some countries, (1), (2), (6)

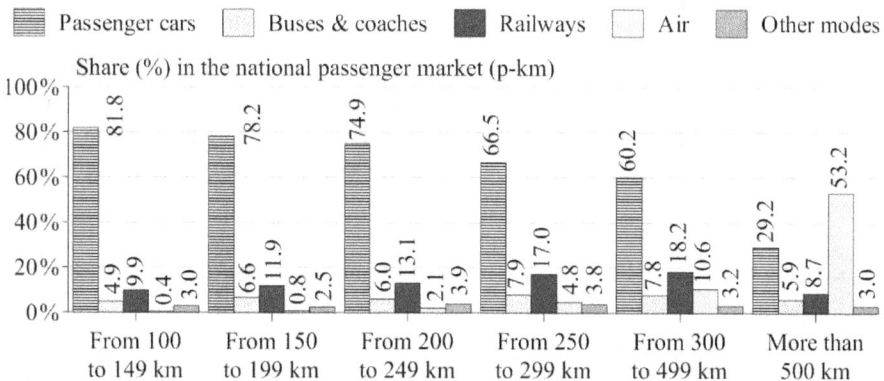

Fig. 1.15. Variation of share in the passenger transport market of medium distances for various transport modes for the EU-15 countries, (2)

17

1.7. A panorama of freight traffic of railways around the world

1.7.1. Suitability of railways for some categories of freight

Freight transport differs from passenger transport, as it aims at providing a variety of products and goods between the production and final consumption points (often distanced thousands of kilometers). As explained in section 1.10.6, freight transport is just one component of the logistics chain which includes collection, transport, distribution-delivery, and for a number of products transshipment and storage. We usually distinguish the various types of freight into four categories: bulk freight (grains, minerals, stones, gravel, steel, coal, cements, oil, gas), specialized freight (cars, chemicals, containers), general merchandise (products consumed in supermarkets), and express freight (products of small volume and high value), (9).

The specific characteristics of each transport mode make it more or less suitable for some of the above categories of freight. Thus, railways are more suitable for bulk freight and specialized freight. For these categories of freight, sea transport and inland waterways could also be a suitable solution. Trucks are suitable for any of the categories of freight, but principally for general merchandise. Air transport is suitable for express freight. In relation to the value of a product, for products of a value $\$0.01 \div 0.02/kg$ sea transport and inland waterways are most suitable, for products of a value $\$0.06 \div 0.20/kg$ railways and trucks are more suitable, and for products of a value higher than $\$1/kg$ air transport is more suitable, (9).

1.7.2. Evolution of freight traffic of railways around the world

Figure 1.16 illustrates evolution of freight traffic carried by railways world-wide during the last four decades. Rail freight traffic worldwide had slow growth rates between 1980 and 1990, negative growth rates between 1990 and 1995 (a consequence of restructuring of the economies of formerly socialist countries in Europe, for which most freight was previously transported by rail), high growth rates between 2000 and 2008 (due to increased freight traffic in China, the USA, and Russia), negative or stagnating growth rates since 2008.

1.7.3. Freight traffic of railways for some countries of the world and evolution over time

Figure 1.17 illustrates the evolution of freight traffic of railways for some countries of the world over the last four decades. Until 2008 rail freight experienced very high growth rates in the USA, China, Russia (after 1998), moderate growth rates in India and Canada, whereas it recessed or stagnated in the EU-28.

Rail freight traffic worldwide (billion t-km)

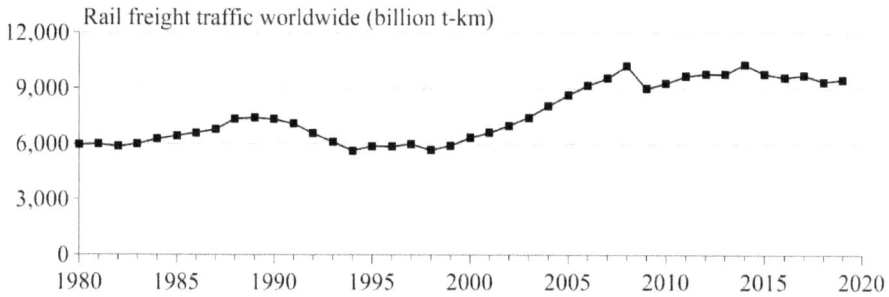

Fig. 1.16. Evolution of freight traffic of railways worldwide, (1), (2), (6)

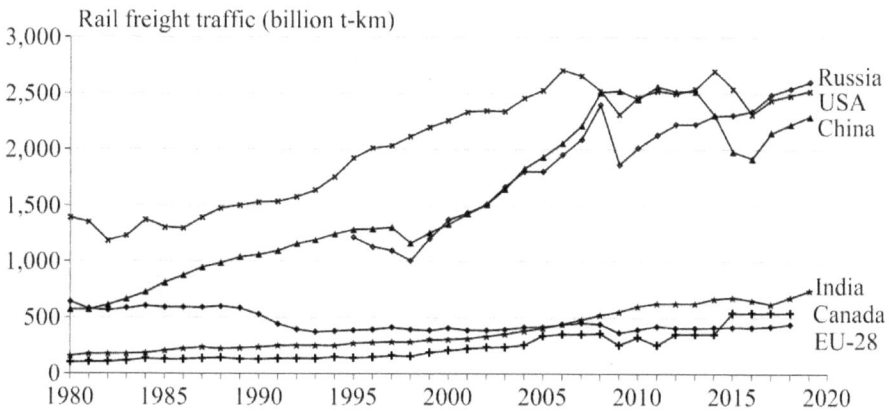

Rail freight traffic (billion t-km)

Russia
USA
China

India
Canada
EU-28

Fig. 1.17. Evolution of freight traffic of railways for some countries of the world, (1), (2), (6)

Figure 1.18 illustrates freight traffic of railways for the year 2018 for some countries. In 2018, 91% of world railway traffic was realized in the USA, China, Russia, India, Canada, Ukraine, and the EU-28.

1.7.4. Comparative evolution of freight traffic for railways and other transport modes

Figure 1.19 illustrates a comparative evolution between 1995 and 2018 of freight traffic of railways and other transport modes for the EU-28 countries. During that period, in the EU-28, railways increased slightly freight traffic (as did inland waterways and pipelines), whereas road and sea transport had a greater increase in their freight traffic.

1.7.5. Share of railways in the national freight transport market

Figure 1.20 illustrates the evolution of the share of railways in the freight transport market for some countries of the world, with a high share in Russia, Switzerland, India, USA, China and a low one in Japan and the UK. High levels of railway freight shares reflect the existence of high volumes of products suitable for rail transport in the specific country, a high traveled distance for rail freight, and appropriateness of the specific railways to compete with road trucks concerning tariffs, capacity, and delivery time.

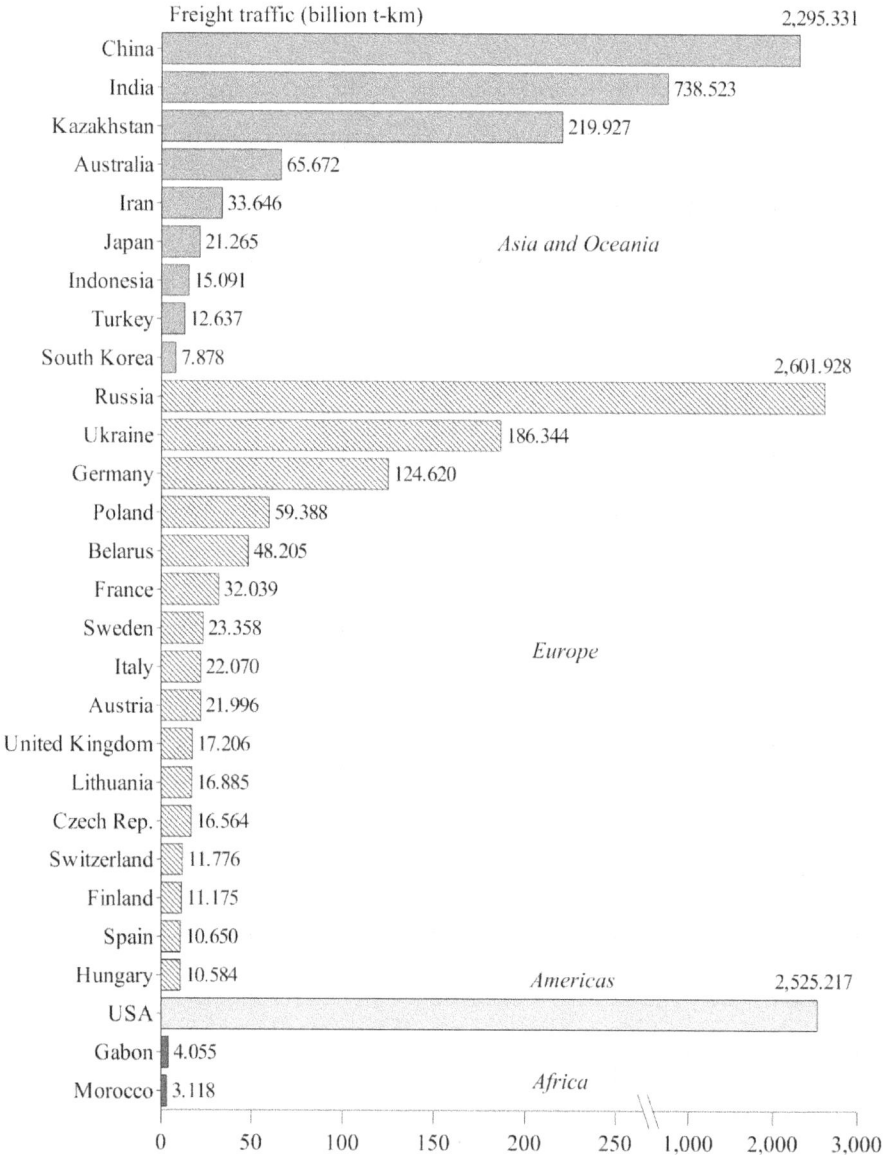

Freight traffic (billion t-km)

China	2,295.331
India	738.523
Kazakhstan	219.927
Australia	65.672
Iran	33.646
Japan	21.265
Indonesia	15.091
Turkey	12.637
South Korea	7.878
Russia	2,601.928
Ukraine	186.344
Germany	124.620
Poland	59.388
Belarus	48.205
France	32.039
Sweden	23.358
Italy	22.070
Austria	21.996
United Kingdom	17.206
Lithuania	16.885
Czech Rep.	16.564
Switzerland	11.776
Finland	11.175
Spain	10.650
Hungary	10.584
USA	2,525.217
Gabon	4.055
Morocco	3.118

Asia and Oceania

Europe

Americas

Africa

Fig. 1.18. Freight traffic of railways in various geographical areas and countries around the world (2018), (1), (2), (6)

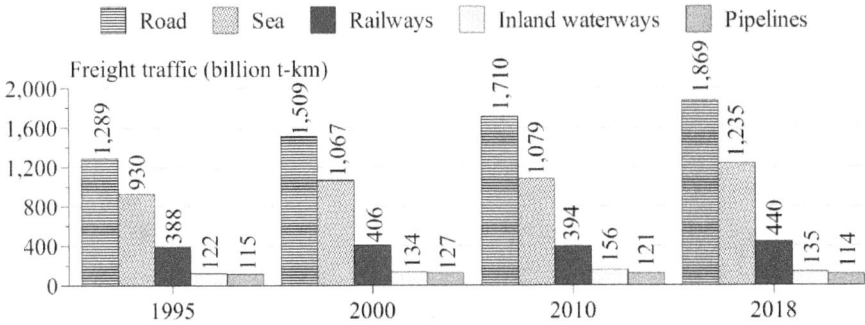

Fig. 1.19. **Evolution of freight traffic for various transport modes for the EU-28 countries, (2)**

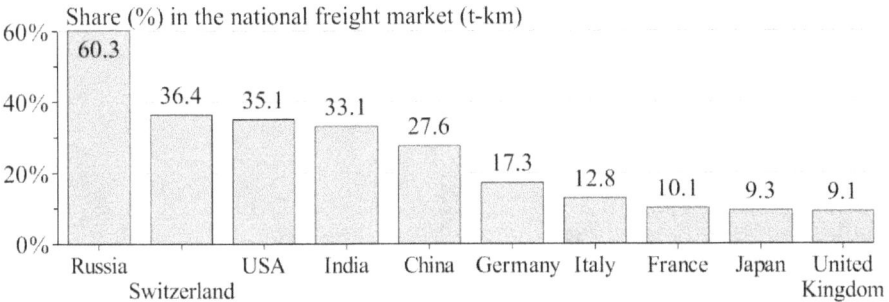

Fig. 1.20. **Evolution of share of railways in the freight transport market for some countries of the world (data for the year 2018), (2), (6)**

Figure 1.21 illustrates the evolution of the share of railways in the freight market for the EU-28 countries. Railway freight share dropped from 13.6% in 1970 to 11.3% in 2017, as most European railways did not manage to efficiently meet the requirements of the transport market. The situation is inverse in the USA, (Fig.1.22), where after a drop by half between 1930 and 1970, rail freight share has during the last two decades moderately increasing tendencies.

1.7.6. Growth rates of freight traffic of railways

Figure 1.23 illustrates growth rates of freight traffic of railways for the periods 1970÷2007 and 2007÷2016. Specifically for the period 1970÷2007, rail freight traffic had high growth rates in India, China and the USA, it was almost stagnating in the EU-15 and Russia, and declining in Japan. However, in all these countries (with the exception of India and Russia) rail freight traffic stagnated or declined for the period 2007÷2016, principally as a result of the financial-economic crisis of 2008÷2012.

21

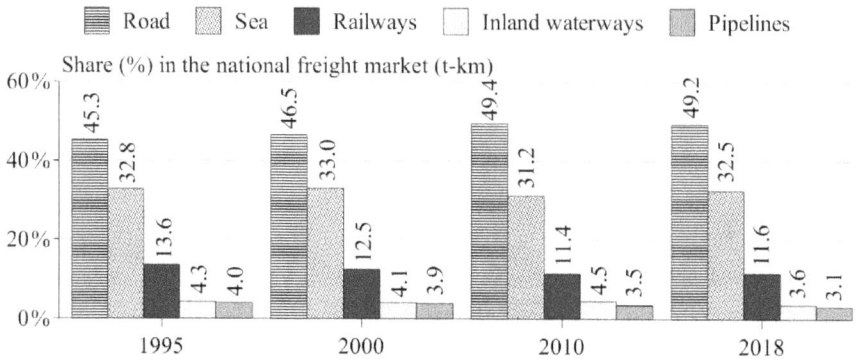

Fig. 1.21. **Evolution of share of various transport modes in the freight transport market of the EU-28 countries, (2)**

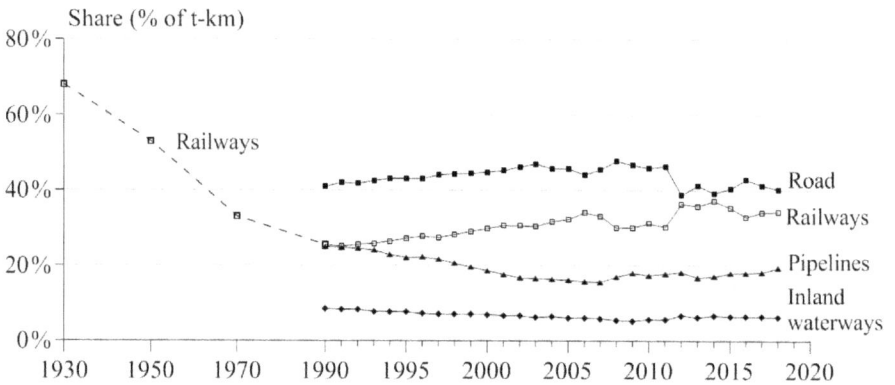

Fig. 1.22. **Evolution of share of various transport modes in the freight transport market of the USA, (1), (6)**

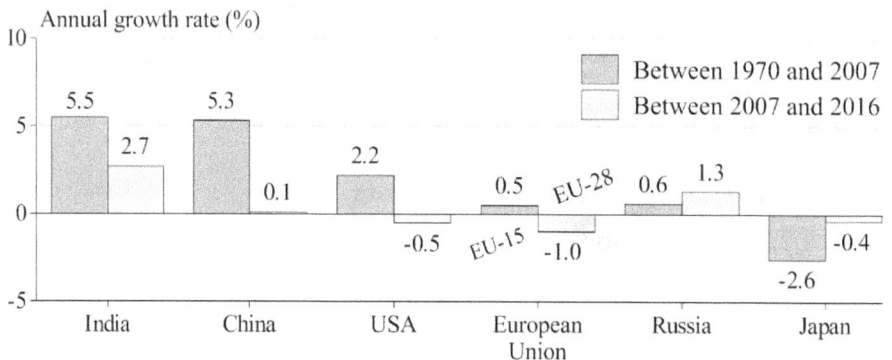

Fig. 1.23. **Growth rates (1970÷2016) of rail freight traffic for some countries (1), (2), (6)**

1.8. Railway traffic, length of lines, staff, and productivity of railways

There are worldwide more than 1.3 million kilometers of railway lines in operation in 2018 and railways realize a worldwide traffic of 3.7 trillion passenger-kilometers and 9.5 trillion tonne-kilometers per year. Table 1.1 illustrates for several countries all over the world for the year 2018, (1), (2), (6):

- length of railway lines,
- length of electrified railway lines,
- passenger traffic (in passenger-kilometers),
- freight traffic (in tonne-kilometers),
- staff numbers,
- railway productivity (in passenger-kilometers + tonne-kilometers per employee).

Length of railway lines of a country reflects the importance attributed by that country to its railway network (public investment, etc.), density populations, geographical characteristics (central, peripheral), and locations of production centers. Length of electrified lines is also a consequence of: public investment for railways in the specific country, density of railway traffic, environmental considerations, cost of production of electric energy. Thus, in Switzerland almost all the railway network is electrified, whereas the situation in the USA is just the opposite.

Levels of employment of railway staff depend on the volume of railway traffic but also on the existence or not of competition, on the degree of state subsidies, on interventionist state policies for unemployment, etc.

Passenger and freight traffic of railways are the outcome of competitiveness of services provided by railways, of the existence or not of state subsidies and competition, and of the level of tariffs and quality of service provided by other transport modes.

Productivity assesses the rates of a production activity and depends on a number of factors, such as labor, capital, etc. When we take into account only one input factor for the measuring of productivity, then we have a partial (or single factor) productivity. The most common indicator to assess partial productivity of railways is labor productivity, which is the ratio of the output volume to the total number of employees.

As output volume for the railway productivity is used the total number of traffic units realized and is usually calculated by adding the volume of passenger-kilometers to that of tonne-kilometers of rail traffic. Thus, railway productivity is usually expressed in passenger-kilometers + tonne-kilometers per employee. Railway productivity can be increased by combining some of the following measures: investment in new equipment, innovation, new organization,

**Table 1.1. Railway traffic, length of lines, staff, and productivity
for railways of several countries all over the world
for the year 2018, (1),(2),(6)**

Country	Length of lines worked (km)	Length of electrified lines (km)	Average staff (thousands)	Passenger traffic (billion p-km)	Freight traffic (billion t-km)	Railway productivity *
Australia	10,651	1,418	21.1	5.157	65.672	3,357
Austria	4,953	3,583	43.0	13.205	21.996	820
Belgium	3,602	3,112	33.2	10.710	7.280	542
China	99,460	77,177	1,619.2	1,461.211	2,295.331	2,320
Czech Rep.	9,406	3,213	41.2	10.220	16.564	650
Denmark	2,519	642	6.9	6.195	2.594	1,279
Finland	5,925	3,331	8.9	4.535	11.175	1,759
France	27,594	16,686	131.8	97.100	32.039	980
Germany	39,299	20,347	322.4	98.100	124.620	691
Hungary	7,752	3,183	36.6	7.770	10.584	502
India	67,415	34,319	1,227.0	1,157.174	738.523	1,545
Indonesia	5,483	471	28.9	16.932	15.091	1,108
Iran	9,306	181	11.6	14.874	33.646	4,183
Ireland	2,045	53	4.0	2.281	0.089	588
Israel	1,599	55	3.8	3.579	1.241	1,268
Italy	16,781	12,016	84.1	55.493	22.070	922
Japan	19,123	11,756	113.2	264.125	21.265	2,521
Kazakhstan	16,061	4,238	130.7	19.110	219.927	1,829
Netherlands	3,223	2,321	36.1	18.895	7.023	718
Norway	4,134	2,770	5.6	3.759	3.970	1,378
Poland	18,536	11,909	48.9	21.043	59.388	1,645
Portugal	2,546	1,696	5.1	4.570	2.765	1,438
Russia	85,497	44,067	710.6	133.381	2,601.928	3,849
Saudi Arabia	2,939	-	1.6	0.135	9.781	6,198
South Korea	4,111	2,990	27.5	23.002	7.878	1,123
South Africa	23,193	10,345	38.3	13.865	140.002	4,017
Spain	15,893	10,016	29.8	28.420	10.650	1,311
Sweden	10,906	8,078	9.1	13.547	23.358	4,060
Switzerland	4,032	4,032	33.9	20.590	11.776	955
Turkey	10,378	5,070	23.4	10.997	12.302	996
Ukraine	21,626	10,266	258.6	28.615	186.334	831
United Kingdom	16,294	6,125	79.2	69.706	17.206	1,097
USA	149,406	-	146.8	10.426	2,525.217	17,275

* thousands of traffic units (passenger-kilometers + tonne-kilometers) per employee.

24

new management, competition, motivation for efficiency (better salaries, bonuses, etc.), training. Productivity is a physical quantity and does not take into account financial terms (revenues, labor costs). As shown in Table 1.1., high railway productivity for several countries is a result of some of the following factors: a high traveled distance, size of the employed staff at the required minimum level (and not above), modern equipment and efficient management, strong competition in the sector, motivated staff. Average railway productivity worldwide was in 2018 2.42 million traffic units (passenger-kilometers + tonne-kilometers) per employee, (25).

1.9. Priority to passenger or freight traffic

Behind the numbers of rail share and traffic presented in the previous paragraphs, a big dilemma concerning many railways is hidden: should railways give priority to (and therefore facilitate the development of) passenger or freight trains? Almost in every part of the world, with the exception of the USA[*], passenger trains have been given priority (concerning departure-arrival times, investment, etc.). On the contrary, in the USA, the priority of railways is freight traffic, with a share of about 34.2% in the land freight transport market in 2018, whereas rail passenger traffic is rather marginal with a share of about 0.7% (2 generations of most Americans have never taken a train in their life). Share of air transport in the USA was 12.1% in 2018, that of buses 6.4%, passenger cars' share 80.4% and motorcycles' share 0.4%, (1), (2), (6).

While freight traffic volumes were stagnating in Europe, (see Fig. 1.19), freight traffic in the USA increased from 1,000 billion t-km in 1970 to 2,257 billion t-km in 2000 and to 2,525 billion t-km in 2018, a number that should be compared to a volume of 32.04 billion t-km for French railways and a volume of 124.62 billion t-km for German railways for the year 2018.

However, differences of rail freight share in the USA and Europe are also due to a number of other reasons, (13), (19), (25):

- traveled distance (around 1,500 km) by freight trains in the USA is much greater than in Europe, Japan, etc. Indeed, a freight train runs on average 1,800÷2,900 km in western USA, 650÷1,000 km in eastern USA, against only 160÷320 km in Europe, (19), (25),
- freight tariffs in the USA are far lower than freight tariffs in Europe,
- productivity in the USA rail freight sector increased from 2 million t-km per employee in 1970 (with 566,000 employees in total at that time) to 14.6 million t-km per employee in 2010 (with only 169,280 employees), which reflects an increase of productivity of 630% in 40 years,

---

[*] The situation is similar in Canada, where railway tracks are primarily used for freight traffic.

♦ state subsidies are granted in the USA only for regional passenger traffic and not for freight, whereas in many parts of the world railways receive subsidies not only for passenger but also for freight traffic. The situation is (at least theoretically) similar in the EU, where states can neither allocate subsidies to freight traffic nor tolerate cross-subsidies between passenger and freight traffic,

♦ investment for railways in the USA was oriented to the freight sector, with 100 billion US$ invested between 1970÷2000 in order to, (19):
  – extend loading gauge, (see section 7.10), so as to permit two-level freight wagons transporting containers to run the American rail network,
  – renew and innovate freight rolling stock, while reducing between 1970÷2000 the total number of traction machines by 30%, wagons by 25%, and lines in operation by 40%,

♦ liberalization of both road and rail transport resulted in the reduction of 30 big rail companies in 1970 to only 8 in 2010 and 6 in 2016, thus providing the possibility for economies of scale,

♦ with the exception of the North-East Corridor (Washington-Philadelphia-New York-Boston), where the real public interest is to avoid saturation in the airports of cities served, railway lines are used principally for freight traffic in almost every other place in the USA,

♦ when considering land transport (road and rail), shipments for distances higher than 1,000 km are realized by rail in the USA at a percentage of 70%, against only 15% in Europe,

♦ the operating revenue per tonne-kilometer is for European freight trains 3 times higher than for American freight trains,

♦ in the USA, there is competition as a result of deregulation, (see section 3.2.3). In the EU, both member-states and the European Commission are trying to force competition through regulation,

♦ as most railway tracks in the USA are dedicated to freight traffic, the maximum value of the axle load, (see section 7.5.1), is 30÷35 tonnes, against 22.5 tonnes for Europe (due to the coexistence of passenger and freight trains),

♦ a freight train has on average a length of 3,500 m in the USA, against only 750 m in Europe,

♦ as a result of higher length of freight trains and increased axle load, net tonnes transported per typical freight train are in the USA 9,000÷12,000 against 1,200÷2,000 in Europe.

Thus, the disappearance of passenger trains in the USA gave freight trains the possibility to be fully developed without any restrictions and cutoffs concerning departure-arrival times. Would this be realized, if passenger trains were still operating in the entire American rail network? The answer is that until a certain level of freight and passenger traffic (which is the case in the great majority of railway lines all over the world), freight and passenger trains can coexist without major inconvenience. For heavy traffic lines, however, authorities must decide whether their priority is passenger or freight traffic, (25).

1.10. Position of railways in the transport market, comparative advantages, and transport services with good prospects for railways

1.10.1. Competition in the transport market and comparative advantages of railways

Position and importance of railways in the transport market depend on the evolution of a number of factors which are the driving forces for transport: increase in population, economic output and increase in GDP, organization of the transport market (monopoly, oligopoly or competition, deregulation, liberalization), technological evolutions and innovations (internet, automations, digitalization, GPS), lifestyle and social attitudes, environmental measures (in view of a decrease of CO_2 emissions), low-cost products and services, relocation of companies, completely integrated logistics services, (9).

Competition between railways and other transport modes (private cars and trucks, airplanes, sea transport, pipelines, inland navigation) is commonplace all over the world. This form of competition is often characterized as inter-modal. However, in some countries (e.g. USA, Canada, EU, etc.) there is competition among many railway operators running on the same track (intra-modal competition). In other countries (China, India, etc.) a railway operator has a monopoly over a specific route. In any case, the transport market becomes all over the world more and more competitive. Within such a competitive environment, railways should promote their competitive advantages: high speeds, possibility to transport high volumes of passengers and freight, extensive applications of automations and intelligent technologies for the provision of just-in-time logistics services.

Predicting the position of railways in the transport market for a number of years ahead (5÷20) involves the assessment of the likely occurrence of events and excludes any possibility to take into account major, abrupt, or revolutionary changes. For instance, an eventual new sanitary crisis, or a new economic crisis, or an escalation of degradation of earth's climate (followed by a common decision at world level to make less use of airplanes and private cars and more use of electrified railways and other sustainable transport modes) are not excluded but cannot be predicted with a high degree of accuracy (when it will happen, to what extent, under what conditions).

1.10.2. Railways and high speeds

Cost and travel time are the two major factors considered by a traveler before deciding the selection among alternative transport modes. Private cars and buses have a technical barrier in the increase of speed beyond the level of 130 km/h, due to the technical constraints in the braking distance and the human limits regarding reaction time in case of danger. On the contrary, railways can achieve higher running speeds up to 350 km/h in 2021, while ensuring safety and high level of comfort. As railways reach city centers, where railway stations are located (airplanes cannot, as airports are situated in remote areas from the city center), they are easily accessible by bus or metro systems and can achieve door-to-door travel times far lower than other land transport modes. For distances up to 500÷600 km, high-speed railways can achieve better door-to-door travel times in comparison to airplanes, provide a level of comfort similar or even better compared to airplanes, ensure reliability of departure and arrival times (airplanes, private cars, and buses can suffer from traffic congestion), and offer satisfactory frequencies and standard intervals between successive trains.

For a 600 km trip, a high-speed rail passenger consumes 6 liters of fuel and emits 8.1 kg of CO_2, against 31.5 liters of fuel and 67.4 kg of CO_2 in the case of private car and 43.1 liters of fuel and 93.0 kg of CO_2 in the case of airplane, (38). In addition to lower emission of CO_2 and less consumption of energy, high-speed trains relieve congestion problems, facilitate regional development, and boost economic activity; these advantages can justify the financing of construction of new high-speed lines (or the upgrading of existing tracks) by the tax payer, that is by the state budget, as is the case in EU, China, Japan, South Korea, etc. On the contrary, in the USA there is still the point of view that the financing of new high-speed lines must be done (at least partially) by the beneficiary (the traveler) and not by the tax payer (the state or federal budget). This fact, combined with the low return rate of investment on high-speed railways, is the principal reason of reluctance of state and federal authorities in the USA for financing the construction of high-speed railway lines.

1.10.3. Urban rail services

In an era of exploding traffic problems, railways can decisively contribute to alleviate them through the large carrying capacity they provide, (Fig. 1.24). Many neglected railway lines connecting city centers to the suburbs are accordingly being modernized and used for urban rail services, thus relieving the traffic problems of many cities, (28).

Carrying capacity (in thousands of passengers per direction and hour)

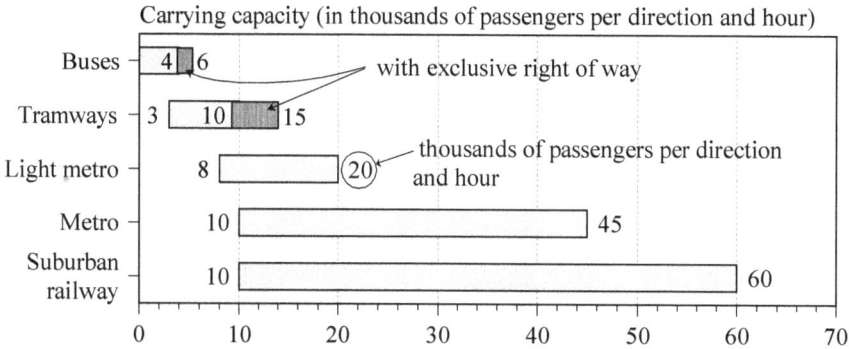

Fig. 1.24. Carrying capacity of various transport systems, (28), (34)

1.10.4. Bulk loads – Rail freight corridors

The metal-to-metal contact of the wheel of a railway vehicle with the rail results in considerably lower rolling resistance for rail freight wagons than for trucks. Thus, railways are in an advantageous position for the transport of bulk loads, such as raw materials, coal, petroleum, grain, and other agricultural products. Railway competitiveness in bulk load transport depends, among other matters, upon the marshaling yards facilities, where freight trains are disassembled and reassembled, and where long and (often) unjustified waits are occurring, (see section 15.14), (22).

Bulk freight of railways is very often transported unpacked in large quantities. Railway vehicles are usually open in their upper side, where loading and unloading is taking place. When bulk loads must be protected (from rain, dirt), covered vehicles are used. Covered railway vehicles can have opening doors on the sides or underside to discharge the load, and in this case they are called hopper vehicles. Liquids (petroleum, chemicals) and compressed gases are carried by rail in tank vehicles.

Punctuality, fluidity, regularity, reduced cost, safety are among the exigencies of the freight market. In the USA, where most railway tracks are used only by freight trains, fulfillment of the above exigencies is an issue of good management. However, in Europe, China, India, and elsewhere, freight trains run on the same track with passenger trains, the latter always having priority. For this reason and for lines with high density of rail freight traffic, it has been suggested that some railway tracks (known as dedicated rail freight corridors) are to be used exclusively for freight transport. The average speed of freight trains in Europe is lower than 20 km/h and this is a result of long waiting times in marshaling yards and frontiers and not of a maximum speed of freight trains of 100÷120 km/h. With dedicated rail *freight corridors*, it is hoped to increase in Europe average speed of freight trains up to 50 km/h and thus provide similar travel times compared to trucks, (15). Freight from Asia to Europe is forwarded almost exclusively by sea transport (90%) in 40 days, air transport in 1 day, rail transport (only 1%) in 15 days. The Eurasian rail freight corridor aims at increasing this very low share of railways between Asia and Europe.

1.10.5. Combined transport

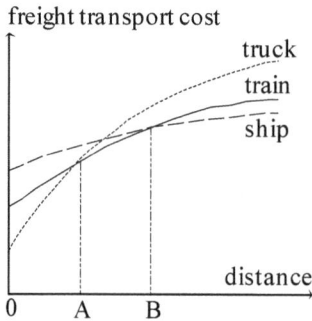

Fig. 1.25. Costs of freight transport as a relation of distance for various transport modes, (16)

The various transport modes present comparative advantages concerning freight transport costs in relation to distance, (Fig. 1.25). Thus, for short distances, trucks have a comparative advantage, for intermediate distances railways have an edge, while long distances favor the use of ships. Increasing competition in the area of freight transport, however, makes the search for the lowest cost compulsory.

The evident solution for a specific traffic itinerary is to combine more than one transport modes, by exploiting the advantages of each mode in relation to the distance covered. Combined transport can be defined as a composite transport process involving at least two consecutive transport modes (e.g. truck-ship, train-ship, truck-train). When no handling of goods is taking place in successively used transport modes, we use the term inter-modal transport. Some techniques developed for combined or inter-modal transport are the following:

– *Containers*, used in road, rail, and sea transport. The tendency is to use containers as large as it is allowed by the existing loading gauge, (33). External dimensions of containers are as follows:
  • 20 feet type: 6.058 m long by 2.438m wide,
  • 40 feet type: 12.192 m long by 2.438m wide,
  • 60 feet type: 13.716 m long by 2.438m wide.
  Due to their structure, containers can be stacked up to six in one pile.

– *The Ro-Ro* (Roll on–Roll off) technique, whereby whole trucks or truck bodies with freight are loaded on a train or ship, so that only a small part of the transport is covered by road. According to EU regulations, the maximum dimensions of freight vehicles for combined transport are: height 4.0 m, width 2.5 m, weight 40 tonnes. Instead of the term Ro-Ro, the terms Piggyback or Swap body are used for the above process.

– *Rolling road* or *highway*, in which trucks can drive straight onto the train, by using mobile ramps. During rail transport, truck drivers travel in seating or sleeping accommodation. This technique is used, among others, in the Channel tunnel for freight transport between the UK and France.

Since combined transport requires transshipment of freight from one mode to another (with associated expenses), it is necessary to determine the minimum distance beyond which combined transport becomes cost-effective. The answer to this question is not simple, since it depends on the cost of labor, energy, the mechanical equipment for transshipment, etc. Therefore, European conditions place this minimum distance at 700÷900 km, whereas in the USA it is around 1,500 km, (15), (32).

The development of combined transport necessitates the existence of a satisfactory level in road and rail networks and modern transshipment equipment. Figure 1.26 illustrates the cost components for rail-road combined transport for the economic conditions of Western Europe.

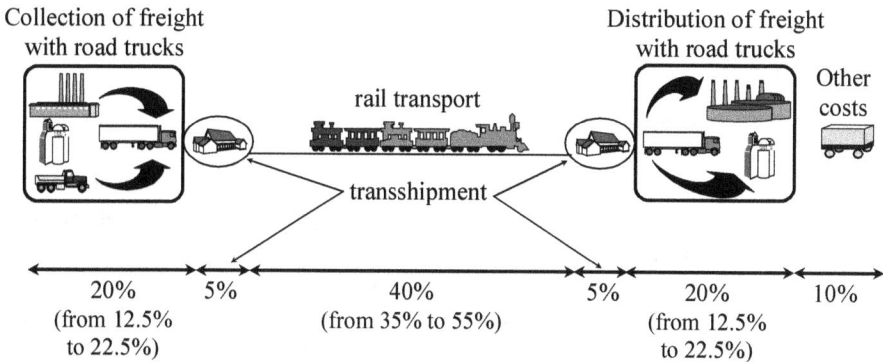

Fig. 1.26. Cost components for rail-road combined transport for the economic conditions of Western Europe, (15), (24)

1.10.6. Rail freight transport and logistics

Freight transport by rail was limited until some decades ago to carrying goods. The dynamics of modern transport, however, have broadened the scope of the transportation process. Reliable and speedy carriage is no longer sufficient. It must also be accomplished at the lowest possible cost, ensuring that a certain quantity of goods be made available at the required place and time with a direct responsiveness upon demand. An important contribution to this effect has been achieved during the last four decades with so-called logistics, which involves the whole process encompassing timely information on the need to make available a certain item at a specific place and time, reliable and speedy transport, possible storage, and final delivery to the recipient, (Fig. 1.27). It is therefore clear that in this sense the transportation process has a much broader meaning.

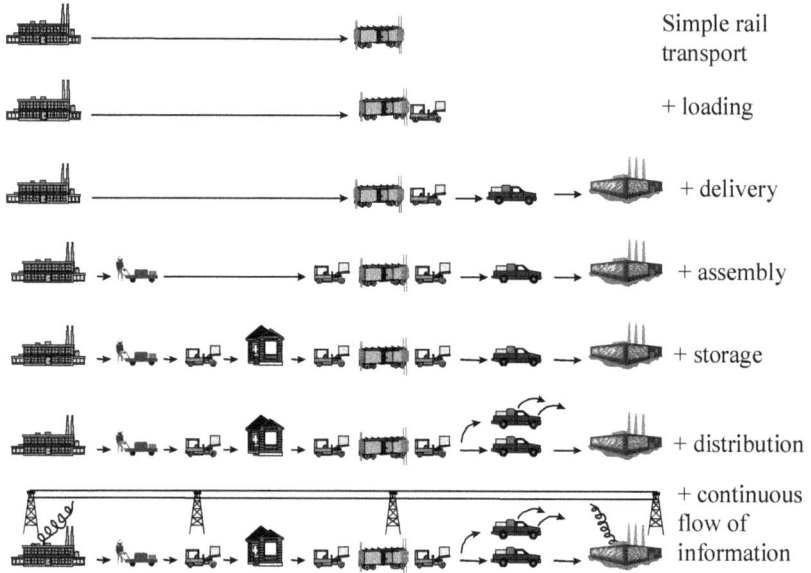

Fig. 1.27. From simple rail transport to logistics, (16)

The progressive computerization of all transport processes, the development of telematics, automations, and the increasing competition in the transport market lead to the consideration of logistics as a unified service (Logistics as a Service , LaaS) within a unified system from production to distribution and not as a succession of separate services and providers. The role and importance of railways within this new conception of LaaS may become in the future more significant than in the past.

1.11. Railways and air transport: competition or complementarity

1.11.1. Fields and conditions of competition and complementarity

Competition among the various transport modes is in most of the cases the rule. The choice of a traveler for a specific mode is the outcome of a complex evaluation, during which an assessment of the various characteristics of each travel mode is conducted: total travel time (door-to-door), total cost (if more than one transport modes are used), quality of services offered, reliability, and frequency. All the above characteristics are incorporated in the generalized cost of transport, (see section 5.1.5). When selecting a transport mode, a traveler aims at minimizing (in relation to the value of time) the generalized cost and consequently at maximizing his personal utility and obtaining the greatest benefits, (9). However, transport modes can cooperate in order to provide a composite inter-modal product that is the best solution for a traveler, who uses successively more than one transport modes (rail-air, rail-road, air-road) that function complementarily. Thus, for distances shorter than 500÷600 km and with travel times between terminal stations of less than 3 hours, railways have an advantage over the airplane, since they reach directly the center of cities served. On the other hand, for distances more than 1,000 km, the airplane has practically no competitor, as even the high-speed train with a maximum speed in the range of 300÷350 km/h cannot travel 1,000 km (with a number of stops in railway stations) in less than 3h 30min, (see Fig. 2.4).

For distances between 600 km and 1,000 km, rail and air transport are in competition and the rail share depends on door-to-door travel time (compared to airplane), cost, frequency, quality of services, etc, (Fig. 1.28).

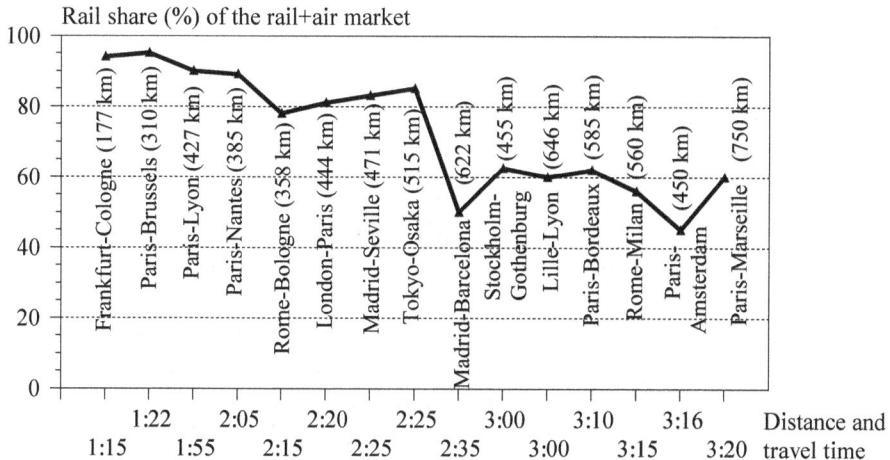

Fig.1.28. Rail share for some high-speed routes in relation to travel time and distance, (5)

Nevertheless, there are two fields where railways and air transport can cooperate complementarily: rail links to airports and medium distance rail connections from airports to other (than the city served) regions.

However, rail and air transport cannot work and cooperate efficiently unless a number of conditions are met, (5), (7):
- physical interconnection of the railway network with the airport, which means that the railway station reaches the airport with direct access to the air terminal and with facilities for people with disabilities,
- coordination of the railway timetables with those of the airline companies,
- eventually combined air / rail tickets with linked fares and simultaneous reservations (i.e. integration of the railway services into the computerized airline system),

– registration of luggage right to the final destination, which involves over-coming the difficulties associated with safety control. However, some airports (e.g. Heathrow, Gatwick, Madrid) which had remote luggage check-in facilities have withdrawn these, while Frankfurt, Abu Dhabi, Swiss, and other airports still offer such facilities.

1.11.2. Rail links with airports

Accessibility of an airport towards its catchment area with fast and reliable land transport connections at a reasonable cost is among the factors of success of the air transport industry. As road transport can suffer from unpredictable conditions of traffic congestion and usually costs more than rail transport, all major airports have efficient rail links to the center of the cities served. In 2021, even in the USA, where railways have a fair share of passenger transport, 15 airports have rail links to the cities served, with a lower cost compared to taxi services, (5), (18).

1.11.3. Rail connections of airports with remote areas

In an era of merciless competition among airlines and airports about extend-ing their catchment area, railways (and particularly high-speed ones), which take customers at the very heart of conurbations, may be a decisive factor for an airport in winning traffic from another city or region than the one served directly by the airport, (5), (7).

Thus, Brussels can be reached through Paris-Charles de Gaulle airport by high-speed train in 1h 20min. Similarly, through Frankfurt airport, Cologne (in 1h 02min) and Stuttgart (in 1h 17min) can be reached by high-speed trains.

In their aggressive commercial approach, airlines give preference to long-haul flights. Thus, railways (and particularly high-speed ones) can become a key element for the increase of the airline traffic by serving medium-size cities, which are not adequately served by air but lie on a principal railway line.

1.11.4. Low-cost air transport and railways

Low-cost is a fairly new conception for the design of products and services, by offering very competitive prices in the market, while providing to the client the essential features of the product or service. Low-cost airlines focus on the essentials of an air transport trip (in terms of travel time, reliability) and not on the non-essential functions, such as meals, drinks, easiness and space of seat, facilities of the airport. Low-cost airlines offer tickets at lower prices (compared to full service airlines), they are based in many cases on remote and less congested airports (where they pay lower landing fees), they provide neither meals nor drinks free of charge during the flight, they oper-ate only direct (and not connecting) flights, they allow only a small luggage free of charge, and they exploit at the maximum possibilities of the internet instead of employing staff. The model of their operation-organization focuses on the principle that a property of secondary importance of the air transport product ceases to be offered free to the clients, who are asked to pay for this property in order to have it.

Low-cost airlines compete harshly with high-speed trains and often offer lower tariffs. Nevertheless, as high-speed trains reach city centers, they achieve lower door-to-door travel times (and consequently a lower generalized cost), while offering higher comfort during traveling and possibility to take along an unlimited volume of luggage, and thus they succeed to retain their competitive position and higher shares in specific markets.

Implementing the principles of low-cost air transport to railways and creating a *low-cost high-speed rail service* is only partially possible because of the peculiarities of railways: technical rigidity, labor legislation, centrally situated railway stations, fixed costs difficult to reduce. However, some rail labor costs can be reduced by minimizing human involvement in purchasing a ticket, which can be reserved online and scanned by the ticket holder in the station. In this way, high-speed railways can introduce the low-cost concept in their services by generalizing an extensive use of automations and thus lowering their costs and tariffs, while offering a less stressful and more comfortable traveling environment (compared to low-cost air transport).

1.12. The sanitary crisis of 2020 and 2021 and its effects on rail transport

In the years 2020 and 2021 the world experienced an unprecedented sanitary crisis due to a very contagious virus (named Coronavirus, Covid-19), which outbroke in December 2019 and within three months it spread out all over the world. Governments were obliged to force citizens to stay home and telework or not to work at all (by indemnifying them), passenger transport was banned (not only between countries but also between cities within the same country), schools, universities, restaurants, commercial activities (out of supermarkets) were closed for several months; all these measures were taken in an effort to minimize personal contacts, as the virus was very spreading among individuals without masks at a distance less than 2 m, particularly in closed spaces, low temperatures (around 8°C), and high humidity.

Compared to the year 2019, air passenger traffic (expressed in passenger-kilometers) was in 2020 worldwide by 66% lower. Rail transport also suffered, particularly intercity travel, as citizens were not permitted to travel to cities out from their home city. Commuter traffic suffered also (though less), as citizens could move out of their home but only for specific reasons (work, if permitted, visit to supermarket, doctor); most leisure trips were banned between March-May 2020 and November 2020-May 2021. Compared to the second quarter of 2019, the number of rail passengers in the second quarter of 2020 was reduced for some European countries as follows: Ireland 90%, Netherlands 79%, France 78%, Spain 78%, Italy 77%, Poland 64%, Germany 59%, Sweden 56%, Finland 52%. Rail freight traffic was also reduced in 2020 and 2021, compared to 2019, but to a lesser degree, as most logistics services continued to operate, albeit an economic environment in recession. The GDP was shrank in 2020 by 4.2% worldwide, by 4.3% in the USA, by 6.0% in Germany, by 9.7% in France, by 9.8% in the UK, by 4.1% in Russia.

The situation started to ameliorate after May 2021 (though worsened again in December 2021), thanks to the development of vaccines, protecting vaccinated persons against the virus, and the vaccinations of large parts of world population. It is estimated that it will take around 3÷4 years after 2021 for the economic and transport activity to reach levels of 2019.

This sanitary crisis unveiled a number of inherent weaknesses of the rail (and more generally the transport) sector. Indeed, seats of passengers are very close to each other, whereas the distance required between individuals to reduce the risk of contagion is 2 m. Standing passengers (when permitted, e.g. commuter transport) can easily squeeze, as it is permitted to have up to 4 standing passengers/m^2 of space of rolling stock (rail cars). Wearing a mask was (and still is) a strange situation for most passengers. It will take some years (perhaps decades) to realize the long-term effects (changes of habits, methods of work, etc.) of the sanitary crisis of the years 2020 and 2021.

A similar crisis, due probably to another virus, cannot be excluded in the near or far future. In view of this eventuality, new regulations will appear and will confront issues such as:

♦ daily disinfection of rolling stock,
♦ availability of free masks for travelers,
♦ eventually obligation for any traveler, before boarding on the train, to present a sanitary certificate attesting that a vaccination has been done; however such a measure will raise issues related to human rights and liberties,
♦ adaptation of space and design of rolling stock, with more space available between seats of passengers, more spacious seats (similar to the first class); inevitably this will increase the cost of transport (for all modes). Redesigning and refurbishing all the existing rolling stock is a complicated task and time will tell what is the best way dealing with this. For this reason, most rail and transport companies put during the sanitary crisis an upper limit of the occupancy rate (load factor), that is the ratio of the number of occupied seats to the total number of available seats, at a level of 60% instead of 100% usually. That means that under sanitary cri-

sis conditions a seat will remain vacant between two adjacent occupied seats,
♦ stricter guidelines in stations, so as to avoid close contacts of individuals when queuing up, boarding on the train, waiting in seating areas, etc.

Railway managers must keep in mind that teleworking and professional (but also personal) meetings through virtual conferences will be more frequent in the future and this will be detrimental for transport in general and the railways in particular.

1.13. International railway institutions

International railway cooperation is realized within the frame of the following international institutions:

1.13.1. The International Union of Railways (*UIC,* from the initials of its French name 'Union Internationale des Chemins de Fer'), which was established in 1922 and had 204 members in 2021 (73 active, 67 affiliate, 64 associate), these being rail authorities or companies from various countries all over the world of both the infrastructure and the operation activity of railways. The general objectives of UIC are, (4), (15):
– promoting rail transport at world level, meeting current and future challenges of mobility and sustainable development, and improving the image and the overall coherence of the railway activity at world level,
– developing international railway cooperation and transactions, planning and implementing measures permitting railway services across national borders, ensuring quality in both passenger and freight traffic, improving environmental performance and competitiveness of railways,
– standardization and design of technical specifications concerning all components of railway activity (e.g. ballast, subgrade, electrification, rolling stock, etc.). Specifications and regulations of UIC are known under the name *code* or *leaflet* or (recently) *international railway solution,*
– informing international organizations, decision centers, and public opinion on the usefulness and advantages of transport by rail.

Within this general framework, UIC activities cover the following sectors:
♦ statistics and general information about railways. The data base RAILISA[*] of UIC provides statistical data covering 45 variables of the railway activity (e.g. length of tracks, passenger and freight traffic, number of rolling stock vehicles, financial results, etc.) of almost all railways around the world,
♦ research of new technological advances concerning track, rolling stock, etc, conducted principally within its own research units,

---

[*]Rail Information System and Analyses.

- planning the optimization and rationalization of the railway technical equipment, exploitation methods, data processing, etc., by using the appropriate UIC specifications related to the classification of locomotives and their axle arrangements, coaches, goods wagons, etc.,
- developing interoperability (technical but also commercial, see sections 1.15 and 21.10),
- advancing signaling systems and equipment of safety and security,
- optimizing environmental performance of railways and particularly reducing CO_2 emissions,
- allocating revenues and offsetting debit between railway operators for international traffic of railways crossing more than one country.

The work of UIC is provided at 3 levels: strategic, technical-professional, support services.

1.13.2. The *European Institute of Rail Research (ERRI)* known also with the initials *ORE* of its former French name ('Organisme des Recherches et d'Essais'), which was until 2004 an agency of the International Union of Railways aiming to organize and coordinate research and test procedures that advance railway technology. Topics investigated are divided into the following five categories: A (traction, signaling, telecommunications), B (rolling stock), C (interaction between rolling stock and track), D (track, bridges, tunnels), E (materials' technology). Nowadays, research within the UIC is conducted within each technical department and is coordinated transversely by a research and innovation unit.

1.13.3. The Community of European Railways and Infrastructure Companies (CER) of the European Union member-countries, candidate-countries, and also Norway and Switzerland. The CER aims at establishing common positions and policies of European railways covering all aspects of the rail transport activity, such as infrastructure planning, passenger and freight services, the environment, etc. Among the many initiatives to promote railway cooperation within the EU area is *ERRAC* (European Rail Research Advisory Council), a joint initiative of the European Commission and the railway industry.

1.13.4. The European Union Agency for Railways (ERA). Though UIC covers (at world level) design and specifications of technical issues of railways, the EU decided to create its own agency for such issues. Thus, the European Union Agency for Railways (ERA) was created with the mission to coordinate European railways on issues related to safety, interoperability, and quality of service, establish common policies and strategies, set mandatory requirements for European railways and manufacturers, and since June 2019 deliver safety certificates, authorization of vehicles that can move on the railway network of the EU, and approval of trackside signaling systems.

1.13.5. The European Committee for Standardization. Another authority that exercises key influence on specifications and regulations related to the various components of the railway system is the European Committee for Standardization (CEN, from its French initials 'Comité Européenne de Standarisation').

1.13.6. The International Transport Forum (ITF), which is an inter-governmental organization integrated within the Organization for Economic Cooperation and Development (OECD). ITF operated till 2007 under the name European Conference of Ministers of Transport (ECMT). ITF has its own data base and publishes regularly scientific reports on issues such as transport policy, financing of infrastructure, logistics, traffic congestion, safety, CO_2 emissions, sustainable development, and social inclusion.

1.13.7. The railway section of the *World Bank,* which has its own data base for railways and publishes reports on special issues of railways, such as lending of railways, competition and regulation, investment and financing, rail's impact on mobility and sustainability.

1.13.8. Transport organizations of international institutions, such as the *United Nations,* which aim at the promotion of international railway transport and cooperation. Among them we should mention the United Nations Economic and Social Commission for Asia and the Pacific (ESCAP) and the Organization for Cooperation of Railways (OSJD), which was a similar to UIC organization of former socialist countries of Eastern Europe (Russia, Poland, etc.) and China, Vietnam, Mongolia.

1.14. The rail industry worldwide

The rail industry may refer to the following components of the railway system:
- infrastructure (subgrade, subballast, ballast, sleepers, fastenings, rails, electrification equipment),
- engineering systems (signaling, control/safety, telecommunications),
- rolling stock (locomotives (diesel or electric), passenger vehicles, freight vehicles, high-speed vehicles, metro vehicles).

The worldwide market of industries of railway technologies (infrastructure, engineering systems, and rolling stock) is estimated at 115 billion US$ for the year 2006, 150 billion US$ for the year 2010, 182 billion US$ for the year 2014, and 163 billion US$ for the year 2017. The share of each of the above components for the period 1990÷2010 was as follows: infrastructure 50%, rolling stock 39%, and engineering systems 11%. However the financial-economic crisis of 2008÷2012 resulted in the delay, postponement, or even cancellation of the construction of a number of new railway lines (particularly in Europe). The rate of construction of new lines in China has also declined since 2012. The combined effect of these two phenomena was a change in the share of the three railway components, which was worldwide for the year 2014 as follows: infrastructure 29%, rolling stock 60%, and engineering systems 11%. Conventional, urban, and high-speed railways had the following shares (in 2014) at world level in the rail technologies market: conventional railways 74%, urban railways 14%, and high-speed railways 12%, (4), (9), (14).

More specifically, the world rolling stock market was dominated in 2018 by the following constructors: Bombardier 23%, Alstom (which absorbed Fiat Ferroviaria) 14%, China South Locomotive & Rolling Stock Corporation (CSR) 14%, Siemens 11.5%, China North Locomotive and Rolling Stock Industry Corporation (CNR) 11%, General Electric 7.5%, Kawasaki 5%, Construcciones y Auxiliar de Ferrocarriles (Spain) 5%, Transmashholding (Russia) 4%, other companies 5%.

Following the example of Airbus, efforts began two decades ago for a closer cooperation of European companies in view of a construction of a high-speed train combining the advantages of the French TGV and the German ICE. However, an attempt to merge Alstom and Siemens in one company was blocked by the competition authorities of the European Commission in February 2019.

Though Europe was for many decades the most significant market for the rail industry, in recent years construction of many new tracks (particularly high-speed) and new metro systems in China, South Korea, Taiwan, and elsewhere shifted the center of interest of the rail industry from Europe eastwards.

1.15. Railway interoperability

The railway industry and railway companies have been presenting many differences in their products for many decades, due to state protectionism and subordination of railways to national industrial targets and regulations. Thus, a great variety of gauges (see section 7.4), electrification systems (see section 20.6, Fig. 20.4), signaling and traffic control systems (see section 21.10, Table 21.3) render railway cooperation difficult.

For instance, a train cannot have a continuous route from Lisbon to Paris because of difference of gauge, electrification, and signaling systems, from Paris to Amsterdam because of incompatibility of electrification systems, from China to Kazakhstan because of difference of gauge. However, a future-oriented rail transport system, such as that currently taking shape in Europe, Asia, and other continents, should break free once and for all from national boundaries and regulations. *Interoperability*, in its strict sense, can be defined as the technical compatibility between the railway rolling stock or track equipment of different countries or industries, allowing the safe and uninterrupted movement of trains, while accomplishing the required levels of performance. Principal technical obstacles that interoperability tries to tackle concern: track gauge, electrification systems, signaling systems, loading gauge of trains, platform length and height, train length, axle load, systems that pertain to communication, control-command, and safety. The above described characteristics refer to the *technical* component of interoperability, which is analyzed in more detail in section 21.10.

Apart from confronting technical incompatibilities, the goal of railways should be an unbroken international service to their customers, meeting the same high quality standards throughout, whatever the length of the journey or the number of countries crossed. Thus, in addition to the harmonization of technical systems, interoperability should aim to similar levels of quality of service and more easily accessible (passenger and freight) information and distribution systems for customers worldwide. These characteristics refer to the *operational* and *commercial* components of interoperability, which are also analyzed in the section 21.10.

A third component of interoperability is the legal one. In contrast to air and sea transport, which have been developed at an international level and thus follow international standards, railways have been developed within national frontiers and interests and thus lack international standards legally compatible. *Legal* interoperability does not entail unique and common forms of organization, standards, or structures, but a compatibility of the legal and institutional framework of railways between the various countries. A typical example of legal interoperability is the COTIF convention for the international carriage by rail, with its various annexes (CIV for the international carriage of passengers, CIM for the international carriage of goods, etc.), (see also section 3.13). Legal interoperability should provide a compatible legal railway regime for the transport contract documents and consignment notes, for the liability and the charges of rail operators for the use of infrastructure, etc. However, interoperability can be considered within a *broader sense* as a property of a product, service, or system with a clear definition and known interfaces with other systems, permitting the specific product, service, or system to work efficiently and continuously (at present and in the future) without any kind of restrictions, (25).

1.16. Applications of GPS in railways

Some critical issues for the operation of railways are the following: knowledge of the accurate position of a train, its speed, the expected arrival time, whether and where other trains are moving on the same or converging tracks, how many seats are vacant on a train at any moment, whether more trains can run on the specific track. Accounting for all the aforementioned issues will result in reducing delays and operation costs, preventing accidents and collisions of trains, providing real-time information to passengers for delays, arrival, and departure times, increasing carrying capacity of track, exploiting the margins of the carrying capacity of a train, increasing customer satisfaction. Confronting the above problems can be easily conducted today with the help of applications of the Global Positioning System (GPS), a satellite-based navigation system which can provide at any moment accurate and real-time information about the position and speed of any moving or fixed point on the earth.

GPS is based on data provided by 24 satellites flying on a circular move around the earth at a height of 20,200 km with a period of rotation of 24 h. Satellites send out signs that are received on earth by the appropriate receivers, (Fig. 1.29). If we know the three coordinates (latitude, longitude, altitude) of a point, then we can define its accurate trace on earth. This is achieved with the help of GPS, when a receiver has in view at least four satellites. Indeed, the three unknown coordinates of the position of a point can be calculated by using information that is provided by three different satellites. Additional information drawn from a fourth satellite is required in order to assess the error and accuracy (of the order of 20÷30 cm) obtained from information received from the three satellites.

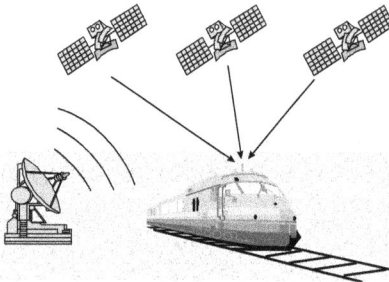

Fig. 1.29. GPS applications in railways

Applications of GPS cover almost any aspect of everyday human activity, such as the monitoring and tracking of the position of any moving point (buses, cars, animals, pets tracking, even depicting stolen cars or cars taken by adolescents from their parents, etc.). GPS are the basis for driverless road vehicles, driverless rail vehicles, and intelligent transport systems.

By installing a GPS in a locomotive of a train, we can know its real-time position at any moment. Having available the accurate position of a train between two consecutive time moments, we can calculate its speed, the required time to reach a specific railway station, and eventual delays. This information is very useful for the passengers on the train or those waiting in the stations, but also for the railway staff in charge of safety and control of speed of trains. In addition, a train driver can have full knowledge whether the track is free in front of him, of the position of other trains (in front or behind) moving on the same track or on converging tracks, of the existence of a train in the opposite direction, of eventual obstacles on the track. This information is also critical for all warning and intervention systems in order to avoid collisions and increase safety. Furthermore, a train driver can communicate directly with drivers of other trains (equipped also with GPS) moving around him and thus improve both safety and operation conditions. Knowledge of the available space between consecutive trains gives the possibility to the train operation center to increase carrying capacity of the track and permit other supplementary trains to run. GPS on the trains can also provide information about the available free seats and thus optimize the existing capacity of a train. All the above contribute to the reduction of operating costs and the increase of customer satisfaction, (see also section 19.8, where other than GPS satellite technologies are also analyzed).

Railways can achieve with the help of GPS a control of operations similar to airplanes, which have a complex but very safe system of navigation based on centralized control. Inspired from the success of Eurocontrol, the European system of management of air traffic, international institutions such as the ERA or the UIC should aim to install an international system for the management of railway traffic and the continuous monitoring of rail vehicles in greater regions such as Europe, central and eastern Asia, etc.

1.17. Big data, Internet of Things, Artificial Intelligence and railways

Operation, construction, and maintenance of railways are based on a number of data, which should be accurate, reliable, and provided in real time. Until some decades ago, data were collected almost exclusively manually and then were used in computations, conducted either manually (till the appearance of calculators) or with the help of computers (the last $4 \div 5$ decades). However, evolutions in electronics and information technologies give the possibility to collect data automatically, without any human intervention. Thus, it is possible nowadays to handle at high speeds big volumes of data coming from various sources and then turn these data into information useful for clients, decision makers, etc. This is what we call big data, (9).

Big data is a new and revolutionary set of advanced techniques (combining software and hardware), which give the possibility to capture, manage, and analyze very large volumes of extremely varied data collected electronically with the help of the internet, mobile telephones, GPS, and various other devices, (27). Big data constitute a seismic shift in our way to record and understand the surrounding world and are the effect of a sharp decrease of costs related to the collection, storage, and processing of information. Principal characteristics of big data are the volume, velocity, and variety of data that can be recovered and processed. In the new digital world, people and goods leave a digital trace in their everyday activity and thus everybody knows about others' activities, while losing a significant number of degrees of freedom at the same time. Big data should be used by respecting scrupulously privacy rights of citizens, as they are in force in a specific state.

Big data change radically the complex nature of railways (both capital and work intensive). The digital railway infrastructure and rolling stock are loaded with electronic devices which generate a lot of data, while providing information about the accurate position of a train, its speed, the accurate mechanical and geometrical characteristics of rail infrastructure, the level of comfort (through information related to vibrations), whether doors are open or closed, etc.

In the following chapters it is explained how big data can be used for specific rail problems. Thus, big data can contribute to:

– providing information on arrival-departure times and on eventual delays (by monitoring at any moment the accurate position of the train),
– minimizing train delays and optimizing train schedules,
– adapting tariff policy at any given moment by knowing how many seats are vacant and available and by offering these for sale to clients at attractive prices,
– increasing carrying capacity of infrastructure; when knowing the position of successive trains, it is easy to calculate whether an additional train can be put on the track,
– enhancing safety by detecting any obstacle or danger on the track and by controlling and adapting the speed of the train,
– avoiding collisions of trains and accidents in level crossings,
– optimizing maintenance of both the track and the rolling stock by accurately monitoring values of defects of the track (see section 16.4), the rail (see section 10.9), the wheel (see section 19.2), the springs of the primary and secondary suspension (see section 19.4),
– managing efficiently vehicles of the fleet of a company(by knowing their accurate position),
– ensuring energy efficiency (by adopting continuously the conditions of operation, such as speed, acceleration, etc.),
– supervising driver performance, behavior, and eventual errors,
– facilitating the use of driverless rail vehicles,
– improving the life cycle of all materials and items (through accurate monitoring of their evolution over time),

Sets of big data that include storage sites, computing infrastructure, various platforms, and software services constitute what we call a *cloud computing*. Cloud computing that covers large parts of the world forms what we call *Internet of things* (IoT), (9). Once the IoT is applied to railways, it can create the IoT to railways, through which trains become interconnected communication hubs that can transmit data to network control centers (and through them to clients), fleet managers, drivers, etc.

Another revolutionary evolution of the last two decades which will be extended and generalized in the current decade is artificial intelligence (AI), that is the science of making machines or systems do things that would otherwise require human intelligence, (9). There is a great variety of AI systems, all of which are programmed to think like people, to act like people, and to think rationally and reasonably. Applications of AI for railways are given in the following chapters, more particularly for the forecast of rail demand (section 4.10), for condition-based management (section 16.12), for the optimization of the fleet of rail vehicles (section 19.11), for signaling systems (chapter 21), for safety (chapter 22), and other, (27).

Even there is a misperception that AI will make railways (and other sectors of the economy) operate automatically and independently of any human intervention, this does not hold true. Though less railway staff will be needed in the future, there will always be a need for more qualified specialists in charge of conception and operation of AI systems, processing of data, interpretation of results, etc. However, additional efforts will be required to homogenize AI systems and reduce differences concerning methods of censoring, monitoring, and communicating among them.

2 High Speeds, Magnetic Levitation, and Hyperloop

2.1. The evolution of high speeds on rails

2.1.1. Definition of high-speed trains and evolution of speed

High-speed trains (HST) were the response of railways to the transport market requirement for reduced travel times. However, there is no universally accepted top speed, beyond which a system can be characterized as HST system. It is generally accepted that existing conventional railway technology, along with improvements in the track and rolling stock, can accommodate top speeds not exceeding 200 km/h. Beyond this speed, additional capital costs are needed to meet the requirements of more stringent design features and sophisticated system components. Thus, we consider HST when V>200 km/h. This definition of HST is included in the European legislation, namely in Directive 49/1996 and Regulation 1315/2013. High speeds were pioneered by two railway networks:

- the Japanese railways, which operated in 1964 the 'Shinkansen' high-speed line between Tokyo and Osaka, with a top speed of 210 km/h, increased in 1985 to 240 km/h and later up to 300 km/h, depending on the section of the line,
- the French railways, which operated in 1981 the TGV[*] high-speed train between Paris and Lyon, with a top speed of 260 km/h, increased to 270 km/h in 1983 and to 300 km/h in 1989.

Both lines were built on heavily traveled routes showing signs of saturation. Faced with improving the existing infrastructure or building a new high-speed line, the latter was opted, (37), (48).

Figure 2.1 illustrates the evolution of maximum speed of trains, in operation and in test runs.

---

[*] TGV: Train à Grande Vitesse.

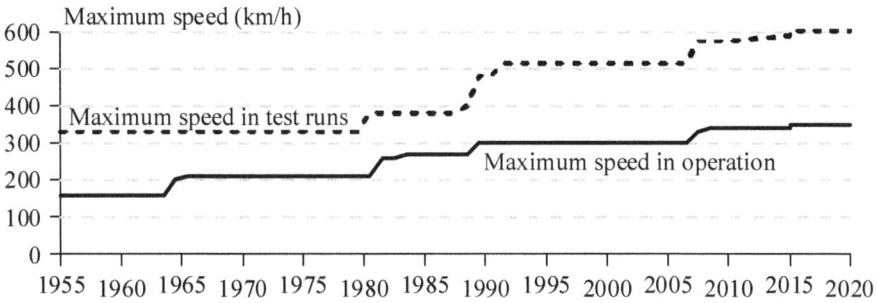

Fig. 2.1. The evolution of maximum speed of railways, (37)

2.1.2. *Panorama of high-speed lines around the world*

High-speed lines were constructed from 1964 to 2021 in the following countries:

- Japan (Tokyo-Osaka-Fukuoka-Kagoshima, Takasaki-Nagano-Kanazawa, Tokyo-Aomori-Hakodate, Tokyo-Niigata) with V_{max}: 320 km/h.
- France (Paris-Lyon, Paris-Bordeaux, Paris-Marseille, Paris-Lille-Calais, Paris-Strasbourg, Paris-Rennes) with V_{max}: 320 km/h.
- Germany (Hannover-Würzburg, Mannheim-Stuttgart, Hannover-Berlin, Aachen-Cologne-Frankfurt) with V_{max}: 320 km/h.
- Italy (Rome-Florence, Rome-Naples, Turin-Milan-Bologna-Florence) with V_{max}: 300 km/h.
- Spain (Madrid-Barcelona, Madrid-Valladolid, Madrid-Cordoba-Seville, Cordoba-Malaga, Madrid-Valencia) with V_{max}: 320 km/h.
- Belgium (Brussels-Lille) with V_{max}: 320 km/h.
- The Netherlands (Amsterdam-Brussels) with V_{max}: 300 km/h.
- The United Kingdom (London-Dover) with V_{max}: 300 km/h.
- Russia (Moscow-St. Petersburg) with V_{max}: 250 km/h.
- Turkey (Ankara-Istanbul) with V_{max}: 250 km/h.
- South Korea (Seoul-Busan) with V_{max}: 300 km/h.
- Taiwan (Taipei-Kaohsiung) with V_{max}: 300 km/h.
- USA (Northeast Corridor (Washington-New York-Boston) with V_{max}: 240 km/h.
- China (Beijing-Shanghai, Ningbo-Xiamen, Zhengzhou-Xian, Nanjing-Wuhan-Guangzhou-Shenzhen, Beijing-Zhengzhou-Wuhan-Guangzhou) with V_{max}: 350 km/h.

Table 2.1 illustrates total kilometers of high-speed lines around the world (in operation (2018), under construction (2018), and planned), with the corresponding maximum speed in each case. A total of 42,978 kilometers of high-speed railway lines with $V \geq 250$ km/h were in operation worldwide in 2018 (around 4% of total railway lines all over the world).

Table 2.1.

High-speed railway lines with $V \geq 250$ km/h (in operation (2018), under construction (2018), planned in various countries all over the world), (1), (38)

Continent and country	Kilometers of high-speed lines		
	Kilometers in operation (2018)	Kilometers under construction (2018)	Kilometers planned (short-term plans)
Europe			
Austria	268	281	0
Belgium	209	0	0
France	2,814	0	0
Germany	1,620	147	81
Italy	896	53	0
Netherlands	90	0	0
Poland	224	0	484
Russia	0	0	770
Spain	2,852	904	1,061
Sweden	0	11	0
Switzerland	144	15	0
United Kingdom	113	230	32
Total Europe	9,230	1,641	2,428
Asia			
China	27,684	10,026	1,268
Taiwan	354	0	0
India	0	0	508
Iran	0	0	1,351
Japan	3,041	402	194
Saudi Arabia	453	0	0
South Korea	887	0	49
Turkey	594	1,153	2,230
Total Asia	33,013	11,581	5,600
Other countries			
Morocco	0	200	0
Brazil	0	0	0
USA	735	192	1,710
Total other countries	735	392	1,710
Total world	42,978	13,614	9,738

Though many European countries had planned a number of new high-speed rail lines, the economic crisis of 2008 combined with competition from low-cost airlines in some of these countries led governments to delay or even cancel a number of such projects. Thus, China is the country where high-speed rail lines are increasing most rapidly. Indeed, although China is building highways rapidly, it is impossible to maintain highway traffic or private car ownership at the current level, due to high rates of economic growth. Thus, in order to support mobility in China, HST may appear as the only cost-efficient and viable solution, (8).

Figure 2.2 illustrates evolution of the total length of high-speed lines in the various continents of the world. In 2018 71.2% of high-speed railway traffic (in passenger-kilometers) is realized in China, 10.8% in Japan, 5.9% in France, 3.3% in Germany, 1.7% in Spain, 1.6% in South Korea, 1.6% in Italy, 1.2% in Taiwan, and the remaining 2.7% in other countries of the world.

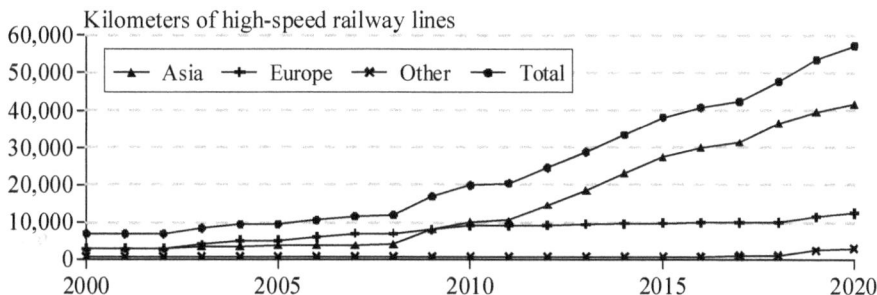

Fig. 2.2. Evolution of the total length of high-speed railway lines with V≥200 km/h for the various continents around the world, (1), (38)

In the USA, a number of routes (Table 2.2) have been suggested as candidates for new high-speed lines. It has been difficult, however, to devise a trustworthy funding model. Most European and Asian high-speed lines have been constructed by public funding. Such a model cannot work in the USA, where a balance and a compromise should be targeted among the private sector, the states, and the federal government. The most promising high-speed line to be realized in the USA was the one connecting Los Angeles to San Francisco with V_{max}: 350 km/h. But, though a part of the project was under construction, it was scaled back in 2019. Thus, the suggested high-speed routes illustrated in Table 2.2 are still in 2021 at the stage of projects.

2.1.3. High speeds for only passenger or mixed traffic

Before deciding the construction of a new high-speed line, we can distinguish two approaches, (45), (48):

Table 2.2.
Suggested routes in the USA for new high-speed rail lines, (8)

Corridor	Length of line (km)	Corridor population in 2050 (million)	Corridor trips in 2050 (million)	Infrastructure costs (million US$ of year 2009)
California (Sacramento-San Francisco-Los Angeles-San Diego)	1,751.0	54.1	101.0	35,904÷63,104
Pacific Northwest (Vancouver-Seattle-Eugene)	751.5	14.5	12.3	7,005÷9,340
Florida (Tampa-Orlando-Miami)	769.3	31.6	28.9	7,170÷26,768
Chicago Hub (Minneapolis-Chicago-Detroit-Cleveland-Pittsburgh-Kansas)	3,497.1	39.1	66.0	49,151÷74,795
South Central (Dallas-Austin-S.Antonio, Dallas-Oklahoma, Dallas-Little Rock)	1,934.4	33.0	63.9	14,424÷52,888
Southeast (Birmingham-Atlanta-Jacksonville-Raleigh)	2,670.0	33.2	84.4	29,862÷49,770
Gulf Coast (Houston-New Orleans-Mobile)	1,648.0	22.0	21.6	18,432÷30,720
Northeast Corridor (Washington-New York-Boston)	735.5	54.5	35.0	11,425÷26,049
Keystone (Pittsburgh-Ney York)	782.1	16.6	9.9	11,178÷17,010
Empire (Buffalo-Boston)	1,013.9	28.1	22.6	12,600÷17,010
Northern New England (Boston-Montreal)	1,070.2	15.3	9.9	13,300÷17,955
Total	16,623.0	342.0	455.5	210,451÷385,409

♦ in the first approach, only passenger trains run on high-speed lines, with low axle loads, very small tolerances of track defects, and large longitudinal gradients (up to 35÷40 ‰). Running HST on dedicated tracks, only for passenger trains, presupposes a high rail passenger traffic to make the construction and operation of the new line cost-efficient,

♦ in the second approach, the new high-speed lines are run by both passenger and freight trains, the coexistence of which entails higher maintenance costs and requires lower values for the longitudinal gradient (up to 12÷15 ‰). A number of high-speed lines have been designed for mixed traffic (both passenger and freight trains).

In any case, for a specific HST system, top speed represents a compromise among the additional capital investment required to achieve a top speed, the higher operating cost, and the travel time savings.

HST run in 2021 with a maximum speed of 350 km/h in China, 320 km/h in Japan, France, Germany, 300 km/h in South Korea, Taiwan, Italy, and elsewhere. However, the Beijing-Shanghai high-speed line was designed for a maximum speed of 380 km/h, but due to high operating costs maximum speed was reduced to 350 km/h. What is more, a further increase of speed beyond 380 km/h seems difficult to be realized, due to the following inherent limitations of the rail technology, (44), (48):

- difficulty in collecting electric power,
- reduced adhesion between wheel and rail at higher speeds, causing wheel slip,
- greater size and weight of on-board equipment.

2.2. High-speed trains and their impact on the rail market

2.2.1. High speeds and population concentrations

High speeds require new lines or major improvements on existing lines. The high construction and operation costs, (see also Table 5.1), cannot be justified, unless a large number of rail trips is realized daily. A first index for the justification of a new high-speed line may be population concentrations on both ends or along the line, (Fig. 2.3). For a new high-speed line to be economically justified, a population of ten million people at the one end and four million people at the other may be considered as a rough first criterion. Otherwise, high-speed lines may become a non-profitable activity, (37).

2.2.2. Impact of high speeds on the reduction of rail travel times

The reduction of travel times has been a constant goal of railways, as can be seen in Table 2.3. Only with HST, however, were railways able to achieve on 500÷600 km routes door-to-door travel times equal to or better than air transport and thus compete efficiently with airplanes.

Indeed, HST capitalize on their advantage to reach city centers and thus make travel times from the center of a city to the center of another one far shorter than for private cars and even, in many cases, shorter than for airplanes, (Table 2.4).

Thus, HST reduced travel times in the Madrid-Barcelona (622 km) route from more than 6h to 2h30min, in the Berlin-Hamburg (286 km) route from 2h20min to 1h55min, in the Milan-Rome (560 km) route from more than 4h to 2h55min, in the Taipei-Kaohsiung (345 km) route from 4h to 1h30min, and in the Beijing-Shanghai (1,318 km) route from around 10h to 4h24min.

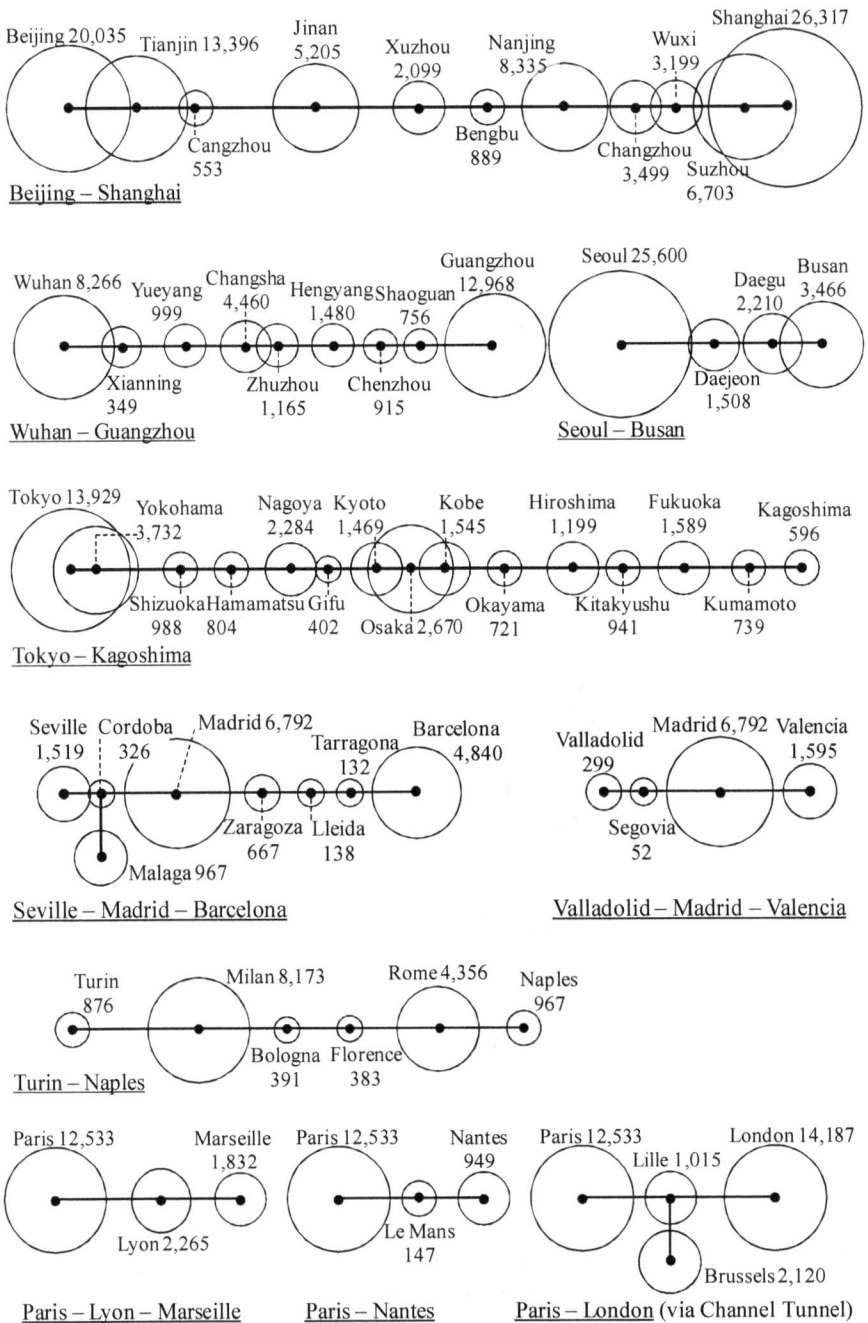

Fig. 2.3. **Population concentrations (in thousands) along major high-speed lines around the world. The greater area of each city is considered**

52

Table 2.3.
Rail travel time reduction on certain high-speed routes

	1960	1980	1983 (TGV line, length 427km)	1987	1999	2021
Paris-Lyon (511 km)	4h00min	3h50min	2h00min	1h50min	1h50min	1h56min

	1963	1965 (Shinkansen)	1990	1999	2021
Tokyo-Osaka (515 km)	5h30min	3h10min	2h30min	2h30min	2h21min

Table 2.4.
Comparison of travel times from the center of a city to the center of another one for railways, airplanes, and private cars (case of the Paris-Lyon route), (48)

HST travel time (1h56min) + access time to and from railway stations	Airplane [1]	Private car [2] (on highway at a top speed of 120 km/h)
3h	3h30min÷4h	5h

[1] The time indicated is the sum of the flight time, travel time from the city center to the airport, and check in time and retrieval time.

[2] The time indicated is the time from the center of a city to the center of another, i.e. it takes into account the time necessary (about 30 min) for a car to reach the highway from the city center.

Figure 2.4 illustrates travel times before and after the operation of HST for the most traveled high-speed routes around the world.

If we try to correlate high-speed rail shares with travel times, a linear correlation may be established, (Fig. 2.5), with a satisfactory value for the coefficient of determination R^2 (R^2=0.76), (see also section 4.4.1). The correlation is less satisfactory between rail share and traveled distance (R^2=0.67), (Fig. 2.5).

2.2.3. High speeds and new rail traffic

Another result of high speeds was the increase of traffic, either as diverted demand from air and road transport or as totally new demand (generated demand). Figures 2.6 and 2.7 illustrate high-speed rail traffic in countries with high-speed lines. More particularly in China, yearly high-speed ridership was reported to be 290 million passengers in 2010, 1,161 million passengers in 2015 and 2,001 million passengers in 2018.

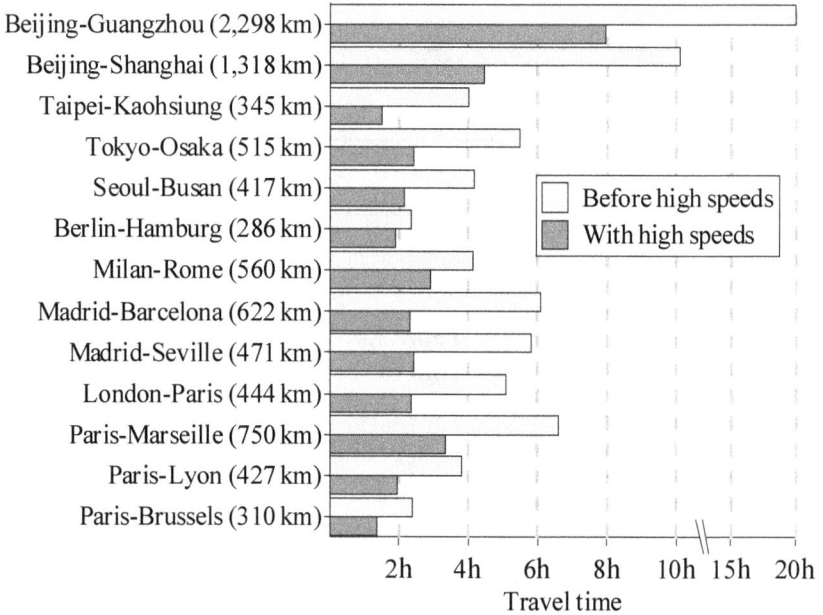

Fig. 2.4. Travel times before and after the introduction of high-speed trains, (1), (38)

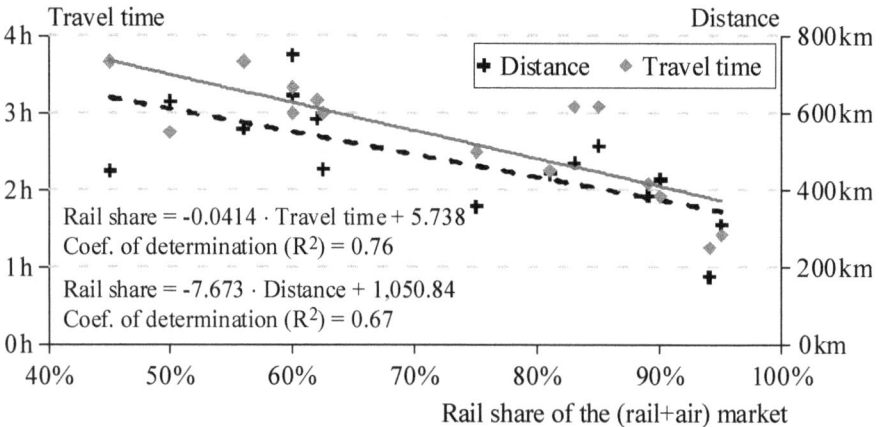

Fig. 2.5. Rail share in relation to travel time (— line) and distance (- - - line), (5)

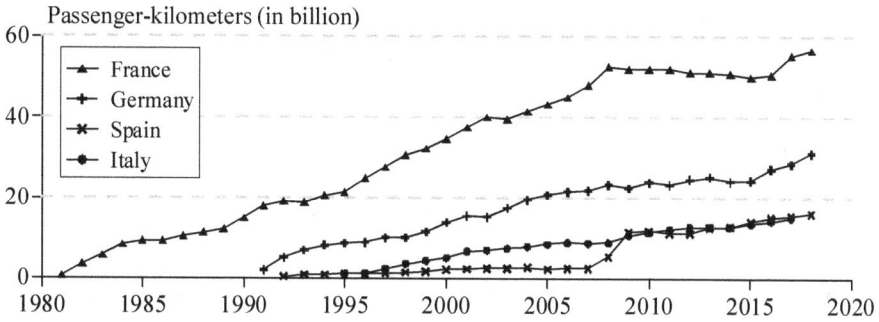

Fig. 2.6. Evolution of high-speed rail traffic in Europe, (1), (2)

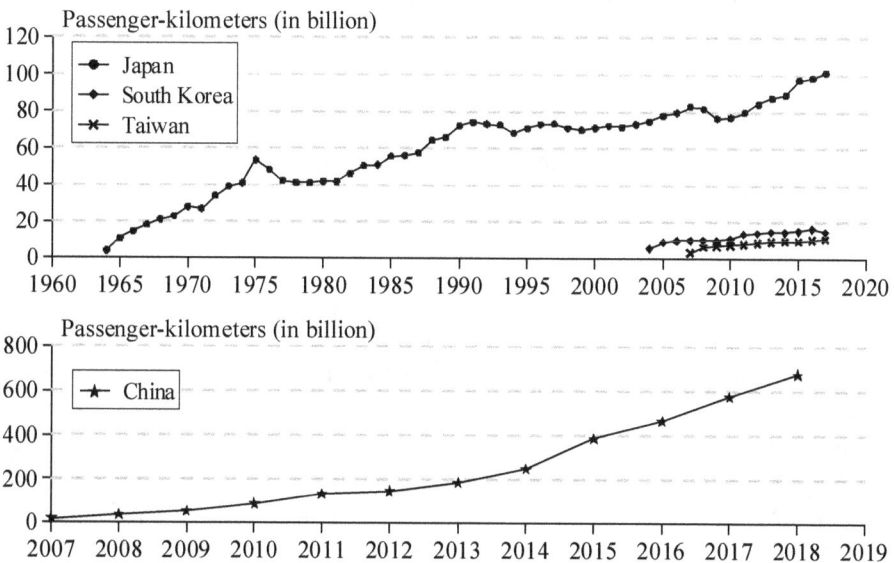

Fig. 2.7. Evolution of high-speed rail traffic in Asia, (1), (2)

High speeds, therefore, attract back to the railways part of the passenger traffic lost in the past or generate new traffic. For this purpose, however, a speed increase is not enough: station accessibility should also be improved through efficient bus or metro systems. In many instances, the connection of railway stations serving HST to airports can contribute to an efficient (in terms of time and cost) air-rail trip, as explained in section 1.11.

However, the success of HST is due not only to the reduction of travel times, but also to the following characteristics:

– frequency of service,
– regular-interval timetables,

55

- a high level of comfort,
- a pricing structure adapted to the needs of customers,
- complementarity with other means of transport,
- more on-board and station services,
- accessibility of railway stations, which implies connection with metro and bus systems, parking facilities for cars and bicycles,
- increased levels of safety and security.

A high-speed rail system should be designed to incorporate the whole range of services which the customer has come to expect when traveling on HST, including both pre-travel services (information, ticket purchasing, seat reservation, etc.) and post-travel ones (after sales services).

2.3. Technical features of high-speed railway lines

2.3.1. Technical characteristics of high-speed lines

Table 2.5 illustrates the technical characteristics of the most traveled high-speed rail lines. Important differences regarding longitudinal gradients and electric traction systems are observed.

2.3.2. Track characteristics for high speeds

HST operation requires that the track be built and maintained to much more demanding specifications and closer tolerances than conventional and lower-speed tracks. Continuous welded rails of type UIC 60, concrete sleepers (monoblock or twin-block), and elastic fastenings have been used to improve ride quality, stability, and safety of the track. Indeed, HST have outstanding safety records, due to exclusive rights-of-way (in some cases), fencing, computerized train control, and extremely good maintenance. Instead of ballasted track, some countries (Japan and Germany, among others) have used slab track for HST, (see also chapter 17).

2.3.3. Rolling stock for high speeds

Rolling stock used for high speeds comprises either lightweight, streamlined and electrically powered locomotives handling passenger coaches or simply trains of self-propelled multiple-unit cars. Their light weight minimizes horsepower and braking effort requirements, wheel wear, and track degradation. High-speed trains have low values of the axle load, (see section 7.5.1), between 11 and 17 tonnes, instead of 22.5 tonnes for non high-speed railways. Their traction power is high, $11 \div 24$ kW/tonne. Traction motors are normally carbody-mounted rather than axle-hung, in order to reduce unsprung masses (i.e. masses below the primary suspension system, see also section 7.11.2), (38), (44).

Table 2.5.
Technical characteristics of high-speed rail lines, (43), (47), (48)

Country	Japan	France	Germany	Italy	Spain	S. Korea	China
Line	Tokyo-Osaka (515km)	Paris-Lyon (427 km)	Hannover-Würzburg (327 km)	Rome-Florence (260 km)	Madrid-Barcelona (622 km)	Seoul-Busan (417 km)	Beijing-Shanghai (1,318 km)
Maximum speed V_{max} (km/h)	260÷300	300	250	250	350	350	380
Travel time	2h27min	1h56min	2h	1h21min	2h30min (for V=320km/h)	1h55min	4h24min (for V=350 km/h)
Radius of curvature R_{min} (m)	2,500	4,000	7,000	3,000	4,000	7,000	7,000
Maximum longitudinal gradient (‰)	20	35	12.5	8	30	25	20
Traction power supply	25 kV 50 Hz, 60 Hz	25 kV 50 Hz	15 kV $16^2/_3$ Hz	3 kV	25 kV 50 Hz	25 kV 60 Hz	25 kV 50 Hz
Distance of axes of two tracks (m)	4.2	4.2	4.0	4.2	n.a.	5.0	5.0
Superelevation (mm)	200	180	150	160	n.a.	n.a.	n.a

2.3.4. Power supply at high speeds

Power is supplied to HST from wayside substations through overhead catenary wires and is collected through pantographs mounted on the locomotive or power vehicle roofs. At high speeds, the catenary tension must be maintained at a constant value to minimize variations of sag. The French HST has a two-stage pantograph in order to minimize pressure (uplift) and to maintain excellent current collection characteristics at high speeds. The majority of other pantograph systems use much more rigid and more complex catenaries with a higher tension, resulting in less sag, (see section 20.8).

2.3.5. Economic data for high-speed trains

Compilation of cost data from a number of high-speed lines and trains in Europe suggests (in values of year 2018) a construction cost of 15÷40 million € per kilometer of high-speed line and a maintenance cost of 90,000 € per year and per kilometer, (see also section 5.2.2). The cost of purchase of a high-speed train (with 350 seats) is 30÷35 million € and its maintenance cost amounts to 1 million € per year, (38).

As high-speed railways are sophisticated and complex systems, each component of the railway activity should ensure synergy with the other ones and thus should be considered within a systems approach. Any project of a new or improved high-speed track should entail improvements in railway stations (space, facilities, aesthetics) and a high level of accessibility with metro and bus systems.

Pricing systems of high-speed trains are similar to those of airlines and are based on yield management techniques, in order to maximize revenues. Thus a high-speed train traveler who buys the ticket a short time (few hours or days) before the scheduled travel time risks to pay more, compared to a traveler who bought the ticket many days or months ahead. On the contrary, conventional railways usually apply a flat value of ticket price, no matter when the ticket was purchased.

2.4. The Channel Tunnel and high speeds between London and Paris

2.4.1. Technical description

The governments of the United Kingdom and France decided in 1986 on a permanent railway link between the two countries, which was realized entirely by private financing. For this purpose the Eurotunnel Consortium was created with responsibilities to construct the Channel Tunnel and operate it for 55 years, which was extended later to 99 years, (40).

The project of a total length of 50.5 km consists of two rail tunnels (one per direction) with a diameter of 7.6 m plus a third tunnel (of a diameter of 4.8 m) for maintenance purposes, emergency incidents, etc. The principal tunnels are connected to the auxiliary one at 375 m intervals. The rail level is situated 25÷40 m below the seabed level.

The entire construction cost was initially underestimated at 4.2 billion €, changed many times, was finalized at 7.4 billion €, and is allocated as follows:
- ◆ 50% for the tunnel construction,
- ◆ 10% for the rolling stock,
- ◆ 40% for tracks, signaling, electrification, etc.

2.4.2. Travel times

Full operation through the Channel Tunnel began in autumn of 1994. Three types of services are provided:

- high-speed trains, named 'Eurostar', with a running speed in the tunnel of 160 km/h and outside of the tunnel of 320 km/h, joining London to Paris in 2021 in 2h 16min and London to Brussels in 2h 01min. Eurostar trains have a maximum capacity of 894 passengers (672 in second class and 222 in first class). Eurostar has a share approaching 90% in the (rail + air) transport market between London and Paris and a share higher than 65% between London and Brussels. Both routes offer a punctuality higher than 90%,

- conventional trains, night trains, freight trains, with a usual speed of 100÷120 km/h in the Channel Tunnel,

- shuttle passenger trains, named 'Le Shuttle', transporting cars, trucks (of a maximum weight of 44 tonnes), and buses. Passengers remain in their seats and maximum speed in the tunnel is 140 km/h. The shuttle trains require extensive terminal facilities at each end, around 1.3 million m^2 in the British side and 7 million m^2 in the French side, the size of a major international airport like Heathrow, (43).

More technical details about the Channel Tunnel are given in the relevant chapters (soil mechanics in section 9.2.5, forecast of demand of Eurostar trains in section 4.4.2, etc.).

2.4.3. Method of financing and forecasts of demand

As said previously, not a public British penny (=0.011 € in February 2021) has been given for Eurotunnel, which was totally financed by the private sector. Forecasts for Eurostar estimated demand for the year 1995 at 11.5 million passengers, whereas the real number was only 2.92 million passengers, and for the year 2003 at 18.9 million passengers, against only 6.31 million passengers traveled that year and 10.3 million passengers traveled in 2017. Demand of shuttle passenger trains was 4.2 million passengers in 1995, increased to 10.1 million passengers in 2000 and then stagnated to 10.4 million passengers in 2017. Total freight transported through the Channel Tunnel was 1.8 million tonnes in 1995 and 2.9 million tonnes in 2017. Figure 2.8 illustrates evolution of the various categories of traffic through the Channel Tunnel.

Overestimation of demand and underestimation of costs led to a real financial disaster, which is reflected in the value of Eurotunnel share (issued at £3.50 in 1987, increased to £11.00 in 1989) in the stock markets, which dropped down 4 times lower in late 1994 compared to its initial value and increased 27 years later in February 2021 to £12.05.

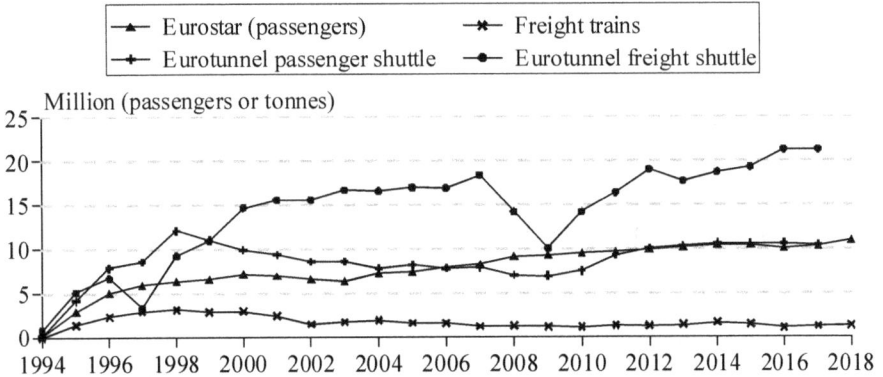

Fig. 2.8. Evolution of the various categories of traffic through the Channel Tunnel (1), (2)

2.4.4. Operation, safety, and maintenance

The ultimate capacity of Eurotunnel can reach up to 30 train movements per hour in each direction. Eurotunnel employs around 3,500 people in 2021, 12% of which are in charge of maintenance.

The Channel Tunnel has a huge cooling system and a signaling system that incorporates full automatic train protection. The system proved rather efficient in the fire of November 1996 on one of the freight shuttles, without any casualties, but with serious damages which resulted in closing the affected tunnel for 7 months for repairs.

2.5. Tilting trains

HST require new layouts and new tracks and are often an expensive solution, which is feasible only for very high population concentrations (as illustrated previously in Fig. 2.3) at each end or along the line. However, the usual reason for the restriction of speed is the small radius of curvature.

The railway industry devised a solution which permits to increase the speed in curves without the necessity to improve the existing layout. The extremely interesting solution is called *tilting* train, due to the fact that the vehicle body tilts when negotiating a curve and thus gives an additional superelevation.

Tilting trains are not, strictly speaking, HST but have succeeded reduction of travel times up to 33% compared to conventional trains. However, average reduction of travel times by tilting trains range around 12÷20%. Tilting trains are in use in Italy, Spain, the United Kingdom, Sweden, Finland, the USA, and elsewhere. The tilting technology is analyzed in detail in sections 14.2.3 and 19.9.

2.6. Aerotrain

The aerotrain technology is based on guided transport (like conventional railways), but avoids any contact between the moving vehicle and the bearing substructure on which transport is taking place, whereas conventional railways rely on the metal (wheel) to metal (rail) contact. The aerotrain is a vehicle running on a concrete bearing substructure in the shape of an inverted 'T', (Fig. 2.9).

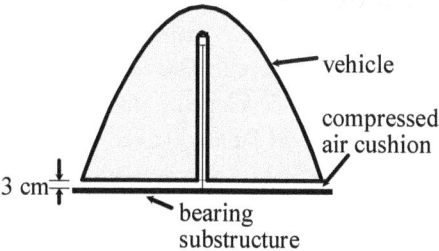

Fig. 2.9. The aerotrain principle

Propulsion in aerotrain is achieved, without any wheel system, by a compressed air cushion blown between the vehicle and the bearing substructure. Thus, the aerotrain replaces the adhesion forces, necessary to propel conventional trains, by compressed air layers, (48).

This technology was developed in the 1960s in France and achieved in 1974 the speed of 430 km/h. Even though there were various plans for aerotrain construction (e.g. Paris – Orleans, where 18 km of bearing substructure had even been built, Brussels – Luxembourg, etc.), they were abandoned in the 1970s for various reasons, the main ones being, (48):

♦ the new technique was not compatible with conventional railways,
♦ construction proved to be much more expensive than a new conventional railway line, without the additional cost being offset by the much lower maintenance cost of the aerotrain compared with conventional railways,
♦ energy consumption (due to the air turbine used for aerotrain propulsion) was much higher than for conventional trains,
♦ the carrying capacity of the aerotrain was low (64÷96 passengers in the prototype, but up to 160 passengers in two-vehicle aerotrains).

Secondary reasons, such as passenger safety considerations (possible fire in the vehicle which rides 5 m above ground), noise, and questionable overall aesthetics have contributed to the abandonment of the project.

2.7. Magnetic levitation

2.7.1. Technical description

In magnetic levitation systems, contact between the bearing substructure and the vehicle is avoided, propulsion being ensured by magnetic phenomena, (Fig. 2.10). The bearing substructure is a concrete slab in the shape of an inverted 'T' (or of a 'U'). Suitably located magnets and coils generate the

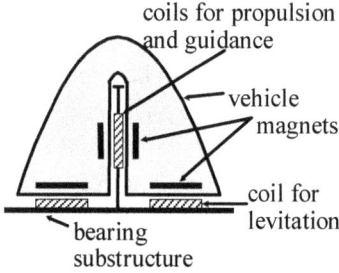

coils for propulsion
and guidance

vehicle
magnets

coil for
levitation

bearing
substructure

**Fig. 2.10. The magnetic levitation
principle**

forces required for levitation, propulsion, and guidance. However, a super-conducting magnet, fulfilling the above three requirements, has been constructed. The maglev technology was developed in the 1970s in Germany and Japan. During test runs, a speed of 517 km/h was attained in 1979, of 581 km/h in 2003, and of 603 km/h in 2015, (44), (46).

Many of the handicaps of the aerotrain invention are still valid for the maglev as well, such as the greater difficulty involved in penetrating city centers, in contrast to conventional railways. In addition, pressure from the rail industry and political goals had delayed applications of the maglev invention, (41).

Two different maglev technologies have been developed: Attraction or electromagnetic suspension technology, developed by Germany, and repulsion or electrodynamic suspension technology, developed by Japan. The basic features of the two maglev technologies are listed in Table 2.6, (44).

Table 2.6.
Basic maglev technology features, (44)

Feature	Attraction or electro-magnetic suspension (Germany)	Repulsion or electro-dynamic suspension (Japan)
Vehicle floats on air gap of	1.3 cm	10.2 cm
Presence of wheels	None	Auxiliary wheels for support at low speeds (less than 80 km/h)
Type of magnets	Iron-core electromagnet	Superconducting coils
To achieve stable levitation	Requires electronic sensing of the vehicle-infrastructure gap and continuous control of magnetic current	Requires reaching a sufficient forward velocity (about 80 km/h)
Propulsion	Long-stator iron-core linear synchronous motor	Long-stator air-core linear synchronous motor

2.7.2. Comparison of magnetic levitation with conventional railways

Advantageous features of the maglev technology are, (41):
– it can achieve high speeds in the range of 400÷500 km/h,

- there is no loss of traction at high speeds, since there is no vehicle-track contact,
- there is no wheel-rail friction. The only resistance to overcome is aerodynamic drag,
- there is no hindrance from rail, ice, or snow.
- it is very quiet because of few, if any, rotary or sliding parts in maglev vehicles,
- there is no possibility to derail from the guideway, since the vehicles are intimately coupled to the guideway,
- there are low maintenance costs both for the vehicle and the track, due to the absence of mechanical contact,
- it can negotiate sharp curves and steeper grades, since wheel friction is not a factor in propulsion.

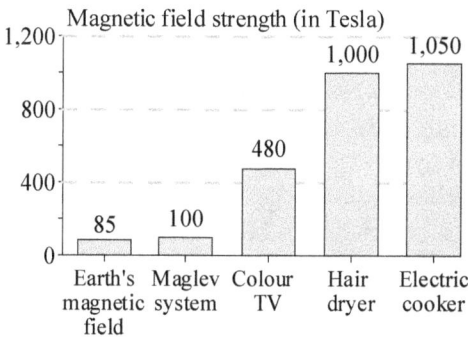

Fig. 2.11. **The magnetic field of a maglev system in comparison to other exposures of the human body, (41), (42)**

Thus, maglev systems can climb steep grades up to 100‰ and negotiate curve radii of 2,250 m at a speed of 300 km/h. On the other hand, the maglevs use approximately 30% less energy than a conventional train traveling at the same speed.

The effect of the magnetic field of maglev systems is slight, far lower compared to a color tv, a hair dryer, an electric cooker, (Fig. 2.11). However any negative effect on passengers with pacemakers is completely ruled out.

Lastly, it should be noticed that due to the absence of rolling noise, maglev systems are much quieter than conventional railway ones, (see also section 23.3), (42).

2.7.3. Applications of magnetic levitation

Though many projects for maglev applications have been planned in the past, the following maglev systems were in operation in 2021:

- the maglev system connecting the airport of Shanghai with the center of the city (a distance of 30.5 km) in 7 minutes, with a maximum speed of 431 km/h and a punctuality of 99.97%. The project had a cost of 1.2 billion US$ and is in operation since 2004,

– the maglev system near the city of Nagoya in Japan, which was inaugurated in 2005. Its maximum speed is only 100 km/h, since the system was designed as an alternative to metro systems. Maximum carrying capacity is only 4,000 passengers per hour and per direction, the minimum radius of curvature is 75m, and the overall construction cost per km (rolling stock included) was 100 million US$,

– the maglev systems in Daejon and Incheon, South Korea, serving the neighboring airport since 2016,

– two maglev systems in China, with a top speed of 110 km/h, in operation since 2017, used principally as alternative to metro systems.

2.8. Hyperloop technology and systems

The hyperloop technology (Fig. 2.12) consists of one tube (of a diameter of approximately 2 m), through which special vehicles (often referred to as pods or capsules) are traveling at speeds that can reach 1,200 km/h. Propulsion of vehicles is assured by magnetic levitation forces and accelerators, which are planned along evacuated (airless) or partially evacuated low-pressure tubes (at a pressure of 1 millibar). Tubes of hyperloop systems are placed over the ground on a series of pylons. Hyperloop vehicles get their initial velocity from an external linear electric motor and they can be powered from solar panels installed on top of the tube, (39).

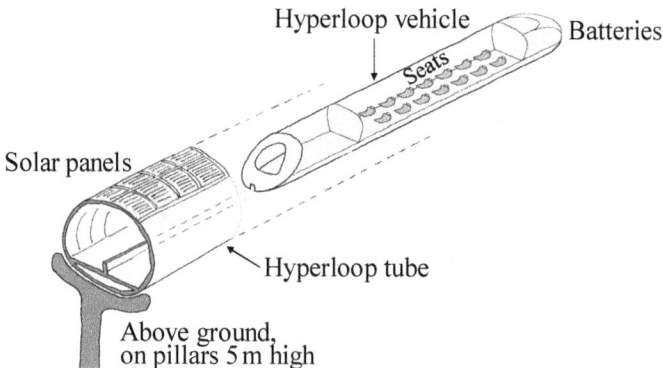

Fig. 2.12. Hyperloop technology and system

Hyperloop technology overcomes the two principal problems of rail systems: friction and air resistance. The capacity of a hyperloop vehicle (8.5 m long) is estimated to be 28÷40 passengers. Emissions of hyperloop systems are expected to be lower by 60% compared to conventional high-speed railways. Hyperloop can be considered as a new form of ground transport, in floating vehicles within low-pressure tubes, faster than air travel, with the same transport capacity as a bus, solar powered, with low emissions, independent from weather conditions, capable to provide trips at frequencies similar to metro systems.

Hyperloop technology was until 2021 on an experimental stage. It aims both the passenger but also the freight market by attracting parts of the current air freight market. A serious criticism against hyperloop technology is that passengers are transported in an unpleasant, windowless environment, under high acceleration forces and noise levels, (39).

There are many ambitious projects to apply the hyperloop technology around the world:

- USA (Los Angeles-San Francisco, Chicago-Cleveland, Washington-New York). For the Los Angeles-San Francisco route (a distance of 608 kilometers) hyperloop promoters estimate a cost of 11÷17 billion US$. Their opponents argue that the real cost can be many times higher. The planned travel time is 30 min and a one-way ticket is estimated to cost 20 US$. On the contrary, a high-speed railway line Los Angeles-San Francisco has a cost of 77 billion US$, (39)
- Dubai-Abu Dhabi, a distance of 150 km, estimated to be traveled in 12 min with 10,000 passengers/h in both directions,
- Stockholm-Helsinki, through a tunnel under the Baltic Sea. It is estimated that the distance of 500 km can be traveled in 30 min,
- Russia (Moscow-Saint Petersburg), India (Chennai-Bengaluru), China (Tongren-Mount Fanjing), and other.

Under the optimistic scenario, the Hyperloop technology will not be in operation before 2025, provided that it will overcome a number of technological, economic, and political obstacles. Procedures and certifications for hyperloop systems should be similar to those of aviation.

Table 2.8 recapitulates the principal technical, operating, and economic characteristics of hyperloop technology in comparison with high-speed trains, maglevs, and air transport, (46). Indeed, hyperloop technology achieves the higher speed, the lower terminal-to-terminal travel time, high frequency, and is solar powered. However, as hyperloops cannot reach easily city centers and require transfer of passengers, so that other transport modes (metro, buses, railways) can be used in a frame of intermodality, door-to-door travel times are less impressive and in some cases comparable with those of high-speed trains. There is still a question about comfort and traveling environment in a hyperloop vehicle and tube. Promoters of hyperloop technology argue for far lower construction and operation costs (compared to HST), but a strong criticism has been addressed that the real final construction cost of hyperloop will be many times higher than the estimated one.

Table 2.7.
Technical, operating, and economic characteristics of hyperloop technology in comparison with high-speed trains, maglev systems, and air transport, (46)

	High-speed trains	Maglevs	Hyperloop	Air transport
Maximum speed (test runs)	575 km/h (2007)	603 km/h (2015)	1,200 km/h	Up to 2,200 km/h (hypersonic aircrafts)
Maximum speed (for the specific project)	– 350 km/h (China, 2021) – 320 km/h (Japan, France, Germany, Spain, Belgium) – 300 km/h (South Korea, Taiwan, Italy, UK)	431 km/h (Shanghai)	Only projects under study. Scheduled speeds: – 1,200 km/h through regular course sections – 500 km/h through urban areas and difficult course sections	Up to 1,000 km/h for aircrafts with a range of 500÷800 km
Travel time (terminal to terminal / distance traveled)	– London-Paris: 444 km / 2h16min – Tokyo-Osaka: 515 km / 2h30min – Seoul-Bussan: 350 km / 1h55min – Beijing-Shanghai: 1,318 km / 4h24min	Airport-city of Shanghai: 30.5 km / 7min	Scheduled for: – Los Angeles-San Francisco: 608 km / 35min – Dubai-Abu Dabi: 150 km / 12min	Around 1h for a distance of 600 km between airports
Accessibility	High (stations in city centers)	Low (stations outside of city centers)	Low (stations outside of city centers)	Low (airports in remote areas)
Time for security controls	Low	Low	Low	High
Frequency	5 minutes headway	15 minutes headway	30 seconds to 2 minutes headway	1 minute headway in runway
Comfort	High, providing productive use of travel time	Unknown for long distances	Unknown, but in a windowless environment under high accelerations and noise levels	Satisfactory, but less productive use of travel time than with HST
Cost of construction	Around 15÷40 million €/km of new double track line in Europe, around 40 million US$/km in the USA (California), (rolling stock not included)	25 million US$/km	11÷17 million US$/km (including vehicles), 40 million US$/km for underwater track	1.2÷2.0 billion US$ for a new runway
Carrying capacity	Up to 1,000 passengers per train	574 passengers per train	28÷40 passengers per vehicle	Up to 555 passengers per aircraft
Energy consumption	Electric- 605 BTU/p-km	Electric- 730 BTU/p-km	Solar powered	Jet fuel
Interoperability with other transport modes	High	Low	Low	Low

3 Policy and Legislation

3.1. The competitive international environment and the evolution of the organization of railways

The organization of railways began in the late 19th and early 20th centuries in the form of small private enterprises. The strategic importance of the railways for the economy and the security of various countries, combined with the deficits of railway companies which had already begun to appear, led most governments, between 1935 and 1960, to nationalize their railways. Therefore, most railways became part of the state administration or were under state control (1950s÷1980s period).

Changes in the transport market during the 1980s and 1990s (mainly the gradual liberalization and deregulation of transport activities from the regulating framework under which they had been operating for four decades or more) compelled railways to show more flexibility in the organization of their services, reduce costs, adapt to new technologies, exploit their comparative advantages, and modernize, in order to become competitive in the transport market. Some countries, like Japan, the United Kingdom, Sweden, have already privatized their railway operators. In the transport market, neither technology nor innovation will have a reason to exist in the third decade of the 21st century, unless they are financially efficient and competitive, compared to services offered by other transport modes (road vehicles, airplanes), (12), (15), (36).

An important step towards the liberalization of railway activities in Europe and the introduction of intra-modal competition in the rail sector (i.e. competition of many railway operators running on the same track) was the separation of infrastructure from operation, thus putting an end to a monolithic organization (for almost 150 years) of railways. Separation entails the creation of a number of independent units out of an integrated railway. Separation can be horizontal, when the existing rail operator is split into specialized operators such as passenger, freight, rolling stock or vertical when activities from infrastructure are separated from those of operation. If no specific mention is made to the term separation (horizontal or vertical), then it indicates vertical separation. Thus, a vertically integrated railway company owns, maintains, and operates the infrastructure, the rolling stock, and the passenger and freight transport activities. In contrast, a separated railway company limits its field of action into one or more (but not all) of the above activities, (52).

Fundamental objectives of a rational rail policy should include higher productivity and efficiency performances of the rail sector, lower prices for the consumer (while increasing quality of service), lower public subsidies, higher new investment (coming not only from the public sector), increased levels of safety and security, (13), (15).

There is no clear model for the new organization of railways that can ensure that all the above objectives of railways are fulfilled. Thus, many new organizational schemes and forms are observed in the various countries of the world, with clear differences in the new structure of railways (integrated or not), the degree of regulation, the level of public subsidies, the degree of involvement of the private sector.

3.2. The dual nature of railways: business and technology

3.2.1. Weaknesses inherited to railways

In the competitive environment of the transport market, railways should search for their comparative advantages, which they should develop with the help of the necessary technological modernization and innovation. On the other hand, they should operate as enterprises, governed by the same rules of competition applied to other businesses and relinquishing the umbrella of state protectionism sheltering them for decades.

Railways, however, inherit serious handicaps as a result of decades of state protectionism, such as, (52), (68), (69):

♦ administration and organization inflexibility. For decades, railway management dealt only with current affairs and was involved principally in technical matters. Important issues were decided by the supervising ministry, often based on political criteria,

♦ accumulation of personnel in routine tasks and staff shortages for administration, organization, and technological upgrading positions,

♦ high costs, often the result of obsolete operating methods and staff overcrowding,

♦ rolling stock often difficult to operate, offering services at a level which does not meet transport requirements in some cases,

♦ maintenance expenses of railway infrastructure undertaken by the railways, as contrasted to road and air carriers which contribute only a small part of the maintenance costs of road network and airports respectively, (15). Separation of infrastructure from operation was an important step to overcome this situation,

♦ obsolete infrastructure (which was overdimensioned in many cases), often as a result of the absence of serious investment for many decades,

♦ obligation to operate lines with a low traffic without sufficient compensation, which, had the line been operating by private enterprise criteria, would not have sustained its operation (public service obligation (PSO)).

3.2.2. Comparative advantages of railways

The aforementioned compendium of disadvantages risks giving the impression that railways have nothing but problems. However, railways contribute to the development of both transport and the economy, since they, (15):

• provide an integrated system of services for both passenger and freight transport, with programmed schedules regardless of day, season, and weather conditions, a fact resulting in network economies,

• pollute the environment minimally in contrast to other transport means, (see chapter 23, Table 23.1 and Fig. 23.11),

• contribute decisively to relieve congestion in peak travel periods in central thoroughfares, because of their huge transport capacity, (see section 1.10.3),

• consume much less energy for the same traffic, compared to any other transport mode, (see section 23.4),

• provide reduced fares for large segments of the society, particularly for social reasons (e.g. students, the elderly, etc.), who can thus travel more easily.

3.2.3. Strategy and restructuring measures

The transport sector in Europe and worldwide is presently oriented to a gradual liberalization and deregulation[*], with emphasis on competition between the various transport modes (inter-modal competition) but also between many companies which use the same transport mode and infrastructure (intra-modal competition). The government-owners of the railways are under the obligation to ensure a real autonomy for the railways, to gradually reduce subsidies to

---

[*] Liberalization is an economico-political vision of the organization of the economy suggesting that the state does not interfere in economic affairs and that the real ruler should be market forces. Deregulation is an economic technique suggesting the withdrawal of state interventions in the market. Deregulation is usually a measure towards liberalization, other measures being to avoid any control of prices and salaries, anti-trust techniques, etc. Let us notice that at the railway field, characterized by a gradual liberalization, regulation is still in force. In any case, even in a totally liberalized transport market, a minimum regulation concerning safety, level of service, and financial capacity of involved companies is necessary.

rail undertakings (used to cover deficits), to establish a regime of transparency in rail operations, and to create a framework in which other rail operators can use the railway infrastructure and enter the rail transport market. Within such a framework, the railways should aim at, (12), (15), (36), (69):

- market oriented activities and the eventual abandonment of unprofitable services,
- greater flexibility in the organization and development of operational criteria for the various initiatives, e.g. investment,
- personnel allocation on the basis of the requirements of the particular transportation task and staffing of the various departments by specialized personnel. It is not to exclude, particularly for management and specialized tasks, the use of high-quality specialists from other sectors,
- trying to reduce drastically costs in order to make rail services more competitive in the transport market. The reduction of costs may come from the application of information technologies, internet and other new technologies, in addition to the rationalization and inevitable reduction of the current personnel levels,
- systematic maintenance and renovation of the rolling stock and infrastructure, enabling the railways to meet the requirements of their clients,
- infrastructure modernization with important investment (for the most part this can be covered by the state, the EU, the World Bank, and in some cases by the private sector). It should be stressed here that modernization does not refer to any particular project, but to those that will enable railways to coexist competitively with other transport means. For the more attractive projects, financing can come from the private sector too, as was the case of the Channel Tunnel project and of many other rail projects using the financial technique of public-private partnership (PPP),
- clear definition of public service obligations in the passenger sector, which, if the only consideration of the railways were business profit, would not have been undertaken to the same extent or degree (e.g. operation of lines with small traffic). The authority enforcing a mandatory public service (e.g. the Ministry of Education for reduced student fares) should refund lost income to the railway operator,
- adequate compensation of the railways for not polluting the environment and not causing traffic congestion. A quantitative and financial evaluation of the effects of the various transport modes on the environment is already available, (see section 5.7). The prevailing view is to subsidize railways with an amount corresponding to what would have to be spent to combat the pollution and traffic congestion, which would have been caused, had the operation of the railways been discontinued,

– gradual reduction of deficits. The ratio revenues/expenses, which represents the ability of a company to survive or not without any subsidies, should not be lower (or far lower) than 1.0. However, as deficits of railways are covered by the state budget, that is by the citizens (who may be not users of railways), a strong pressure to the railways is exerted and will continue to do so for the reduction of deficits,

– commercial and tariff policies which increase revenues, assure high degrees of load factor, and respond to requirements of clients and the society,

– fulfilling financial, commercial, and technological targets, which should be clearly defined. Factors that can measure the global result of a railway undertaking can be the degree of adaptability and the operating costs, which present huge differences and very contrasting situations among the various countries.

3.2.4. Railways and transport requirements

Any transport activity is not an end in itself, but exists in order to fulfill specific needs of transport of persons and goods. Railways should try to offer more efficient and competitive services and must take into account the following, (15), (69):

– the evolution in the transport market, resulting from globalization of the economy, liberalization, and increasing deregulation,

– competition and the need for reduction of costs,

– the obligation for harmonization, known as interoperability, (see sections 1.14 and 21.9), of the various railway technologies (e.g. track gauge, electrification, and signaling systems), to permit a global railway service,

– the need for a long-term operational profitability and a market oriented strategy focusing on profitable segments.

Survival in the evolving and highly competitive international environment demands efficient, accessible, and competitive rail transport systems which must fulfill economic and social expectations, whilst ensuring objectives of wider environmental protection, efficiency of resources, and safety. Moreover, rail development should allow for maximum synergy with other transport modes, thus responding to modern and door-to-door requirements for seamless transport and sustainable mobility.

3.3. Globalization and liberalization of the rail market

Since more than two decades, the world is characterized by an increasing globalization (of economic and commercial activities), which can be described as a procedure of opening national markets to products and services and reducing state subsidies and costs. Globalization requires a competitive environment and a liberalization of the transport market. More particularly for the rail sector, liberalization means the withdrawal of any obstacles concerning issues such as entrance of new operators in the rail market, commercial and tariff policy of the railway undertakings, management, strategy, investment, etc.

Rail liberalization generates both *opportunities* and *threats* for railways, (15):

– intra-modal competition (from other rail operators) will press for reduced rail tariffs and increased quality of rail services,
– state subsidies will be reduced and should be allocated under the terms of open tendering. In any case the continuation of operation of a rail activity causing deficits should be appropriately justified; thus railways will be pressed to curtail costs with as an inevitable effect the loss of some jobs,
– new investments will incorporate new technologies, which will increase quality of services, create new products (concerning particularly international rail services, combined transport, high speeds, etc.),
– a more customer oriented commercial and tariff policy will permit railways to gain segments of the market, such as business travel, transport of groups, transport of bulk or dangerous loads, etc.

However, even in a liberalized rail market, the role of the state remains critical and should assure, (58):

• high standards of safety,
• a certain level of quality of services,
• that only new operators with a sufficient financial capacity and technical performances can enter the rail market,
• fair conditions for inter-modal (with other transport modes) and intra-modal (with other rail operators) competition,
• transparency and accountability in the use of public funds,
• a stable economic environment for long-term investments and technological innovations,
• prevention of pricing abuses,
• an appropriate environment for the reduction of costs, while avoiding social unrest,
• further development of international rail services, which presumes the simplification of customs procedures and efficient cooperation between infrastructure managers and rail operators,
• a new organization that avoids unnecessary fragmentation and limits the risks and costs concerning both finances and safety in managing the inter-

face between train operations and infrastructure, with clear identification of the responsibilities of each part,
- writing off (totally or partially) past debts.

Nevertheless, liberalization should be clearly distinguished from privatization, i.e. the property regime of an undertaking. It is possible to have a privatized rail operator with monopolistic rights (the case of many rail operators in the United Kingdom) or a state-owned rail operator in a competitive context (the case of some national railways), (55), (66).

3.4. Separation of infrastructure from operation and the new challenges for railways

3.4.1. Separation as an incentive for competition

The railway of the past was in most cases a natural monopoly with a specific status for its staff and had two levels of contact with the external environment: on the one hand the state owner and on the other the passengers and freight forwarders.

In section 3.2.3 we described some measures that railways should undertake on their initiative. Many railways have been reluctant to restructure, even when they have been pressed to take some measures (such as the closure of lines or reductions of personnel).

Therefore, the question is whether it is possible to reverse the decline of the railways by means of reforms based on competition, which can become the catalyst for radical changes in the oligopolistic transport sector, (13), (58).

The answer lies on the ideas of contestability: the threat that a potential competitor can enter the market is a critical and sufficient motivation for the existing operator to behave as if competition existed, (15), (69).

The tendency of the period $1985 \div 2020$ to restrict monopolistic activities and introduce competition in all sectors of transport will continue and become stronger, as this is encouraged by the international economic environment. Air transport is fully liberalized in many countries of the world and national carriers, who enjoyed monopoly and state protection for years in both international and domestic routes, are struggling in a very competitive environment with private carriers (often low-cost) who entered the market. The question concerns whether this model should also apply to the railways and under what conditions.

It has been argued that there can hardly be any competition if railways keep their old organization, with one state-owned company in charge of both infrastructure and operation. Thus, the separation of operation from infrastructure appeared as a first step to introduce competition within the railway market. Infrastructure will be the responsibility of an authority separated (at least financially and organizationally) from operation and every railway operator will pay charges for using a track, in relation to the time of the journey, the distance traveled, the kind of railway operation, etc. Charges should not have any discrimination against new entrants, (58).

However, there is a counter-argument whether separation of infrastructure from operation is a prerequisite for competition. In the USA, Japan, and elsewhere, rail operators own their infrastructure, whereas another operator can run on an infrastructure that it does not own by paying appropriate charges.

3.4.2. Competition and new challenges for railways

In this new organization of the railways, many fundamental challenges are arising, (15), (20), (51), (Fig. 3.1):
- *Culture*. Is the difference between a service-rendering operator and an engineering company understood?
- *Technology*. Do systems fit to new objectives and requirements?
- *Human resources*. Do employees have the right skills within the new organization?
- *Competition*. Will there be many and competing operators on the infrastructure and what will the impact be? Will rail operators be partners or clients of infrastructure?

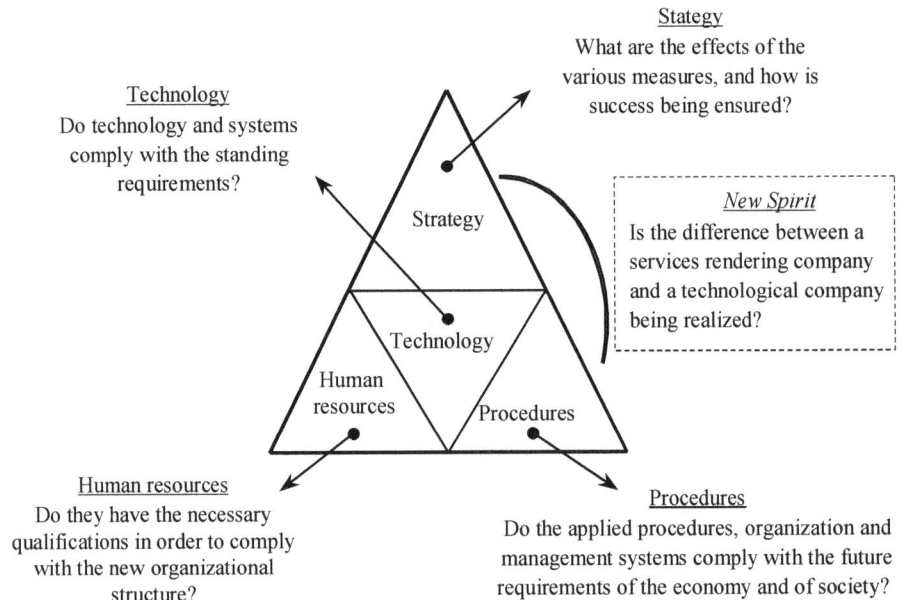

Fig. 3.1. New challenges for railways (15), (20)

- *Investment*. Where will new investment come from? Is a public-private partnership feasible?
- *Debt*. How will the accumulated debt be reimbursed? Mostly from the state (case of Germany), partly from the new entrants (case of Japan), and how?
- *Organizational responsibility*. As multiple operators run on the same infrastructure, this will necessitate a strong independent infrastructure entity, which will be responsible for path allocation, traffic management, etc.
- *Role of the state*. In a liberalized rail market, the state should assume the role of Regulator. There are many forms and degrees of regulation, (see below section 3.6).
- *Reason for existence*. Is railway transport either technically necessary or more efficient to assure a further increase in mobility of persons and goods? What is the added value of railways in modern economies?

3.4.3. Various forms of separation

However, as stated previously, a counter-argument against separation and for keeping the status quo of the integrated organization of the railways is that in some parts of the world (USA, Canada for freight, Japan for passengers), a railway operator can be the owner of infrastructure and this fact by itself does not exclude other operators from running on the specific track by paying appropriate charges.

Thus, intra-modal competition can exist without separation of infrastructure from operation. However, separation can be the catalyst to introduce competition or to facilitate the entrance of many rail operators.

In relation to the degree of separation we can observe various forms of separation:
- *full separation* (e.g. Sweden, United Kingdom, etc.). All rail operators are separated from the infrastructure provider,
- *no separation but intra-modal competition* (e.g. USA, Japan). The infrastructure manager controls and provides the dominant rail operations, but other rail operators can run on that infrastructure by paying appropriate charges,
- *no separation and no competition* (e.g. China, India, etc.). There is only one fully integrated railway company, that is one infrastructure provider which is also the operator.

In relation to the legal structure, there are many types of separation:
- *accounting*. Railway activity is integrated, only accounts of infrastructure are separated from those of operation,
- *institutional*. Infrastructure and operation are totally independent companies, both financially and legally,

75

♦ in a *holding structure*. The former integrated company is split in two or more companies, which are merged into a holding system with a common board and chairman.

A major concern in the transition from the fully integrated railway to the various separated forms is safety. The railway system is complex and safe transport is based on a synergy of its various components; this synergy should continue (it should even be strengthened) after separation. Another issue is transaction costs and appropriate management after separation. It has been argued that in the USA an integrated freight railway could have a 20÷40% cost advantage over a vertically separated railway. Vertical separation of railways in the EU may have caused an increase of operating costs of EU railways of 5.8 billion € per year. Complete separation of infrastructure from operation in the UK is estimated to result in an infrastructure management 20÷40% less efficient in the UK, compared to many EU infrastructure managers, (51).

Separation may serve many objectives, (58), (65):

- clarify roles of government and the degree of its involvement in the railway activity,
- encourage a stronger participation of the private sector,
- promote competition, intra-modal (with other rail operators) but also inter-modal (in the market, with other transport modes),
- focus business on parts of railway activity (e.g. freight),
- establish clear terms for rail infrastructure provision,
- reduce public subsidies to the rail sector,
- afford more customer-oriented rail services.

3.5. A definition of railway infrastructure

A definition of railway infrastructure is given by European Union Regulation 851/2006 (which replaced Regulation 2598/1970) and comprises routes, tracks, and fixed installations necessary for the safe circulation of trains.

Railway infrastructure consists of the following items, (70):

- *Ground area* and the *line of route*. It comprises the subgrade itself, (see chapter 9), including in particular embankments, cuttings, geotextiles, drainage channels and trenches, masonry trenches, culverts, lining walls, planting for protecting side slopes, etc.
- The *track* and *track bed*, (see section 7.2, Fig. 7.1), which consist of the rails, sleepers, fastenings, ballast, and subballast.
- *Switches* and *crossings,* (see chapter 15).

- *Engineering structures*: bridges, culverts and other overpasses, tunnels, covered cuttings and other underpasses, retaining walls, etc.
- *Level crossings*, including appliances to ensure the safety of road traffic.
- Passengers and goods *platforms* and *access ways*.
- *Safety*, *signaling*, and *telecommunications* installations which include fixed signals, track circuits, (see section 21.3.2), train control equipment, signal cables or wires, signal boxes and control systems, and (particularly for high-speed lines) cab signaling systems.
- *Electricity power* supply, which includes catenaries and supports or third rail, substations and power supply cables, and control equipment.
- *Lighting installations* for traffic and safety purposes.
- *Buildings* used by the infrastructure department and without any connection with transport activities.

Stations, *marshaling yards*, and *warehouses* may be owned either by the infrastructure manager or by the train operator.

3.6. European Union rail legislation

European Union legislation aims to introduce competition in the rail market, rationalize and reduce public subsidies, reduce costs and transform railways to customer oriented businesses, achieve interoperability, strengthen safety, boost high speeds, and take advantage of the friendly to the environment performance of the railways, (12).

The various successive steps in the EU rail legislation are often referred to as the five railway packages, (58), (65):
- railway package zero (Directive 440/1991),
- first railway package (Directives 12, 13, 14/2001), aiming to open the railway market,
- second railway package (Directives 49, 50, 51/2004), aiming to create a legally and technically unified European railway area,
- third railway package (Directives 58, 59, 137/2007), aiming to open up international rail passenger services,
- fourth railway package (Regulation 796/2016, Directives 797, 798/2016, Regulations 2337, 2338/2016, Directive 2370/2016) aiming at removing all remaining legal, institutional, and technical obstacles in order to achieve the single European railway area and thus revitalize and make the rail sector in the EU more competitive.

The European Union rail legislation can be summarized as follows, (12), (49), (65):

♦ Separate infrastructure from operation (at least at the accounting level), (Directives 440/1991, 12/2001, 14/2004, 51/2004, Regulation 2337/2016). Furthermore, separate the activities of passenger and freight transport at the accounting level and avoid any cross subsidies between them (Directive 13/2001).

♦ Determine the minimum conditions (about safety, finances, etc.) to be met, for a rail operator to run infrastructure (Directive 18/95). If these conditions are met, the rail operator can apply for a License, valid within all EU countries. The Licenses issuing Body should be independent from any railway operator (Directive 13/2001). However, in order to operate in a specific country, the operator must possess in addition to the License a Safety Certificate, valid only in this specific country (Directive 14/2001). The fourth railway package (Regulation 796/2016, Directive 2370/2016) changes radically jurisdiction of national authorities over railway safety. Indeed, the EU Agency for Railways undertakes all duties related to authorizations for railway vehicles, Safety Certificates for train operators and trackside control-command and signaling systems. From June 2019, all responsibilities of member-states of the EU related to Licenses, Safety Certificates and vehicle authorizations are transferred to the EU Agency for Railways (Regulation 796/2016) and are incorporated within a unified online portal (Directive 798/2016), (65).

♦ Fulfill safety conditions. Any rail operator should possess a Safety Certificate, which is issued by each state (and is valid only within the specific state) and examines whether the rolling stock is checked and approved and if the personnel (particularly drivers) are properly trained (Directives 14/2001, 49/2004, 59/2007, 110/2008). From June 2019, jurisdiction of national authorities for issuing Safety Certificates is transferred to the EU Agency for Railways (Regulation 796/2016).

♦ Establish specific rules for interoperability aiming to assure excellent compatibility between the characteristics of the infrastructure and those of the rolling stock and of operation, in order to increase performance levels and safety, improve quality of services, and reduce costs (Directives 50/2004, 57/2008 and 16/2001). However, Directive 797/2016 gives authorization (from June 2019) to the EU Agency for Railways to check compatibilities of trackside equipment with interoperability requirements, (see also sections 1.14 and 21.9).

♦ Provide methodology of path allocation and calculation of infrastructure charges (Directives 14/2001, 49/2004, 10/2008, 57/2008, 34/2012 and Regulations 913/2010, 909/2015). Infrastructure charges should take into account the nature of service, the time of its supply, the market situation, the quality of railway infrastructure and should prevent congestion. Comparable services should be subjected to the same charges. Sanctions for delay and bonuses for punctual arrival can be provisioned.

♦ Encourage a greater participation of the private sector in railway activities. This participation can take various forms of PPP, (see section 6.3.5), regarding funding of infrastructure construction, maintenance, and modernization.

♦ Permit a further separation of the operation into separate business units (e.g. passenger, freight, commuter, etc.), (12). However, EU legislation does not impose any rules on ownership of transport undertakings (article 222 of the Treaty of Rome), leaving the possibility to the states to privatize or not either parts or the entire railway undertaking, (58).

♦ Determine the duties of the Infrastructure Manager, who should be fair and avoid discriminations (Directive 12/2001). In addition, governance of rail infrastructure should give new entrants in the rail market fair and non-discriminatory access to rail related services and facilities (Directive 2370/2016).

♦ Introduce a Regulator, who will settle disputes in the playing field and particularly decisions of the Infrastructure Manager (Directive 14/2001). The regulatory body can be solely and exclusively responsible to regulate rail affairs (Austria, Bulgaria, Denmark, Greece, Latvia, Poland, Romania, Slovakia), or integrated within a wider transport regulatory organization (Belgium, Czech Republic, Finland, France, Hungary, Ireland, Italy, Portugal, Sweden), or integrated within a wider organization responsible for all types of regulated industries of the specific country (Croatia, Estonia, Germany, Lithuania, Luxembourg, the Netherlands, Slovenia, Spain).

♦ Ensure transparency in finances, without any possibility of state subsidies for freight transport, a clear justification of state subsidies for passenger transport (through public service obligations), and the possibility of subsidies for the Infrastructure Manager (Directive 12/2001).

♦ Open up national and international rail freight services on the European rail network (since 2007), (Directive 51/2004).

♦ Aim at the creation of a European high-speed railway network which guarantees safe and uninterrupted travel (Directives 48/2004 and 50/2004):
  – at a speed of at least 250 km/h on lines specially built for high speeds, while enabling speeds of over 300 km/h to be reached in appropriate circumstances,
  – at a speed of the order of 200 km/h on existing lines, which have been or are specially upgraded,
  – at the highest possible speed on other lines.

♦ Open up international passenger transport (since 2010), (Directive 58/2007).

♦ Introduce progressively a mandatory tender of PSO for domestic passenger transport services by rail (Regulation 2338/2016). However, member-countries can continue to award directly PSO contracts (that is without any tendering process, but under contractual requirements for service quality, frequency, and capacity) until the end of 2022. This deadline gives practically EU member-countries the possibility to award long-term PSO contracts by the year 2037, if they were awarded in 2022, since maximum duration of public service contracts for rail passenger services is 15 years (Regulation 1370/2007).

♦ Take measures in view of a full liberalization in the future of all rail services (including cabotage rights concerning domestic rail passenger transport (Directive 2370/2016)). Though theoretically such measures aim at boosting competition within the railway market, open-access operators will be permitted to run services in competition with the current operator of a PSO contract on domestic routes only under the condition that the new entrant does not compromise the economic equilibrium of a PSO contract (Regulation 2338/2016), (65).

♦ Strengthen security (Directive 49/2004, Regulation 796/2016) and allocate more responsibilities to the EU Agency for Railways, with duties on security, interoperability, coordination of policy and strategies.

♦ Ensure basic rights of rail passengers concerning insurance, ticketing, delays, etc. and those of passengers with reduced mobility (Regulation 1371/2007).

3.7. Some representative models of separation of infrastructure from operation in European railways

Some representative organizational models of separation of infrastructure from operation in Europe are the following, (58):

3.7.1. The Integrated model

This model is applied in Luxembourg, Ireland, and elsewhere; it is based on the will to maintain the integrity of the railway activity. The model consists in the creation of business units for infrastructure and operation, with management independence but without legal status, under a common executive board and a common chairman, within a single unique legal structure, (Fig. 3.2). In compliance with the EU regulations, the Bodies responsible for path allocation and infrastructure charges have been created outside the railway company.

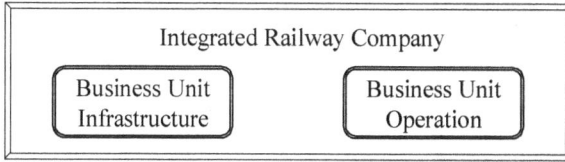

Fig. 3.2. The Integrated model

3.7.2. The Semi-integrated model with apparent organic separation

This model was applied in France. It is based simultaneously on an institutional separation of responsibilities, assets, and liabilities, leading therefore to separated balance sheets and operating accounts between the Infrastructure Manager (IM) and the Railway Operator, (Fig. 3.3). However, if responsibilities, objectives, strategies, and financial issues concerning infrastructure management are devolved on the IM, the maintenance of infrastructure is carried out by the Operator, a consequence of a public subcontract signed between the IM and the Operator, the latter acting as the subcontractor and the IM fixing the rules in this matter. Concerning investments, if the IM is normally the contracting authority defining the scope, consistence, and the objectives of infrastructure investments, the Operator is entitled to work as an executor. Consequently, this organization of responsibilities did not bring about any separation of the manpower force within the old vertically integrated railway company, some of the workers working on operation's issues and others as subcontractors of the IM. As far as the application of EU Directive 12/2001 is concerned, any responsibility on access charges issues and path allocation has been transferred to other bodies formed within the IM.

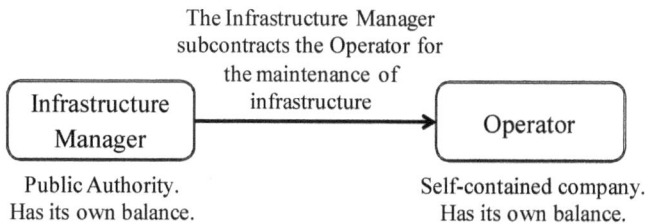

Fig. 3.3. The Semi-integrated model with apparent organic separation

This model, which only pretended to apply the European legislation, while keeping intact the unified railway system, was condemned by the European Union Court of Justice in April 2013 and was changed in France in 2015 in a holding form: a real Infrastructure Manager in charge of maintenance and operation of infrastructure and an Operator in charge of operation of trains.

3.7.3. The Holding model

This is the most popular model within the EU countries and is applied in Germany, France, Italy, Austria, and other countries and has led to a legal separation of business responsibilities, (Fig. 3.4). Consequently, every sector is regarded not just as an independent business unit but also as a legal entity, therefore having separate accounts, balances, and financial results. Independence between sectors is therefore better assured than in previous models. All these legally independent companies are amalgamated into a holding system with a common executive board and a chairman. In theory, the chairman cannot give any orders to the Infrastructure Manager, which must remain completely independent and impartial towards operators, with no discrimination whatsoever. However, this assumption of independence has not been verified in practice in many holding models; thus, many forms of discrimination in favor of the state Operator have been observed. The activities and responsibilities of the Railway Operator and the Infrastructure Manager within the holding company should be completely separate. More particularly, the formerly vertically integrated German Railways were transformed in a holding company, composed of two companies:

- one in charge of infrastructure (Deutsche Bahn), with four subsidiaries in charge of: safe exploitation of the railway network and infrastructure, passenger railway stations, electric power and fuel supply, construction-maintenance-exploitation of tracks,
- one in charge of operation (DB Mobility Logistics), with six subsidiaries in charge of: long-distance rail passenger transport, logistics, rail passenger transport in urban areas (metros, trains, buses), provision of various services such as cleaning of trains, maintenance of rolling stock, information technologies, etc., (49).

Holding Company

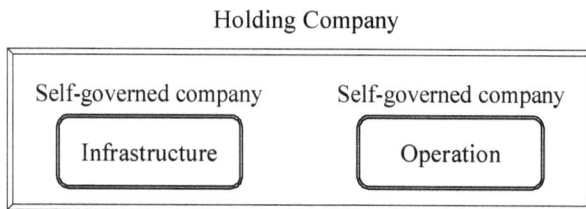

Fig. 3.4. The Holding model

Path allocation and access charging issues have been so far kept within the domain of the infrastructure manager, who must demonstrate a non-discriminatory approach towards every operator.

Such a holding organization has generated the creation of a lot of subsidiaries and a rapid liberalization of the railway market. There were more than 500 new rail operators in 2017 in Germany, having a share of 15% in the rail freight market and about 6% in the passenger market. Though the holding model has been strongly criticized, the European Union Court of Justice in a decision of 28 February 2013 (case C-556/European Commission against Germany) judged that the holding model is compatible with the EU legislation. Major advantages of the holding model are synergy of the system, economies of scale in the various costs, high coordination, and absence of conflicts of interest between infrastructure and operation.

3.7.4. The Separated model

This model is based on a complete institutional separation of the former integrated railway company between the infrastructure and the operation, (Fig. 3.5). Path allocation and access charging issues can remain within the domain of the Infrastructure Manager. The separated model is applied in Spain, the UK, the Netherlands, Belgium, Sweden, Finland, Denmark, Portugal, Greece, and elsewhere.

Company of Infrastructure	Company of Operation

Fig. 3.5. The Separated model

3.7.5. The Separated model along with further separation in infrastructure

This model has been applied in the Netherlands and is based on a complete separation between the Infrastructure Manager, on the one hand, and the various activities concerning operation of the former Dutch railways on the other, (Fig. 3.6). In addition, there was a further separation in the organization of the Infrastructure Manager into three independent parts, the first being responsible for path allocation and access charging, the second for maintenance activities, and the third for the planning of infrastructure activity. However, many disputes broke out between these different bodies, whose goals and limits were not clear. In order to resolve the matter, the Dutch government decided to settle the issue by appointing an Infrastructure Manager, a fact that brings the Dutch model closer to the fully separated one.

Both infrastructure and operation companies were in 2019 under state control and ownership in the Netherlands. No competition has been adopted till 2019 for rail services at national level, while competitive tendering was introduced progressively for rail services on most regional lines.

Company of Infrastructure

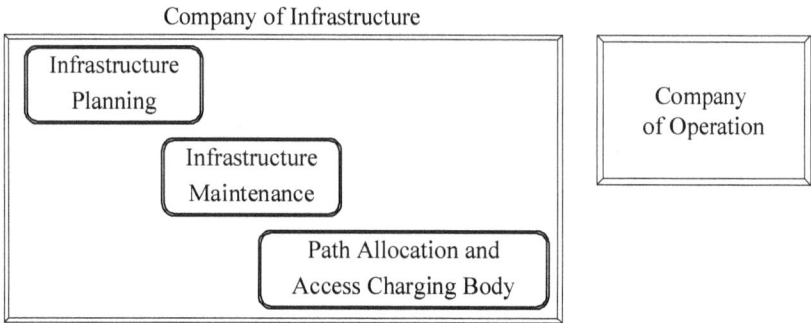

Fig. 3.6. The Separated model with further separation in infrastructure

3.7.6. The Separated model along with privatization

This model is applied in the United Kingdom and it is inspired by the Swedish model of a complete legal separation of the Infrastructure Manager's responsibilities and those of operation, with in addition a privatization of all the activities concerning, (Fig. 3.7), (55):

i. the Railway Undertaking, that was split up into a number of operators (called TOCs[*]) in the passenger sector and a number of operators in the freight sector. But even as private companies to which regional franchises have been allocated through calls for tender, these TOCs have survived so far thanks to important subsidies granted by national authorities and whose level is part of the contract, (see also section 6.10.7),

ii. the Infrastructure Manager, with the formation of Railtrack, which has tried to render the rail infrastructure activity profitable by drastically cutting maintenance and operation costs. Serious financial problems of Railtrack led the British government to renationalize rail infrastructure, whose duties have been taken over by Network Rail.

It has been argued that the privatization of all sectors has been chosen in the UK principally to render deregulation irreversible. Thus, British railway reform has placed the Stock Market at the heart of the new railway organization.

There were in 2020 in the UK 24 passenger train operating companies (20 franchised, 4 open access) and 7 freight operating companies. However, the largest UK freight company was acquired by the German Railways. In 2017, the British government contributed £1.55/passenger journey in England, £6.08 in Scotland, and £8.82 in Wales. Public subsidies for the passenger sector in 2017 represented 36% of total revenues from fares, (60).

---

[*] TOCs: Train Operating Companies.

Recent evolutions concerning the Infrastructure Manager have been characterized by a new involvement of the state, something that proves that it is hard for the private sector to operate efficiently and safely the railway infrastructure. Though UK is still the most liberalized rail market in Europe, problems arising at the interface infrastructure-operation lead to thoughts that some future rail services (e.g. the new railway line Cambridge-Oxford) may well be vertically integrated, that is without any form of separation between infrastructure and operation.

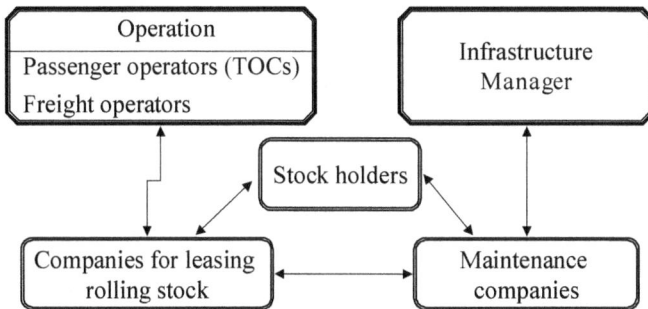

Fig. 3.7. The Separated model with privatization

3.7.7. Assessment of the various models

The integrated models correspond more to railways that have experienced interventionist government policy for a long time and emphasize more on cooperation. The changes as to the current situation are few and in any case not fundamental.

The holding and separated models correspond to a competitive transport market with the entry of many new railway operators, thus encouraging competition. These models presuppose fundamental organizational changes and can boost the establishment of a new competitive railway that will be easily adapted to market requirements. There is competition in the freight and long-distance passenger markets, whereas local passenger services are usually awarded by public biddings.

The separated models with further separation can be considered as a variation of the separated one. However, the split of infrastructure in many units may prove inefficient.

The separated model with privatization aimed at a drastic reduction of costs and subsidies. Each railway operator is a private company; it monopolizes railway services in specific routes and is subsidized by the state. This model did not encourage competition considerably and serious problems of cooperation among operators have emerged.

Figure 3.8 illustrates for the railways of the various European countries the degree of separation, the form of organization and legal structure, and the functions of the Infrastructure Manager (IM).

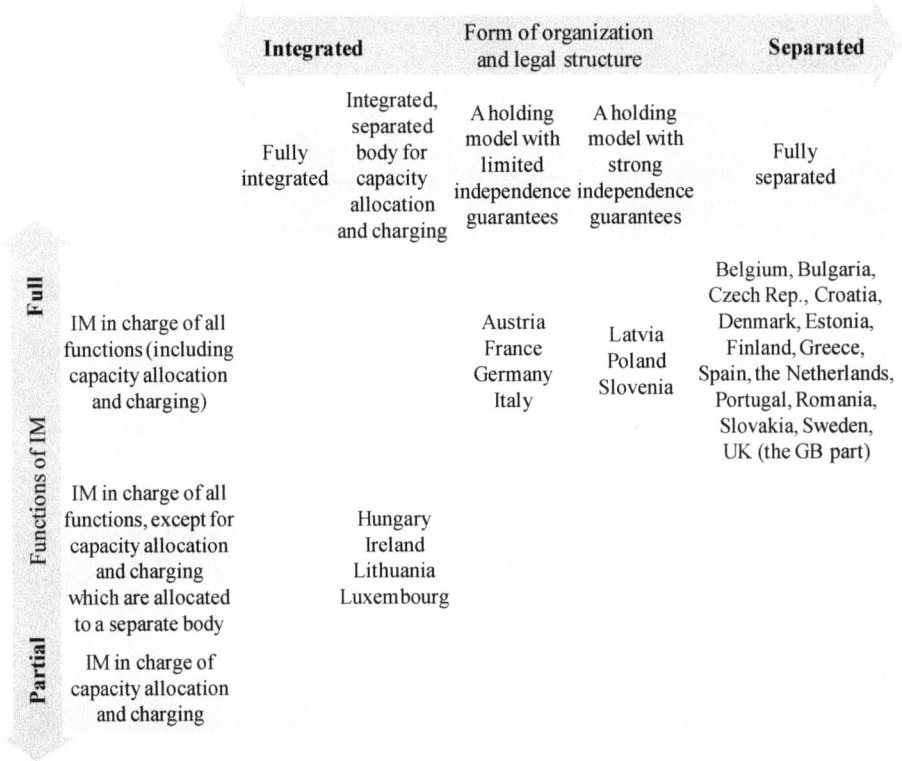

Functions of IM	Integrated		Form of organization and legal structure		Separated
	Fully integrated	Integrated, separated body for capacity allocation and charging	A holding model with limited independence guarantees	A holding model with strong independence guarantees	Fully separated
Full — IM in charge of all functions (including capacity allocation and charging)			Austria France Germany Italy	Latvia Poland Slovenia	Belgium, Bulgaria, Czech Rep., Croatia, Denmark, Estonia, Finland, Greece, Spain, the Netherlands, Portugal, Romania, Slovakia, Sweden, UK (the GB part)
IM in charge of all functions, except for capacity allocation and charging which are allocated to a separate body		Hungary Ireland Lithuania Luxembourg			
Partial — IM in charge of capacity allocation and charging					

Fig. 3.8. Degree of separation, form of organization and legal structure, and functions of the Infrastructure Manager (IM) for the railways of the various European countries, (100)

However, any evaluation of reforms should take into account historical, geographical, political, and economic particularities of each country.

Therefore, the degree of liberalization and segmentation of the sectors of operation and infrastructure of railways can be categorized as follows, (12), (13):

♦ Operation:
  – One national operator.

- One principal operator + regional operators.
- Many operators by segmentation of the network.
- Many operators – Open access.

♦ Infrastructure:
- Business unit within an integrated railway company.
- A separated infrastructure company either amalgamated within a holding system or totally separated.
- Infrastructure company totally privatized (but the British experience has proved that this model cannot work efficiently).

No country in Europe, however, has adopted the extreme case of open access with a total privatization of each sector.

3.7.8. Assessment of the impact of railway reforms

A first evaluation of the impact of railway reforms (taking into account that the railway is a heavy industry and it takes some time for reforms to bring about results) can be conducted by comparing performances (traffic, productivity, personnel) of a more liberalized model (Germany) with a less liberalized one (France), (Fig. 3.9). In fact, passenger traffic increased more rapidly in France than in Germany. In contrast, freight traffic increased in Germany but collapsed in France, a result of different restructuring patterns in these two countries. However, personnel reduction and productivity had higher rates in French railways compared to German railways. Nevertheless, it is difficult to calculate how much of the results can be attributed to the reforms. On the other hand, the quality of service for passengers does not seem to have changed dramatically as a whole. Some argue that no significant correlation between perceived quality of service and market opening policies can be established, while others report that customer satisfaction increases are the result of the introduction of competition, (59).

There is scepticism as to whether reforms in the organization of railways (and particularly the separation of infrastructure from operation) may have contributed to stabilize the modal share of rail passenger transport and possibly even stimulate a modest increase, (Fig. 3.10). However, the Community of European Railways argued in 2017 that it is not possible to establish a trustworthy statistical correlation between changes in the rail share and the degree of separation in the railway industry.

Experiences from the deregulation of other sectors, such as telecommunications and electricity, show that common results are mergers and concentration. This evolution is slowly emerging in the rail sector.

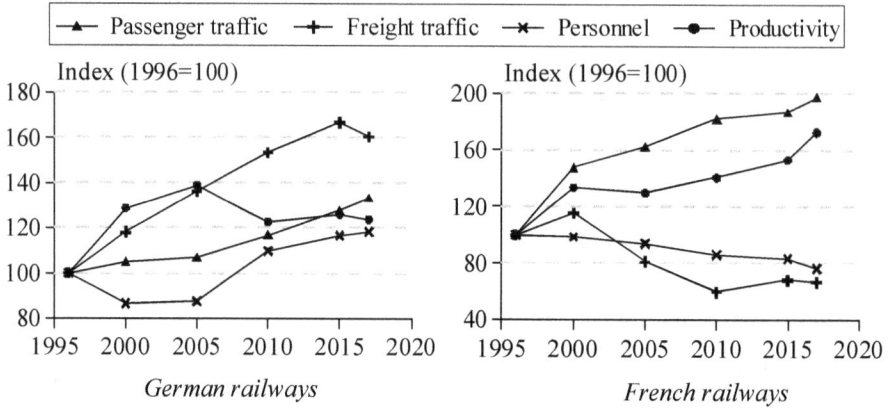

Fig. 3.9. Evolution of traffic, personnel, and productivity before and after the separation of infrastructure from operation in German and French railways, (1)

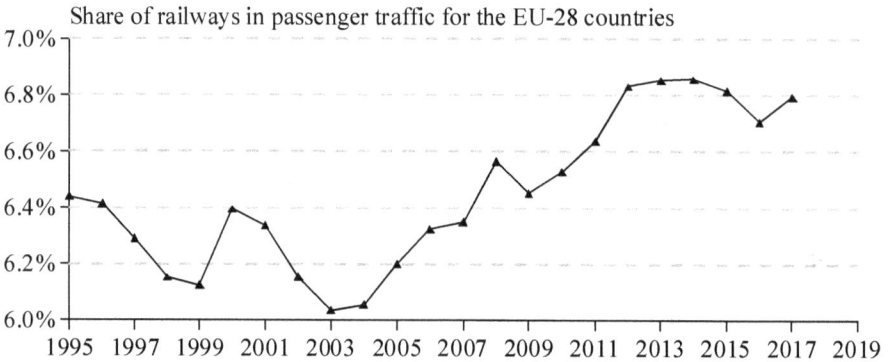

Fig. 3.10. Evolution of the market share of rail passenger transport in the EU-28 countries between 1995 and 2017, (1)

In conclusion, competition is still partial in the EU rail sector, regulation proves difficult, and there are few new entrants in most countries. There is no clear model that can ensure a more efficient railway, that is a railway with lower tariffs, higher quality of service, higher productivity, and less subsidies. The organizational and legal structure to be adopted (integrated or separated, totally separated or in a holding form, public or private) and the degree of regulation vary greatly from one country to another. Separation should not be considered as an end in itself but just as one of the many tools to convert railways in modern and efficient transport companies.

3.8. Rail legislation in the USA and Canada

As analyzed in section 1.9, railways have a marginal role in the passenger market in the *USA* (with a share of 0.7% in 2018) but play an important role in the freight sector (with a share of 35.1% in 2018 in the land transport market of the USA). American rail legislation should be examined within the context of the North American Free Trade Agreement (NAFTA) between the USA, Canada, and Mexico.

Another characteristic of the American rail market is that the principal rail operators can own the tracks they are running on. As competition is the rule in the American economy, legislation tried to assure rights of rail operators to run on infrastructure owned by another (and often competitor) operator.

Railways operated in both the passenger and freight sector as private companies until 1970, when the national passenger railroad corporation (Amtrak) was formed, a federally owned corporation subsidized by the federal government. Amtrak owns the track infrastructure it uses in the northeast of the USA and has the right to operate over all other tracks under negotiated access agreements (subject to adjudication in the event of dispute with the infrastructure owner).

During the 1970s, about 20% of the railway industry faced bankruptcy in the USA. Thus, the Consolidated rail corporation (Conrail), owned by the federal government, was formed in 1976, resulting from the consolidation of the bankrupt companies in the Northeast and Midwest of the USA. The option was to make Conrail viable and then to sell it. If it were not possible to make it viable, it would be liquidated. Conrail was sold in 1987 (having suffered considerable losses).

The more important measure in response to the continuing financial crisis in the rail industry in the USA was deregulation, which was introduced by the so-called *Staggers Act* of 1980, with the objective to achieve a balance between the financial viability of the rail sector and the interests of the shippers.

Regulation of the transport sector in the USA was conducted by the Interstate Commerce Commission, which was created in 1887 and was replaced in 1995 by the Surface Transportation Board, with jurisdiction covering all railways operating within the USA and the following duties, (53):

- ensure that rail carriers have trackage rights to operate on another carrier's infrastructure,
- reduce tariffs, particularly when complaints for market dominance and power abuse have been addressed. In 2011, it was reported that the Staggers Act resulted in a reduction of 51% in average freight tariffs,
- address quality,
- control exit, under specific circumstances, from the market,

– approve or decline mergers in the rail industry or impose conditions (i.e. trackage rights) on the merger, in order to promote competition.

This legislation seems to have worked well in preserving competition overall; however, some cases of disputes revealed the many, more or less, subtle ways in which the owner of infrastructure can create barriers to the entry of another operator, when access exists in theory.

Deregulation and liberalization of rail legislation in the USA is a good example of how the health of a transport company can be affected by a change in policy, organization, and shift of an activity from the umbrella of the public sector to the private sector. Railways in the USA concentrated on the freight transport of bulk materials, which was the most profitable and the least subject to competition activity. Operating costs were reduced by staff reductions (from 1.7 million employees in 1910 to 0.18 million in 2018), (56).

A major debate in the USA in the third decade of the 21st century focuses on whether the rail market is sufficiently deregulated and re-regulation measures should eventually be taken, which is the position of the American railroad corporation.

Similarly, railways in *Canada* have principally freight traffic (with a share of 67.9% in their national freight traffic), which is realized at a percentage (concerning tonne-kilometers) of about 95% by the two main trans-continental railways, Canadian National and Canadian Pacific.

Canada began to deregulate railways before the USA with the national transportation act of 1967, which changed in 1987. Subsidies were terminated in 1996 and labor productivity after deregulation increased by 93% from 1988 to 1997.

In both countries (USA and Canada) railway pricing is determined by market forces. Competition is not considered as an end in itself, but is balanced against other public interest considerations. Regulatory barriers are low concerning entry of a company in the rail market. However, a prerequisite in both countries for operating or maintaining a railway is the possession of a railway operating certificate, issued by the national regulator.

A fundamental difference in the rail policy between USA and Canada consists in that if a railway company wishes in Canada to run over the lines of another railway it can ask the national railway regulator to approve such rights and settle any dispute. US rail policy does not include any similar term, (56).

3.9. Rail legislation in Japan

Strong densities of population in Japan (with 334 people living in 2019 per km^2 of land area, against 123 in France, 274 in the United Kingdom and 34 in the USA) favor rail passenger traffic. On the contrary, rail freight has in Japan a rather marginal share. Japanese National Railways (JNR) started facing in the mid-1960s serious fiscal problems, which had not been overcome for two decades, in spite of four restructuring plans.

In order to pursue the fast economic growth and rising personal income of the country, Japanese railways invested huge amounts of capital. However, rail infrastructure is extremely costly with a low rate of return. Debt augmented greatly, increases in fares resulted in fewer customers, and privatization and segmentation were seen as the only way to revitalize the Japanese railways.

The whole rail network was split in 1987 into 6 regional passenger companies (each one owning its own infrastructure) totally privatized (JR Hokkaido, JR East, JR Central, JR West, JR Shikoku, JR Kyushu). Another company, Japan freight railways, which pays fees to the 6 rail passenger companies for using their tracks and other facilities, took freight traffic.

Before deregulating and privatizing the rail market, the Japanese government undertook specific measures: the transfer of JNR's long-term debt to the JNR Settlement corporation, the reduction in excess labor (from 400,000 in 1980 to 191,000 in 1994), and the abandonment of unprofitable local lines.

Table 3.1.

Ratio of labor and capital costs to fare income before and after privatization of Japanese railways, (64)

	Labor costs (%)	Capital costs (%)
1985	73.6	53.8
1987 Privatization	31.6	22.2
1994	35.1	33.0

The ratio of labor and capital costs to fare income decreased greatly after privatization, (Table 3.1), (64). The number of passengers increased between 1986 and 1993 by 20%. The rolling stock kilometers traveled also increased about 20% after privatization.

Since 2006, all shares of JR East, JR Central, and JR West have been traded in the Stock Market. On the other hand, all shares of JR Hokkaido, JR Shikoku, JR Kyushu, and JR Freight have been transferred and are owned by the Japan Railway Construction, Transport, and Technology Agency, an independent administrative institution of the state.

In contrast to the EU, most of the passenger railways in Japan own their infrastructure and thus maintain a vertically integrated structure. However, in order to encourage competition in the rail market, rail legislation in Japan distinguishes three types of railway licenses, (66):

- companies that provide passenger/freight services, while holding their infrastructure,

- companies that provide passenger/freight services by using rail infrastructure owned by another organization,
- companies that own rail infrastructure only for one purpose, to give access to a rail operator without infrastructure, who pays the appropriate charges.

More than three decades after the deregulation and privatization Act of the former unified Japanese railways, a great number of rail operators are competing in the rail market of Japan: the 6 major JR companies, 16 major private railway companies, 128 mid- to small private railway companies, 5 semi-major companies, 11 public authorities, 33 rail transit companies, (66).

3.10. Rail legislation in China and India

In both China and India railway activity is totally regulated.

Railways in *China* are owned by the state and controlled by the Ministry of Railways of China. Infrastructure belongs to the state and major issues concerning tariffs, service planning, and investment are taken by the government. With 2 million employees, Chinese railways are geographically split into 16 districts, each one covering a particular region of China. There are fears, however, that competition from the road sector and low rail profit rates combined with a highly regulated environment may lead to low rail profitability. However, in 2013 the policy and planning responsibilities of Chinese railways were transferred to the Ministry of Transport.

Indian railways is a state-owned company in charge of railway and metro systems, owned and operated by the government through the Ministry of Railways. Principal railway legislation can be found in the Railways Act of 1989, which is amended regularly. With 1.3 million employees, Indian railways are divided into several zones (16 in 2020), which are further subdivided into divisions.

3.11. Rail legislation in Russia

Russian railways remain a vertically integrated company, which manages infrastructure and operates passenger and freight services. However, management and operation of railways were separated from policy functions.

In the late 1990s, competition was seen in Russia as an efficient way to attract private capital, achieve cost reduction, and improve profitability. The new legislation in early 2000s aimed at a preliminary restructuring, privatization of some railway supply industries, and introduction of competitive tendering to procurement processes. Just afterwards, legislation introduced a separation of policy-regulatory functions from commercial-business activities.

Influenced by EU policies, rail legislation in Russia has opted for a separation of infrastructure from operation in the future. Within this policy frame, some measures were the following, (52):

– separate policy-regulatory functions from commercial-business functions,
– transfer of rail activities to a joint stock holding company,
– transfer of intercity passenger services to a separate company,
– facilitate private investment in rolling stock in order to renew the fleet, though the national freight carrier still keeps owning locomotives and controlling freight flows,
– split existing companies to a number of subsidiaries and then sell some of them to the private sector.

3.12. Rail legislation in Australia and New Zealand

Railways in *Australia* have differences from one state to another concerning both gauge and organization. The gauge problem was resolved partly by the conversion of all interstate tracks to standard gauge in 1995. Rail passenger transport is limited outside of the major cities, because of long distances in a sparsely populated country of 25.3 million people (in 2019). Thus, railways in Australia focus principally on freight.

Until the early 1990s, railways operated as vertically and horizontally integrated public sector monopolies. After 1995 and the creation of the Australian competition and consumer commission, policy aimed at the introduction of competition in rail operations with as a key factor the conditions of access to rail infrastructure, which can be done in 3 ways: declaration under the national access regime, certification of the state regime, and authorization of an undertaking from an infrastructure provider. Concerning safety, a national rail safety law and a national safety regulator were established in 2009 (they are operational since 2013) and the specific law was afterwards detailed in the various Australian states. The Railways Act of 2005 imposes in Australia that any rail operator or access provider must hold a license.

The states of Western Australia and Queensland own the vertically integrated railway systems and have created separate business units and separate accounts. The states of Victoria and New South Wales have separated infrastructure from operation and freight from passenger.

Legislation in *New Zealand* was very protective for railways and until 1961 carriage of goods by road was limited to distances up to 50 km, which was raised to 67 km in 1961 and to 150 km in 1977. Liberalization of the road haulage in 1983 pressed railways toward restructuring. Infrastructure and rolling stock were treated as separate accounting units supported by an internal pricing structure. The labor force was reduced, traffic was maintained, and productivity increased.

3.13. International rail law – The COTIF convention

Conditions, obligations, rights, and liability of all parties involved in rail transport within the frontiers of a country are described in its national legislation, they have usually the character of public law, and are valid only within the frontiers of a specific country. However, international rail transport needs a uniform system of law and regulations, applicable to all countries through which international rail transport is realized. The problem was first confronted in 1890 by means of the international convention for the transport of goods by rail CIM (from the initials of its French name 'Convention Internationale pour les transports ferroviaires internationaux des Marchandises') and next in 1924 by means of the international convention for the transport of goods by rail CIV (from the initials of its French name 'Convention Internationale pour les transports internationaux par chemin de fer des Voyageurs'). Later in 1973, another convention related to the liability of railways for death and personal injury of passengers was added.

Conventions CIM and CIV were unified in 1980 in one convention under the name COTIF (from the initials of its French name 'Convention pour les Transports Internationaux Ferroviaires'). COTIF has some general dispositions and the modified CIM and CIV conventions. Liberalization of rail transport, competition, electronic transactions, separation of infrastructure from operation, and the need to increase safety made necessary a radical modification of international rail legislation. Thus, the new COTIF convention was signed in 1999, is applicable in more than 50 states, on 250,000 kilometers of track, and on 17,000 kilometers of shipping lines and inland waterways around the world. The COTIF convention has the characteristics of private law, thus it is subject to prescriptions of public law of the various countries. The COTIF convention valid in 2020 has some general provisions and is accompanied by 7 appendices as follows, (62):

♦ CIV: General conditions of carriage for rail passengers. CIV confronts, among others, conditions and contract of carriage of passengers, wording of tickets and reservations, passengers' obligations before and during the journey, conditions of carriage of luggage and animals, rights of passengers in case of delays and cancellations, liability for personal injury and loss and damage of property, actions in cases of claims, complaints, and disputes.

♦ CIM: General conditions of carriage of goods by rail. CIM confronts, among others, wording of the contract and the consignment note, the responsibility of particulars involved in the consignment note, payment of costs, loading and unloading of goods, packing and delivery issues, liability of carriers, successive carriers and substitute carriers, compensation for loss, damage, or exceeding the agreed transit period, and liability for rail-sea traffic.

♦ RID: Regulation concerning the international carriage of dangerous goods by rail. RID confronts, among others, a definition of dangerous goods, training of employees, safety obligations of all involved parties, checks to ensure compliance with safety requirements, internal emergency plans for marshaling yards.

♦ CUV: It refers to contracts of use of vehicles in international rail traffic and confronts, among others, forms of signs and inscriptions on the vehicles, liability in case of loss of a vehicle or of damage to a vehicle or caused by a vehicle.

♦ CUI: It refers to contracts of use of infrastructure in international rail traffic and confronts contents, form, and mandatory law of such contracts, obligations of the rail operator and the infrastructure manager, liability of all involved parties, damages in case of death or personal injury during the travel.

♦ APTU: It refers to the validation of technical standards and the adoption of uniform technical prescriptions to railway material in international traffic.

♦ ATMF: It refers to the technical admission of railway material in international traffic.

It is noteworthy to outline that the COTIF convention itself gives the possibility to member-states to conclude agreements with derogations from the uniform rules described above. Railways which have adopted the COTIF convention manifest greater legal reliability, simplification and homogeneization of customs formalities and other administrational procedures, savings in costs and time, improvement of the image of railways in relation to competing modes (air and road), which for many years have put together solid legislation and rules for international traffic.

95

4 Forecast of Rail Demand

4.1. Purposes, needs, and methods for the forecast of rail demand

Forecast is an effort to foresee and anticipate developments in the future. It is a complex procedure that must take into account the expectations and tendencies of a particular society, the situation in the industry, economic and political factors, human fears, and the psychology of the human being. Rail demand is the function which relates the amount of rail passenger and freight traffic that people will choose for a specific price and other quality conditions of the rail transport activity under consideration. Forecast of future demand is a prerequisite for many railway projects and actions, such as:

- construction of a new railway line (or station),
- opening (or closing) of a new railway service (e.g. high-speed services, commuter services, etc.),
- purchase of rolling stock vehicles,
- programming of the necessary staff in trains and stations,
- revenue estimation,
- commercial and pricing policy,
- management strategies.

Conducting a forecast is a difficult task. Parameters affecting demand are of both technological and human nature, the latter ones being difficult to foresee; the form of their intercorrelation is complex and statistical data concerning past demand are often insufficient or inaccurate.

All rail transport forecasts are based on a kind of rail demand model. A *model* can be defined as a human effort (through a simplified representation) to understand, explain, and foresee the evolution of a physical, human, or social phenomenon. It tries to investigate whether a causal interrelationship can be found between the phenomenon under study (e.g. number of high-speed train passengers between London and Paris) and some of the parameters affecting it (year of the study, cost of rail transport and of competing modes, travel times by rail and competing modes, quality of service, Gross Domestic Product of countries at the origin and destination of a trip, etc.).

Once some form of correlation is established and the statistical and logical validity of the model are checked, then the model can be used for the forecast of future rail demand.

A model can be based on different methodologies of the phenomenon under study and thus we can distinguish the following categories of methods (with a variety of models suggested for each method), (9):

- *qualitative* methods: they are based on perceptions, evaluations, and assessments made either by transport specialists or by the users of a transport service,
- *statistical* methods: they are based on historical data of demand which can be extended to the future. The fundamental assumption that should be fulfilled in any kind of statistical method is that all factors which affected rail transport demand in the past will remain the same in the future, with their degree of influence unchanged, while the characteristics of supply remain also unchanged. The only variable in statistical methods is time. Trend projection of statistical data and time-series are the statistical methods mostly used for the forecast of rail demand,
- *causal* methods: they establish a causal correlation between demand (the dependent variable) and its generating forces, such as time (period of analysis), costs, economic indices, travel times, etc. (the independent variables). Causal methods of rail transport demand are usually based on *econometric* methods which establish a correlation (in the form of equation) between rail transport demand and the most important factors (not all of them) that affect demand. Causal methods are sometimes based on *gravity* methods, an analogy of the gravitation law of mechanics,
- methods based either on *artificial intelligence* or on *fuzzy numbers*. Artificial intelligence makes machines do things that would require human intelligence; these machines are programmed to think like people do, to act like people do, and to think rationally and reasonably. On the other hand, fuzzy numbers do not refer to one single value (as an ordinary real number) but to a set of plausible numerical values within a specified domain and around a central value.

When forecasting rail demand, we distinguish the short-term level (6÷24 months), the medium-term level (2÷5 years), and the long-term level (5÷10 years). A model can be appropriate for short- or medium-term forecasts but totally inappropriate for long-term ones and vice-versa.

However, a model can explain only a small category of specific problems. It is the result of a number of assumptions and should not be extended, generalized, or used for cases for which the assumptions, upon which it is based, do not hold. Rail specialists should avoid such inappropriate extensions and generalizations of models.

A forecast model should not necessarily have a complicated form, just the opposite. Within the acceptable limits of accuracy, the simplest form of model should be the target.

Any demand forecast has inherent weaknesses and uncertainties and should clearly address assumptions on which it is based, the degree of accuracy of the forecasts, and the frame and conditions within which the forecast can be used.

On the other hand, forecasting presumes a minimum of stability. The forecaster considers many parameters that will continue to evolve under the influence of the past. The forecaster can forecast what is likely to occur, but he cannot predict the unpredictable (like the shrinkage of rail traffic, due to the sanitary crisis of 2020 and 2021).

4.2. Driving forces and parameters affecting the various categories of rail demand

4.2.1. Driving forces affecting rail demand

The need for rail transport is not an end in itself but a means to satisfy other human needs. The demand for transport is derived, it is usually linked with other human activities. With the exception of sightseeing, people travel to satisfy another need (work, leisure, meeting other persons, shopping, etc.) at their destination. The need for transport would not have existed, if all these activities had been located and developed in neighboring areas, something that did not happen even in the first forms of organization of human societies. As a result of economic and social activities, transport is strongly influenced by economic factors, which can be expressed by the GDP or the purchasing power or the income.

Rail transport is also strongly influenced by the spatial distribution of human activities. High concentrations of populations and goods are favorable for rail transport.

Social attitudes and lifestyle, characterized by increasing individualism and search for more comfort and ease, can affect rather subjectively the choice of an individual for a specific transport mode against another.

Travel times (along with costs) affect critically rail demand. Travel times may be calculated from station to station or from door to door, in which case they should consider access time to (or from) the station (on foot, by bicycle, bus, private car, or metro). Railways have the advantage of reaching the center of cities and the disadvantage of not providing direct door-to-door transport (as private cars do).

Rail transport has a dynamic character and differs from day to day and from hour to hour.

The institutional framework of organization of the transport market has a critical impact on rail demand. For many decades, railways had monopolistic control of the rail market, facing only external (inter-modal) competition from buses, private cars, and airplanes. However, internal (intra-modal) competition, that is operation of many railway companies on the same line, has been recently introduced in some countries, as analyzed in chapter 3.

The sensitivity of citizens to the protection of the environment has as a result that environmental issues should also be taken into account among parameters affecting demand.

Technological developments may be critical for rail demand, as they affect directly the operating characteristics of rail transport. On the other hand, fuel prices have a direct impact on rail costs of operation.

The performance and characteristics of other transport modes also affect rail demand.

4.2.2. Effects on rail demand of the principal parameters of rail transport

4.2.2.1. Passenger rail demand

Passenger demand is divided into intercity demand (among cities) and commuting demand (from the central part of a city to its suburbs and vice-versa). Intercity demand can have as a motivation either business activities or leisure. The same applies to commuting, but all components of commuting demand have similar characteristics and for this reason there is no distinction in commuting between business and leisure.

Business, leisure, and commuting demand are affected in a different way by the various parameters of rail transport, which are: cost of travel, travel time, frequency of services, quality of services, and punctuality. Thus travel time, punctuality, frequency, and quality of services are critical for business rail demand. Cost is critical for leisure, whereas cost, punctuality, and frequency are greatly influencing commuting, (Table 4.1). The reason for the differences illustrated in Table 4.1 is rather simple: the company (and not the person who travels) undertakes usually the cost of business travel, whereas for leisure and commuting it is the traveler who pays the cost of the ticket.

4.2.2.2. Freight rail demand

Freight demand is, in most cases, part of the industrial process. Parameters that influence freight demand are, (see also section 1.7.1), (9), (22):
- type of goods: characteristics and nature of materials and of final products,
- geographical parameters: location, vicinity with a port, density of population,
- socio-economic parameters,

Table 4.1.
Parameters of rail transport and their degree of influence
for business, leisure, and commuting demand

Parameter affecting rail passenger demand	Category of rail passenger demand		
	Intercity		Commuting
	Business	Leisure	
Travel time	+++	+	+++
Cost of travel	++	+++	++
Frequency of services	++	+	++
Quality of services	++	++	+
Punctuality	+++	+	++

Legend: +++ high influence, ++ medium influence, + small influence

– legislation and road traffic restrictions,
– price. The pricing policy for freight is more flexible than for passengers and includes usually negotiations with clients,
– seasonality for some types of goods,
– terminal and combined transport equipment.

We usually distinguish rail freight in bulk quantities (oil, cereals, etc.) from isolated small items.

Many surveys among shippers and transport forwarders have revealed the reasons of preference for rail freight, which are: high volumes, low cost, non-availability of road vehicles, safety. The reasons of non-preference of rail freight are: high shipment times, bureaucratic procedures, high costs, and uncertainty of time of delivery of goods, (24). It is clear that some shippers consider rail tariffs low, whereas others consider that they are high. The reasons of preference of road transport by shippers are: speed of shipment, door-to-door transport, simple procedures and flexibility, accuracy of shipment, low cost, and the availability of vehicles. If railways want to increase their freight traffic, they must overcome all these handicaps and weaknesses, as are revealed by several surveys, (15), (22), (91).

4.3. Qualitative methods

4.3.1. Market surveys

The qualitative method most commonly used for the assessment and forecast of rail demand is the market survey, which however requires a certain time (from some days to months) and has a high cost. The market survey is the only method that can be applied when there are no statistical data (e.g. the opening of a new railway station or the construction of a new railway line) or when railway managers try to identify the reactions of their customers to eventual changes in existing rail services or to the supply of new ones, (92).

Transport market surveys, besides the determination of passenger characteristics, can (by means of appropriate questions) identify passenger intentions. Indeed, up to the 1980s, transport market surveys were about questions over trends and choices that had already taken place. Such surveys are characterized as surveys of *revealed preference.*

During the past five decades, however, market surveys include questions of a hypothetical nature (e.g. "how often would you use the train if the ticket's price were reduced by 20% ?"). Thus the intentions of the person questioned are identified and some indications are provided regarding the development of future demand. Such surveys are characterized as surveys of *stated preference,* (92).

Any survey has unique characteristics and though the questionnaire can be tailored on the basis of other similar questionnaires, it should focus on the targets of the specific survey, the prevailing conditions, the conclusions to be reached, the practical decisions to be taken. Some issues raised in surveys for rail services are the following, (9):

- purpose of traveling (work-business, tourism-leisure, visiting friends or relatives, education),
- type of train (commuter-suburban, intercity, high-speed),
- degree of satisfaction from railway services already provided: travel time, punctuality and reliability, frequency, cleanliness, behavior of staff in trains and in stations),
- means of transport used to access the station (on foot, by bicycle, motorcycle, private car, bus, taxi, metro),
- travel time to access the station (<15 min, 15÷30 min, 30÷60 min, >60 min),
- assessment of facilities (cleaning, internet, air conditioning, space of seats, etc.) and services (restaurant, bar, newspapers, etc.) provided in the train,
- assessment of services provided in the station (ticket sales, cleaning, waiting areas, restaurants, bars, air conditioning, access to platforms),
- frequency of traveling by train,
- reasons that deter from using trains (delays, many stops, cleanliness, long travel times, personnel behavior in the train and in the station),
- whether passengers possess or not a private car,
- questions of stated preference: whether passengers would use more frequently the trains if travel times or fares were reduced, if trains were put on for specific slots, etc.

The construction of the *questionnaire* is the first critical aspect in any market survey. Table 4.2 illustrates a questionnaire used in a market survey addressed to passengers of intercity trains.

The second critical aspect in a market survey is the calculation of the size of the *sample,* which is a small part of the whole population that will be investigated. A correctly designed sample should represent as accurately as possible the population under study. However, any survey has an inherent degree of error, as recorded values of the survey will never coincide with actual values that will be realized. The error in the values of the survey is a function of the size of the sample.

Among the various sampling methods, we can distinguish probability methods (random, stratified, cluster) and nonprobability ones (convenience or opportunistic, quota). In most rail surveys, random sampling is used and selection of participants of the survey is made purely by chance, (9).

The size n of the sample (the number of people that will be questioned) is calculated in relation to the size N of the population (the number of customers) from the formula, (9):

$$n \geq N \cdot \left(1 + \frac{N-1}{p \cdot (1-p)} \cdot \left(\frac{d}{z_{\alpha/2}} \right)^2 \right)^{-1} \tag{4.1}$$

where   p: a probability parameter, usually given the value p= 50%,
         d: the margin of error, with usually accepted values for d in the range of 1.0%÷10.0%,
         $z_{\alpha/2}$: a parameter related to the confidence level c, which measures how certain we can be that the sample accurately reflects the population within the accepted value for the margin of error.

In most rail surveys we accept the values for d and c given below. We take the example of a survey of passengers of Eurostar trains (N=10.399 million in 2019) and we can select:

- margin of error d=10%, confidence level c=90%. From statistical nomograms it is deduced that $z_{\alpha/2}$=1.645 (where a=1−c), and from formula (4.1) n=68 questionnaires,
- margin of error d=5%, confidence level c=95%, $z_{\alpha/2}$=$z_{0.05/2}$=1.960, and from formula (4.1) n=384 questionnaires,
- margin of error d=1%, confidence level c=95%, $z_{\alpha/2}$=1.960, and from formula (4.1) n=9,565 questionnaires.

It is clear that equation (4.1) is strongly nonlinear. Thus, doubling the size of the survey does not mean doubling also the degree of accuracy.

Table 4.2.
Questionnaire for a market survey on intercity trains, (80)

QUESTIONNAIRE SURVEY ON INTERCITY TRAINS

Date: [/   /2022]   *Name of researcher:* [   ]

1. Class: [   ] 1st class   [   ] 2nd class

2. What is the purpose of your trip today?

   [   ] Work
   [   ] Family reasons
   [   ] Leisure – Tourism
   [   ] Studies
   [   ] Other reason   Please specify: [   ]

3. Did you find the scheduled departure time convenient?

   [   ] Yes   [   ] No

   If no, what departure time for the service would you wish? [   ]

4. How often do you travel by rail?

   [   ] Rarely   [   ] Up to 10 times a year
   [   ] Up to 2 times a year   [   ] Up to 20 times a year
   [   ] Up to 5 times a year   [   ] More than 20 times a year

5. Which of the following reasons led you to choose rail for your travel?
   (passengers can select more than one reasons)

   [   ] Travel time   [   ] Service frequency
   [   ] Safety   [   ] Low fare
   [   ] Comfort   [   ] Non availability of other means of transport
   [   ] Punctuality   [   ] Other reason (please specify)

6. Should you not use the train, which means of transport would you use?

   [   ] Private car   [   ] Bus   [   ] Airplane

7. Did any of the following cause you concern?
   (passengers are able to select more than one reasons)

   [   ] Delays   [   ] Train personnel behavior
   [   ] Too many stops   [   ] Station personnel behavior
   [   ] Cleanliness   [   ] Other (specify)
   [   ] Long travel time [   ] Nothing

8. Are you satisfied with the means of transport from the origin point to the departure station?

    ☐ Yes   ☐ Moderately   ☐ No

9. What means of transport did you use to get to the departure station?

    ☐ Private Car   ☐ Bus   ☐ Taxi   ☐ Metro

    ☐ Motorcycle   ☐ Bicycle   ☐ On foot

10. Does the departure station meet your expectations?

    ☐ Yes   ☐ Moderately   ☐ No

If no, what do you think should be improved? _____

11. Do you use the train's restaurant and bar services?

    Restaurant: ☐ Yes ☐ No   If no, why? _____

    Bar: ☐ Yes ☐ No   If no, why? _____

12. Do you own a car?

    ☐ Yes   ☐ No

13. Would you be interested in a car transport rail service?

    ☐ Yes   ☐ No

14. Should you use a night train, would you be interested in a sleeper service?

    ☐ Yes   ☐ No

15. Passenger sex:

    ☐ Male   ☐ Female

    (The traveler escorts a child under the age of 12 ☐ Yes ☐ No)

16. What is your profession?

    ☐ Public servant   ☐ Pupil

    ☐ Private sector employee   ☐ Pensioner

    ☐ Freelance professional   ☐ Military

    ☐ University or College student   ☐ Unemployed

    ☐ Housewife

17. Passenger age (estimated by the researcher):

    ☐ Less than 26 years old ☐ 26÷65 years old ☐ Over 65 years old

18. Passenger nationality: ☐

4.3.2. Scenario writing method

The search for alternative qualitative approaches for long-term forecasts resulted in the development of the scenario writing method. This method can be used for medium- and long-term forecasts.

A general definition of the method is that with the writing of scenarios, one attempts to present the pattern through which certain conditions will evolve in the future, by using as a point of reference and comparison the existing situation, which is described by means of a series of events and conditions, (9).

A scenario is a set of facts, events, and plausible actions that could lead to a future situation. Thus, scenario writing is not in principle a forecasting method but merely a tracking of pathways and processes that either will lead from the present situation to a future one (prospective scenario) or a plausible or desired future situation will be the outcome of the present one (projective scenario). Scenarios complement the deficiencies and limitations in the forecasting ability of other methods or try to establish relations between causes and effects for complex situations and multiple or uncertain or vague processes and conditions. Scenario writing requires identification of the problem and of its key factors, formulation of the various scenarios as clear and distinguishable alternatives with detailed description of conditions for each one, comparative analysis of the various scenarios, assessment of the more suitable one (eventually with the help of some form of quantification).

The scenario writing method is among the methods used for long-term forecasts of the European Commission and many others institutions as well as in studies with a long-term range.

4.3.3. Executive judgment method

The executive judgment method is based on the evaluation and assessment made by rail executives (specialists or managers) over a specific topic, for which they have a broad and in-depth knowledge. The method exploits both the scientific background and the empirical knowledge of specialists, who can consider in the analysis a plethora of factors that other methods fail to take into account.

4.3.4. Delphi method

The Delphi method is an evolution of the executive judgment method and is used to forecast certain medium-term events and to calculate the probability of their happening in the future, by seeking advice from a pool of experts in successive rounds. More specifically, the Delphi method has three separate stages: the preparatory stage, the stage of controlled feedback mechanism, and the final stage of conclusions and forecast.

Within the three stages of the Delphi method, we can distinguish the following steps: definition of the problem and of topics to investigate, choice of experts (upon their knowledge, experience, willingness, and availability), design of the questionnaire of the survey, first round of completion of the questionnaire by the experts, evaluation of the first round, completion of a new questionnaire by the experts, finalization of the procedure until a consensus is reached by the experts (usually after three, but in some cases even six, successive rounds of completion of the questionnaires), (9).

Participants in a Delphi method are encouraged to justify their position and eventually bring some kind of quantification. Whether consensus is reached or not among experts can be assessed with the help of the so-called Kendall's coefficient of concordance, (9).

As an example, the Delphi method is usual at the scheduled meetings of international rail institutions (such as the International Union of Railways), where future policies (and the probability for them to occur) are discussed among experts.

4.4. Method of trend projection of statistical data

4.4.1. Theoretical background and conditions of applicability

Trend projection of statistical data of the past is the statistical method most commonly used among railways for a quick estimation of future demand. Based on statistical data (which should cover at least 10 years), a projection of past trends into the future can be conducted, which requires from some hours to 1÷2 days of work and has a low cost to prepare it. The method provides adequately reliable forecasts for a period of 2÷5 years and up to 10 years beyond the year of forecast, as long as no unpredictable events take place (such as a sudden change in the economic situation or in the conditions of competition, accidents, a sanitary crisis, etc.) and supply remains unchanged.

The method is based on the assumption that all parameters affecting rail transport demand over a specific route for both the railways and their competitors (travel times, fares, income, elasticities, etc.) will continue over the course of time to affect rail transport in the same manner. This can be acceptable on a short-term or medium-term level, but hardly on a long-term one, (71), (74).

Many railways begin their forecasts by using trend projections of statistical data of the past, which can be improved with the results of a market survey or an econometric model.

Data are set on a demand (Y-axis) – time (X-axis) diagram (Fig. 4.1). This diagram provides a first indication whether the development of the phenomenon is linear or exponential.

106

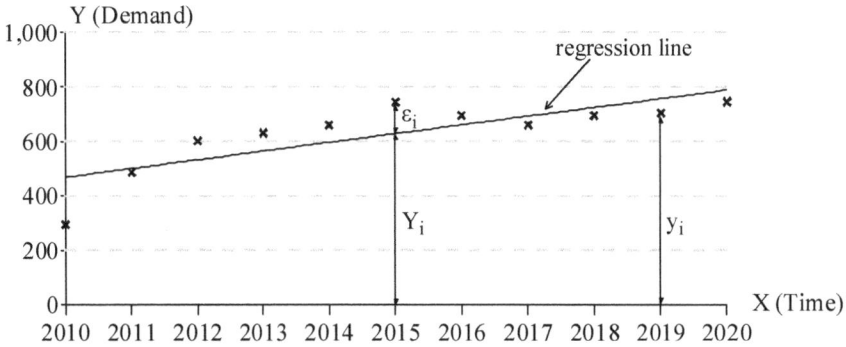

Fig. 4.1. Statistical data (y_i), regression line, calculated values (Y_i), and errors (ε_i) of the forecast

Thus, if the phenomenon develops linearly, demand Y_t for the year t will be:

$$Y_t = a + b \cdot t \tag{4.2}$$

If the phenomenon develops exponentially, demand for the year t will be:

$$Y_t = c \cdot (1 + d)^t \tag{4.3}$$

It should be noted that rail transport demand may have a linear development over the beginning of a new rail service and an exponential development later on, or an exponential development on the beginning followed by phenomena of saturation (an asymptotic development) later on. In this case, a combination of the aforementioned formulas (4.2) and (4.3) should be applied, (71).

Calculation of parameters a, b, c, d of equations (4.2), (4.3) will be the result of a correlation (either linear or exponential) of the dependent variable Y_t with regard to the single independent variable t. Therefore, we can determine the straight line (for linear development) or curve (for exponential development) from which the various points of Figure 4.1 are the least distanced. This is achieved by employing the method of least squares. The correlation (formula or line) between two (or more) variables which is derived on the basis of a series of statistical data is known in statistics as *regression*.

Let y_i be the various values given by statistical data, $\bar{y}$ the average of values y_i, and Y_i the values provided by the curve of Figure 4.1. Whether the curve is satisfactorily adjusted to the statistical data of the past will be dependent upon the value of the *coefficient of determination*, R^2, which is defined as:

$$R^2 = \frac{\sum_i (Y_i - \bar{y})^2}{\sum_i (y_i - \bar{y})^2} \tag{4.3}$$

107

The coefficient of determination multiplied by 100 gives how closely the forecasted values approach statistical data. Values of R^2 approaching 1.0 show that the regression curve is satisfactorily adjusted to the statistical data of the past, whereas R^2 values approaching zero show that no satisfactory correlation in the past's statistical data, which present irregular fluctuations, can be found. For most demand forecasts, values of $R^2 > 0.85$ are considered satisfactory. This means that if $R^2 = 0.85$, the forecasted values will diverge from values that will be realized by 15% on average, (9), (71).

A plethora of computer software allows a quick and easy calculation of the parameters a, b (or c, d) of the regression equation[*]. These software allow the use of various forms of functions (linear, polynomial, or exponential), for which the various statistical indices (coefficients of independent variables, data averages, sample variance, coefficient of determination, etc) are determined.

A number of questions are raised regarding the methodology of trend projection by using existing statistical data, (71), (77):

- what is the time period that the available data should cover? As derived by a number of analyses, 10 years is the minimum, provided that the statistical data collected represent sufficiently the evolution of demand,
- how far into the future can the past's data be projected? It was concluded that the projection period could not exceed half the analysis period, as long as is valid the fundamental assumption that the parameters affecting demand in the past will remain the same in the future and with their degree of influence unchanged, whilst the characteristics of supply remain unchanged.

4.4.2. Example of a projection of statistical data

Consider for instance an effort in year 2020 to try a projection of past statistical data for the forecast of demand of passengers for Eurostar trains, (see section 2.4.3), between London and Paris. The first step is to collect data that should be reliable. Points in Figure 4.2 are the annual numbers of passengers between 1995÷2019, collected from the internet site www.eurotunnel.com.

Ordinates of demand of Eurostar trains, as illustrated in Figure 4.2, do not make clear whether the phenomenon under study develops in a linear or nonlinear form. For this reason, both a linear and a nonlinear (more particularly a 2nd degree polynomial) regression will be attempted.

First we try a linear regression (Fig. 4.2, line —) and we enter the statistical data of Figure 4.2 in a computer software in order to calculate the coefficients of the linear regression:

$$D_t = 273{,}050 \cdot t - 5.398 \cdot 10^8 \tag{4.4}$$

where D_t : demand for the year t.

---

[*] For example the software Microsoft Excel, Grapher, Microfit, Eviews, and Harvard Graphics are, among others, suited for this purpose.

Passengers of Eurostar trains

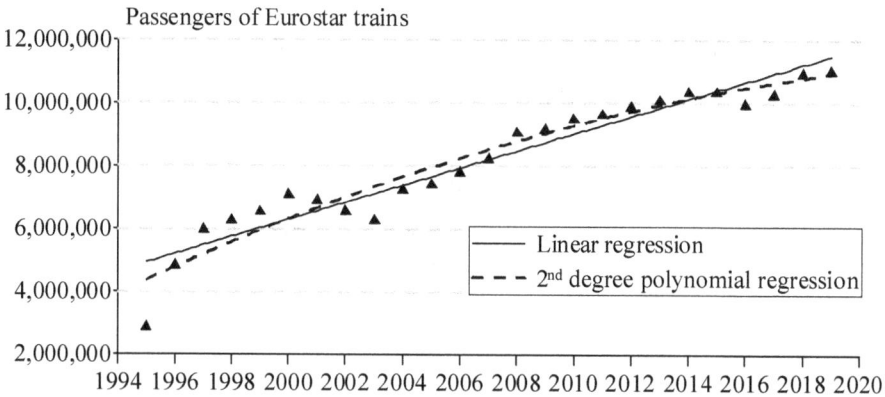

Fig. 4.2. Data of annual demand of passengers of Eurostar trains and linear (—) and 2nd degree polynomial (– – –) regression curves

The coefficient of determination for the linear regression of equation (4.4) is found to be $R^2 = 0.92$, which is a quite high value ($R^2 > 0.85$) and allows a trustworthy forecast.

Next we try a 2nd degree polynomial regression (Fig. 4.2, line – – –), for which future demand D_t can be calculated according to the equation:

$$D_t = -6.097.47 \cdot t^2 + 24,748,290 \cdot t - 2.51 \cdot 10^{10} \tag{4.5}$$

The 2nd degree polynomial regression renders an equally high value for the coefficient of determination $R^2 = 0.93$.

Which curve should the forecaster choose? There is no a priori answer to this question. Of course, the forecaster will look for a curve with the greater value of the coefficient of determination R^2. But in our example both curves (linear and 2nd degree polynomial regression) have almost exactly similar values for R^2. Thus the forecaster should consider which of the curves of equations (4.4), (4.5) is closer to the nature of the phenomenon under study, as witnessed from past experiences. Indeed, a linear evolution presupposes constant rates of yearly increases, which of course cannot continue forever. A 2nd degree polynomial regression, which begins exponentially and then turns asymptotically, may be closer to reality. Therefore, it is apparent that in addition to a good statistical analysis, other qualities of the forecaster such as experience, intuition, and imagination are also essential in order to put together a good forecast.

However, if the forecaster examines the data carefully, he/she can remark that the first full year of operation (1995) has a very low demand. For this reason, the data for this year could be omitted.

How long can this forecast be used? To achieve statistically accurate forecasts, the period of forecast should not be more than half the period for which statistical data are available, that is around 10 years and provided that all factors that affected demand of Eurostar trains in the past (such as tariffs of Eurostar trains and air transport companies, travel times, etc.) will continue to be the same in the future and with the same degree of importance.

4.5. Time-series models – Box-Jenkins method

Time-series is defined as a series of successive observations, which are sufficient for a description of the phenomenon under study. The independent variable in time-series models is time t.

The simplest form of a time-series analysis is a trend projection of past statistical data, which, however, due to its simplicity, is usually presented separately and has been already analyzed in section 4.4.

Time-series models try to identify the form of development of the studied phenomenon in the past, to investigate whether (and under what conditions) this evolution could continue in the future, and finally to forecast what could be expected in the future. Any time series is composed of one or more of a number of components: trend, seasonal, cyclical, random.

The most popular time-series model refers to the names of Box and Jenkins, who devised techniques allowing the choice among specific patterns of evolution of a phenomenon, while trying to simulate and to identify either the whole phenomenon under study or part of it. Simulation of the various forms of patterns can be achieved by means of a variety of regression curves, known as processes, with most frequently employed the following: AR (Autoregressive), ARI (Autoregressive Integrated), MA (Moving Average), IMA (Integrated Moving Average), ARMA (Autoregressive Moving Average), ARIMA (Autoregressive Integrated Moving Average), and SARIMA (Seasonal Autoregressive Integrated Moving Average with Seasonality), (75), (85).

The Box-Jenkins model is not often used for railway problems, as it requires lots of data (ranging from 50 to 100), is complicated, and cannot assure that a calibration (that is a sufficient adjustment of calculated values to statistical data) will be achieved. When it is used, the rail forecaster should look for the appropriate pattern, which suits the evolution of demand.

4.6. Econometric models

4.6.1. Definition, domains of application, and successive steps for the construction of an econometric model

Econometric models can provide a causal correlation between demand (dependent variable) and the driving forces (independent variables) affecting it. Econometric models require time (from some days to $1\div2$ months), exper-

110

tise, and are costly; therefore, they are used only by large railway authorities, state services, or university institutions. Econometric models are necessary either when we do not have any statistical data of rail demand or when we have statistical data, but we need a causal correlation of rail demand to the driving forces generating it. Thus, we use econometric models in the following cases: construction of a new railway line, a new station, terminal and logistics facilities, marshaling yards, purchase of new railway vehicles, rail master plans and business plans, commercial and pricing policy, withdrawal of a rail service and transfer to another one, organization of train schedules and timetables, etc.

Figure 4.3 illustrates the successive steps for the construction of an econometric model for the forecast of rail demand. Rail specialists can either look for existing econometric models and adapt them to their specific case and problem or try to construct their own econometric model. An econometric model leads to satisfactory results as long as there is strong dependence between dependent and independent variables, the independent variables are not inter-related among them, and accurate values for the independent variables can be provided.

4.6.2. Statistical tests for the validity of an econometric model

The statistical validity of an econometric model is tested by means of a number of statistical and diagnostic tests, which are, (9):

- stationarity test, testifying that the first year (or time) of statistical data, upon which the model is based, does not affect the pattern of the model and its evolution over time. Such tests are the correlogram of the autocorrelation function or alternatively the unit root tests,
- tests examining the correlation of the dependent variable (rail demand) with each one of the independent ones. Such a test is the Pearson correlation coefficient test,
- tests ensuring a strong dependence of the dependent variable with each one of the independent ones. This is the t-statistic test,
- tests checking the functional form of the econometric equation. This is the F-statistic test,
- multicollinearity tests, ensuring that those variables that are considered as independent are indeed independent and there is no interdependence among them. When this is sought, the Pearson correlation coefficient among the various pairs of independent variables can be applied,
- tests checking whether the residuals (differences between actual and calculated values) are normally distributed. These are the skewness and kurtosis tests,

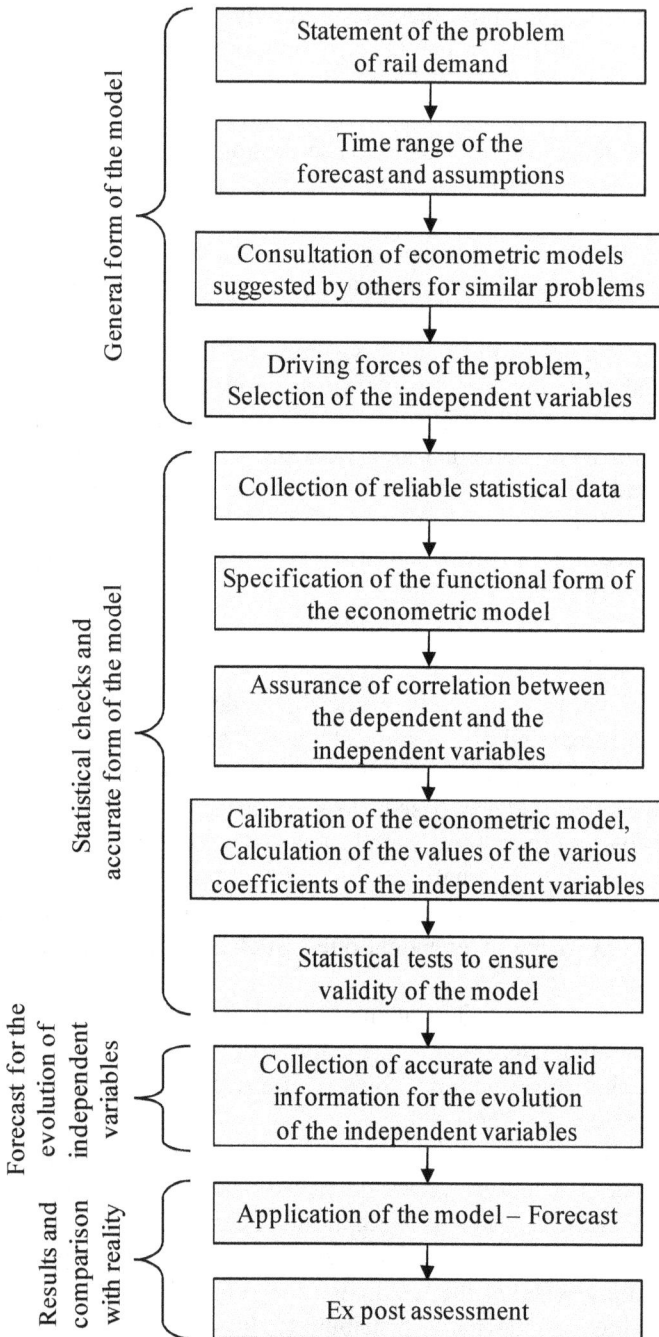

Fig. 4.3. Successive steps for the construction of an econometric model for rail demand, (9)

- tests examining whether the residuals of a specific time period are not correlated with the residuals of another time period. This is the Durbin-Watson test,
- heteroscedasticity tests, testifying that the residuals do not depend on the values of the independent variables. Such is the Breusch-Pagan and other tests.

4.6.3. Examples of some econometric models for the forecast of rail demand

We will first analyze an econometric model, which was suggested for the forecast of *annual* railway passenger *demand* in Greece. The analysis period spans over the years 1960÷2014. The specific case presents some interesting characteristics such as a great increase of rail fares (up to 50% in 2003), a significant reduction of frequency of services (route cuts and in some cases route cancellations), a serious economic and debt crisis (between 2009÷2018), the organization of a major event (the Olympic Games in 2004).

Variables referring to monetary units were adjusted in constant 2014 prices according to the annual consumer price index. All variables are incorporated into the model as indices that have the value 100 for the year 1985 and then logarithmized.

The econometric model's equation is, (77):

$$\ln D_{pass} = -0.9865 \cdot C_r - 0.1989 \cdot I_{co} + 0.8104 \cdot F_r + 0.2494 \cdot GDP + 5.1985 \qquad (4.6)$$

where:

D_{pass} : rail trips per capita,
C_r : unit cost of transport by rail (per passenger-kilometer),
I_{co} : car ownership index,
F_r : frequency of rail services (vehicle-km),
GDP : Gross Domestic Product of Greece per capita,

The model's adjustment to real data is satisfactory with a coefficient of determination $R^2 = 0.94$. Figure 4.4 illustrates the econometric model's results compared to data (actual values).

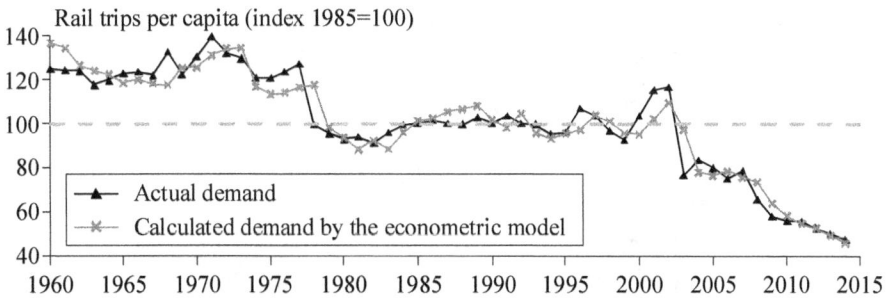

Rail trips per capita (index 1985=100)

Fig. 4.4. Comparison of results of the econometric model with real data, (77)

In a similar approach to forecast demand for *interurban rail travel* in Ireland, the following independent variables were selected: rail fares, income, car ownership index, quality of service, consumer expenditure, seasonality. The econometric model had two forms: one linear and one logarithmic. However, the coefficient of determination was higher in the linear than in the logarithmic approach, (93).

Other econometric models for the forecast of demand of *local rail services and stations* have identified the following independent variables: rail fares, rail service level, journey times, frequency, costs and service levels of competing modes, and economic activity (GDP), (89).

For the forecast of rail *freight* demand, econometric models of the following form have been suggested, (94), (9):

$$\ln D_{freight} = a + b_1 \cdot \ln(\text{freight}) + b_2 \cdot \ln(\text{EXR}) + b_3 \cdot \ln(\text{GDP}) \qquad (4.7)$$

where:

$D_{freight}$: rail freight demand,

freight : the freight (in volume or monetary value) of the specific product to be transported,

EXR : the exchange rate of currencies of origin and destination countries,

GDP : Gross Domestic Product.

4.6.4. Exogenous and endogenous variables in rail econometric models

Independent variables in an econometric model may be divided into two categories:
♦ *exogenous,* which are not affected by the rail industry,
♦ *endogenous*, which are affected by the rail industry.

Exogenous and endogenous variables can be identified as follows, (72):
– exogenous variables: GDP or employment, population, car ownership index, car fuel costs, car journey times, bus cost, bus journey time, bus headway, air transport cost, air headway, and metro cost,
– endogenous variables:
  • rail fares,
  • rail generalized journey times (incorporating in-vehicle time, frequency, and eventual interchange),
  • rail quality of service,
  • non-timetable related service quality (station facilities, rolling stock facilities, and environment).

4.7. A statistical method of forecast for highly diverging data

Forecast of rail demand by projecting into the future statistical data of the past is based on the minimization of the sum of squared errors between actual values (provided by statistical data) and calculated values (derived by the regression analysis). The method can work as long as no extreme and diverging values of statistical data are observed, that is values very distanced in relation to the regression curve. These distanced values are often called outliers.

Problems of rail demand with a high number of outliers can be encountered when non predictable events (such as an economic crisis, a war, a sanitary crisis, etc.) occur and statistical data of the past do not follow a more or less continuous pattern. In such cases, neither trend projection and time series nor econometric methods can ensure a forecast within the acceptable limits of errors. However, there are methods that limit the effects of outliers on the forecast, while pursuing the natural evolution of the problem, as it is represented by statistical data.

Thus, the "Least median of squares (LMS)" regression permits trustworthy forecasts for situations where non normal distributions, extreme outliers, or both are observed. LMS is based on the minimization of the median of the squared residuals (the difference between actual and calculated values) and not on the average (as in ordinary least squares method). At this point, it is worthwhile reminding that the average of a set of numbers is the sum of these numbers divided by the number of items in that set, whereas the median of a set of numbers is that number for which half the numbers are higher and half the numbers are lower.

A characteristic feature of the LMS method is that the breakdown point corresponds to a 50% percentage of outliers among observations, which is the highest possible for a statistical method, (Fig. 4.5). The breakdown point is the maximum percentage of extreme observations in the sample, so that they do not affect the outcome of estimation, (78), (86).

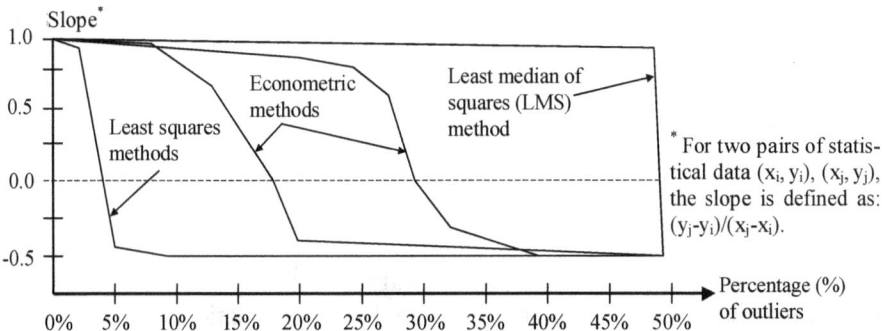

Fig. 4.5.   The percentage of outliers for which various statistical methods break down, (78), (86)

An example of an application of the LMS method in railway problems can be given by taking into account data of yearly demand illustrated in Figure 4.3. Application of LMS method gives results illustrated in Figure 4.6, (78).

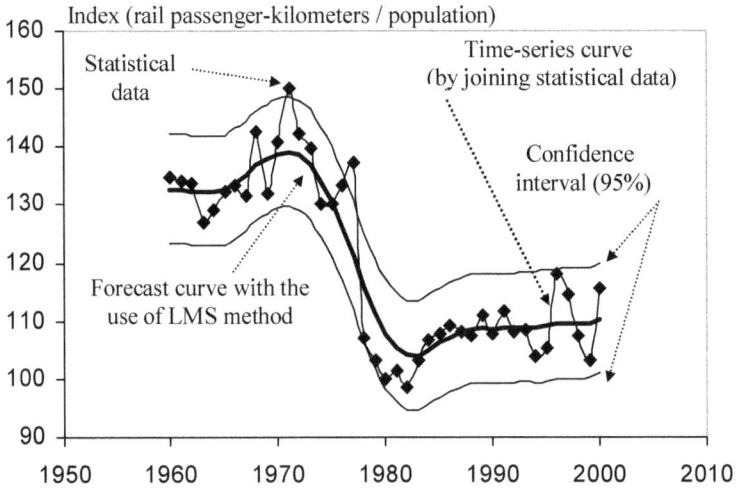

Fig. 4.6. Forecast of demand with the use of LMS method, (78)

4.8. Gravity models

Gravity models are the equivalent for transport of the gravitational law of mechanics and they are used for the forecast of rail demand principally in the case of the construction of a new railway station and sometimes in the case of the construction of a new railway line. In both cases no statistical data of rail demand are available, as the project is new.

Thus, for the forecast of demand of a new railway line between cities i and j, it was suggested that the gravitational law be specified as follows, (92):

$$D_{ij} = k \cdot \frac{P_i \cdot P_j}{d_{ij}^2} \qquad (4.8)$$

where: D_{ij}: rail demand between cities i and j,
A_i: population of city i,
A_j: population of city j,
d_{ij}: distance between cities i and j,
k : proportionality factor.

Equation (4.8) presents a simplistic analogy with the gravitational law. For this reason, it has been improved by replacing population with the total transport demand of each city and distance with the generalized cost of rail transport. Thus, eq. (4.8) can be written:

$$D_{ij} = k \cdot \frac{A_i \cdot A_j}{C_{ij}^a} \tag{4.9}$$

where: D_{ij}: rail demand between cities i and j,
  A_i: total transport demand of city i,
  A_j: total transport demand of city j,
  C_{ij}: generalized cost of rail transport between cities i and j,
  a : parameter of calibration. Various studies estimated values of the parameter a to be between 0.6 and 3.5, (90),
  k : proportionality factor.

In addition to rail passenger, gravity models have been suggested for the forecast of rail freight. In this case, rail freight demand D_{ij} from point i to point j can be expressed by the following function, (22), (91):

$$D_{ij} = f\left(O_i, P_j, e^{-b \cdot C_{ij}}\right) \tag{4.10}$$

where: O_i : Production of product in point i,
  P_j : Demand of product in point j,
  C_{ij} : Generalized cost of freight transport between i and j,
  b : Parameter of calibration.

Generalized cost C_{ij} of freight transport is the sum of the monetary cost paid for freight transport and the monetary value of freight travel time and quality of freight transport; it can be expressed as, (16):

$$C_{ij} = f_{ij} + b_1 \cdot S_{ij} + b_2 \cdot \sigma S_{ij} + b_3 \cdot W_{ij} + b_4 \cdot P_{ij} \tag{4.11}$$

where:

f_{ij} : fare for freight transport from origin point i to destination point j,
S_{ij} : total travel time from origin point i to destination point j, (transshipment included),
σS_{ij} : variance of total travel time,
W_{ij} : waiting time from the moment demand has been manifested till the beginning of the transportation procedure,
P_{ij} : probability of losses, alteration of products, robberies, etc.

4.9. Fuzzy models

4.9.1. Fuzzy numbers and fuzzy logic

Fuzzy means not clear enough, not well-known. A fuzzy number does not refer to one single number (as an ordinary number, e.g. 1.582) but to a set of plausible numerical values within a specific domain. Our logic of under-standing and interpreting a phenomenon is based on the principle of Aristotle, which accepts only true or false statements (Fig. 4.7.a) and is expressed in computers through the binary system that employs only two symbols (0 and 1).

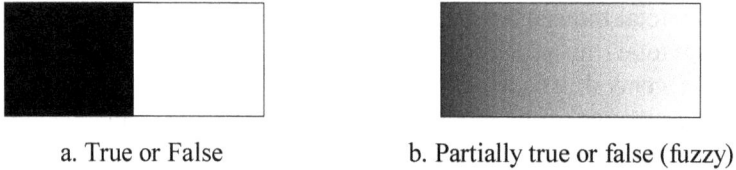

a. True or False   b. Partially true or false (fuzzy)

Fig. 4.4. From true or false statements to partially true or false statements (fuzzy logic)

A fuzzy number is characterized by one or more central values and a spread through which it varies. Triangular fuzzy numbers have only one central value, while trapezoidal fuzzy numbers have more than one central values. Fuzzy numbers used in applications for transport problems are usually of a triangular form (Fig. 4.8) and they are specified as $A= (r, c)_L$, where r is the center, c the spread, L a reference function, and μ the value of the membership function μ_A describing the boundaries of the fuzzy number $A= (r, c)_L$, $(0 \leq \mu \leq 1)$

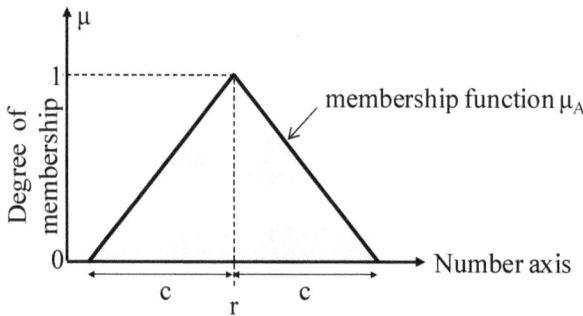

Fig. 4.8. Characteristics of a triangular fuzzy number

4.9.2. Fuzzy regression analysis

The basic tool in most methods of forecast of rail demand is regression analysis, which relates the dependent variable Y (demand) with a number of independent variables (X_i) according to the equation:

$$Y= a_0 + a_1 \cdot X_1 + a_2 \cdot X_2 + ...+ a_n \cdot X_n \qquad (4.12)$$

where $a_0, a_1, a_2, ..., a_n$ are real numbers that are called regression coefficients.

If in equation (4.12) we use fuzzy numbers instead of common real numbers for the regression coefficients, then we have a fuzzy regression equation of the form:

$$Y= A_0 + A_1 \cdot X_1 + A_2 \cdot X_2 + \ldots + A_n \cdot X_n \qquad (4.13)$$

where $A_i = (r_i, c_i)_L$, $i=0,1,2,\ldots,n$ are fuzzy numbers.

Fuzzy numbers and fuzzy equations are more suitable than usual ordinary numbers for the analysis of phenomena which present variations or oscillations around a numerical value (such as abrupt economic and social changes affecting critically and irregularly rail demand), vague statements and assessments (such as the reasons for the choice of a traveler between a high-speed railway and a low-cost airline), etc.

Employing fuzzy analysis and constructing a fuzzy model is less complicated than it appears to be. A number of computer software[*] permit a rather easy application of fuzzy methods to deal with railway problems. Rail specialists have to choose the form of fuzzy number (e.g. triangular isosceles), to give values in a number of coefficients which describe the ambiguity of the problem, and to select and apply the appropriate software.

4.9.3. Example of a fuzzy model

We will illustrate the above by applying the fuzzy regression equation (4.13) with the use of statistical data presented in Figure 4.4. The analysis period spans over the years 1960÷2014, triangular fuzzy numbers are chosen, and the analysis is executed with the use of Maple V4 software. Now the fuzzy regression equation (4.13) has two terms: the first gives the center of fuzzy regression and the second (added or subtracted) the upper and lower limits. The fuzzy regression equation will be, (77):

$$D_r = (r_0 + r_1 \cdot C_r + r_2 \cdot I_{co} + r_3 \cdot F_r + r_4 \cdot GDP) \pm (c_0 + c_1 \cdot C_r + c_2 \cdot I_{co} + c_3 \cdot F_r + c_4 \cdot GDP) \quad (4.14)$$

where $r_0, r_1, \ldots, r_4$ and $c_0, c_1, \ldots, c_4$ are coefficients which are derived by the fuzzy regression analysis, (Table 4.3).

Table 4.3.
Values $r_0, r_1, \ldots, r_4$ and $c_0, c_1, \ldots, c_4$ of the fuzzy model

r_0 :	6.3945	c_0 :	0
r_1 :	-0.9587	c_1 :	0
r_2 :	0.5841	c_2 :	0.2565
r_3 :	-0.2376	c_3 :	0
r_4 :	0.2296	c_4 :	0

Values of Table 4.3 permit to finalize the form of the fuzzy econometric model for the evolution and forecast of rail demand and to conclude the curves which are presented in Figure 4.9. If we compare results of the fuzzy

---

[*] Such as Matlab, Maple, Mathematica, FisPro, etc.

model with actual (observed) values of rail demand, we can witness that the center curve of the fuzzy model follows the series of statistical data and particularly the turning points. The bounds of the fuzzy model are the equivalent of the errors or divergencies of usual statistical methods.

What is the use of employing the fuzzy methodology instead of usual statistical methods? For some problems (as for the one presented previously) a fuzzy method can result in less ambiguity and better accuracy than a usual econometric model.

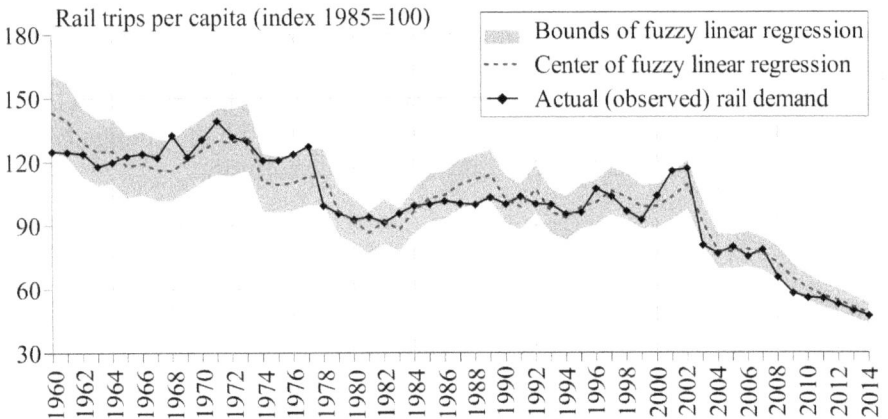

Fig. 4.9. Results of a fuzzy econometric model and comparison with actual values, (77)

4.10. Artificial Neural Networks (ANN) models

4.10.1. Artificial neural networks and biological neurons

The method of artificial neural networks (ANN), also called neural network method or simply neural method, is one of the techniques of artificial intelligence, (see section 4.1). ANN is a machine learning method which tries to imitate in a simplistic form the way that the human brain works and processes information. Indeed, each one of the approximately 100 million neurons of our brain allows human beings to understand, remember, think, and apply previous experiences. An artificial neuron simulates these operations of a natural neuron by imitating them; however, it is strictly impossible to replicate them, (82).

The fundamental function of a biological neuron consists in receiving stimulus and impulses either from the external environment or from another neuron in the form of electric signals, which are the transmitters of any outside information or impulse. These electric signals move in a kind of flow from one neuron to another and can either accelerate or decelerate. Whether an electric signal will be taken into account or ignored, accelerated or decelerated depends on a threshold value of the specific electric signal (which is different for each individual, and even for the same individual it becomes different in relation to time), below which the signal is not transmitted and over which the signal is transmitted to a neighboring neuron. This process permits the human brain to record, store, and learn from the various sources of information.

4.10.2. Artificial neurons and how they operate

An artificial neuron (Fig. 4.10) imitates in a very simplistic way some of the operations of a biological neuron. Thus, an artificial neuron can receive a number of inputs x_i with a relative weight for each input w_i, resulting in a total input $\xi = \sum w_i \cdot x_i$. If this total input is greater than the neuron's threshold limit value for activation b, then the artificial neuron will be activated and the output y_i, which is a function of the total input ξ, will be transmitted to another neuron or to the environment. For a total input lower than the activation value b, no transmission takes place and any effect of inputs x_i is ignored. The whole process is repeated in successive calculations, through which the output value y_i is optimized from one step to another or approaches a targeted value Y. What changes in successive calculations is adjusting the values of the weight factors w_i (Fig.4.10). Depending on the nature of the problem, there are many forms (linear, binary, sigmoid, Gaussian, hyperbolic) for the transfer (called also activation) function f, which transforms the various inputs x_i to outputs y_i.

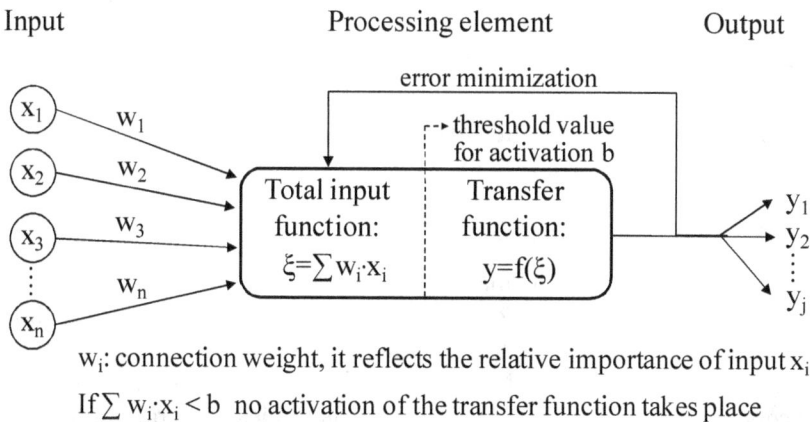

w_i: connection weight, it reflects the relative importance of input x_i

If $\sum w_i \cdot x_i < b$ no activation of the transfer function takes place

Fig. 4.10. Mechanism of operation of an artificial neuron and transformation of external inputs x_i (through connection weights w_i) to outputs y_i, (9)

4.10.3. Input, output, and hidden layers of ANN

The process illustrated in Figure 4.10 is materialized in practice by ANN models with the help of the following, (Fig. 4.11), (9):

- an input layer, which incorporates all input variables, that is variables receiving signals from the external environment,
- an output layer, which incorporates all output variables, that is variables sending signals to the external environment,
- one or more hidden layers, not connected to the external environment, and through which optimization of weight factors (called connection weights w_i) is achieved.

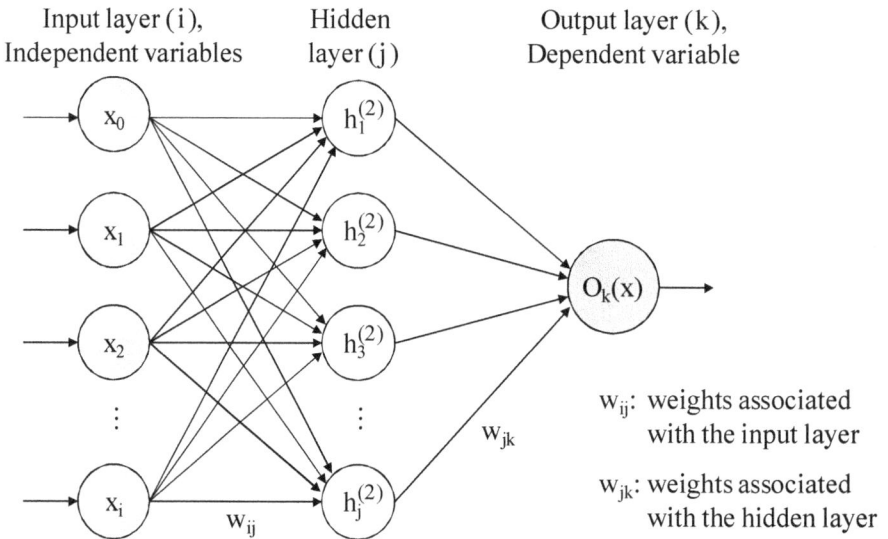

Fig. 4.11. Input layer, output layer, and hidden layer of an artificial neural network, (9)

4.10.4. A variety of ANN models

In the successive steps of calculations of an ANN, optimization in search of output y_i can take place either only forward (feedforward ANN) or both forward and backward by interacting continuously with input variables and the starting situation (recurrent ANN). The connection weights w_i are modified in successive calculations and this is known as the learning rule of the ANN model and can be implemented according to one of the following algorithms: error-correction, competing, perceptron, Boltzmann, Hebb, (9). Thus, a number of available statistical data (around 70%) are used for training the ANN model, a smaller number (around 15%) for validating it, and a third one (around 15%) for testing it. A number of computer software[*] employing ANN methods are available in the market, (88).

4.10.5. Suitability and areas of applications of ANN

The big handicap of ANN models is that though they can produce very satisfactory numerical results, they do not understand neither the mechanism nor the process governing the problem under study. ANN imitate and execute what is asked from them, without understanding it and consequently without interpreting it.

However, ANN models can provide solutions in a number of problems for which traditional methods based on mathematics and statistics fail to do so and particularly for problems, (9):

- with a nonlinear evolution,
- with a dynamic and changing behavior or without a specific pattern,
- so complex that a number of factors affecting them is omitted and not taken into account in usual analyses,
- based on assumptions that limit the degrees of freedom and flexibility of the problem under study by transforming it to a different problem than the initial one,
- requiring analysis of large amounts of data (like big data).

Therefore, ANN operate empirically and do not provide any kind of interpretability, but they can take into account nonlinearities, great amounts of data, fluctuating and irregular evolutions; in addition, ANN can provide real-time and accurate results. All usual methods based on mathematics and statistics simplify a problem (and sometimes even distort it) to a rather different one, which is easy and simple to study. ANN analyze the real problem with its particular characteristics and irregularities.

In conclusion, ANN are efficient and suitable for problems for which we are not interested neither in the mechanism nor in the interpretation of what happens, when large amounts of data are available or some data are missing, for nonlinear problems, when lots of assumptions simplify to a great extent the real problem and restrict the validity and applicability of solutions.

ANN are applied nowadays in all sectors of science. Many applications of ANN are already available for the transport sector: self-driven vehicles, driver behavior and road safety, routing and scheduling of freight transport, condition-based maintenance for all transport infrastructures, assessment of effects of unpredicted events.

---

[*] Among them Matlab, Neurosolutions, R, and other.

ANN models are particularly suitable for the analysis and forecast of rail demand, as they can simulate problems with irregular evolution of demand, even missing values for some years, nonlinear evolution of demand, large amounts of demand data.

4.10.6. Example of application of ANN for the analysis and forecast of rail demand

We will illustrate now how ANN can be used for the analysis and forecast of rail demand. We will use the statistical data of rail demand given in Figure 4.4. The input variables will constitute the input layer (Fig. 4.12) and will be the same with the independent variables of the econometric equation (4.6): unit cost of transport by rail (C_r), car ownership index (I_{co}), frequency of rail services (F_r), Gross Domestic Product (GDP) per capita. The output layer will consist of one output variable which is the dependent variable of the problem, in this case rail demand (expressed by the number of rail trips per capita). The hidden layer (where numerical calculations take place) includes 9 nodes. The available statistical data are used as follows: 70% for training, 15% for validating, and 15% for testing.

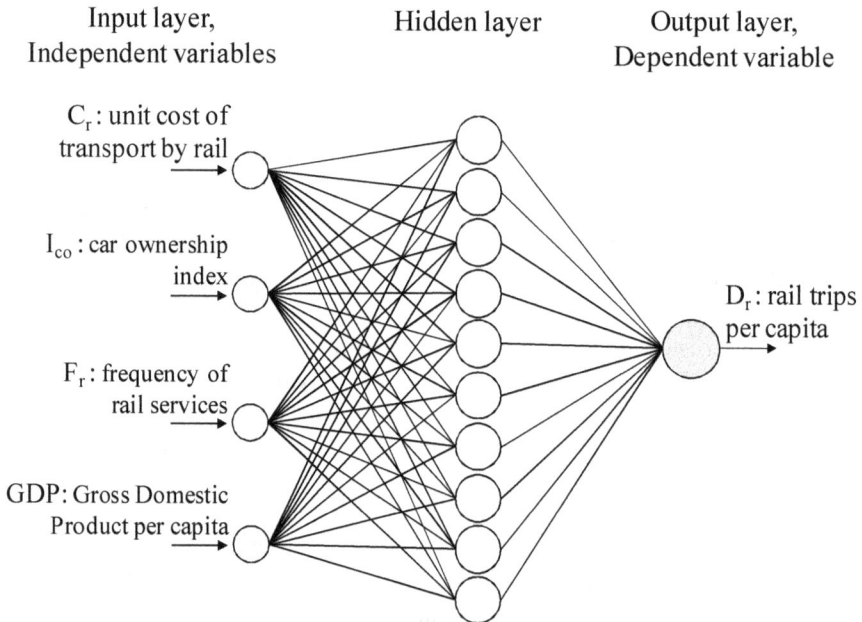

Fig. 4.12. Architecture of an ANN model for the forecast of rail demand, (9)

The computer software Matlab was used and a feedforward algorithm was selected for the execution of the various calculations; the coefficient of determination between calculated and actual values is found to be extremely high, $R^2 = 0.97$, thus testifying a high correlation of the dependent variable to the independent ones.

Figure 4.13 illustrates the results of the ANN model in comparison with actual values, that is values of past statistical data. We can notice how close to actual values are the calculated ones, which in addition pursue the pattern and turning points of the curve of statistical data.

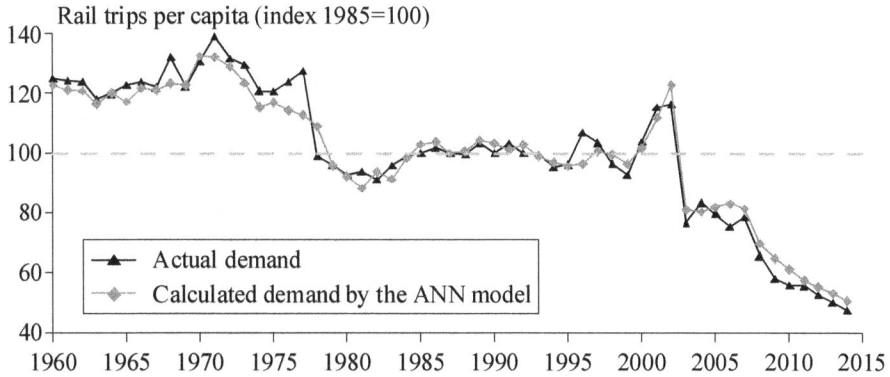

Fig. 4.13. Results of an ANN model for rail demand and comparison with actual values, (9)

4.11. Evaluation of the forecasting ability of a model for the forecast of rail demand

Any of the previously presented models for the forecast of rail demand is based on efforts to minimize errors, that is the differences between actual values and values calculated by the model. The usual criterion for assessing how close the calculated values are to the actual ones is the coefficient of determination R^2. The closer the values of R^2 are to the value 1.0, the better the adaptation of the forecast curve to the data of the problem is. Values of R^2 greater than 0.85 are a good index that the forecast curve is suited to the data of the problem. However, railway specialists should not take for granted any forecast, but they should continuously verify the goodness of the forecast with ex post analysis by comparing the forecasted values to values that are realizing, (71). There is a number of criteria for assessing the forecasting accuracy either of the model itself or among various models. Such criteria are:

- the mean absolute deviation (MAD), which is the average of the absolute values of errors,
- the mean squared error (MSE), which is the average of squared errors,
- the root mean squared error (RMSE), which is the square root of MSE,

- the mean absolute percentage error (MAPE), which is the average of the absolute errors divided by the actual values,
- the Theil's inequality coefficient (also known as Theil's U), which is a measure that emphasizes large errors (as the MSE criterion), while providing a dimensionless measure for comparison among the various forecasting methods.

A good forecast is assured when all the above criteria (other than the Theil's inequality coefficient) take the smallest value.

When the Theil's U of a model is calculated equal to zero, then the model's forecasting ability is perfect, whereas when Theil's U is calculated equal to one, the model lacks any forecasting ability. In practice, values of Theil's U of 0.50 or less are considered to be good and values less than 0.10 are considered to be excellent, (77), (9).

A comparison has been conducted to evaluate the forecasting ability of the econometric and fuzzy models, given by equations (4.6) and (4.14) respectively, and the ANN model; all three models describe the same phenomenon. The Theil's U takes the value of 0.007 for the econometric model, 0.008 for (the center of) the fuzzy model, and 0.005 for the ANN model. Thus, the model which best describes the phenomenon under study is the econometric one, while all three models have values of Theil's U very close to 0.0.

Figure 4.14 gives the comparative performance of the various models analyzed previously (Econometric, Fuzzy, ANN) and comparison with real data.

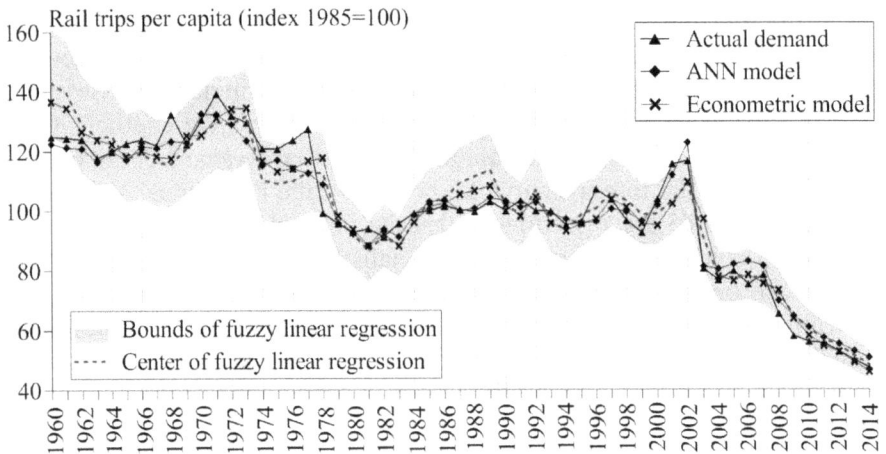

Fig. 4.14. Comparative performance of econometric, ANN, and fuzzy models for the forecast of rail demand

4.12. A comparative analysis of performances of each method and selection of the appropriate one

With such a plethora of available methods, the question arising concerns the choice of the most appropriate one; however, the answer depends on the following, (Table 4.4):

– *Nature* and *range* of the forecast: For short- (1÷2 years), medium- (<5 years), and medium-long (up to 10 years) term forecasts, trend projection can be a first and rather suitable method. The results of the trend projection can be complemented by and correlated with the Delphi method or an executive judgment. If the forecaster wants to normalize unpredictable events concerning past data, then besides the trend projection, he can try the LMS method. Yet, medium-long term (>5 years) and very long-term (>10 years) forecasts will require a causal technique, usually an econometric model. Besides usual regression analyses, the forecaster can use applications of the fuzzy method in order to determine a more accurate range for the forecast and relieve the forecast from the impact of unusual events in the past. For forecasts with high variations and unusual patterns of evolution, ANN may be very useful.

In the case of a new line or a new station, a market survey, which is time-consuming and costly, is necessary and must be complemented by a gravity or an econometric model.

Table 4.4.
A comparative analysis of performances of the various forecasting methods of rail demand

		Qualitative methods				Trend projection	Quantitative (causal) methods			
		Executive judgment	Market survey	Delphi method	Scenario writing		Econometric	Gravity	Fuzzy	ANN
Suitability for forecasts till	1 year	H	H	H	H	H	H	H	H	H
	2÷3 years	M	H	M	H	H – M	H	H	H	H
	5 years	L	M	M	M	M – L	H	H	H	H
	10 years	L	L	L	M	L	M	M	M	M
Special expertise required		H	M	H	H	M – L	H	H	H	H
Time required to conduct the forecast		L	H	L	M	L	H	H	H	H
Data requirements		M	M	M	M	H – M	H	H	H	H
Cost requirements		M	H	M	L	L	H	H	H	H

Legend	Symbol	H	M	L
	Meaning	High	Medium	Low

- *Expertise* required. Executive judgment, Delphi method, scenario writing, econometric, gravity, fuzzy, and ANN methods require a qualified forecaster with a high level of expertise, while trend projections and market surveys can be conducted by less qualified personnel.
- *Time* available. For a forecast within some hours or days, a trend projection accompanied, if possible, by executive judgment or Delphi method is the only accurate approach. If the forecaster has time availability for the completion of the forecast, regardless of the cost, he/she can try more time-consuming methods, such as an econometric model, ANN and fuzzy methods, a market survey, etc.
- *Data* available. Trend projections and all causal methods require accurate data for a rather long period of time. If such data are unavailable or unreliable, then the use of qualitative methods is suggested.
- *Cost* of the forecast. Econometric methods, ANN, fuzzy methods, and market surveys are costly compared to trend projection, scenario writing, executive judgment, and Delphi methods.

Finally, any forecast, most particularly a medium- or long-term one, should be regularly checked and updated with new data.

However, the longer the period of forecast the lesser the accuracy that can be expected. Forecasts for more than 15 years ahead should be used only as an indication of what may occur. Forecasts within the range of 5÷15 years have an inherent uncertainty, related to non-predictable evolutions of the exogenous parameters of the problem (such as the economic indices), even if no abrupt changes take place.

5 Costs and Pricing

5.1. Definition of railway costs

5.1.1. Construction, maintenance, and operation costs

Understanding the structure of railway costs is essential and crucial for all railway activities. Construction of a new railway line will be strongly based on an accurate knowledge of costs. Operation of a railway service also needs the most accurate and detailed knowledge of costs. Pricing of infrastructure requires knowing the values of maintenance costs. Establishing tariffs for passenger and freight traffic requires knowledge of both operation costs and elasticities.

Cost can be defined as the amount of available resources spent in conjunction with the construction or operation of a railway activity, (116). Railway costs can refer to the construction of a line, in which case they are called *construction* costs, to the maintenance of a line (*maintenance* costs), or to the operation of a railway service (passenger, freight, combined, terminal), in which case they are called *operation* (or *operating*) costs.

When a separation of infrastructure from operation exists, we also discern the infrastructure cost, which is the sum of costs related to the provision and use of railway infrastructure. These costs include maintenance and operation costs pertaining to subgrade, ballast, sleepers, rails, signaling, telecommunications, electric traction installations, lighting, police inspection, as well as to station[*] installations and the staff needed to operate the infrastructure, (116).

5.1.2. Fixed and variable costs

Fixed costs refer to those costs which do not vary with the level of traffic. In contrast, *variable* costs relate to the quantity of traffic transported. *Total* costs are the sum of fixed and variable costs.

---

[*] If stations belong to the infrastructure.

5.1.3. Marginal cost

Marginal cost is the additional cost when increasing traffic by one unit. More strictly, marginal cost is defined as the unit cost resulting from an increase or decrease in traffic volume (progressive or regressive marginal cost respectively). When the variation in traffic is infinitely small, marginal cost is the derivative of cost in relation to traffic, (112).

Marginal cost may refer either to a given production capacity (*short-term* marginal cost) or to a changing production capacity (*long-term* marginal cost), which usually needs a longer period to occur.

Development cost is the total of all costs incurred to implement a new facility or project (e.g. new rolling stock, station, line, etc.)

The costs of a railway line in relation to traffic can be illustrated as in Figure 5.1. We can distinguish the following components of costs, (16), (115):

a. general fixed costs (section OA of the curve), related to the expenses of administration, maintenance, etc.,
b. fixed costs related to a specific category of traffic. If, for instance, passenger trains stop operating on a line run by passenger and freight trains, then the additional component of maintenance costs related to passenger traffic will cease to exist,
c. marginal costs (section BC of the curve), related to fuel, maintenance of rolling stock, and necessary personnel in the trains. Railways are characterized by the fact that traffic usually does not reach capacity and thus railway infrastructure is in most cases underutilized,
d. development costs (section DE of the curve), related to the purchase of new rolling stock or to the construction of a new line or facility in order to respond to a demand that cannot be served by the existing infrastructure or rolling stock.

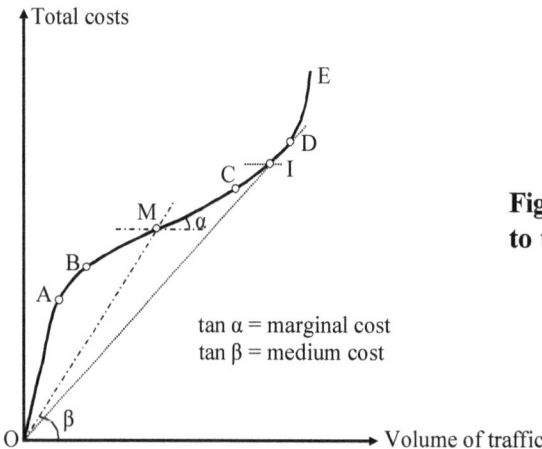

$\tan \alpha$ = marginal cost
$\tan \beta$ = medium cost

Fig. 5.1. Total costs in relation to traffic for a new railway line

5.1.4. External costs and marginal social cost

External costs are the costs that the use of a transport system imposes on non-users of the system, (95), (16). Thus, the use of a private car causes emissions that pollute the air and cause damage to the health of other people driving along the same route or living in the neighboring area. These emissions are a typical case of external cost: a negative effect is experienced by several people and is caused by somebody who does not endure any consequences for his action. It is widely accepted that air pollution, noise, climate change, accidents, congestion, worsening of the living environment are principal components of external costs that all transport modes generate during their operation.

Social cost is the total cost to the society and is the sum of construction, maintenance, operation, and external costs.

Marginal social cost of infrastructure is defined as the total cost entailed by the running of an additional train on a particular infrastructure and is composed of the following, (116):

i. a marginal cost related to infrastructure, which measures the increase in maintenance and renewal costs resulting from an additional train running,

ii. a marginal congestion cost, expressing in monetary terms the value of delays and constraints imposed on the rest of the traffic by an additional train running,

iii. a marginal external cost, representing the increase in other costs to the society incurred by the running of an additional train. This cost component measures principally the variation of costs of accidents, pollution (air and sound), climate change, etc.

5.1.5. Generalized cost – Monetary value of time

A person's choice between two modes of transportation is made by taking into account three parameters, (16), (112):

♦ the *direct monetary cost*, which in the case of railways is the sum of the train ticket cost, plus the cost to reach the departure station, plus the cost from the arrival station to the destination point, (Fig. 5.2),

♦ the *value of total travel time* h (from origin to destination),

♦ the *quality of service* q.

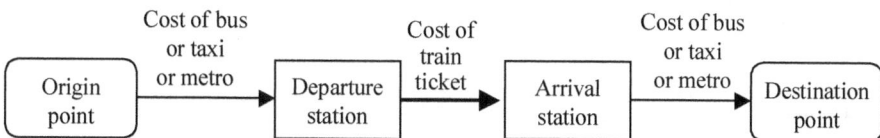

Fig. 5.2. Direct monetary cost from origin to destination point

131

The *generalized* cost (GC) takes into account the above three parameters and is defined as the sum of the direct monetary cost (DMC) paid by the traveler, plus the monetary value of total travel time, plus the monetary value of quality of service (q):

$$GC = DMC + h \cdot T + q \qquad (5.1)$$

where: h : the time from origin to destination,
        T : the monetary value of a man-hour.

When railways increase speed and reduce travel times, then for a specific value of man-hour, the generalized cost is reduced and some traffic can be diverted from airplanes, buses, and private cars to the railways, (Fig. 5.3). The volume of diverted traffic is a relation of the monetary value of a man-hour for the various categories of passengers.

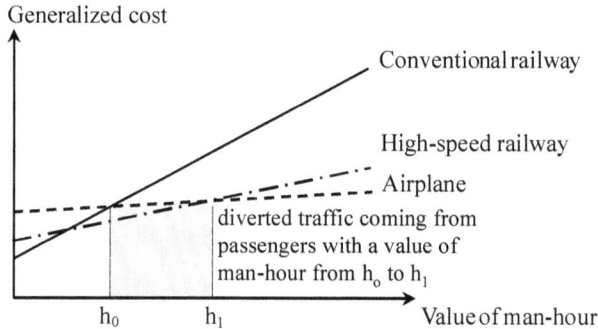

Fig. 5.3. Diverted traffic when reducing generalized cost

Reduced rail travel times may lead to increased rail revenues. Indeed, passengers with a high value of time (e.g. businessmen) may be willing to pay a higher rail tariff, if rail travel times are substantially reduced (compared to competing modes) and result in lower generalized costs. Thus, in an attempt to minimize the generalized cost of transport, individuals try to maximize their *utility* when selecting among many alternative transport modes, (9).

A critical assumption in the calculation of generalized cost is the monetary value of a man-hour (for passengers) or a tonne-hour (for freight). The following values (converted in € of year 2020, based on initial values and inflation rates) of travel time have been used in some studies for railways, (9), (105):

– business travel 15÷31 €/man-hour,
– leisure travel 7.2÷8.25 €/man-hour,
– freight transport 1.25÷1.55 €/tonne-hour.

5.2. Construction costs of a new railway line

5.2.1. Factors affecting construction costs of railways

The construction cost of a new railway line is influenced by several factors:

- layout characteristics, mainly the number and size of bridges and tunnels. It should be noted that in railway lines of comparable levels of service, the existence of many civil engineering structures (tunnels, bridges) may double, triple, or even quadruple the construction cost, (see below Table 5.2),
- expropriation cost, which, especially in urban areas, may considerably increase construction costs,
- costs related to works necessary for the protection of the environment,
- number of switches and crossings per kilometer of track,
- existence or not of electrification of the line,
- labor costs, which vary from country to country (and often within the same country).

The use of cost data based on information drawn from the analysis of projects of other countries should therefore serve only as a rough estimate of the various cost parameters, while always keeping *proportions* in mind.

5.2.2. Construction costs for new high-speed lines

Cost data derived from high-speed lines constructed during recent years can give a first estimation of the construction cost of a new high-speed railway line. All monetary values which will appear below are that of year 2020. This means that initial values of construction costs were actualized in values of year 2020, based on inflation rates of all the previous years. Initial values of construction costs are derived either from information of the author over the specific project or from data of UIC and other institutions. These initial values of construction costs did not take into account modifications (during construction) of costs.

The high-speed line 'TGV Méditerranée' of French railways, in operation since 2001, with $V_{max}=350$ km/h, on ballast, with 6.5% of tunnels and 12.7% of bridges (reported to the total length of the line), has a construction cost per kilometer of line* of 20.4 million €. A more recently constructed French high-speed line is 'Sud-Europe-Atlantique' between Tours and Bordeaux, in operation since 2017, with $V_{max}= 320$ km/h, and a construction cost per km of 27.3 million €.

---

* Unless otherwise stated, costs per km of line refer to double track railway lines and do not include rolling stock.

The Spanish high-speed line Madrid-Barcelona, in operation since 2003, with $V_{max}= 270 \div 350$ km/h, on ballast, had a first part with 26.8% of its length in tunnels and 3.4% in bridges and a second part with only 2% in tunnels and 2.7% in bridges. The construction cost per kilometer in the case of the first part was almost double compared to the second part. Globally, the whole high-speed line Madrid-Barcelona-frontier with France has an average final construction cost per km of 15.2 million €.

The German high-speed line Cologne-Frankfurt, with $V_{max}=300$ km/h, on concrete slab, with 26.5% of tunnels and 4.2% of bridges (reported to the total length of line) has a construction cost per km of 26.7 million €.

The Italian high-speed line Rome-Naples, with $V_{max}=300$ km/h, on ballast, with 17.8% of tunnels and 24.0% of bridges per km of line, has a construction cost per km of 26.1 million €.

The Japanese high-speed lines, with $V_{max}=300$ km/h, have an average construction cost per km (rolling stock included) in the range of $21.0 \div 32.0$ million €, (111).

The Korean high-speed line, with $V_{max}=300$ km/h, on concrete slab in tunnels (a cross-section of 107 m^2) with a length greater than 5 km, on ballast elsewhere, has a construction cost per km (including 46 units of high-speed rolling stock) of 52.2 million €.

The Chinese high-speed lines present a great range of values of construction costs per kilometer. The following values are based on construction costs calculated in the Chinese currency and converted in Euros according to the exchange rate on 30.06.2010 (1€ = 8.29 Chinese Wan). Chinese high-speed lines constructed between $2005 \div 2010$, with $V_{max}=250$ km/h, have construction costs per km in the range of $6 \div 15$ million €, with an average value of 9.0 million €. High-speed lines constructed in China between $2005 \div 2012$, with $V_{max}=350$ km/h, have construction costs per km in the range of $15 \div 21$ million € with an average value of 16.5 million €. The high-speed line Beijing-Shanghai constructed between 2008 and 2011, with $V_{max}=380$ km/h, has an average construction cost per km of 19.3 million €.

The high-speed line Mecca-Medina (which has a distance of 450 km) in Saudi Arabia, with $V_{max}=300$ km/h, has a construction cost per km of 26.6 million €, (111).

The Indian high-speed line Mumbai-Ahmedabad, with $V_{max}=320$ km/h, has an estimated (in the year 2015) initial construction cost per kilometer of 26.7 million €.

Bearing in mind the economic conditions in Europe in 2020, a rough estimate for the construction cost of 1 kilometer of a new high-speed line is in the range of $15 \div 40$ million €, (38). It should be noted that lower construction costs for France and Spain compared to Germany and Italy are partly due to the fact that high-speed lines in France and Spain are dedicated only for passenger trains, whereas in Germany and Italy are run by both passenger and freight trains.

Any decision to construct a new high-speed railway line aims at the reduction of travel time, thus it is of interest to know the cost of reducing travel time by one minute. Table 5.1 (3rd column) illustrates the cost per minute saved (from 35 million to even 369 million € per minute saved) for some recently constructed high-speed lines in Europe.

A usual problem in the construction of any rail project is that the initially calculated values of construction costs are modified and almost always are increased during the construction of the project. This is due to either initial approximate calculations of costs or to changes and modifications of the final layout in relation to difficulties encountered in situ. Table 5.1 illustrates the difference between initial and final construction costs. Therefore, railway engineers, managers, and economists should be very wary when trying to assess ex ante the construction costs of a railway line.

Table 5.1.
Initial and final values of construction costs for some high-speed rail lines in Europe and cost per minute saved in relation to pre-construction travel times (according to the European Court of Auditors in 2018)

Routes	Initial construction cost (million €/km)	Final construction cost (million €/km)	Cost per minute saved (million €)
Berlin-Munich	12.4	21.9	104.87
Stuttgart-Munich	6.4	44.7	368.69
Rhin- Rhône	14.9	18.8	34.51
LGV Est Européenne	12.4	16.5	51.63
Madrid-Barcelona-French border	11.0	15.2	39.70
Ele Atlantico	12.5	15.7	34.61
Madrid-Leon	11.8	15.7	57.00

5.2.3. Allocation of construction costs to the various rail components

The allocation of construction costs of a new railway line to the various components of the railway system differs greatly and depends on the peculiarities of each particular situation. Figure 5.4 illustrates the average values from data of France, Spain, Germany, and Italy for lines with no major civil engineering structures.

Civil engineering works
(subgrade, expropriations, tunnels, bridges)
55%

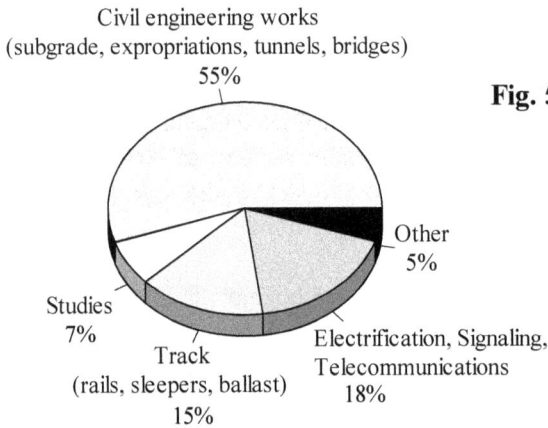

Fig. 5.4. Allocation of construction costs of a new railway line to the various components of the railway system (compiled from field data)

Other
5%

Studies
7%

Track
(rails, sleepers, ballast)
15%

Electrification, Signaling,
Telecommunications
18%

5.2.4. Construction costs of civil engineering works

Construction cost of civil engineering works of a new railway line depend principally on three factors:

- the ground conditions (degree of difficulty of topography). A line in a plain requires evidently fewer tunnels and bridges than a line in mountainous areas,
- the maximum speed. Lower speeds require lower values for the radius of curvature, which results in a lower length of the line and in fewer tunnels and bridges,
- whether the line has one or two tracks. However, as the width of a double track is not the double of a single track, (see Fig. 12.8 and 12.9, section 12.6), construction costs of a double track are not the double compared to a single track (all other parameters being unchanged).

Table 5.2 illustrates the range of rough average values of construction costs of civil engineering works of a new railway line in relation to topography (easy, average, difficult), the speed, and the number of tracks (single, double), (110). In case of difficult topography, the construction cost increases due to the number of tunnels and bridges that will have to be constructed. Values of Table 5.2 should be considered only as orders of magnitude and reference points.

Average values for the boring of a railway tunnel are for a single track line 15÷75 million €/km and for a double track line 30÷100 million €/km, depending on soil mechanics conditions. Average values for the construction of a railway bridge or a viaduct are for the case of short spans and easy foundation 15÷30 million €/km of track and for long spans and difficult foundation 30÷75 million €/km of track (all values of year 2020), (110).

Table 5.2.

Construction costs of civil engineering works (values of year 2020) of a new railway line in relation to topography, speed, and number of tracks, (110)

Type of track	Maximum speed (km/h)	Cost of a new line/km (million €)		
		easy topography	average topography	difficult topography
single	100	1.5÷4.5	4.5÷22.0	22.0÷58.0
double	100	1.5÷6.0	4.5÷29.0	29.0÷73.0
double	300	3.0÷8.7	8.7÷44.0	a great range and dispersion of values

5.2.5. Construction costs of track superstructure

Construction costs of a typical track superstructure (rails UIC 60, concrete sleepers, ballast) amount to 0.45÷0.75 million €/km of track, (110).

5.2.6. Construction costs of electric traction

Construction costs of electric traction differ in relation to the system of electrification and the distance of substations, (see section 20.5). We can consider separately the cost for catenaries (overhead contact wires) and the cost for substations:
– catenaries costs. For alternating current systems (25 kV, 50 or 60 Hz and 15 kV, 16 ⅔ Hz) and low speeds they amount to 0.15÷0.30 million €/km of track, whereas for high speeds they amount to 0.22÷0.45 million €/km of track. For direct current systems (3 kV, 1.5 kV) they amount to 0.22÷0.45 million €/km of track, (values of year 2020), (110),
– substation costs. For alternating current (25 kV, 50 Hz) substation costs amount to 0.22÷0.45 million €/km of track, for alternating current (15 kV, 16 ⅔ Hz) to 0.3÷0.75 million €/km of track, and for direct current (3 kV, 1.5 kV) to 0.3÷0.75 million €/km of track, (110).

5.2.7. Construction costs of signaling

Costs of railway signaling systems, (see chapter 21), include, (110):
♦ cables (for signaling and communications): 0.07÷0.15 million €/km of line, in relation to the traffic of the line,
♦ automatic block system: 0.15÷0.75/block section, depending on whether signaling permits movement of trains on one or both directions,
♦ automatic or advanced train protection: 0.015÷0.06 million € per signal,

♦ cab signal (automatic train control with transmission by track circuits or by cables): 0.3÷0.45 million €/block section,

♦ radio links: 0.015÷0.03 million €/km of track,

♦ installing wayside equipment on the track for the European Rail Traffic Management System (ERTMS) Level 2 has diminishing values of costs, which are expected to stabilize in mid-2020s around 100,000 €/km of track; installing ERTMS equipment on a rail vehicle has also diminishing values, which are expected to stabilize around 100,000 €/vehicle, (see also section 21.10.5).

5.2.8. Costs of installing level crossings

Level crossings are the points where a railway line crosses a road at the same level. Safety measures to avoid collision and ensure safe circulation of trains and road vehicles may include installing one of the following types of level crossings, (see also section 22.9.2 and section 22.9.4):

• level crossing with light and acoustic signals without barriers (for train speeds up to 140 km/h). Average costs of equipment amount to 30.000 €,

• level crossing with light and acoustic signals and in addition with either automatic half barriers (for train speeds up to 160 km/h) or automatic full barriers (for train speeds 160÷200 km/h). Average costs of equipment for the full barriers solution amount to 150,000 €, (see also section 22.9.4).

5.3. Maintenance and operation costs of rail infrastructure

5.3.1. Maintenance costs of rail infrastructure

Whether a railway is unified or separated, it is essential to know the maintenance and operation costs of rail infrastructure. Infrastructure maintenance costs comprise:

♦ maintenance and renewal of track (rails, sleepers, ballast) and subgrade,

♦ maintenance of electrification, signaling, telecommunications facilities, and substations,

♦ maintenance of tunnels and bridges,

♦ maintenance of platforms (in stations).

A maintenance cost per year of 54,000 €/km of track was reported for France and a cost of 73,000 €/km of track was reported for the Netherlands (monetary values for the year 2020). Rail infrastructure maintenance cost is allocated as follows to the various maintenance components:

– 65% for track and platforms,

– 30% for electrification, signaling, telecommunications, and substations,

– 5% for bridges and tunnels.

Maintenance costs of infrastructure for new high-speed lines are reported in 2020 to have an average value per year of 90.000 €/km of line, (38).

Specifically for the track and in relation to the traffic of the line, average annual maintenance costs are estimated at 45.000 €/km of track for lines classified UIC group 2, (see Fig. 7.6), for a speed of 100 km/h and 60.000 €/km of track for a speed of 300 km/h. For UIC group 4, annual maintenance costs of track are estimated at 28.000 €/km for a speed of 100 km/h and 30.000 €/km for a speed of 300 km/h, (110).

5.3.2. Operation costs of rail infrastructure

The operation costs of rail infrastructure include traffic management (92% of total operation costs) and schedule planning (8% of total operation costs), and are estimated per year at 1.60 €/train-km.

5.4. Costs of purchase and of maintenance of rolling stock

5.4.1. Costs of high-speed rolling stock

As there is a variety of contracts for the purchase of rolling stock, significant differences can be observed among the principal rolling stock constructors or even for the same constructor among various markets. Since most of new high-speed rolling stock orders refer to mid-2000s, values of purchase costs will be provided for that period of time.

The cost per seat for the purchase of new rolling stock (values for the year 2008) was 75,100 € for the Spanish high-speed train (named AVE), 64,750 € for the German ICE1 and 73,650 € for ICE2, 67,500 € for the Paris-Brussels-Amsterdam-Cologne train (named Thalys), and 46,600 €÷48,700 € for the French high-speed trains.

If we report the purchase cost per seat-km and per year, then we have the following values: 0.221 € for the Spanish AVE, 0.130 € for the German ICE1 and 0.184 € for ICE2, 0.225 € for Thalys, and 0.116 €÷0.123 € for the French TGV, (Table 5.3).

As airplanes are the principal competitor of high-speed trains, a comparison with the economic data concerning aircrafts may be useful. Thus, the cost per seat was 386,000 € for Boeing 757-200 (with a capacity of 190 seats), 550,000 € for Boeing 767-200ER (with a capacity of 191 seats), and 362,000 € for Airbus A320 (with a capacity of 150 seats), (values of year 2008). Costs reported per seat-km and per year were 0.157 € for Boeing 757-200, 0.167 € for Boeing 767-200, and 0.204 € for Airbus A320. It can be deduced that high-speed trains and airplanes had comparable purchase costs per seat-km and per year, (Table 5.3).

More recent orders of high-speed rolling stock report for the year 2018 a purchase cost per seat of 60.000 € for the French TGV duplex and 87.700 € for the German ICE 4, which should be compared with a cost per seat of 543.000 € for the aircraft A320 of Airbus and 508.000 € for the aircraft B737 of Boeing.

Table 5.3.
Costs of purchase of high-speed rolling stock and of aircrafts
(values of year 2008), (compiled from data of UIC and constructors)

Country	Type of train	V_{max} (km/h)	Cost per seat-km and per year	Cost per seat-km and per year for: Airbus A320	Boeing 757-200	Boeing 767-200ER
Germany	ICE 1	300	0.130			
	ICE 2	300	0.184			
France	TGV	300	0.116÷0.123	0.217	0.167	0.177
Spain	AVE	300	0.221			
Europe	Thalys	300	0.225			

5.4.2. Costs of purchase of ordinary passenger vehicles

The cost of ordinary passenger vehicles (values of year 2020) is estimated to be 1.50÷3.00 million €/vehicle, (110).

5.4.3. Costs of purchase of freight vehicles

The cost of freight vehicles depends on the characteristics of the vehicle (open-covered, flat-hopper, etc.) and averages 75÷150,000 €/vehicle, (110).

5.4.4. Costs of purchase of diesel locomotives

The costs of purchase of diesel locomotives are spread over a great range of values. As a rough approximation, they may be calculated in relation to the power W (in MW) of the locomotive from the empirical formula, (110):

cost of a diesel locomotive (in million €) = W/3 + 1

5.4.5. Costs of purchase of electric locomotives

Similarly, the cost of an electric locomotive can be calculated from the empirical formula, (110):

cost of an electric locomotive (in million €) = 2W+2.

5.4.6. Maintenance costs of rolling stock, signaling, and electrification

Annual maintenance costs (values of 2020) are estimated for rail passenger vehicles at 0.3÷0.5 €/vehicle-km, for freight vehicles at 0.075÷0.19 €/vehicle-km, for diesel locomotives at 2.2÷3.6 €/locomotive-km, for electric locomotives at 0.3×purchase price (in million €)/10^6 €/locomotive-km, (110).

Annual maintenance costs of signaling are estimated at 4% of the purchase price and annual maintenance costs of electrification equipment (catenary-substations) at 2% of the purchase price, (110).

Annual maintenance costs of high-speed trains are estimated at 1 million €/km and are calculated on the basis of maintenance costs of 2 €/km and an average annual traveling of 500,000 km per train, (38).

5.5. Economic life of the various components of the railway system

The various components of the railway system can be used efficiently and safely for a more or less limited period of time, which is called economic life (and sometimes service life or useful life) and depends on the nature, degradation, and scope of the specific component. Thus, in addition to the various costs analyzed previously, it is essential to know how much time any component of the railway system can be used and when it should be replaced.

The *economic life* of a railway component is defined as the period (usually expressed in years) during which the specific component is expected to be usable, with normal repair and maintenance, (16). Economic life is usually shorter than physical life, which is the time until the moment that further use of a railway component may be dangerous and safety is not assured. The depreciation period may coincide with the economic life but due to various uncertainties may well be shorter than the economic life. Any railway material or component should be replaced before the end of its economic life. However, condition-based management, which is supported by a continuous monitoring of the status of each railway component, permits nowadays to conduct the maintenance at the desired (from an economic point of view) or needed (from a safety point of view) time, neither earlier nor later than the appropriate time.

The economic life of a railway material or component depends on both the economic conditions of the country and the level of services of the specific railway network and may differ greatly from one country to another. Average values of economic life of the various railway components and materials for the economic conditions of Europe are given in Table 5.4, (16), (110).

Table 5.4.

Economic life (in years) of various components and materials of the railway system, (16), (110)

Railway component	Economic life (years)
Subgrade (ground area)	100
Tunnels	100
Bridges (concrete, steel)	50
Rails (depending on the traffic load of the line)	20÷50
Ballast (depending on the traffic load of the line and the hardness of the stone)	15÷30
Sleepers (depending on the traffic load of the line and the material of the sleeper)	20÷50
Substations (civil engineering component)	60
Substations (electric equipment)	40
Catenaries (overhead contact wires)	40
Signaling – Safety equipment	30
Electric locomotives	30
Diesel locomotives	20
Buildings – Stations	50
Access roads	50
Passenger vehicles	25
Freight vehicles	20

5.6. Costs of operation and revenues of a railway company

5.6.1. Passenger transport

Costs differ in the various categories of rail passenger traffic: urban and suburban, intercity, regional. Statistics of rail operators usually refer to the whole activity of operation and lack analytical data, which should be available for every specific category of traffic and even more for every route. Railways should introduce analytical accounting techniques in order to have the possibility of an accurate measure of costs for every segment of the market and even for every route.

Costs of operation of railways are the sum of expenses for labor, energy, materials, outsource (external) services (such as cleaning of trains), and depreciation. The energy component of operation cost depends on the fluctuations of the cost of a crude oil barrel (see chapter 1, Fig. 1.2) and averages 5÷10 % of total operation cost for rail passenger traffic and 10÷25 % for rail freight traffic, (9).

With regard to operation expenses for both infrastructure and operation, in the case of European railways, expenses of a railway company, operating both passenger and freight trains, are segmented to 30% for expenses paid for the maintenance and operation of infrastructure and 70% for expenses paid for the maintenance and operation of trains, (100).

In those railways where there is a separation of infrastructure from operation, the rail operating costs for the running of trains are the sum of the following, (5):

- *infrastructure* access cost, which usually has a fixed access charge component and a maintenance access charge component,
- *energy* cost (fuel cost for diesel traction, electrification cost for electric traction),
- *rolling stock* cost (purchase or leasing cost, maintenance, and service cost),
- costs for the use of various facilities in railway *stations*,
- *staff* costs (wages for staff in charge of operations such as drivers, controllers, and hostesses and also wages for staff in management and administration),
- passenger *services* costs (e.g. catering-refreshments, if included in the ticket price),
- service of past *loans*,
- *marketing*, advertising, and promotion costs,
- *outsource services* costs (e.g. cleaning of trains),
- general *administration* costs,
- *other* costs, such as compensation to passengers in the case of delays, cancellations, etc.

Rail operating costs per seat-kilometer vary in Europe in 2020 at the range of 0.08÷0.15 €/seat-kilometer.

Principal revenue sources from operation of European railways are passenger traffic at 40%, freight traffic at 20%, state subsidies at 30%, and other 10%. A useful index for the comparison of revenues among many railways is *yield*, which is the average revenue per unit of transport and is usually expressed in monetary values per passenger-kilometer. Average yields in the passenger sector were for railways of Europe in 2012 0.11 €/passenger-kilometer. However, great differences can be observed from one railway to another, as illustrated in Table 5.5 (see also section 5.11.4), (104).

Average railway passenger yields decreased greatly during the last two decades as a result of a harsh competition from both road and air (for long distances) transport. Just for comparison with values given in Table 5.5, average revenues per passenger-kilometer were in 2003 0.16 € in Germany, 0.19 € in Norway, but only 0.09 € for East Japan Railways and 0.05 € for Amtrak in the USA, (106).

Table 5.5.
Average revenues (in € per passenger-kilometer)
for railways of some countries, (104)

Austria	UK	Germany	France	Netherlands	Belgium	Spain	Italy	Poland
0.17	0.15	0.13	0.125	0.125	0.102	0.082	0.064	0.064

5.6.2. Freight transport

Cost of freight transport is a small component of the total value of freight. For medium and long distances, transport costs represent approximately 21% of the value of freight, (102). This means that only high reductions in the transport cost can have an essential effect on the total cost of the goods transported. For this reason, reliability and on time delivery are essential factors in the very competitive freight transport market, (102), (113).

As explained in section 1.10.5, (Fig.1.25), railways present comparative advantages for medium and long distances. According to the statistics for the 15 countries of the European Union, (Table 5.6), 49.1% of total tonne-kilometers of rail freight are performed for distances from 150 to 500 km, 9.3% for distances from 50 to 150 km, only 2.4% for distances shorter than 50 km, and the remaining 39.2% for distances greater than 500 km, (2), (109).

We usually distinguish in the case of freight transport (as for passenger transport) fixed from variable costs. Rolling stock and staff represent fixed costs, whereas access charges and energy consumption represent variable costs.

Many railway companies choose not to publish data concerning freight costs. We will present an analysis of freight costs, which refers to the labor costs of Italy. Table 5.7 illustrates allocation of operation costs of a freight train of a useful load of 315 t and of 630 t, (109).

Table 5.6.

Share in freight transport, in relation to distance,

of various transport modes for the EU-15 countries, (2), (109)

Distance	Road		Rail		Inland waterways		Total	
(km)	tonnes	t-km	tonnes	t-km	tonnes	t-km	tonnes	t-km
0÷49	53.7%	5.2%	24.1%	2.4%	29.2%	5.3%	52.0%	8.5%
50÷149	22.8%	16.4%	22.7%	9.3%	39.6%	29.1%	23.1%	19.0%
150÷499	18.4%	41.9%	40.4%	49.1%	28.9%	54.1%	19.5%	46.0%
>500	5.1%	36.5%	12.8%	39.2%	2.3%	11.5%	5.4%	26.5%
Total	100.0%	100.0%	100.0%	100.0%	100.0%	100.0%	100.0%	100.0%

Table 5.7.

Allocation of operation costs of rail freight for freight trains

of a useful load of 315 t and of 630 t, (109)

Costs	% of total freight cost for a train with a useful load of 315 t	% of total freight cost for a train with a useful load of 630 t
Staff	66.8	59.4
Energy	7.3	12.9
Depreciation	20.4	18.1
Other	5.5	9.6

Average yields in the freight sector were for railways of Europe in 2012 0.05 €/tonne-kilometer, with, however, great differences from one railway to another, (Table 5.8), (see also section 5.12).

Table 5.8.
Average revenues (in € per tonne-kilometer) for railways
of some countries, (104)

France	Austria	Italy	Germany	Portugal	Spain	Romania	Belgium
0.290	0.120	0.040	0.030 (0.025÷0.050)	0.030	0.020	0.020	0.015

5.6.3. Combined transport

The cost of combined transport differs in relation to the total distance, partial distances traveled by rail and road, the terminal equipment, etc., and it has been illustrated in Figure 1.26 (section 1.10.5).

5.7. Quantification of external costs in monetary values and internalization policies

5.7.1. Quantification of external costs in monetary values

For many decades, a crucial issue concerning the various components of external costs was their accurate and objective quantification in monetary values. This work has been conducted and applied to data of the year 2016 and refers to the 26 EU countries (Malta and Cyprus do not have railways), plus Norway and Switzerland, (95).

The various components of external costs are: accidents, noise, air pollution, climate change, habitat damage, congestion, well-to-tank emissions (the last being understood as the emissions released to the atmosphere from the production, processing, and delivery of fuel). All these components have been identified and described as can be seen in Table 5.9. For each one of them, an appropriate method for quantification in monetary values has been developed, (Table 5.9). Congestion costs can be incorporated in total external costs or considered separately. Other (non-cost related) aspects of external costs are presented in chapter 23.

Total external costs (including congestion costs) for the 26 EU countries (with railways) + Norway + Switzerland amount to 987 billion € for the year 2016, which is 6.6% of GDP of these countries. If congestion costs (amounted to 271 billion € in 2016) are not taken into account, then the total external costs (excluding congestion costs) amount to 716 billion € for the year 2016, which is 4.8 % of GDP of the countries under study. Accident costs are the most important category, amounting to 29% of total external costs. Congestion costs are 27% of total external costs, climate change costs 14%, air pollution costs 14%, noise costs 7%, habitat damage costs 4%, well-to-tank emissions costs 5%, (95).

Table 5.9.

Description of the various components of external costs and methods of their quantification in monetary values, (95)

External cost	Cost components	Method of quantification	Leverage points and variability
Accidents	Human costs (pain, suffering), Medical costs, Administrative costs (police, insurances, etc.), Production losses, Material damages	The value of human life is estimated based on the amount that individuals are willing to pay in order to reduce accident risks	Depending on different factors (partly on vehicle-kilometers)
Noise	Damages of human health and costs of annoyance	Willingness of disturbed persons to pay medical costs and risk value due to transport noise	Depending on traffic volume and environmental performance of each mode
Air pollution	Damages of: - human health - material / buildings (non-health costs)	Concentrations of particulates are the basis for the calculation of repair and damage costs	Depending on vehicle-kilometers, energy consumption, and environmental performance
Climate change	Damages resulting from global warming	Avoidance costs to reach Paris Agreement targets* per country or to reach long-term reduction targets	Depending on consumption of fossil fuels
Congestion (only for road transport)	Additional time and operating costs	Time costs (value of a man-hour) and additional operating costs of road users due to congestion	Depending on traffic volume (number of vehicles)
Habitat damage	Effects on nature, landscape, and natural habitat	Based on the infrastructure length (or area) and average cost factors for habitat loss and fragmentation	Depending on length of roads, railways, area of airports, ports, transport demand, density of population
Well-to-tank emissions (energy production emissions)	Costs for energy consumption: extraction, processing, transport, transmission, energy plants	The same as for air pollution	Depending on the quantities of energy consumed

* Paris Agreement targets refer to the Agreement signed by a number of countries in Paris (2015). Central target of this Agreement is to reduce CO_2 and other emissions, so as to keep the earth's temperature increase in the range 1.5°C ÷ 2 °C compared to pre-industrial levels.

Road transport is the mode with the highest share in total external costs (82.9% of total external costs if aviation and maritime transport are included; 97.5% if aviation and maritime transport are not included). Maritime transport causes 10% of total external costs and aviation 5%. On the contrary, railways have a small share (1.8%) and inland waterways even smaller (0.3%). More than 2/3 of external costs are caused by passenger transport, against less than 1/3 by freight transport, (95).

Figures 5.5 and 5.6 illustrate the average values (for the year 2016) of the various components of external costs for all transport modes, for passenger and freight traffic respectively, (95).

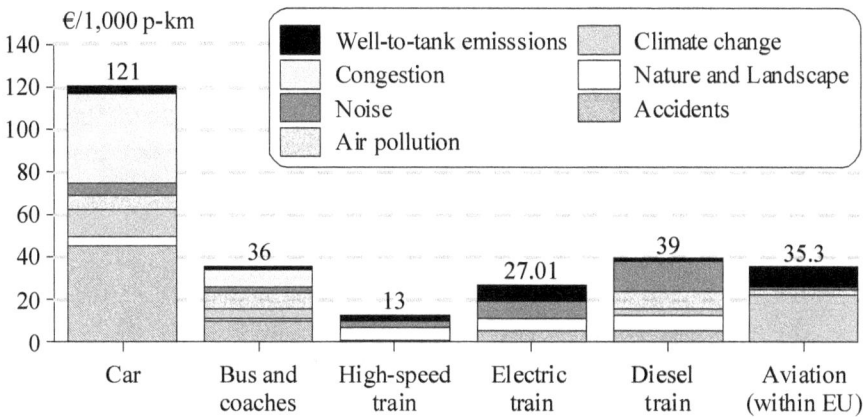

Fig. 5.5. Average external costs of passenger transport for the various transport modes (EU-28 countries), (95)

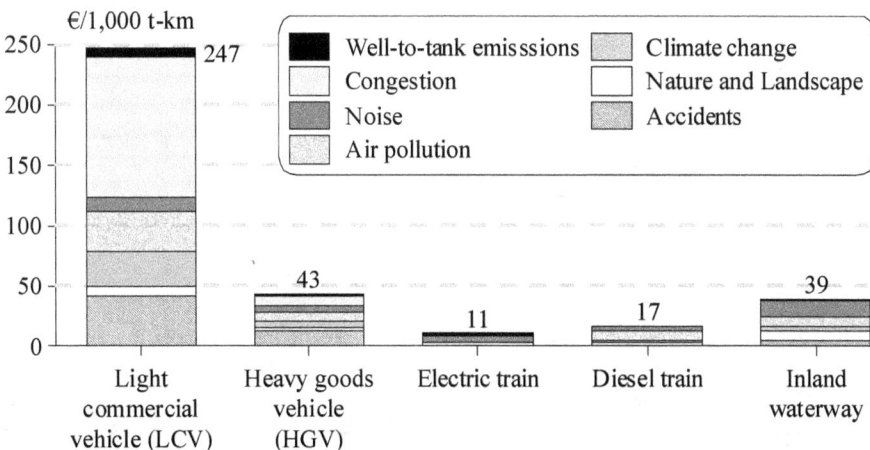

Fig. 5.6. Average external costs of freight transport for the various transport modes (EU-28 countries), (95)

147

5.7.2. Internalization of external costs

A coherent transport policy should be based on pricing policies for the various transport modes which should incorporate into the costs paid by users not only costs of operation but also the real costs of wear of infrastructure and the external costs. Such a strategy is known as internalization and aims at prices of transport which should reflect the real costs to the society. Internalization may include either full infrastructure costs ('the user pays') or external costs ('the polluter pays') or both. Internalization of external costs will make the transport sector more efficient economically, environmentally, and socially and even fairer fiscally for the tax payers.

Table 5.10 illustrates analytical values for each component of external costs for the various transport modes for passenger and freight transport for the EU-28 countries for the year 2016, (95). If internalization of (some or all components of) external costs is conducted on a full cost basis, then values of Table 5.10 can be used. If internalization is conducted on a marginal cost basis, values of marginal external costs should be taken into consideration.

Table 5.10.

Average external costs of passenger and freight transport for the various modes of the EU-28 countries – Values in cents of €

per passenger-kilometer or per tonne-kilometer, (95)

Cost component / Transport mode		Accidents	Air pollution	Climate change	Noise	Congestion	Well-to-tank emissions	Nature and landscape	Total
Passenger transport	Private car	4.5	0.7	1.2	0.6	4.2	0.4	0.5	12.1
	Bus-Coach	1.0	0.7	0.5	0.3	0.8	0.2	0.1	3.6
	Motorcycle	12.7	1.1	0.9	9		0.5	0.3	24.5
	High-speed train	0.1			0.3		0.3	0.6	1.3
	Electric train	0.5	0.01		0.8		0.8	0.6	2.71
	Diesel train	0.5	0.8	0.3	1.4		0.1	0.8	3.9
	Aviation (within EU)	0.02	0.2	2.2	0.2		0.9	0.01	3.53
Freight transport	Light commercial vehicle	4.1	3.4	2.8	1.1	11.6	0.8	0.9	24.7
	Heavy goods vehicle	1.3	0.8	0.5	0.5	0.8	0.2	0.2	4.3
	Electric train	0.1			0.6		0.2	0.2	1.1
	Diesel train	0.1	0.7	0.2	0.4		0.1	0.2	1.7
	Inland waterway	0.5	0.8	0.3	1.4		0.1	0.8	3.9
	Maritime	0.001	0.44	0.16	n.a.	n.a.	0.06	n.a.	0.661

5.8. Pricing of infrastructure

5.8.1. Principles for the pricing of railway infrastructure

The crucial question in economics of transport infrastructure is to what extent the user of an infrastructure undertakes the various costs (construction, maintenance, operation, external) and whether a part or some components of infrastructure costs can be undertaken by the state (by the tax payer instead of the user). However, users of road and air transport cover only a part of total costs of roads and airports. Pricing of railway infrastructure emerged as an important issue when competition was introduced in the railway sector and thus different railway operators could make use of the same track. Pricing of railway infrastructure and decision of levels of access charges are among the critical factors for balancing finances of both the infrastructure managers and the railway operators. Pricing of infrastructure should incorporate the following principles: simple, transparent, stable, fair, non-discriminatory, and efficient. In addition, it should take into account:

- the essential characteristics of the specific infrastructure (speed, availability of departure and arrival slots (specific time intervals), electrification, signaling),
- train characteristics (length, axle load, required power, etc.),
- efficient use of infrastructure and consistency with general transport policy objectives.

It is common practice that the basis for the pricing of infrastructure is in most of the cases (see below Table 5.12) the number of train-kilometers run.

5.8.2. Objectives of infrastructure pricing

Any infrastructure pricing model should clearly establish its objectives and rank them by priority, (20), (96):

- cover, in whole or in part, the operating and maintenance costs of railways. Table 5.11 recapitulates the various assets and costs of infrastructure. If infrastructure account is not balanced, then a public subsidy is necessary to cover the deficit,
- favor the best possible use of rail infrastructure,
- promote some categories of traffic (urban, regional, intercity, freight),
- reflect the level of infrastructure quality provided to the rail operators,
- take into account external costs and thus compensate and encourage the use of transport modes (like railways) which are more friendly to the environment,
- contribute to the costs of developing the rail network through making investment self-financing,
- contribute to a balanced regional development.

Table 5.11.
Rail infrastructure assets and costs, (99), (101), (116)

Type of asset	Description of asset	Capital expenditure	Major renewals	Maintenance costs	Operation costs
1. Line of route (subgrade and structures)	Land, embankments, cuttings, bridges, tunnels	Original route construction	Major upgrading of subgrade, tunnels, Replacement of steel bridges	Culverts, vegetation, steel structures, bridges and tunnels	
2. Track	Rails, sleepers, fastenings, ballast, switches and crossings	Construction of rails, sleepers, and ballast	Renewal of rails, sleepers, and ballast	Inspection, ballast tamping, partial repair of track	
3. Signaling	Fixed signals, track circuits, signal boxes, control equipment, ERTMS	Construction of signals boxes, etc.	Replacement of signals boxes, etc., usually with more modern equipment	Maintenance of signals and control equipment	Staff for signal operations, power, and other supplies
4. Electricity power supply	Overhead line or third rail, substations, cabling, power control equipment	Electricity power supply and associated supply equipment	Replacement of supply equipment	Maintenance of supply equipment	Staff for electric power, costs of electric energy
5. Stations, marshaling yards, depots	Platforms and buildings	Construction of stations, marshaling yards, depots	Renewal or replacement of buildings	Maintenance of structures	Labor, energy, and other costs incurred by the Infrastructure Manager

As certain of the above objectives are contradictory to some extent, the infrastructure pricing model should establish a compromise and address the ranking of priorities, which should be characterized by cohesion.

5.8.3. Financial consequences of infrastructure pricing

Infrastructure charges can be high, which is beneficial for public finances but detrimental to the finances of railway operators running on the specific infrastructure. On the other hand, if infrastructure charges are low, this is beneficial for railway operators but detrimental to public finances.

5.8.4. A commercial approach of infrastructure pricing

Rail infrastructure can be considered either as a commercial product (that is, a product to sell) or part of the public estate (that is, a public utility). Even in the second case, however, the pricing of rail infrastructure should be characterized by a commercial approach and should incorporate the following principles:
- flexibility of prices (without, however, any discrimination),
- possibility of discounts for rail operators with heavy traffic,
- allocation procedures, when many rail operators are seeking for the same departure or arrival slot.

Figure 5.7 illustrates some of the parameters, which could be taken into account in a pricing model for rail infrastructure, (97), (108).

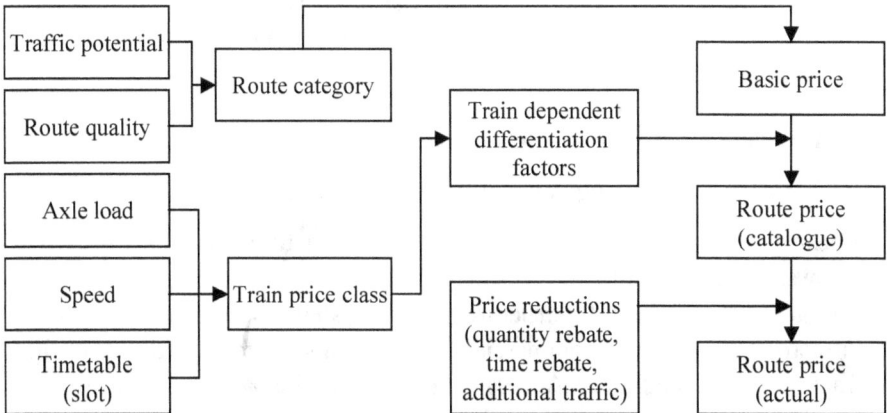

Fig. 5.7. Factors affecting a pricing model of rail infrastructure, (97), (108).

5.8.5. *Theoretical and practical infrastructure pricing*

There are many theoretical alternatives for infrastructure pricing, leading to pricing according to a variable of infrastructure cost, which may be:

♦ marginal cost (short-term),
♦ marginal social cost, including external costs,
♦ marginal social cost of investment (long-term), including the cost of renewal investment,
♦ total cost.

Economic theory suggests that optimal pricing should be based on the long-term marginal social cost or development cost, (114). Some countries (e.g. Germany, United Kingdom) combine the principle of coverage of total costs with the financial potential of the railway undertakings. Thus, the state subsidizes the theoretically high charges that the rail operators are unable to pay.

Pricing in some countries is based on the marginal social cost, whereas in other countries only on marginal cost. Both pricing methods result in deficits that are covered by state subsidies.

5.8.6. *Structure of infrastructure pricing*

We can distinguish two great structures of infrastructure pricing models:

• *one-part* models, with a single component based on variable cost and the following weighting factors: speed, axle load, equipment of track, electrification, specific route, time of the day (slot), type of commodity, etc.
• *two-part* models, with one component based on variable cost and another fixed part, which can reflect capacity to be used and path allocation, without however any discrimination.

5.9. Infrastructure pricing models in some countries

5.9.1. *Infrastructure pricing according to European Union legislation*

European Union principles for pricing rail infrastructure can be summarized as follows, (20), (100), (101):

– The accounts of the Infrastructure Manager shall balance revenues from infrastructure charges, commercial activities, and state subsidies with infrastructure expenditures. Access charges should be paid to the Infrastructure Manager by any operator when running a train on the track. Charges must be non-discriminatory for different railway undertakings that perform similar services. Charges cannot be smaller than marginal costs.

– Any discount shall be limited to the actual saving of the administrative cost in cases of long-term contracts. Limited discounts can be granted to encourage the development of new rail services or the use of underutilized lines. For similar services, similar discount schemes shall be applied.
– A component of prices may be reservation charges as an incentive for an efficient use of capacity and should reflect that capacity, which, if requested, should be paid (at least partially) even if not used.
– Charges may take into account the following parameters: distance, nature of traffic, composition of train, speed, axle load, specific day and time of use of rail infrastructure (slot), electrification and signaling installations, whether or not infrastructure is underutilized, eventual congestion on some tracks, level of noise from trains, etc.

The information on infrastructure charges given in the following paragraphs is taken from the Network Statement of each Infrastructure Manager of EU (data of 2020). Since 2016, member-states of the EU have the option of delivering data to the European Commission by using a data portal (www.rne.eu) and following a standard format for the Network Statement, (51). Thus, it is possible for anyone interested to conduct cross-country comparisons and actualize in the course of time data provided in this book.

The legal basis for the pricing of railway infrastructure in EU countries consists of the following: Directives 34/2012, 49/2004, 10/2008, 57/2008, and Regulations 913/2010, 909/2015. The minimum access charges for the essential railway services (track, electrification (if any), signaling, stations) should reflect the cost of these services. The rail pricing models for some countries are analyzed below.

5.9.2. France

The French pricing model of rail infrastructure takes into account the following categories of lines, (51):
• lines with *ordinary passenger* traffic (conventional lines),
• lines with *high-speed passenger* traffic,
• lines with *freight* traffic.

Rail infrastructure charges are composed of the following components:
♦ *access charge*, owed by the state on a fixed flat rate and covering part of fixed costs,
♦ *traffic (or running) charge,* calculated according to the transport service provided and the actual use of the line; it is allocated on the basis of marginal cost of maintenance, operation, and renewal of the line. For passen-

ger trains it is calculated by 85% on path-kilometers and by 15% on train-kilometers. For freight trains, both net and gross traffic are taken into account in relation to the UIC group of the line, (see section 7.5.2),

♦ *market charge,* allocated on fixed costs and calculated according to market size and population (important, medium, low) and the conditions of competition (high, average, low) of rail transport of the line with road and air transport. Lines enjoying some kind of state contract are segmented from other lines, which are further segmented in international and domestic (classified in 5 categories) routes,

♦ *electric traction charge*, calculated according to the actual use of the electric facilities and allocated on the basis of marginal cost of maintenance and renewal of electric facilities. Related to electric power supply is another charge, aiming to covering the *losses* in electrical systems, which is allocated on the basis of marginal cost of providing the electric energy, so as to compensate for losses in electrical systems from substations up to train detection points,

♦ *congestion charge,* as a financial incentive for the good use of a specific line, allocated as a unit price per train-path along sections of the lines which are declared congested and for the specific congestion times,

♦ *special charges,* for the use of specific facilities (combined transport and other) and sections of track.

Charges are differentiated for peak periods (06.00-08.00, 16.00-20.00), normal periods (14.00-16.00), slack periods (20.00-00.00, 00.00-06.00, 08.00-14.00) of a typical day as well as for crowded week-ends and national holidays.

5.9.3. Germany

The first German pricing model for rail infrastructure in early 2000s focused on infrastructure equipment configuration, whereas the more recent one puts emphasis on the type of train running on the track and is more oriented towards the market. In the general structure of the model, a number of surcharges and discounts can be accepted. Framework agreements can be concluded for long-term use of tracks. No deduction can be applied to the standard usage charge for the noise related component of charge, (51).

The German rail pricing model has a differentiation in relation to the category of traffic and the various lines are segmented as follows:

• *local* passenger (urban, suburban, regional), when maximum speed does not exceed 130 km/h, maximum traveled distance is less than 50 kilometers, and journey time is less than 1 hour. Charges are a responsibility of federal states ("Bundesländer"),

- *long-distance* passenger: it includes all (except local) lines run by passenger trains. Charges are a relation of speed, slot, location of the line, priority, flexibility, frequency, network connections,
- *freight.* Charges are a relation of train weight, train path length, flexibility, priority, whether dangerous goods are transported or not.

The German rail pricing model has a component of *track* access charges and another one of railway *station* charges. It gives incentives for the usage of low-noise freight wagons by providing discounts or surcharges in relation to the level of the emitted noise. Cancellation fees are applied for slots reserved and cancelled later as follows: 15% of the agreed charge for cancellation 5 to 15 days before departure, 30% for cancellation 1 to 4 days before departure, and 80% for cancellation less than 24h before departure.

Charges are calculated in relation to the train-path kilometers run. Each of the three categories of traffic (local passenger, long-distance passenger, freight) is further segmented into a number of subcategories. Discounts are provided in order to promote the development of new railway services; surcharges can be applied for congested railway lines.

5.9.4. United Kingdom

The British rail pricing model has two components, a fixed one (applied only to passenger services under public service obligations) and a variable one which aims to cover the short-run marginal costs of running on the network. The variable component of charges is composed of a variable usage charge (aiming to recover operating, maintenance, and renewal costs that vary with the volume of traffic), an electrification asset usage and traction electricity charge, a freight specific charge, charges for access to service facilities. Financial incentives are provided for long-term framework agreements, (51).

5.9.5. Italy

The Italian rail pricing model takes into account the following:
- use of infrastructure, which is calculated by taking into account train speed, departure time, composition of train, density of circulation,
- access to stations, which is null for regional stations.

Reductions are afforded for low speeds, circulation in non-peak hours, high volumes of traffic, etc.

155

The Italian rail charging model has two components, an access charge (specific to a category of traffic) and a running charge related to the volume of traffic.

5.9.6. Spain

The Spanish rail pricing model has the following components, (51):
- *access charge*, allocated on the total annual train-kilometers running,
- *reservation charge*, allocated on the basis of fixed cost of maintenance and operations of the track. It is dependent on the time of the day, the route category, and the type of railway service,
- *traffic charge*, which is a relation of the period of the day, the route category, the type of railway services. It is calculated on the basis of the value of seat-kilometers offered,
- *running charge*, allocated on the basis of variable cost,
- *charge* for the use of *stations* and other facilities,
- *charge* for the use of *platforms*.

5.9.7. Poland

The Polish rail pricing model has the *operating charge* component (allocated with regard to costs of maintenance, renewal, rail traffic management, depreciation), the *reservation/cancellation charge* component, and *additional charges* for specific services.

5.9.8. Sweden

The Swedish rail pricing model has the following components: *track charge* (in relation to axle load), *train-path charge* (in relation to traffic), *passage charge* (for congested links).

5.9.9. Austria

The Austrian rail pricing model is segmented in passenger traffic (commercial, public service long-distance, heavy short-distance, light short-distance) and freight traffic. Incentives are provided if delays are reduced and if the emitted noise is below a particular level.

5.9.10. Belgium

The Belgian rail pricing model identifies 4 major market segments: commercial passenger traffic, passenger traffic under public service obligations, high-speed traffic, freight traffic.

5.9.11. Denmark

The whole rail network of Denmark is under public service obligations. An environmental subsidy is provided by the state for intermodal traffic making use of rail infrastructure.

5.9.12. A comparison of rail infrastructure charges

Rail infrastructure charges in Europe present great differences from one country to another and reflect, in most cases, interventionist state policies, usually in order to protect the historical state-owned rail operator.

Table 5.12 summarizes the characteristics of rail access charge systems for the various European countries. The component of charges expressed per train-km or gross train-km refers to traffic that is realized. The component of charges expressed per train-path-km represents requested infrastructure capacity for a certain period of time between two places and can be part of either the running charge or the reservation charge. Some systems charge the use of stations separately, (51).

In the United Kingdom, charges are different from one rail operator to the other, and though they appear to be high they are greatly subsidized by the state.

Similarly, in Switzerland, theoretical charges for freight traffic are also high, but almost two thirds of these charges are compensated by state subsidies.

High differences can be observed between charges for passenger and freight trains. Some countries have higher unit charges for freight trains, other have higher unit charges for passenger trains, while some countries have a similar level of unit charges, both for passenger and freight trains.

In conclusion, concerning rail pricing models, two great categories can be observed:
- models based on short-run marginal cost,
- models based on long-run marginal cost.

The tendency, however, is that charges should reflect as much as possible the real cost of maintenance and operation of infrastructure as well as a real equity for all operators. This means that the critical components of charges should be variable costs (related to the traveled distances), whereas components of fixed costs should be greatly reduced.

Outside Europe, there is also a great range of infrastructure charges, which reflect different objectives of cost recovery, different balances between passenger and freight, network complexities, and the intensities of traffic.

Table 5.12.
Characteristics of access charge systems for the various European countries, (complied from data of Network Statements for the year 2020), (51)

Country	Reservation charges	Running charges			
		per train-km	per gross train-km	per train path-km	per station stop
Austria		√	√		√
Belgium		√			
Bulgaria	√	√	√		
Croatia		√			
Czech Rep.		√	√		
Denmark		√			
Estonia			√		√
Finland			√		
France		√		√	√
Germany				√	√
Greece		√			
Hungary		√		number of paths	√
Italy		√			√
Lithuania		√	√		
Latvia		√	√	√	
The Netherlands	√	√		√	√
Norway	√		√		
Poland		√	√		
Portugal	√	√		√	√
Romania		√	√		
Slovenia			√		
Slovakia		√	√		
Spain	√	√ (+per seat-km)		√	√
Sweden		√	√		
Switzerland		√		√	
United Kingdom		√	√		√

Figures 5.8 and 5.9 illustrate infrastructure charges in Europe for passenger (suburban and regional, long-distance, high-speed) and freight (1,000 tonnes, 1,600 tonnes, 6,000 tonnes) traffic for the year 2016. Average values

of infrastructure charges are for passenger trains around 2 €/train-kilometer and for freight trains 2÷4 €/train-kilometer, (104).

Just for comparison, in the USA, Amtrak has paid in 2009 2.7 US $/train-kilometer for using rail infrastructure of other American rail operators.

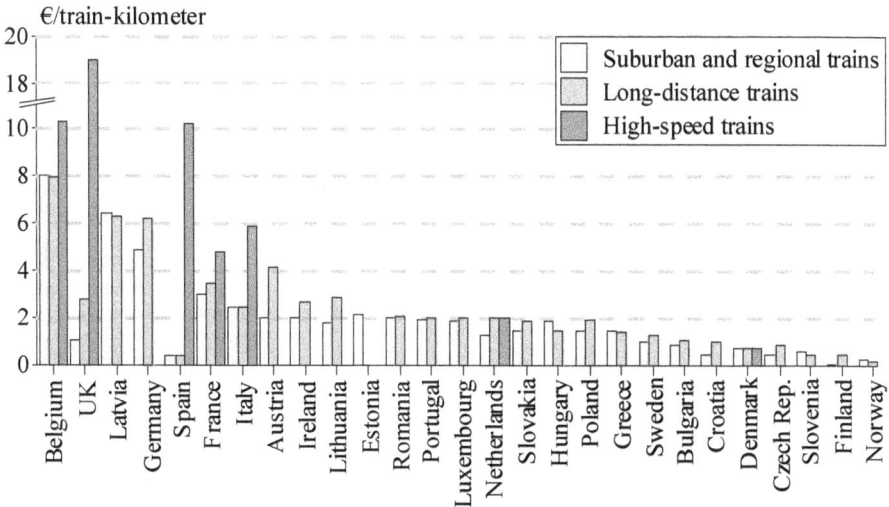

Fig. 5.8. Infrastructure charges (€/train-km) for the three categories (suburban and regional, long-distance, high-speed) of *passenger* trains in various European countries for the year 2016, (104)

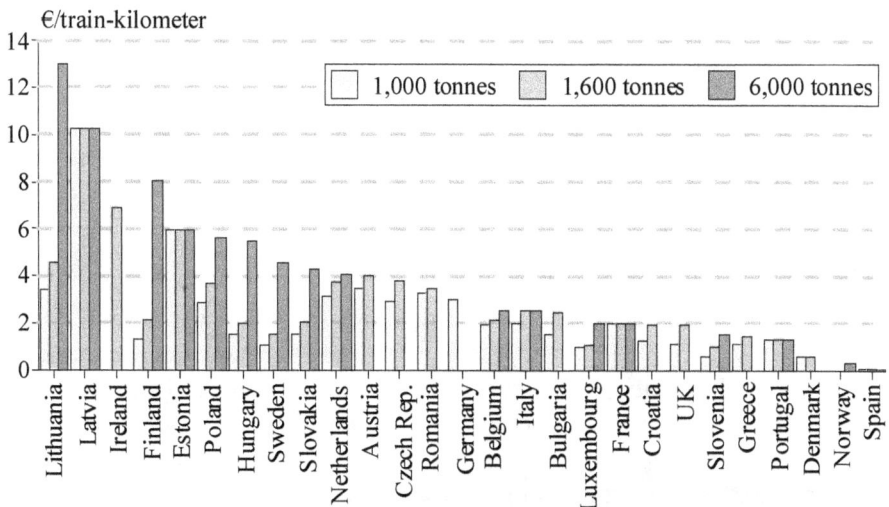

Fig. 5.9. Infrastructure charges (€/ train-km) for *freight* trains in various European countries for the year 2016, (104)

5.10. Pricing of operation

5.10.1. Targets of pricing of operation

The pricing of operation (which leads to the calculation of tariffs for passengers and freight) should cover the expenses of the rail operator, while at the same time assuring the financing of the necessary investment for renewal and modernization of its equipment (rolling stock, etc.) and a margin of profit. *Tariff* is defined as the charge paid by the user of a rail service. Tariffs aim at:

♦   part or total coverage of expenses,
♦   orienting clients to those services which are more beneficial either for the rail operator or for the society.

A rational pricing should take into account existing or offered capacity, cost, demand forecasts, price elasticity of demand, and cross elasticities, (see section 5.10.3), with competing modes.

5.10.2. The traditional method of pricing

Cost $C(x)$ of rail transport is usually expressed as a sum of two components (equation 5.3): one $(B \cdot x)$ depending on the distance traveled (x) and the other (A) being constant for each specific route and representing expenses which are not a relation of the volume of traffic,

$$C(x) = A + B \cdot x \qquad (5.3)$$

Based on this structure of cost, rail companies have used for many decades a similar formula for pricing passenger and freight services:

$$P(x) = a + b \cdot x \qquad (5.4)$$

Variable b is not constant, but is usually differentiated in relation to the range of distance it refers to.

5.10.3. Effects of elasticities

The term elasticity records and reveals whether (to which degree and with what consequences) a human need (such as the need for transport) can be reduced or not. Elasticity in general is the responsiveness of demand for a transport mode to a change in one of its determinants (price, income of customers, travel time, etc.) or determinants of a substitute transport mode (car, bus, airplane). Thus, elasticities of rail transport may refer to changes of rail tariffs (price elasticity), of customers' income (income elasticity), of tariffs of another transport mode (cross-elasticity), etc.

Price elasticity of demand (or simply price elasticity) helps to assess the extent to which demand is affected as a result of a change in price and is defined as:

$$e_p = \frac{\dfrac{\Delta q}{q}}{\dfrac{\Delta p}{p}} \qquad (5.5)$$

where: e_p : price elasticity,
q : demand when price is p
Δq: change in demand when the price changes from p to p±Δp.

Price elasticity may refer to short-term or long-term changes. It is a relation of distance, the purpose of trip, and the specific conditions for each case (existence and fares of competing modes, etc.). As a rough estimate, rail passenger price elasticities can be given values in the long-term between -1.0 and -1.4 for leisure and between -0.5 and -0.7 for work and in the short-term half as the previous long-term values. For rail freight transport, long-term price elasticities are between -0.25 and -0.35, (9).

When price elasticity is close to zero, then an increase in tariffs of 1% has no effect in demand and consequently is beneficial for the rail operator. This can be encountered in some urban or suburban rail services.

When price elasticity is between 0 and –1, an increase in tariffs of 1% will cause a reduction in demand to a percentage less than 1% and an increase in revenues between 0 and 1%. The global effect of such a strategy may or may not be beneficial for the rail operator.

When price elasticity equals –1, then an increase in tariffs of 1% causes a reduction of demand of 1% and thus revenues remain invariable.

When price elasticity is less than –1, an increase in tariffs of 1% causes a reduction of demand of more than –1% and thus revenues will be reduced.

Income elasticity reflects how a change in real income affects demand and is defined as:

$$e_{in} = \frac{\dfrac{\Delta q}{q}}{\dfrac{\Delta I}{I}} \qquad (5.6)$$

where: e_{in} : income elasticity,
I : real income, which is the inflation-adjusted income (the ratio of nominal income to the inflation rate) and reflects the buying power of the nominal income.
ΔI : change in real income.

Income elasticities for rail interurban trips in Europe have values between 0.8 and 1.0, reaching as much as 2.0 for high-speed railways.

The elasticity of rail freight transport in relation to GDP has been found to be close to 1.0: a little lower than 1.0 for developed economies and a little higher than 1.0 for developing economies, (9).

Cross-elasticity of demand measures how the demand for one transport mode changes, when the price of another mode (competitor or substitute) changes, and is defined as:

$$e_{i,j} = \frac{\dfrac{\Delta q_i}{q_i}}{\dfrac{\Delta p_j}{p_j}} \tag{5.7}$$

where: $e_{i,j}$: the cross elasticity of demand for transport mode i (e.g. rail) in relation to a change in the price of mode j (e.g. private car),

q_i : the demand of transport mode i when its price is p_i and the price of transport mode j is p_j,

Δq_i : the change in the demand of transport mode i when the price of transport mode j changes from p_j to $p_j \pm \Delta p_j$.

Cross elasticity of rail demand with respect to the cost of use of private car has in Western Europe values around $0.20 \div 0.25$, (9).

5.10.4. Pricing and competition

Traditional methods of pricing based on distance (and given by formula (5.4) of section 5.10.2) are not currently an efficient tool for pricing, since they ignore competition in the transport market, which comes either from other modes (road, air) or is intra-modal competition coming from other rail operators running on the same infrastructure.

However, the choices of clients to use a specific transport service are based on two critical parameters, tariff and quality of services (which is a relation of travel times, comfort, just in time arrival, etc.). Evidence has shown that if clients are satisfied with the quality of services, then a moderate increase in tariffs may have practically no impact on demand, (15). According to a survey in South Korea, including both business and leisure travelers, decision factors and their degree of importance when considering long-distance travel were as follows: fare 32.8%, safety 22.5%, accessibility 18.5%, travel time 15.3%, comfort 6.8%, frequency 4.1%, (103). Another survey conducted to passengers of high-speed trains in the UK, France, and Spain, in which respondents could select more than one criterion for their choice of mode of transport, revealed the following degrees of importance for the various deci-sion factors: fare 80%, travel time 69%, timetable 33%, reliability 31%, safe-ty 21%, flexibility 19%, lost (wasted) time 18%, transfers 15%, accessibility 14%, luggage 14%, security 11%, on-board services 8%, environmental im-pact 5%, (38).

Competition puts pressure on rail operators to take into account tariffs applied by their competitors and thus to abolish the traditional pricing method based on distance.

5.11. Pricing of passenger traffic

5.11.1. The existence (or not) of public service obligations

Public service obligations (PSO) are these rail services that, if the only con-sideration of the railways were business profit, would not have been underta-ken to the same extent or degree (e.g. the operation of lines with small traffic, low tariffs for some segments of the market, etc.).

PSO can refer either to certain categories of traffic (the elderly, students, etc.) or to the region served (isolated or non-accessible areas). In both cases, the authority imposing a PSO must subsidize the lost revenues of the rail op-erator. The justification for PSO lies on theories of regionality and on the fact that every citizen should have a minimum level of accessibility and mobility, which should be assured by more than one transport modes.

PSO aim (at least theoretically) at maximization of the public benefit; they usually refer to passenger transport and only rarely to freight transport.

5.11.2. The strategic dilemma: profit or increase of traffic

The decision at the dilemma of choosing between profit and increase of traffic is the responsibility of the state policy (if the state subsidizes some or all rail-way services) and of the rail operator.

Pricing strategies aiming at profit have been adopted in the United King-dom and Germany, (see Fig. 5.10 below), among others, and have led to the abandonment of many secondary lines. Surprisingly and in spite of high unit tariffs, this strategy has led to an increase of traffic in the United Kingdom and elsewhere.

Pricing strategies aiming at an increase of traffic have been adopted in many countries with an interventionist policy and favor more socio-economic factors and less entrepreneurial spirit. Such policies, however, lead to deficits, which are covered by state subsidies. In any case, a prerequisite for the suc-cess of this strategy is a sufficient quality of rail services; otherwise such a strategy may lead to huge financial losses for the rail operator.

5.11.3. Pricing for rail operators without public service obligations

Rail operators without public service obligations are obliged to balance revenues and expenses. Tariffs should not be less than marginal costs and can be as high as the market can bear.

For similar levels of quality of service, rail tariffs should not exceed a reference value, which is defined by tariffs of competing modes. By setting a target for the traffic of the rail operator and taking into account tariffs of competing modes and elasticities, econometric models, (see section 4.6), can facilitate the calculation of rail tariffs to be applied, (9).

Such a strategy has been used by Eurostar, where, for a similar quality of service rail tariffs are lower by 15% compared to tariffs of the principal competitor, which is the airplane.

5.11.4. Yield management techniques and unit revenues

The everyday problem of a rail operator is to make the maximum profit of the capacity offered, which if not used is lost. Yield management techniques were first used by airlines and later by railways for services other than local, so as to combine the best use of the capacity offered with a maximization of profits.

Yield management relies on the so-called Ramsey pricing technique, which suggests that the sooner the ticket is bought, the higher the offered reduction in tariffs will be. Thus, tickets bought some months before the day of travel may be half or even lower than the price of the ticket bought just some hours before traveling. Any change related to tickets bought at low price is followed by strong financial penalties. A strong criticism has been expressed over this form of pricing and has focused on whether liberty of action of an individual is restricted or not. The pandemic disease that swept the whole world in $2020 \div 2021$ may put in question some aspects of the Ramsey pricing technique, particularly those associated with high financial penalties in case of change of the day of travel for tickets already paid.

Yield management principles lead to a differentiation of tariffs and can take into account the following characteristics:
♦ period of the day, by offering lower tariffs in non-peak hours,
♦ day of the week and season of the year, so as to discourage an extensive use during week-ends or holiday periods.

The differentiation of tariffs leads to a maximization of revenues, permits a penetration to other segments of the market, and can have (at least apparently) the character of a social policy. However, it has also negative effects. For those clients who paid a higher tariff, there is the risk of disappointment, which is offset by offering them other benefits, such as supplementary services, delivery of luggage, etc.

Figure 5.10 illustrates average unit passenger revenues (that is average revenues per passenger-kilometer) for passenger traffic of railways of European countries for the year 2012. The higher the income of the country and the level of railway services, the higher the unit revenue per passenger-kilometer, (100).

Revenue (€) per passenger-kilometer

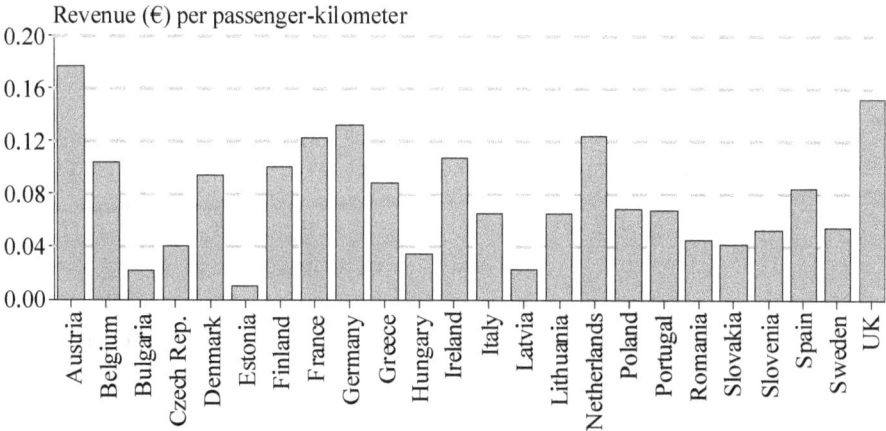

Fig. 5.10. Unit passenger revenues (€/passenger-kilometer) for railways of European countries, (104)

5.11.5. Complementary commercial measures to increase revenues

Commercial policy of a rail operator is based on marketing, advertising, and tariff policy. Commercial measures aim at, (16):

♦ segments of the market, like students, the young, tourists, the elderly (more than 65 or 70 years old), pensioners, personnel of enterprises, travel of groups, foreigners,

♦ strengthening the link between rail operator and clients, by implementing measures like:

   – cards (daily, monthly, yearly) offering unlimited use on rail services for the owners of the specific card,

   – advantages and free tickets for clients who are frequent users of rail services.

5.12. Pricing of freight traffic

As analyzed is section 1.7.1 railways are more suitable for the transport of bulk freight (minerals, oil, grains, etc.) and specialized freight (containers, cars, etc.) for long and medium distances. These conditions are fulfilled in the USA, China, Russia, India and for these countries railways have a high share in the freight market. This is not the case in Europe, as in 2015 the freight market consists of products for half of which railways do not have a comparative advantage: food and leather 15.3%, grouped goods 16.4%, agricultural products 9.8%, small parcels and furniture 7.6%. Only for the remaining part of products transported in Europe, railways have a comparative advantage: coal and metals 9%, containers 5.3%, chemicals 5.7%, wood 5.7%, transport equipment 4.1%, fabricated metal products 4.1%, waste and other raw materials 3.5%, various equipment 2.3%, minerals 4.0%, machinery 3.0%, petroleum coke and other refined petroleum products 2.9%, (83).

Rail freight in most countries does not receive any kind of public subsidies and therefore revenues should cover total costs of the freight sector. Pricing of rail freight must take into account rail freight costs and tariffs of competitors (road transport). Thus, before any effort to increase rail freight tariffs, railways should improve their performances in relation to the following, (15), (117):

- low rail *shipment speed*, which for Europe has an average value around 20 km/h, against an average shipment speed of around 50 km/h for road freight,
- *punctuality*, which is for railways far away from the level of 95% of road freight (delivery of goods in the agreed time, with a margin of a maximum delay of 60 minutes),
- quasi-impossibility for railways to achieve *door-to-door* rail freight transport, unless more rail connections to industrial centers are provided.

Most of the above handicaps are a result of the coexistence on the same track of fast (passenger) and slow (freight) trains, the latter usually being given a lower priority. This handicap does not occur in the USA, where railways are specialized in freight traffic, but is frequent in Europe and Asia. As a remedy, it has been suggested to transform some rail routes with heavy rail traffic to dedicated freight corridors, on which freight trains will run either exclusively or with priority. Dedicated freight corridors are the equivalent for freight of lines dedicated only to high-speed passenger traffic. Technical specifications for such dedicated rail freight corridors could include, (15):

- maximum speed of 100÷120 km/h, the principal target being to achieve an average rail shipment speed approaching 50 km/h. Higher speeds than 120 km/h may be justified only for traffic of products with a high value,
- axle load of 22.5 tonnes, (see section 7.5),
- length of platforms from 600 m to 750 m, allowing the running of long freight trains,

- a loading gauge in tunnels compatible with the loading gauge (known under the initials GC) of UIC, (see section 7.10, Fig.7.13),
- a signaling of the type of version 3 of ERTMS, (see section 21.10.4),
- electrification systems that could feed multi-current locomotives,
- appropriate equipment in marshaling yards in order to minimize long delays in the transfer of goods,
- facilities of combined transport and particularly equipment for the horizontal loading of containers,
- sidings with clients who generate high rail freight flows, thus providing direct rail connections to industrial or production sites.

Figure 5.11 illustrates average unit freight revenues (that is average revenues per tonne-kilometer) for railways of European countries for the year 2012, (104). High rail freight tariffs in France were followed by an important reduction of rail share in freight transport. Austria, in spite of high rail freight tariffs, still keeps a high share of railways in the freight transport market, due to huge transit traffic through the country.

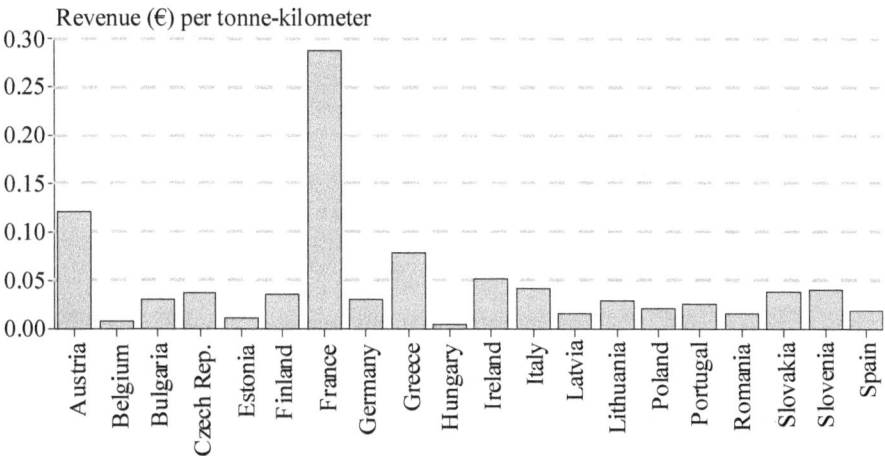

Fig. 5.11. Unit freight revenues (€/tonne-kilometer) for railways of European countries, (104)

167

6 Planning and Management of Railways

6.1. Railways, the society, and the economy

6.1.1. A systems approach for the railways

Considered either as a whole or separated (infrastructure–operation), railways constitute a complex system. Each component (track, traction, operation) has many subcomponents (e.g. for track: rails, sleepers, etc.), the interaction of which is not easy to predict. However, a good synergy of all rail components is necessary in order to achieve the desirable result, i.e. safe, quick, comfortable transport of people and goods at the lowest possible cost. For this reason, railways should always be examined as a *system*.

Application of systems approach in railways is given in the simplified flow chart of Figure 6.1. Even if the problem focuses on a technical need, railway managers should begin from defining the real problem, which could be put as: *'What is the transport need to be satisfied and what are the targets being aimed at?'*. In every step of a systems approach, all alternative solutions should be carefully examined.

6.1.2. Railways and the social and economic environment

6.1.2.1. The internal and external environment

Each railway activity must be examined in relation to its *internal* and *external* environment, (Fig. 6.2). The whole organization of railways must be characterized by the principle of adaptability, that is the ability to adapt to changing situations of its internal and external environment, (122).

6.1.2.2. Strategic and tactical level of decisions

In management, we often distinguish between the strategic and the tactical or organizational level of decisions. To be more specific:

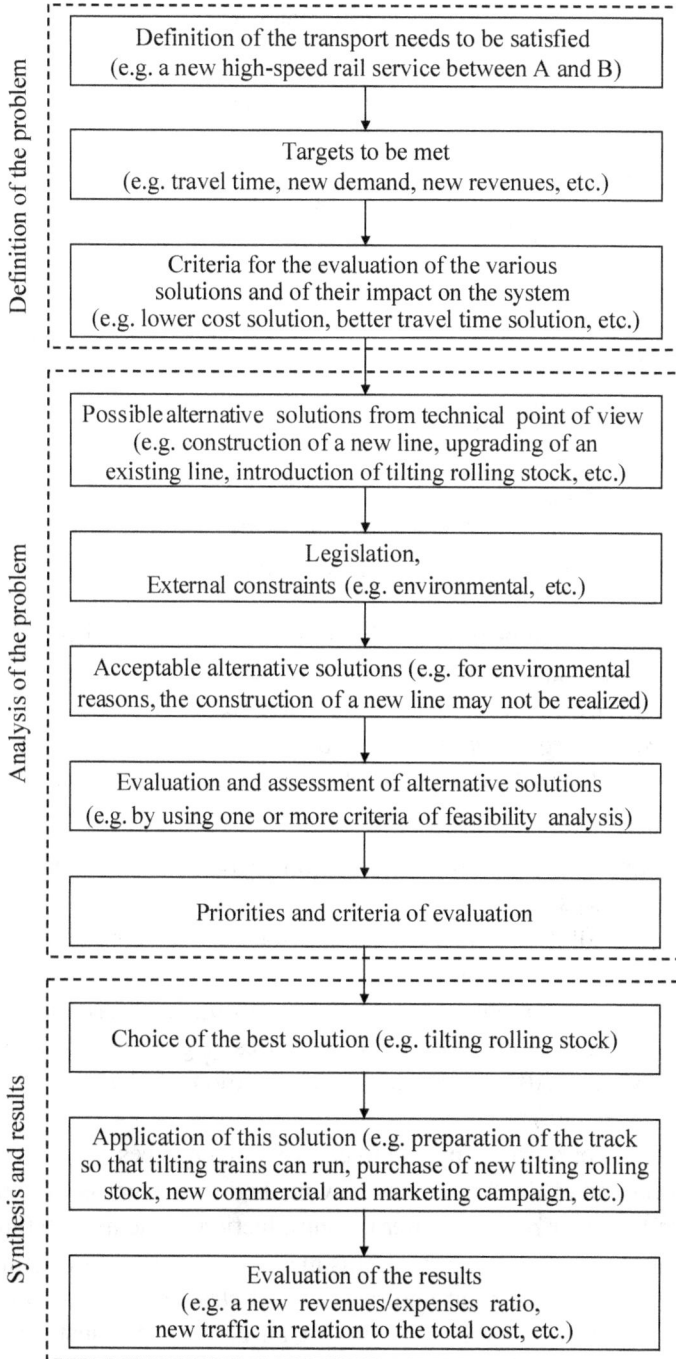

Fig. 6.1. Systems approach applied in railway problems

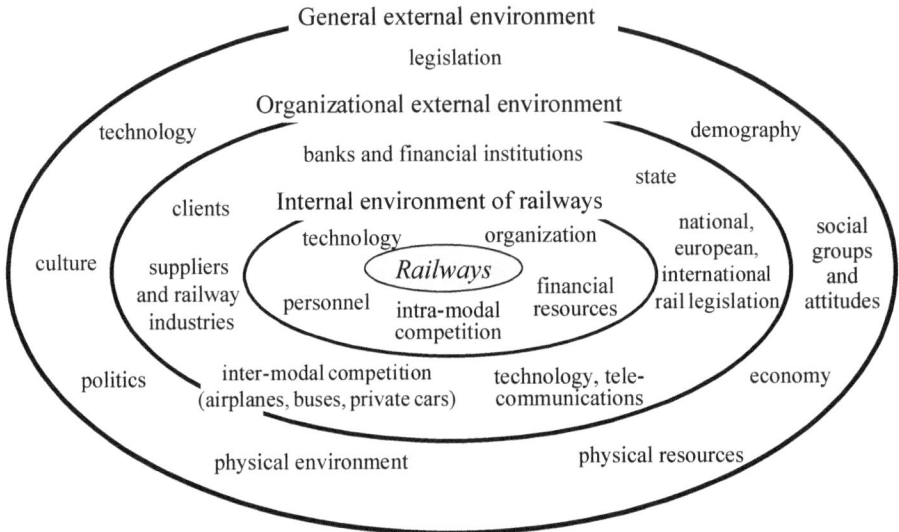

Fig. 6.2. Railways and their internal and external environment

- the *strategic* level of decisions refers to the fundamental orientations of the railway undertaking, such as: revenues/expenses ratio, volume of passenger or freight traffic, level of state subsidies, etc,
- the *tactical* or *organizational* level of decisions concerns the following: introduction of new technologies, changes in human resources, organizational changes, etc.

The adaptability of the railway activity to its environment requires a number of tactical level steps, (36):

- periodic (e.g. all 3 or 6 months) comparison between targets and achieved results (e.g. volume of traffic or revenues, etc.),
- localization of divergences, research of reasons, and formulation of the possible methods to confront the divergences (e.g. in the case of loss of traffic, new methods of marketing, modification of the product offered, new personnel, etc.),
- choice of the appropriate method to confront divergences,
- application and evaluation of this new method (e.g. what is the rate of increase of traffic or revenues after the introduction of the new method),
- if divergences continue to persist, this means that tactical or organizational measures are not sufficient and decisions at strategic level should be undertaken, such as closure or ending of an activity (for instance, freight transport of low traffic, passenger services between low-density population areas), creation of a new service, etc.

6.1.2.3. Separation in business units

As railway transport is extremely complex and is composed of many activities that usually bear no relevance to each other (e.g. activities of marketing and track maintenance), it is mandatory to separate and categorize activities in separate homogeneous units that are called *business units*. A typical categorization of the whole railway activity in business units is:

- ◆ infrastructure (maintenance and operation),
- ◆ rolling stock (maintenance and operation),
- ◆ operation of passenger traffic,
- ◆ operation of freight traffic.

Some of the above business units may further be divided into smaller ones; thus, rolling stock can be divided into two units: one being in charge of maintenance and another in charge of operation.

6.1.2.4. Changes and requirements of the environment of railways

The third decade of the 21st century is characterized by a fast changing environment with the following features, (15):

- the economic and debt crisis of 2008÷2012 and the world sanitary crisis of 2020÷2021 pushed railways to drastically reduce personnel and costs, so as to face efficiently both inter-modal and intra-modal competition,
- economy and society change quickly (at least concerning appearances and exigencies) and ask for new products and services (e.g. just in time delivery of freight, increased quality of service, etc.),
- research and development of new technologies may quickly render existing technologies obsolete and without any value (e.g. electronic ticketing, when introduced in railways, dramatically changes the existing ticketing systems),
- continuous changes and new products of competitors (e.g. low-cost air transport) has obliged railways to drastically change their offer and tariffs in many routes,
- changes in the policy of government and world institutions concerning social security, liberalization of the transport market, consumer protection, pollution, climate change, human rights, etc. (e.g. railways in the case of great delays must compensate clients) impose abrupt changes in strategy and organization,
- frequent and fast changes in the values of the society and the people (concerning safety, environmental effects, etc.) put pressure on railways to change policy and strategy, something that results in additional costs,
- conditions of survival in such a changing environment are systematic marketing surveys, in order to monitor in time forthcoming changes, and adjustment of products and policy of the railways to the external requirements.

6.1.3. Quality control

Verification of achievement of goals cannot be left to an empirical assessment, as was the case in the past. Criteria of assessment should be clear and quantifiable. Quality control aims to assure consistency with certain standards, keeping always in mind the needs of customers. Quality control is usually easier for products than for services (such as rail services). Quality control is of great importance and railways can adopt the ISO (International Standards Organization) or some other certification. Since 2006, railways have adopted ISO 9001, the international standard for a quality management system. In addition, the European norm EN 13816 ('Service quality for public passenger transport') prescribes methods and procedures to assure quality control and achieve an homogeneous (in time and space) rail product. Figure 6.3 illustrates an example of how railway organization and efficiency can be improved through rational successive steps and with continuous quality control.

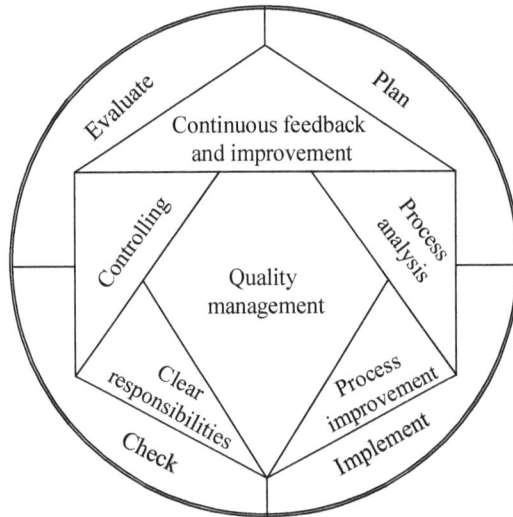

Fig. 6.3. Organization and control of the railway activity, (126)

6.2. Competition and impact on railway management

Railways have operated for many decades as a physical monopoly under the protectionist umbrella of the state, which covered all the deficits that the railway activity was producing. This becomes less and less the case with each passing year. Competition is increasing and can be either external (i.e. intermodal competition from other transport modes) or internal (i.e. intra-modal competition among many rail operators running on the same track).

However, while competition is increasing, regulation still holds. In fact, governments can through regulation subsidize some railway deficits, set and impose tariffs, define regimes of entry-licensing-access for new railway operators, and ask all parties involved for increased levels of safety, (Fig. 6.4).

The structure of railways has direct effects on management. Whether separated or unified, it greatly affects the methods of management to be applied.

In a strongly competitive market, railways should try:
– to have the lowest generalized cost in order to attract new clients,
– to understand and take into account all kinds of elasticities (i.e. price, revenue, cross) in order to react in time and avoid losing markets,
– to present a new image of the specific railway activity,
– to establish a closer and permanent link between clients and the various railway activities.

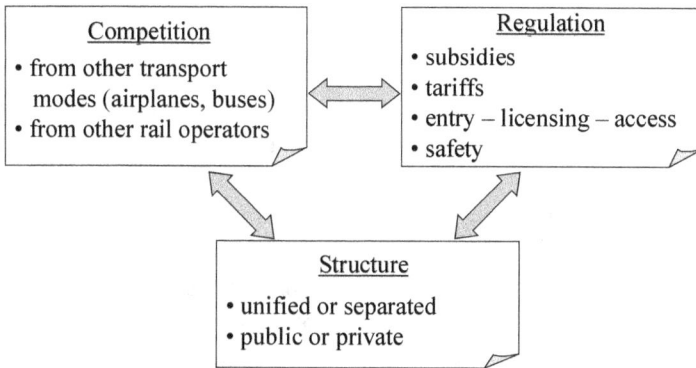

Fig. 6.4. Effects of competition, regulation, and structure of the railway company on the management of railways, (99)

6.3. Feasibility studies and methods of financing

6.3.1. Need for evaluation of any rail project

In the past, some railway lines were constructed as part of a national development plan, or for strategic and security reasons, or for the development of national resources, without detailed economic or financial considerations. As times have changed, such old practices are no longer valid. Thus, even a small railway project must be justified from an economic and financial point of view. Clear and complete answers should be given from the early stages to questions such as: 'Why is this railway line or facility needed?', 'What do we want it to achieve?'. Otherwise, there is the risk of constructing a rail infrastructure with low traffic which will accumulate deficits, while the money spent could have been used for other more useful and efficient purposes.

6.3.2. Benefits – costs in the case of a new railway infrastructure

Feasibility studies compare benefits to costs of a specific railway project.

Cost has two basic components:
– construction cost,
– operation and maintenance cost.

Benefits from the realization of a railway project can be, (16):
• reduction of travel time,
• reduction of operation cost,
• reduction of accidents,
• improvement of the quality of service,
• regional and national development,
• security and national integration.

Among the above benefits the only direct and commercial one is the reduction of operation costs, whereas all other benefits are related to social reasons. Comparison of benefits and costs implies that all benefits should be transposed in monetary terms. This involves:
– For travel times, an assessment of the value of time. Many railway projects have as a primary objective the reduction of travel times. Reduced travel times are the main benefit when considering a rail project and account usually for 2/3 of total benefits. However, what is the monetary value of a man-hour saved? There is no doubt that the value of time is different for a businessman, a public servant, a student, a pensioner, or an unemployed person, (see also section 5.1.5). Each category of traffic has its different value of time. Extremely great differences concerning values of time exist from one country to another.

   Saved travel times related to work activities are taken into account by 100% in feasibility studies. Saved travel times related to leisure, tourism, etc. are taken into account at a percentage of 20%÷35%, (129).
– For regional and national development, assessment of the increase of economic product in the considered region or country,
– For accidents, evaluation in monetary terms in the case of a death or an injury.

6.3.3. Evaluation methods for rail projects

There are many evaluation methods for rail projects, (16), (129):
♦ In the method of Present Value (PV), all expenses (for construction and operation) are calculated for the entire economic life of the project; the alternative solution with the lowest present value is the most economic one.
♦ In the method of Net Present Value (NPV), for each alternative the net present value is calculated according to the following formula:

$$NPV = (B–O) – (C–Y) \qquad (6.1)$$

where:

NPV : Net Present Value,
B : Present value of all benefits,
O : Present value of all operation and maintenance costs,
C : Present value of construction costs,
Y : Salvage value (the project's value at the end of its economic life).

♦ In the Cost-Benefit method, the ratio λ is calculated as follows:

$$\lambda = \frac{B-O}{C-Y} \qquad (6.2)$$

A project is to be realized if $\lambda > 1$. Among many alternative solutions, the one with the greatest value of λ is chosen, (120).

♦ In the Internal Rate of Return (IRR) method, the value of the discount rate is calculated (by the trial and error procedure), for which the present value of benefits equals the present value of expenses. If IRR is greater than the opportunity cost of capital, then the specific rail project has chances to be realized, (129).

♦ Previous methods of evaluation focus on economic parameters, since they assess utility generated by a rail project. In this way, however, important parameters such as quality of service, mobility, noise, pollution, etc., are neglected in the evaluation procedure. *Multi-criteria* methods help to take into account *all* parameters related to a rail project: construction cost, operation and maintenance cost, expected demand, reduction of travel times, increase in national or regional product, quality of service, safety and security, land uses, air pollution, noise, mobility, and accessibility. Each parameter is given a weight factor, which reflects priorities of evaluation. If, for instance, construction cost is given a high weight factor (say of 50%), this will lead probably to the selection of a solution with low cost. If, in contrast, expected demand is given a high weight factor, this will lead to a project serving more people.

However, it should be stressed that the selection of the rail project to be realized is largely a political decision. Evaluation methods just help decision makers to rationalize the procedure of selecting a particular project.

6.3.4. Methods of financing a new rail project

The economic reality is that most of the financing of the private sector is oriented toward industrial projects (particularly in the energy and telecommunication sectors), only a small part to transport projects, and even a smaller one for rail projects. Principal reasons for this situation are the high cost of rail and transport projects, the long period of construction (3÷7 years), and the relatively low expected revenues. As illustrated in Figure 6.5, the cash flow of a rail project becomes positive only within 15÷20 years from the beginning of financing, a long period for bankers and entrepreneurs. In contrast, the cash flow of an industrial project becomes positive within 5÷6 years from its beginning, (128).

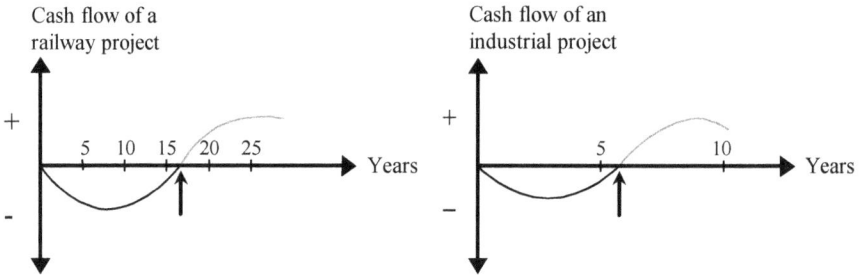

Fig. 6.5. Cash flow of a railway and an industrial project, (128)

Rail projects, in the majority of cases, are an unattractive investment for the private sector. For this reason, they are mostly financed by state funds, and thereafter they operated as a public company. However, the worldwide economic environment exercises pressure on public finances and reduces the possibilities for states to finance unprofitable rail projects. This is why rail managers seek financing (partial or total) from the private sector for a certain part of the required investment, clearly the most attractive from a revenue generating point of view, (see also section 2.4.3).

Some critical issues, when financing a rail project, are the following:

* first of all, who gives the money, the state or the private sector (i.e. entrepreneurs, investors, etc.)?
* who undertakes the risk during construction? For instance, if the initially calculated construction cost were increased by 30%, who would pay the additional money needed?
* when the project is finished, who will run the operation, the railway company or the company which constructed the project? As railway projects are extremely complex, some constructors may be unwilling to operate the projects they have constructed,
* if the constructor of a rail project also undertakes its operation, who is in charge of entrepreneurial risks during the operation? For instance, if real demand were by 25% less than the forecasted demand, who would pay the difference?

6.3.5. Public-Private Partnerships

Depending on the answer to the questions raised in previous sections, there are many schemes of involvement of the private sector at the financing of a rail project, known as Public – Private Partnerships (PPP), (121):

- in the *Build – Operate – Transfer (BOT)* method, a private company undertakes in a public bidding, under state specifications, a rail project, finances (partially or totally) its construction and then operates it for a period, usually 20÷40 years, but even up to 99 years (Channel Tunnel). During the concession period the owner of the project is the public (i.e. the infrastructure manager or the operator). Return of the invested money is achieved through revenues from passenger, freight, or commercial activities.

  A variation of the BOT method is the *Build – Own – Operate – Transfer (BOOT)* method, in which the private partner owns the project during its operation. A variation of the BOOT method is the *Build – Own – Lease – Transfer (BOLT)* method, in which the private partner leases, after construction, the finished project to the rail authority, which pays to the private partner periodic payments for the invested capital.

  Another variation of the BOT method is the *Build – Transfer – Operate (BTO)* method, in which upon completion of its construction by the private partner, the rail company (infrastructure manager or operator) becomes the owner of the project and rents it for a period to the private sector.

  In the *Build – Own – Operate (BOO)* method, the private partner finances, builds, owns, and operates the project for a period during which he can earn revenues from the operation of the project. The method is suitable for investments in railway stations, where facilities may serve many users.

- In the *Private Services Contract: Operations and Management,* the infrastructure manager concludes for a specific facility a contract with a private partner, who is in charge of its operation and maintenance, while the infrastructure manager continues to be the owner and in charge of management.

  A variation of this method is the *Private Services Contract: Operation, Maintenance and Management*, in which the private partner is also in charge of the management.

- In the *Developer Financing* method, an area belonging to the railways is given to a private partner who constructs a rail facility and in return the partner is given the right to construct houses, commercial centers, or industrial facilities. The method is suitable for areas with a high value of land.

- In the *Long-Term Lease*, rail facilities are leased to a private partner who invests money for their modernization and then operates them for a specific period. A variation of this method is the *Lease–Rehabilitate–Operate* method.

177

Selection of the most appropriate method is done in relation to the characteristics of the rail project (e.g. a new track, a new station, a marshaling yard, upgrading of a facility, etc.), the expected demand and revenues, the risks during construction and operation, etc.

6.4. Planning the railway activity

6.4.1. Need and purposes of planning

Human activity is developed in an extremely complex environment. When this activity concerns a complicated system, like a railway, then the need emerges for all components of the system to be constructed and operated in order to achieve the best result. Organization and investment should contribute to this result. A powerful tool to achieve fixed targets is planning and is understood as the process that sets goals, defines actions to be performed, estimates and allocates resources, determines stages in time and deadlines, identifies responsibilities for actions, and defines mechanisms of monitoring and evaluation, (130).

A planning procedure departs from understanding the existing situation, tries to forecast plausible evolutions in the future, anticipates coming problems and evolutions, suggests the investment and organization that will be required, and looks for the necessary funds to finance the suggested actions.

There are many levels of planning, in relation to time:

– long-term planning, which refers to the coming $10 \div 15$ years and describes the whole or sectorial strategies, where investment should be oriented, and the financial strategy to be followed,

– medium-term planning, which refers to the coming $3 \div 5$ years and is an implementation of the long-term planning at medium level concerning strategy, detailed investment, organizational changes and funding, staff requirements, commercial and tariff policies,

– yearly planning, which is detailed in the yearly budget of the railway activity.

It is clear that among the various levels of planning, there should be a consistency of goals and measures. For instance, a decision to lower tariffs and increase traffic may be beneficial in the short term, but may be negative in the medium or long term, if there is no availability of rolling stock to serve increased demand.

Planning is nothing more but the management of change; failure of the railways to respond timely and efficiently to external changes may be disastrous.

Railway planning is a framework, within which the various facilities can operate their separate functions at the highest possible levels of efficiency. Successful planning must be characterized by *flexibility* and *adaptability*.

6.4.2. Master Plans and Business Plans

Master Plans and *Business Plans* are the most current forms of planning railway activity. Master Plans refer to the whole of the railway activity and contain analysis of investment, technical equipment, organization, and finance. However, Business Plans emphasize on organizational, economic, and financial aspects. Each infrastructure manager must have an actualized Master Plan and each railway operator an actualized Business Plan.

A railway Master Plan can be defined as the conception of a planner for the further development of the various components and operations of a railway system. It is the frame within which the planner suggests the evolution and development of the various components of the railway system, i.e. higher efficiency, productivity, and revenues, while reducing costs and respecting environmental rules.

A Master Plan is not an implementation program, as some believe, but a guide and nothing more. It is based on certain assumptions concerning the evolution of the economy (such as Gross Domestic Product, consumer prices, etc.) and of the transport market (share of each mode, prices, elasticities) and it suggests scenarios for confronting future evolutions. A Master Plan should be updated, if possible, on a yearly basis. A Master Plan must clearly set targets, priorities, and methods of implementation.

If in a Master Plan, technical aspects and problems are given less priority, then we have a Business Plan.

Whenever a new investment or expense is suggested in planning, the crucial question *'who pays the bill?'* should be clearly answered.

6.4.3. A brief description of a Business Plan of a railway undertaking

Business Plans vary considerably from one country to another, due to the influence of diverse historical, geographical, sociological, demographic, and economic factors. However, all rail Business Plans should contain a minimum of analysis described below, (15):

a) *External socio-economic environment*: economic growth, agricultural, industrial, and economic production, tourism, population and demography, legislation, state policy.

b) *Railways and the transport market*: evolution of railways' share and traffic, tendencies of the transport market, situation and prospects of competing modes, forecasts for future railway traffic.

c) *Financial situation, costs, and productivity:* Evolution of rail revenues and expenses, costs and tariffs, comparison with costs and tariffs of competing modes, yield analysis, personnel employed and unit costs of services, productivity indices (total and sectorial), comparison with other railways.

d) *Weaknesses of present organization and management.*

e) *Formulation of a new strategy and targets*, that should be quantifiable, such as a new revenues/expenses ratio, increase of traffic and productivity, etc.

f) *New investment required*: description and justification of new investment (e.g. in the sectors of infrastructure, rolling stock, facilities, etc.), estimated advantages and benefits from each investment, sources (state-private) of investment and guarantees of financing, expected return of investment, volume of loans.

g) *Forecast of evolution of the various financial indices* such as revenues, expenses, yields, investment, cash flow, etc.

h) *Human resources changes.*

i) *Sensitivity analyses for all forecasts*, i.e. how will change a forecast (e.g. of revenues) if a basic assumption of the Business Plan (e.g. volume of investment, volume of traffic, etc.) changes.

Planning procedures vary with the size of the railway undertaking. Whatever the size of the railway, however, it is vital that the plans produced have the full support of those levels of the personnel who actually do the work, can manage the changes to which the railway has to respond, or can be made responsible for internal changes.

6.5. Project management for railways

6.5.1. Definition of project management

A project usually starts with an idea, which is then elaborated in a study and finally is developed to the stage of implementation, completion, and operation. A project implies a major capital investment and differs from normal work in several aspects, such as size, cost, complexity, and criticality (concerning partial deadlines and final completion). Any project, however small, needs effective management, if it is to be carried out successfully. A project presents a degree of complexity, rendering it difficult to be carried out by staff who are in charge of daily and routine work.

Project management is the art of directing and administering a project. It covers the necessity to define, formalize, control, and coordinate a wide range of activities. It constitutes of breaking down the whole project into easily understood and measurable work items, so that the tasks and responsibilities of each team unit can be clearly defined and followed up.

The cost of failure of a project is enormous and there are not many volunteers in the administration that would undertake it eagerly, with vague responsibilities (not easily identified) dispersed in various levels. Project management is a wise solution, which permits the administration (e.g. infrastructure manager, ministry, etc.) to efficiently organize the various team units, to optimize methods of work, and monitor, analyze, and follow the progress of work.

6.5.2. Scope, benefits, and costs of project management

It is the responsibility of the administration to assess whether or not its organizational structure, temporarily enlarged as necessary, can efficiently develop, plan, administer, and supervise a particular project, while keeping it on schedule and within budget. This assessment should be as objective as possible and the administration should be assisted in its decision by consultants specialized in management.

Project management has clear advantages in a number of situations such as when, (131):

– the magnitude of the project is large in relation to the administration's management structure and size,
– the scope of work is unfamiliar to in-house personnel,
– if management is conducted by the administration's personnel, even with additional temporary staff, the routine day-to-day functions risk being left behind,
– unpredictable delays arise, thus necessitating rigorous programming,
– high political risks may emerge, if the project is delayed or failed,
– independent and impartial recommendations are required by banks or other funding institutions.

Thus, when project management services are engaged in a project, this offers additional guaranties to the administration for the successful execution of the project and also other benefits, such as, (131):

♦ an impartial, objective, and professional approach,
♦ experience of the project manager, arising from similar projects,
♦ evaluation of all available alternatives,
♦ in time monitoring of deficiencies, testing, and controlling quality of works,
♦ overviewing budget control, financial, and cash flow requirements,
♦ following up the programming of the project and completion on schedule,
♦ economies, due to the opening of the project manager to more competitors,
♦ reduction of the risk of delaying or failing to deliver the project.

Project management services may appear as an expensive solution. However, experience from several projects has proved just the opposite. In addition to assure quality and in-time delivery, project management may contribute to the reduction of costs. Furthermore, it should not be forgotten that if the administration conducts the management, many administration costs are effectively hidden.

6.5.3. Some rail projects that could require project management

Project management may be beneficial or necessary for a number of rail activities. We will mention some of them:

- construction of a *new high-speed line*. Such a project requires a wide range of professionals, among them economists, planners, civil engineers, electrical engineers, accountants, architects, so that project management is quasi-inevitable,

- *track upgrading*. Many railways still have a military discipline and inflexible internal organization. Furthermore, their staff continues to believe that railways are an engineering oriented business. Within such a structure and when the need for upgrading a track arises (e.g. increase of speed, axle load, etc.), there is the risk not to choose the best solution. Therefore, open-minded, objective, open to all solutions project management will then be necessary,

- *a new marshaling yard or other freight facility* will necessitate track and rolling stock specialists, good interface with road freight facilities, operating conditions under low costs. In such situations, the services of a project manager will be in demand,

- *a new tunnel or bridge*. Tunnels and bridges are very expensive projects with a long lifetime (50÷100 years), require many specialists with a high level of expertise, and would need project management services,

- *a new railway station*. Not only architects and civil engineers, but also city planners, transportation engineers, and marketing and advertising specialists could be sought in project management services for a new station,

- *electrification of a line*. The decision of whether to electrify a line or not is principally a matter of cost. It should be based on the volume of traffic, additional cost for its implementation, and the potential of lower operating costs. However, power supply, transmission system, pantograph, insulation, choice among the many alternative technical solutions, and the interface with track and interoperability are all very specialized issues that require a high expertise but also the best coordination and management,

- *signaling and safety installations* are extremely complex systems, whose performance, reliability, impact on safe operation, and technological advances will require project management services.

6.5.4. A description of tasks of project management for railways

As a case-study for the tasks of project management for railways, we will consider the services of a project manager to the infrastructure manager for the construction of a high-speed line.

The activities of project management can be divided in four stages: Organization, Development, Setting up, and Execution. Each stage begins with assessment and conclusions of works of the former stage and ends with the submission of a report to the infrastructure manager, (130), (131):

1ˢᵗ Stage: Organization. It comprises the following tasks:

♦ Definition of the project and of its components (e.g. for a new high-speed line: expropriations, technical studies and surveys, studies and selection of the appropriate materials for subgrade, ballast, sleepers, fastenings, rails, design of tunnels and bridges, design of signaling and electrification equipment).

♦ Definition of requirements and objectives of the infrastructure manager (e.g. cost and time restrictions, eventual deficiencies in personnel and staff of the infrastructure manager, etc.).

♦ Conceptual planning (e.g. the project manager plans its successive tasks: studies, procurement of materials, phases of execution, etc.).

♦ Activity plans, team composition, and resources (e.g. the project manager plans each activity, allocates responsibilities to his personnel and provides the necessary resources (such as funding from the infrastructure manager)).

♦ Determination of physical constraints and approvals (e.g. expropriations, licenses from various authorities).

♦ Cost evaluation and assessment of implications (e.g. the project manager checks and changes the various cost estimates of the infrastructure manager).

♦ Programming of works, funding, allocation of resources (e.g. many computer software can contribute to a rational programming).

2ⁿᵈ Stage: Development. It comprises the following tasks:

− Design and construction standards to be adopted (e.g. does the infrastructure manager have the appropriate specifications or should specifications of another railway authority or institution be followed?).

− Further studies and surveys to be carried out (e.g. in areas of seismicity, more geotechnical investigations may be required).

− Finalization of alternative strategies.

− Finalization of administrative procedures and approvals from the various authorities involved (e.g. ministries, municipalities, state institutions, etc.).

− Proposals for procurement of the various materials (e.g. details concerning bidding procedures, legislation restrictions, etc.).

− Allocation of tasks for each team unit of the project manager.

- Preparation of target cost estimates for the various components of the project.

3rd Stage: Setting up. It comprises the following tasks:
- Procurement plan (e.g. for ballast, sleepers, rails, etc.).
- Quality assurance (e.g. that ballast has the appropriate geometrical and mechanical characteristics).
- Calculation of quantities of the various materials required.
- Cost controls and expected margins of variation.
- Site management organization (e.g. who will be in charge of what).
- Invitation of tenders for the various work components.
- Analysis of offers and results of tenders, recommendations, and suggestions.
- Final estimation of costs and of cash flow.
- Finalization of schedule of works.
- Review and finalization of planning, which may change according to bidders' offers and proposals.

4th Stage: Execution. It comprises the following tasks:
◊ Awarding of contracts and issuing of instructions for work.
◊ Eventual modifications to drawings and execution methods in relation to proposals of the bidders.
◊ Finalization of instructions in order to avoid accidents and ensure health of all personnel working on the construction site.
◊ Establishment of all required medical facilities and personnel.
◊ On-site quality control of materials and suppliers' works.
◊ Appointment and supervision of site staff.
◊ Monitoring and reporting progress against work schedules. Re-scheduling if it proves necessary.
◊ Measurement of quantities, calculation of payments, and adjudication of claims.
◊ Finalization of expenditure against final cost estimates.
◊ Collection of data and records concerning all site works.
◊ Inspection and acceptance of partial deliveries of components of the project.
◊ Test runs and control of operating conditions.
◊ Delivery of the finished project to the infrastructure manager.

6.6. Management of infrastructure

6.6.1. Tasks and objectives for rail infrastructure

The primary tasks of rail infrastructure are:
- to ensure safe operation of rolling stock at the scheduled speed,

- to afford conditions for the highest quality of transport,
- to contribute to a sustainable development.

Principal objectives that infrastructure management should respond to are:
◊ to maintain and increase high levels of safety,
◊ to reduce costs, without however lowering safety standards,
◊ to improve organization, materials, equipment, and personnel's qualifications in order to respond more efficiently to the requirements of operation,
◊ to define and follow a policy which balances expenses to revenues.

The issue of definition of what belongs to infrastructure has been analyzed in section 3.5. Thus, management of infrastructure can refer to the following components:
− maintenance and operation of track,
− maintenance and operation of electrification equipment,
− maintenance and operation of signaling equipment,
− management of rail traffic,
− allocation of paths, when many rail operators are asking for the same slot concerning departure or arrival time,
− computation and collection of charges for the use of infrastructure, when a separation of infrastructure from operation exists.

However, there are different approaches concerning railway stations. Some countries (e.g. the United Kingdom) consider stations as a component of infrastructure. Other countries (e.g. France, Sweden, etc.) consider stations as a component of operation. Certain countries (e.g. Germany, Italy, etc.) have created an independent body in charge of the operation of stations.

6.6.2. A new management approach

Infrastructure managers and staff should get rid of old methods and ideas inherited from the past. Even if engineering aspects may be critical, they should not drive and orient infrastructure's management. Infrastructure exists to serve the operation of trains, which must be the driving factor in all kinds of decisions. Within this view, each investment or technical improvement should respond to specific goals of passenger and freight operation. Thus, the creation of a new entrepreneurial spirit is the first task. Evaluation of needs must begin from zero. Every cost component should be justified and examined in relation to what is more profitable, to be executed by in-house personnel or by external contractors (outsourcing). Even though railways were in control of everything in the past, nowadays the picture has changed drastically, (15).

Infrastructure expenses can no long stay out of control. They should be calculated in detail for each component. The long-term target should be an equilibrium between expenses and revenues, the latter originating from infrastructure charges (paid by operators), commercial activities, and state funding.

Infrastructure charging policies can have two strategic alternatives:

– high charges; this alternative is beneficial for infrastructure finances, but detrimental for operators,

– low charges; this alternative is detrimental for infrastructure finances but beneficial for operators.

The organization of personnel must also change. In the past, critical factors for promotion were qualifications and the number of years of previous service. Today and in the future, promotion is achieved in relation to skills, responsibilities, and productivity.

Infrastructure Managers must have their own Business Plans. The interaction of the various components of infrastructure is illustrated in Figure 6.6.

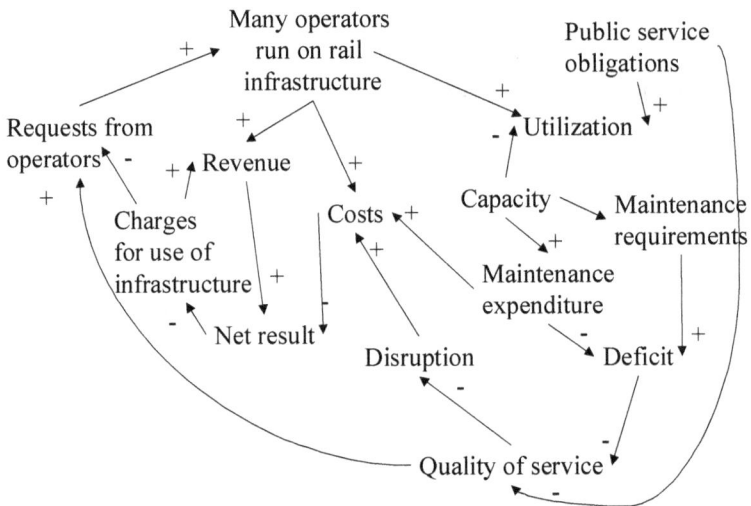

Fig. 6.6. Interaction of the various components of infrastructure, (126)

6.6.3. The issue of outsourcing

Maintenance and operation of infrastructure is a labor intensive activity. In Europe, 30%÷60% of infrastructure expenses are personnel salaries. Working conditions for maintenance are extremely difficult at the available time between successive trains (usually during the night). With one train passing per hour, efficiency of working personnel is 80%, while for two trains passing per hour, this efficiency is reduced to 50%.

Many infrastructure managers follow a policy of outsourcing, which consists in asking for an outcontractor's services (usually through a bidding procedure). Outsourcing policy is applied to activities such as the maintenance of track, tunnels, bridges, etc. Outsourcing policies have permitted high reductions of costs, as the right project is executed at the right time and at the right price. However, outsourcing should not affect safety. Infrastructure managers must impose specifications, conditions of work, and supervision for outcontractors.

6.6.4. The need for homogeneous rail products and services

Railways have been developed on many occasions to meet national targets. Thus, when crossing a frontier, rail infrastructure may have a totally different quality from one country to another. In the era of international railway cooperation, however, infrastructure managers should seek for collaboration in order to achieve a similar quality of infrastructure from origin to destination (which often means crossing many frontiers). This homogeneous rail infrastructure will require:

- a quality of track maintenance, which has as a prerequisite that track defects are of the same magnitude everywhere,
- electrification and signaling systems that permit a continuous running of trains without any interruption for technical reasons. Technical interoperability is the tool to tackle incompatibilities concerning track gauge, electric power, and signaling systems from one railway to another,
- technical facilities that give the client the impression of an homogeneous infrastructure. This applies both to passenger but also to freight (appropriate terminal and transshipment systems).

6.7. Management and policy for rail passenger transport

6.7.1. Tasks and objectives for rail passenger transport

Rail passenger transport must struggle in order to compete in a changing environment, which threatens even the existence of some railway services. The primary tasks of rail passenger transport are:

- safe transport of people at the scheduled time,
- high quality of service, which should be at least similar or even higher compared to the railway's competitors,
- contribution to regional development and to the increase of mobility for certain categories of citizens,
- increase of revenues.

Principal objectives that management of passenger transport should respond to are:
- to increase share of railways in the passenger transport market,
- to increase rail yield, that is unit revenue,
- to reduce costs in order to balance expenses and revenues. Huge deficits cannot continue to exist and in any case they should be clarified. The state can continue to finance some rail services with low revenues, within the frame of public service obligations. In the European Union, for services other than those related to public service, rail operators must succeed in finding a balance between expenses and revenues, which becomes a condition for survival, (15).

6.7.2. A segmentation of traffic

Rail passenger market can be segmented as follows:
- *intercity* traffic. It serves major population centers and customers are very demanding concerning travel times and quality of service. Railways face strong competition in intercity traffic from airplanes and buses,
- *regional* traffic. It serves regional centers, competition comes from buses and private cars, customers' exigencies are lower compared to intercity traffic; many regional rail operators receive public service obligations,
- *commuting* traffic. It serves the suburbs of a city and competition comes from buses and private cars. Commuting traffic is usually strongly subsidized by the state.

Rail managers should consider the expectations of each segment of traffic and conduct policies, which may be different from one category of traffic to another.

6.7.3. A new strategy combining competition, cooperation, and alliances

Gradual liberalization of the railway sector raises opportunities and threats. Within this environment, railway managers must think and act quite differently from what they have been accustomed to in the past.

Competition (both inter-modal and intra-modal) will be the rule. Reduction of costs and increase of quality of service are the least prerequisites to face competition efficiently.

Experiences from liberalization of other sectors of the economy (air transport, telecommunications, electricity) suggest that before expanding their activities, railways must ensure that they control an essential part of their domestic market. Otherwise, in the case that the expansion leads to failure, they risk losing even their previously ensured domestic market.

New opportunities will be given, particularly in international traffic. Railways must prepare new products (such as intercity services in international routes without delays in the frontiers), which should be differentiated from those of their competitors.

However, competitors may be allies in several occasions. Thus, rail and air transport can cooperate at least in two cases, (see also section 1.11):

- short-distance rail services from airports to city centers,
- medium- or even long-distance high-speed rail services from airports to other cities.

Railways can also cooperate with buses, which can transport passengers from railway stations to their final destination, assuring in this way a door-to-door transport, something that railways cannot offer by themselves.

Competition and cooperation will require changes in the structure, organization, and legislation of railways, which usually take time. Railway managers should be prepared for it.

6.7.4. Traditional weaknesses and offer of a new global product of railways

A railway trip is only a part of a more complex trip from origin to destination, including other transport modes such as bus, taxi, or metro, (see section 5.1.5).

Factors and weaknesses deterring citizens from using the railway have been studied in the past and are presented in Table 6.1, (80).

In order to alleviate these weaknesses, railways should try to offer a new and global product to customers by taking a number of measures in order to:

- make railway trip organization and ticketing easier,
- improve facilities in railway stations,
- improve accessibility to public transport networks, such as buses, metros, and taxis.

Thus, many railways have established and maintain web sites with useful and easy-to-obtain information, many of which also provide ticketing services. Concerning ticketing, many stations have been re-organized with the use of queuing systems, have expanded the use of information technologies, and have introduced innovative ways of issuing tickets, such as automatic tellers. To help travelers reach their final destination and increase the ticket's perceived value, some railways have extended the validity of the rail ticket for public transport, (123).

Railways must imperatively reduce distribution costs. For most rail operators, 30%÷60% of expenses have to do with staff costs; thus the introduction of information technologies can greatly improve productivity and reduce costs.

Table 6.1.
Railway usage deterring factors and the grade of discomfort
they cause (5: full satisfaction, 0: null satisfaction), (80)

Part of the trip	Problems arising	Degree of satisfaction
Departure from place of residence or work	*Trip organization*: difficulty to obtain schedule information, uncertainty regarding fares and their probable changes.	2.9
	Trip to the station: the traveler will have either to walk to areas which are usually unsafe at certain times of the day, or to use public transport, or to drive to the station by car, in which case he will be faced with the trouble of finding an empty parking space.	3.9
Departure station	*Inside the station*: ticket issuing, a probable wait in a queue, a difficulty of orientation towards platforms, unpleasant decoration and problematic ascent and descent of stairs.	2.6
	Waiting: at the platform or in the train.	2.0
	Rolling stock: unpleasant exterior and interior, poor cleaning and lack of air conditioning.	2.1
Change of train	*Change of train*: even the most effective and shortest transfer causes displeasure to the passengers, who are forced to carry their luggage and wait. In addition, as many seats are usually taken, passengers face difficulties in finding a seat, they are traveling with unknown people, and they experience a change of environment, all of which are always unpleasant	1.8
Arrival station	*Inside the station*: An environment usually unknown and unfamiliar to the passenger; it may cause doubts about whether it is the right place, while the passenger faces problems handling his luggage.	2.1
	Trip to the passenger's final destination: passengers may experience lack of orientation, a risk of getting lost, and a difficulty in catching a taxi or the appropriate public transport.	2.7
Final destination	*Waiting time*: the risk that the traveler may arrive too early or too late.	2.2

6.7.5. Application of information technologies (internet, SMS)

Some railways have modernized their distribution chain by using the internet or SMS, following a strategy of offering monetary benefits both to the railway company but also to the customers, by creating a gap in the levels of prices of tickets delivered in stations, according to the usual old way, and those delivered with the use of internet or SMS.

Thus, in Germany, prices of tickets purchased on the internet are 5÷10% lower than for the same tickets delivered in stations. In the Netherlands, a surcharge of 0.50€ (and in some cases of 1€) is to be paid if tickets are issued in stations (people with disabilities or older than 60 years do not pay this surcharge).

In some railways, the client can reserve a seat and purchase a ticket with a specific SMS message, which is sent to the appropriate railway call-center. In exchange and after having paid the amount of the ticket with a credit card, the customer receives a specific code. This code will be given to the controller on the train, who will check the validity of the code with a specific pocket computer. In addition, the client can change or cancel his/her reservation.

However, not all clients (particularly old ones) can easily be accustomed to new technologies. For this reason, and in order to avoid the clients' annoyance, it is suggested that railway authorities create reception areas in the stations, with assistants advising and helping people how to use the internet.

6.7.6. Marketing – Customer satisfaction surveys – Creation of a new culture

As explained in section 4.3.1, railway managers, in order to monitor clients' reactions and adapt railway offers to their expectations, must use customer satisfaction surveys systematically. As many marketing campaigns have shown, improving a product and promoting it is not enough; it should be integrated within a new spirit and a new culture of the railway company. Thus, railways can create a new lifestyle: an environmentally friendly transport mode, which respects clients, transports them with punctuality, safety, and security, with less stress and more comfort. Railway personnel must share the values of this new culture, (15).

Within this new spirit, some activities (such as cleaning of trains, maintenance of rolling stock, etc.) may well be outsourced.

6.8. Management and policy for rail freight transport

6.8.1. Tasks and objectives of rail freight transport

The primary tasks of rail freight transport are:
– safe transport of goods and delivery at the scheduled time without any delay or any damage to the content of the freight,
– contribution to a sustainable development, particularly by trying to reduce the noise level from freight trains, which usually operate during the night.

The principal objectives of rail freight management are, (117):
• reducing costs, while continuing to improve safety,
• as in many countries (European Union, USA, etc.) rail freight transport cannot receive public subsidies, revenues from the freight activity should balance freight costs,

- increase of punctuality and reliability. This does not mean necessarily the increase of speed of freight trains but the elimination of waiting times,
- improvement of organization and introduction of appropriate equipment in order to respond to clients' requirements.

6.8.2. A merciless competition

In addition to inter-modal competition (from road trucks), rail freight market is a field of a strong intra-modal competition between the established state railways and private entrants (in the rail freight market), which offer tariffs 10%÷50% lower than the state-owned railways for certain categories of freight.

For many decades, railways have been losing freight traffic, as their competitors were less expensive and more reliable. Even for markets for which railways have a comparative advantage, e.g. massive transport of bulk materials, the risk and the fear of strikes led some of their clients to dismiss the use of railways. In order to reverse this situation, rail freight managers should undertake a number of painful measures, such as, (117):

- ensuring the delivery of goods at the scheduled time by reimbursing clients with amounts high enough in relation to the value of rail tariffs in the case of delays,
- increasing productivity by adopting policies in which train drivers, in addition to driving, offer services of informing the client, assembling the cargo, etc., if this proves necessary,
- promoting cooperation with companies of road trucks in order to offer a door-to-door freight transport.

6.8.3. Integration of rail freight in the logistics chain

Figure 6.7 illustrates how rail freight transport (which is only a segment of a transport from origin to destination) can be integrated in the logistics chain, (see also section 1.10.6), in order to succeed on-time delivery similar to that of road transport.

With the exception of the USA, most other countries have given priority to passenger transport. Thus, freight trains are obliged to operate during the available intervals among passenger trains. To overcome this situation, rail freight managers must put pressure on infrastructure managers to separate slow (freight) from fast (passenger) traffic. For heavy freight routes, the creation of rail corridors dedicated only to freight transport may be examined.

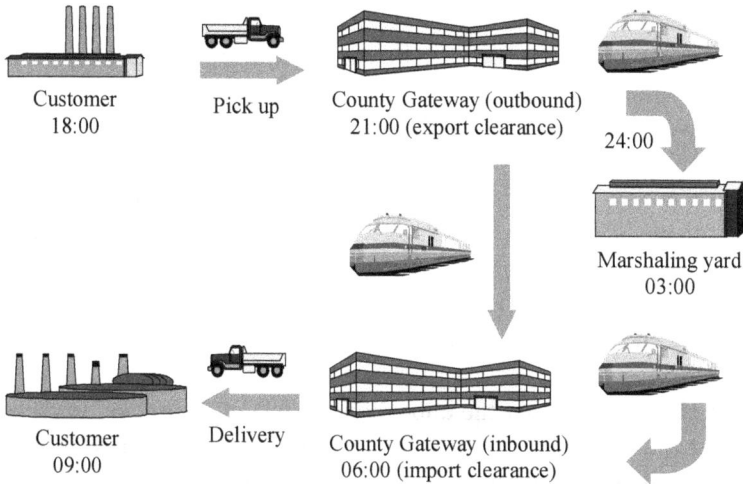

Fig. 6.7. Rail freight and the logistics chain from origin to destination, (126)

6.9. Human resources and their revalorization

6.9.1. The need for a more entrepreneurial approach

For many railways and for many decades, the number of staff, their qualifications, working conditions, and promotion were to a certain degree a result of hierarchy, political, and social considerations. As a consequence, some railways experience the disease of overmanning with unqualified staff, resulting in low productivity and a low quality product. Such a situation is not acceptable today, if railways want really to survive in an extremely competitive environment, where competitors employ people in a more rational way.

Financial constraints press railway managers to reduce costs and to maximize the use of human resources, so that additional resources are added or sought only if they can be justified by contributing significantly to an increase of output. The target should be to succeed an optimal allocation of tasks within a working organization, which attains predetermined production targets by using the minimum levels of staff and working resources at the highest work rate for the largest possible time and for the minimum cost, while respecting technical specifications. Although this is hardly accessible, it should be the target for a rational and efficient allocation based on cost, time, and resource availability, (130).

6.9.2. Allocation of human resources

In a simple business it is easy to identify needs and tasks and allocate the appropriate human resources. The bigger the business, (railways are indeed a big business), the more difficult it is to obtain an optimum allocation solution, since there are so many decisions to be made about a multitude of resources, most of which interact with each other in a complex way.

Allocation of resources can be achieved with the use of one of the following methods: past experience, guessing (not recommended), network analysis, linear programming, simulation and mathematical optimization. With the exception of methods based on experience and guessing, all other methods require computer software* that have been developed either by the railways or by operational research teams.

Concerning the level of resources allocation (national, regional) there are two approaches:

- the first one suggests that the allocation of staff at local level is scheduled in great detail with the use of computer methods, regional managers having small possibilities to alter allocations done by computers,
- the second one uses also computer calculations at national level, but grants higher authority to regional managers for an optimum allocation.

Resources allocation can be done on a short-term or on a long-term basis. A part of train crew and maintenance personnel is employed in some railways on a temporary basis, due to requirements at peak periods.

Resources allocation must take into account other components of the railway activity, such as level of technology, equipment, materials, etc. However, human activity is more and more replaced by machines. Electronic ticketing, for instance, or the use of the internet will greatly reduce the number of staff in charge of selling tickets.

The form of the employment contract depends not only on the needs of the rail company but also on the terms of labor legislation, which in recent years has become more flexible in many countries.

Selection of the appropriate human resources should account for the following: skills, experience, personal incentives, unit costs, possibility of alternative outsourcing (e.g. cleaning of stations), labor legislation, and unions' attitudes.

6.9.3. The art of motivating people to work

Motivation is the process by which staff and workforce are stimulated to work as fast and purposefully as possible. Motivation is triggered by several factors, such as: level of salary, incentives and rewards (e.g. bonuses) for higher productivity, job satisfaction, status and sense of identity, working environment, sense of purpose, opportunities for advancement, working relationships,

---

* such as: Trapeze Rail System, Productive, eResource Scheduler, Float, Hub Planner, Saviom, etc.

social and welfare facilities, performance recognition, and stability of employment. Not all the above factors have the same weight and priority for each working person, since every individual has different motivations for working at a job. However, it is generally easier to recognize symptoms of demotivation such as: time wasting, absenteeism, poor timekeeping, poor quality work, non-cooperation, decline in personal appearance, etc., (130).

Many railway managers do not pay as much attention as they should to the working environment of their staff. They forget that the highest levels of productivity are attained by people well paid, well trained, contented, confident, and equipped with the necessary equipment. Lack of interest of managers in the working environment results to demotivation on the part of their staff.

6.9.4. Increase of productivity

A principal objective of resource allocation is to increase productivity. In fact, productivity relates the traffic produced (passengers, passenger-kilometers, tonnes, tonne-kilometers) to the number of staff used to realize this traffic. Productivity can also relate the work produced to the cost of production or to the equipment (rolling stock, etc.) used to achieve this production. Thus, there are many indices of productivity.

Increase of productivity with the best use of available staff and equipment implies optimization of the following:
- *organization structure*, in order to ensure a minimum of time losses between successive activities (e.g. cleaning and checking of rolling stock, laying continuous welded rails and finishing with ballast laying and compaction, etc.),
- *planning*, in order to minimize time losses among interactive activities,
- *communication* facilities and channels,
- increased *speed* in the rates of work,
- regular *training*, particularly if new technology has been introduced,
- creation of an environment, which ensures *health, safety*, and *welfare*,
- *supervision* and *disciplinary* procedures,
- reduction of *overhead* costs (coming from additional working hours, normally unnecessary in the ordinary production procedure),
- reduction of *unit costs*, which usually incorporates all the above actions.

6.9.5. Restructuring and revalorization of human resources

Organization in many railways is characterized by inflexibility, excess of personnel (usually in routine works), lack of specialized personnel (usually in management, marketing, and operation of new technologies), and a gap between responsibilities and level of skills (a result of the lack of retraining

195

for many years). Labor restructuring and revalorization are among the first priorities of railway managers and comprise the following, (15):

- good understanding of the national and international economic environment,
- estimation of future rail transport demand,
- calculation of required staff and of level of outsourcing activities,
- estimation of excess labor and surplus assets,
- guess of political intentions of state officials to subsidize loss-making services and activities,
- network rationalization and definition of a new culture for rail services offered to clients,
- withdrawal of unprofitable activities,
- description of new services and products,
- restructuring of excess labor. Options that can be deployed are: transfer to other companies or state departments, creation of new activities, dismissal of staff, which should be the last solution,
- organizational changes, such as location of work, hierarchy position, level of responsibility, etc.,
- necessary retraining,
- estimation of cost and time required to implement changes,
- a commercial orientation of all units (including the technical ones) from the lower to the higher level of the organization of the railway activity,
- assessment of the impact of restructuring on productivity, revenues, and production costs,
- creation of a new philosophy and culture of employees, which should place the service of clients at the center of their responsibilities.

In the planning of their restructuring strategies, railway managers should not neglect that:

- institutional factors may seriously constrain the ability of the railways to respond to change. Such factors are: labor law restrictions, unions' defense of current working practices, a political environment supporting social employment policies,
- downsizing of railways is not possible, unless a strong political commitment can be assured.

6.10. Privatization of railways

6.10.1. Prerequisites and targets of privatization

The sole purpose of a private company is to make a profit. All other motivations are of secondary importance and they do not exist unless the primary condition of profit is either a reality or a strong and forthcoming expectation. In view of this, the question is how privatization can work with deeply loss-making railways.

In most cases of privatization of railways, the departure point was of ideological nature: it was suggested that railways should be considered in the same way as other sectors of the economy (telecommunications, airways, etc.), while leaving aside the particularly complex character of the railway system.

Whether ideological or not, the process of the privatization of railways has many targets:

- cut costs, reduce deficits and consequently state subsidies,
- introduce innovations and increase performances of railways,
- attract private funding for investment,
- get rid of inertia and of the political lobbying of railways, though the last is rarely said publicly.

The desire to privatize railways is not enough. A number of prerequisite conditions should be met:

- the privatized activity must generate a kind of profit,
- the privatized railway enjoys full freedom of management and operates in accordance with commercial and labor law,
- public service obligations are either abolished or properly compensated.

These prerequisites are valid whether they concern a state-owned railway or a new entrant in the market.

6.10.2. Privatization and competition

Privatization is not a condition for competition, which may well exist between state-owned railway companies (governed by the same rules as private companies) and new private entrants in the market. It is the responsibility of the state to establish clear conditions of competition in the market. But usually the state fails to do so and presents privatization as a condition of competition, something that is not true, (125).

6.10.3. The problem of debt

In few cases (e.g. Germany), the state has undertaken the accumulated debt of the state-owned railway, thus giving a greater chance for the success of a future privatization. This was, however, not the case in the privatization of Japanese railways in 1987, whose debt was more than 10% of Japan's GDP. In the privatization procedure, 40% of the debt of Japanese Railways was transferred to the three most powerful and promising of the six new railway companies (in charge of both infrastructure and passenger traffic), while the remaining 60% of the debt was assumed by a new government organization. However, the Japanese case is an exception.

6.10.4. The need for a strong Regulator

A liberalized market needs a Regulator. This is even more the case when privatization proceeds. The Regulator should assure the following:
- protect staff and passengers from any danger and ensure healthy and safe conditions of transport,
- safeguard all kinds of interests of users of railway services,
- impose on operators the minimum restrictions concerning performances (e.g. quality of services),
- take all measures for a fair and on equal basis competition,
- enable operators to exercise their activity freely in an environment of a reasonable degree of stability.

6.10.5. Privatization of infrastructure

The principal fear before privatizing rail infrastructure is that the investment choices of the private entrepreneur will be determined solely on the basis of profit and the expected return on the invested capital, while neglecting aspects of maintenance, which may have a catastrophic effect on safety. The British attempt to completely privatize rail infrastructure was finally abandoned, as the minimum standards imposed (by the state) on maintenance did not permit enough benefits, in spite of rather high infrastructure charges paid by operators.

The other extreme, that is keeping the current situation unchanged, is untenable. In most cases, rail infrastructure will remain under state control, but a significant part of its activities may be outsourced to the private sector.

6.10.6. Privatization of operation

Activities of operation that may attract the private sector are:
- high-speed services, especially over long distances or international routes,
- some local passenger services,
- some categories of freight (bulk volumes, combined transport).

Regional and urban traffic are not usually attractive for the private sector, as tariffs are low, unless they are strongly subsidized (through a competitive open bidding procedure) by the state.

A partial privatization of some activities of operation has an inherent danger, to split railway services in two categories:

– the part under private control will enjoy investment and innovation and will increase the quality of services,
– the part under state control will have less investment and modernization and its declining course risks to be further accelerated.

6.10.7. Some cases of privatization of railways all over the world

The most spectacular privatization was that of the former British railways in 1995. Rail infrastructure was assigned to *Railtrack*, a private company which due to economic problems and a series of accidents was renationalized in 2003 (creation of *Network Rail*). The responsibility of Railtrack and Network Rail was the provision of track, stations, and depots to companies wishing to operate trains. Passenger train services in the United Kingdom are provided by a number (24 in 2020) of train operating companies (called TOCs), which successfully bidded for the right to operate services in a specific area, for a period of 7 to 22 years, with an average state subsidy of 25÷40%, (127).

Though the principal driver for the privatization of the rail sector in the United Kingdom was the reduction of public spending for the railways, an ex-post analysis (Fig. 6.8) illustrates the opposite. Indeed, public subsidies to the railways are estimated to be after privatization 2÷4 times higher compared to the situation before privatization. Public spending in Fig. 6.8 includes both operating subsidies but also public spending for rail infrastructure, which increased greatly after a serious rail accident in 2000.

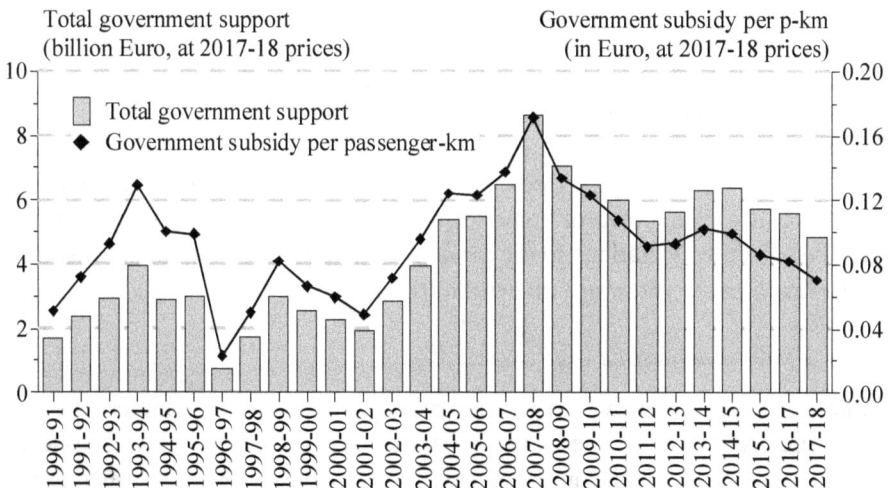

Fig. 6.8. Public spending for the railways in the United Kingdom before and after privatization in the mid-1990s, (127).

Another successful privatization was that of the former *Japanese railways* in 1987, which were taken over by 6 private companies for passenger transport (each one owning its infrastructure) and one company for freight (which pays fees to use the infrastructure of the aforementioned 6 companies).

Privatization of railways in *Australia* aimed principally at political goals: get the government out of any kind of business concerning not only railways but also airlines, ports, banks, etc. Poor financial results, low productivity, insufficient investments in the former state-owned railways facilitated greatly the route towards rail privatization. Almost all freight railways are privatized, whereas PPP schemes have been promoted to many rail passenger services and infrastructure. Globally, the rail privatization experience in Australia can be considered as positive. In the case of failures, the principal reasons were unjustified optimism, overestimated demand and revenues, and underestimation of infrastructure costs.

Privatization of railways in *New Zealand* took place in 1993, without any claim from private operators for state subsidies. However, while profits appeared in the first few years after privatization, later revenues decreased and government decided in 2008 to renationalize railway infrastructure. Railways in *Estonia* were privatized in 2001 and renationalized in 2006 at almost three times the sell price in 2001.

German Railways scheduled to ask for the participation of private capital (through the Stock Market), but the project was postponed after the financial crisis of 2008.

The situation in the *USA* is described in section 3.8. Freight operators (each one owning its infrastructure) are private companies, while Amtrak, the federal passenger rail operator, is strongly subsidized and runs on other operators' infrastructure by paying appropriate charges, (124).

Other cases of privatization of railways around the world are Argentina, Chile, Brazil, Mexico, Ivory Coast, Madagascar, Cameroon, etc. Indian railways have also plans towards some form of privatization, with the auction of a number of railway stations to the private sector and the possibility for some private operators to run trains on certain tracks.

6.10.8. Effects and degree of privatization

In almost all cases of privatization, the quality of services was increased and costs were reduced. In some cases, subsidies were reduced drastically (New Zealand, Australia), while in other cases (United Kingdom) they increased after privatization, (Fig. 6.8). Traffic also increased, but it is difficult to consider this increase as a result of privatization only. The big controversy over safety and the loss of the benefits of the integrated railway system still remains. There is evidence that after privatization some accidents were a result of an absence of synergy among the various components of the formerly integrated railway.

What has been mentioned suggests that privatization should be viewed with caution, while taking into account the particularities and the political environment of each country. There are various degrees of privatization, from full to partial, and the benefits and weaknesses should be carefully examined both for the railways on the one side and the economy and the society on the other.

6.11. Justification and calculation of public service obligations

Clarification of the economics of railways requires a justification and a calculation of public service obligations (PSO). In fact, the state must ensure for each citizen accessibility to local and national centers with at least one transport mode (preferably two). PSO are often founded on theories of regionality, which have a great range of definitions. Theories of polarity relate population centers (e.g. suburbs, villages, towns) to developed poles (industry, administration, leisure, etc.). Other theories are based on generalized cost approaches. Regionality is an inverse function of accessibility.

PSO may refer to:

♦ obligations to operate railway lines, which otherwise would be closed,
♦ obligations to transport some categories of passenger (freight is usually excluded) under certain conditions and tariffs,
♦ obligations to apply lower tariffs, which are imposed by the state.

Analytical accounts are necessary for a detailed calculation of public service obligations. The state can either impose a PSO on a rail operator or choose the lowest cost operator through an open bidding procedure. The state must compensate for each PSO the difference between the additional expenses, caused by the PSO, and the additional revenues generated by them.

In principle, PSO must be awarded through competitive procedures and usually after a competitive bidding. However, even in liberal economies, such as in EU countries, rail PSO can be awarded directly (Regulation 1370/2007) if their annual value is estimated at less than 1 million € or when they concern annually less than 300 kilometers of tracks. If in addition a rail PSO is awarded to a small or medium-size enterprise of no more than 28 rail vehicles, the above thresholds can be increased to 2 million € or 600 kilometers of tracks.

7 The Track System

7.1. The traditional division of railway topics into track, traction, and operation

For many decades, the organization of the unified railway activity has oriented railway science in an interdisciplinary approach which requires competences of the sectors of the civil engineer, the electrical and the mechanical engineer, the economist, and the manager. Thus, following railway network organization, it has become customary to distinguish railway science into three topic areas:

♦ *Track* topics. Subjects of railway infrastructure are dealt with, in order to ensure the safe operation of the rolling stock at the scheduled speed. The superstructure (rails, sleepers, fastenings, ballast or concrete slab) and the subgrade are central subjects of track topics, (Fig. 7.1). Track topics also include layout, stations, switches and crossings, maintenance and safety issues.

♦ *Traction* topics. Subjects concerning rolling stock are elaborated on. Traction topics also include electric traction, telecommunications, and signaling. Certain railways, however, include these latter in the area of track topics, since they are part of the permanent railway infrastructure.

♦ *Operation* topics, which include:
  − *Commercial* operation: marketing, commercial, and pricing policies are analyzed.
  − *Technical* operation: issues concerning schedule organization, optimum use of rolling stock, and traffic safety are examined.

To the above should be added the topics of urban railways (metros and tramways), which constitute a specific railway class of their own of great importance to mass transit in large urban centers.

However, after the separation of infrastructure from operation, track topics, electrification, telecommunications, signaling, and technical operation belong to the responsibilities of infrastructure, whereas rolling stock operation and maintenance and commercial operation belong to the responsibilities of operation. Railway stations may be studied as a component either of infrastructure or of operation, depending on the choice of where stations are belonging, (see section 3.5).

In the next chapters of this book we will deal with all the aforementioned issues, with the exception of station buildings. Differences in track characteristics from one country to another combined with the need to afford accurate specifications for each engineering structure have led international institutions, such as the International Union of Railways (UIC) and the European Commission, and national authorities of various countries to adopt specifications for each component of the railway system. The specifications that will be most used are those of UIC and the European standards, (134), (136), (140).

7.2. The track system and its components

In a railway track, (Fig. 7.1), two discrete subsystems are distinct:
- The *superstructure* (rails, sleepers, track bed (ballast, subballast)), which supports and distributes train loads and is subjected to periodic maintenance and replacement.
- The *subgrade* (formation layer, subsoil), on which the train loads, after adequate distribution in the superstructure, are transferred and which in principle should not be subjected to interventions during periodic maintenance of the railway track.

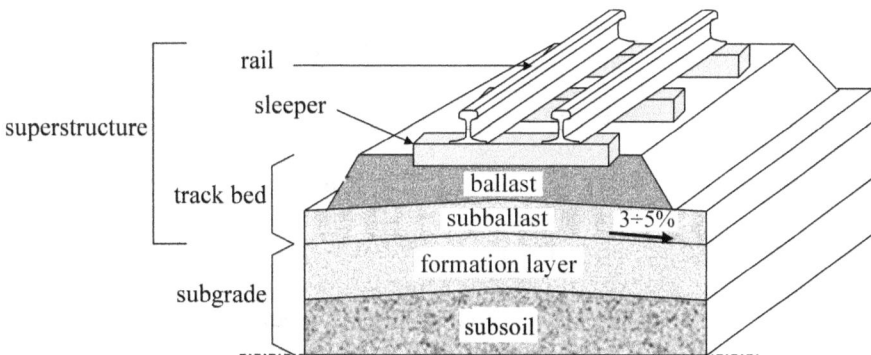

Fig. 7.1. The track (superstructure – subgrade) system

The superstructure is composed of:
- The *rails*, which support and guide the wheels of the train by providing a smooth pathway with little friction between wheels and rails. Loads of the rolling stock are successively transmitted to the rails through axles (mounted often on the same frame, called bogie) and wheels.
- The *sleepers* (also called *ties*, principally in North America) with their fastenings, which keep the rails at a constant spacing, distribute the loads

transmitted from the rails, and give the track the required transverse and longitudinal stability.

♦ The *ballast*, which consists usually of crushed stone and only in exceptional cases of gravel. The ballast should ensure damping of most of the train vibrations, adequate load distribution, fast drainage of rainwater, and a sufficient transverse and longitudinal stability of the track.

♦ The *subballast*, which consists of gravel and exceptionally of sand. The subballast protects the subgrade top from the penetration of ballast stones, while at the same time further distributing external loads and ensuring the quick drainage of rainwater.

In the subgrade the following are distinguished:

– The *subsoil*, which in the case of track laid along a cut consists of on-site soil, while in the case of track laid along an embankment is composed of soil transported to the site.

– The *formation layer*, used on top of the subgrade whenever the subsoil material is not of appropriate quality. Permeable fabrics, called *geotextiles*, are often interposed between the subballast and the formation layer to separate soil from gravel, reinforce soil strength, filter and drain water.

The design of the track system (choice of materials, dimensioning) should ensure safety, passenger comfort, rational construction and operation cost, and the least possible effects to the environment (air pollution, noise, ground vibrations, etc.).

The depth to which mechanical effects resulting from train circulation occur extends to around 2 m below the subgrade top, and this is the depth down to which will henceforth be referred to by the term subgrade, (148).

Resilient pads are placed between rail and sleeper to further attenuate train vibrations, (Fig. 7.2.a). Thicknesses of pads are usually between $5 \div 10$ mm. Elastic pads are composed of some kind of elastic material (rubber, etc.) and in addition to attenuating train vibrations they provide some insulation between rail-sleeper and contribute to a more uniform distribution of external loads. The rail is connected on the sleepers with the help of *fastenings*.

In recently constructed or renewed tracks, however, a baseplate is placed between rail and sleeper, (Fig. 7.2.b). In this case, resilient pads are placed between rail and baseplate and between baseplate and sleeper.

The succession of the various layers of the track system is characterized by a gradual increase of the surface area as we proceed to lower layers and by a considerable reduction of the developed stresses, (Fig. 7.3). We take into account a wheel load of 10t and a track without defects and irregularities. The contact surface between wheel and rail is around 1.3 cm^2, (see section 7.7, Fig. 7.8). As will be explained in section 8.4.8, when a wheel load is applied on a sleeper, the sleeper under load supports 40% of the applied load (against 50% of older theories). Thus, beneath the sleeper, 40% of the applied load will be transmitted, (146). Accordingly, stresses are reduced by more than 15,000 times between the point where the wheel load is applied and the subgrade, (Fig. 7.3). In this preliminary analysis, dynamic effects and track defects, (see section 8.7), have not been taken into account, (152).

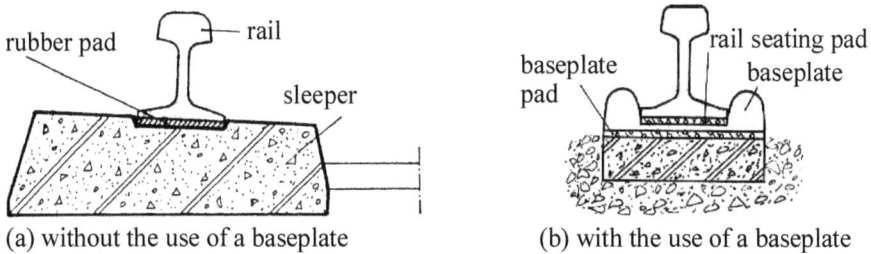

(a) without the use of a baseplate   (b) with the use of a baseplate

Fig. 7.2. Resilient pads between rail and sleeper

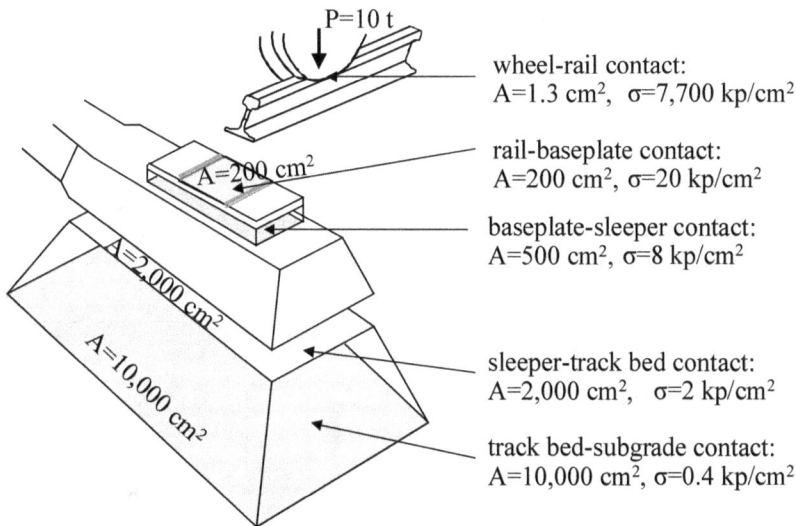

wheel-rail contact:
$A=1.3$ cm^2, $\sigma=7,700$ kp/cm^2

rail-baseplate contact:
$A=200$ cm^2, $\sigma=20$ kp/cm^2

baseplate-sleeper contact:
$A=500$ cm^2, $\sigma=8$ kp/cm^2

sleeper-track bed contact:
$A=2,000$ cm^2, $\sigma=2$ kp/cm^2

track bed-subgrade contact:
$A=10,000$ cm^2, $\sigma=0.4$ kp/cm^2

Fig. 7.3. The base area (A) of each component of the track system and the distribution of train load.

7.3. Track on ballast or on concrete slab

The track usually lies on ballast, in which case we have a flexible support or a ballasted track, (Fig. 7.4.a). However, it is possible that the track lies on a concrete slab, instead of ballast, in which case we have a non-ballasted or slab track, (Fig. 7.4.b). Although a slab track is used in certain countries (e.g. Japan and Germany, among others), it is most effective when employed in tunnels, as it allows a smaller cross-section and facilitates maintenance. In most of the tracks worldwide, a ballasted track is still the case, as it ensures flexibility (an important factor in the event of differential settlements) and much lower construction cost, while at the same time offering a very satisfactory transverse resistance, even at high speeds, (148), (151), (153). The problem of noise, which is much greater with the track on concrete slab than with the track on ballast, should not be disregarded. When a slab track is applied (e.g. in the case of a tunnel), the sudden variation in track stiffness between ballasted and non-ballasted track (felt by passengers as a jolt) is lessened by placing rubber pads of a suitable thickness along the tunnel entrance and exit.

The choice between ballasted and non-ballasted track should be done in relation to construction cost (much greater for non-ballasted track), maintenance cost (much greater for ballasted track), technical requirements (both solutions have advantages and disadvantages), taking into account the level of technological performance and labor cost for each case, (139). Slab track is examined in more detail in chapter 17.

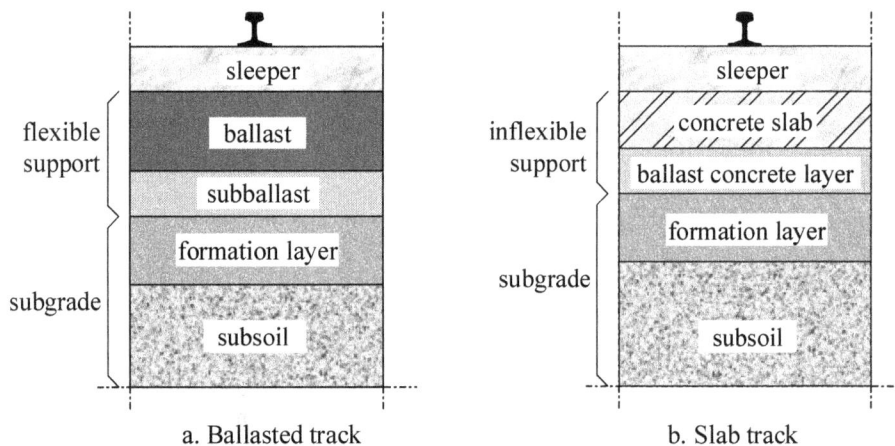

a. Ballasted track   b. Slab track

Fig. 7.4. Ballasted and non-ballasted (slab) track

7.4. Track gauge

The track gauge e is defined as the distance between the inner sides of the heads of the two rails, measured 14 mm below the rolling surface, (Fig. 7.5). Tracks with different gauge values have been laid, as follows:

♦ *Standard* gauge, e=1.435 m. Most lines all over the world have been laid at this standard gauge, which has been found to be the optimal distance of rails for the usual rolling stock dimensions.

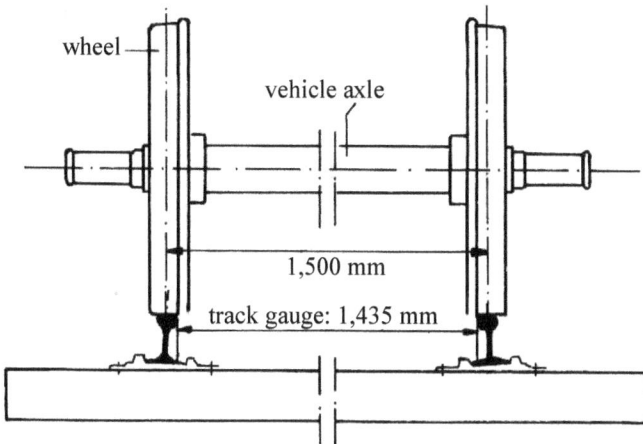

Fig. 7.5. Track gauge (case of a standard gauge track)

♦ *Metric* gauge, e=1.000 m (Spain, Switzerland, Brazil, Argentina, Chile, Bangladesh, Thailand) or e=1.067 m (Japan, Taiwan, Australia, New Zealand, Indonesia, Nigeria). In most cases, secondary lines are laid using the metric gauge. However, metric gauge lines in some railways (Japan, India, South Africa, Australia, New Zealand, South America, and others) operate as principal lines at speeds up to 160 km/h and can support axle loads up to 16÷18 t. The metric gauge was chosen based on the assumption that construction costs would be highly reduced, (138), (140).

♦ *Broad* gauge, e=1.520 m (Russia, Ukraine, Azerbaijan, Baltic countries), 1.524 m (Finland), 1.600 m (Brazil, Ireland), 1.668 m (Spain, Portugal), 1.676 m (Pakistan, India, Bangladesh, Sri Lanka, Argentina, Chile). Though rolling stock running on broad gauge tracks has no essential differences with rolling stock running on standard gauge tracks, broad gauge was chosen to provide a better stability of track and optimal operation conditions of steam engines of the time. Political reasons prevailed also, so that vehicles running on standard gauge tracks cannot trespass into broad gauge tracks.

♦ *Narrow* gauge, e=0.914 m or e=0.760 m.

It should be noted that gauge values had initially been expressed in British measurement units (inches), hence the general irregularity of the above numerical values by their conversion into metric units.

In a total of more than 1.3 million kilometers of railway lines worldwide, 54.9% are laid on the standard gauge, 29.3% on the broad gauge, and 15.8% on the metric gauge. There are few narrow gauge tracks in operation (principally for touristic reasons) and represent less than 0.5% of the total kilometrage of other railway lines.

A delicate issue for trains running on curves with small radii of curvature is the easy passing of the wheels over the rails. To facilitate it, track widening is provided in tight curves. Thus, on curves with a radius R<400 m track widening is provided up to 20 mm for standard gauge tracks on timber or steel sleepers, up to 10 mm for tracks on twin-block reinforced-concrete sleepers and up to 5 mm for tracks on monoblock prestressed-concrete sleepers. For metric gauge tracks, track widening is provided for radii R<500 m and can take values up to 20 mm, (136), (140).

Small tolerances may be accepted between nominal values of track gauge and actual values and are detailed in the relevant specifications. According to the European Technical Specifications for Interoperability (TSI), actual values of the track gauge in relation to train speed must be within the limit values illustrated in Table 7.1.

Table 7.1.
Limit values of track gauge in relation to train speed for standard gauge tracks according to the TSI, (134)

Speed (km/h)	Minimum value of track gauge (mm)	Maximum value of track gauge (mm)
V ≤ 120	1,426	1,470
120 < V ≤ 160	1,427	1,470
160 < V ≤ 230	1,428	1,463
V > 230	1,430	1,463

7.5. Axle load and traffic load

7.5.1. Axle load

The axle load and the traffic load (tonnage) running on the line are critical factors for stresses and fatigue of the materials of superstructure and subgrade of the track. Permitted values of axle load depend principally on track equipment and more particularly on rail, sleeper, and ballast characteristics. Depending on track equipment, different values of axle load may be applied. For *standard* gauge tracks, axle loads have been standardized and classified by UIC into four categories:

    A : Maximum axle load 16 t,
    B : Maximum axle load 18 t,
    C : Maximum axle load 20 t,
    D : Maximum axle load 22.5 t.

Category D was derived by increasing the axle load of category C from 20t to 22.5t, in an effort to reduce the operating cost, especially for freight traffic. This increase was made after years of research and studies, with controversy which did not focus as much on the strength of track (whose components are replaced at regular intervals) as on the behavior of bridges which had been designed for a 20t axle load on the basis of simplified theories of elastic behavior, (145). Research on the elastoplastic behavior of materials, (see section 8.4.4), has shown that bridges designed for axle loads of 20t can withstand axle loads of 22.5t without the need for strengthening, due to strength reserves which the elastic theory did not take into account and left aside, (145).

Railway axle loads for standard gauge tracks were only 10t in 1850 and progressively increased to 12t in 1880, 14t in 1900, 20t in 1930 and 22.5t in the 1980s.

Certain railways with standard gauge tracks, however, use larger axle loads. In the USA (where railways are mainly used for freight transport) the maximum axle load for standard gauge tracks is 25÷32t.

Axle load for *broad* gauge tracks is usually 25t. For *metric* gauge tracks, axle loads are up to 14÷16t (some metric gauge tracks can support axle loads up to 16÷18t), (136), (140).

A series of research has shown that rail fatigue is an exponential function of the axle load Q, and stresses developed within the rail are proportional to the parameter Q^a, where the exponent a takes values in the range of 3 to 4 and closer to 4, (152). Thus, any increase in the axle load results in a much larger increase in stresses of the rail and the other track materials.

7.5.2. Traffic load

Various kinds of rail vehicles are running on a track: passenger vehicles, freight vehicles, locomotives. The algebraic sum of the vehicle loads cannot give an accurate quantification of the running load, because it does not take into account the way in which the load is applied, the running speed, etc. Therefore, a parameter giving an accurate estimate of the passing traffic load is necessary. Railway engineering uses the analogue of the passenger car unit (called also passenger car equivalent) of traffic engineering. In order to determine the traffic load (or tonnage) on a track, the loads of the various trains are first converted into equivalent passenger train loads and then speeds are also taken into account.

For this purpose, a composite rail traffic value is calculated, taking into account both the effects of speed and the relative wear provoked by axle loads. Railway line classification has been standardized by the UIC (Code 714R) and is determined on the basis of a theoretical traffic load T_{th} given by the following formula, (143):

$$T_{th} = S_p \cdot (T_p + k_t \cdot T_{lp}) + S_{fr} \cdot (k_{fr} \cdot T_{fr} + k_t \cdot T_{lf}) \qquad (7.1)$$

where: T_p : the mean daily passenger tonnage hauled (in gross tonnes),

T_{fr} : the mean daily freight tonnage hauled (in gross tonnes),

T_{lp}: the mean daily tonnage of locomotives used in passenger traffic (in tonnes),

T_{lf} : the mean daily tonnage of locomotives used in freight traffic (in tonnes),

k_{fr} : a coefficient taking into account effects of both the load and the wear provoked by freight vehicles and is given, (143):

- normally the value $k_{fr} = 1.15$,
- however, for tracks handling heavy loads, coefficient k_{fr} is given the following greater values:
  - $k_{fr} = 1.30$ for traffic based principally on 20t axle loads (more than 50% of traffic) or for a significant proportion of traffic with 22.5t axle loads (more than 25% of traffic),
  - $k_{fr} = 1.45$ for traffic based principally on 22.5t axle loads (more than 50% of traffic) or for traffic largely consisting of 20t or heavier axle loads (more than 75% of traffic),

k_t : a coefficient which allows to take into account wear resulting from traction locomotives. The coefficient k_t is usually given the value $k_t = 1.40$,

S_p and S_{fr}: coefficients related to the running speed of the train. More particularly, S_p relates to the speed of the fastest passenger trains and S_{fr} relates to the speed of ordinary freight trains. These coefficients are assigned the following values, (143):

S_p, S_{fr} $= 1.00$ for $V \le 60 \mathrm{km/h}$,

$= 1.05$ for   60 km/h $< V \le 80 \mathrm{km/h}$,

$= 1.15$ for   80 km/h $< V \le 100 \mathrm{km/h}$,

$= 1.25$ for 100 km/h $< V \le 130 \mathrm{km/h}$,

$= 1.35$ for 130 km/h $< V \le 160 \mathrm{km/h}$,

$= 1.40$ for 160 km/h $< V \le 200 \mathrm{km/h}$,

$= 1.45$ for 200 km/h $< V \le 250 \mathrm{km/h}$,

$= 1.50$ for $V > 250 \mathrm{km/h}$.

Based on the daily traffic load, the various railway lines are classified, according to the UIC (Code 714R), into 6 groups. The former classification of the UIC (valid until 1989) included 9 groups. As a number of railway authorities still apply the old classification of UIC, Table 7.2 and Figure 7.6 illustrate the new classification (into 6 groups) and the old one (into 9 groups), (143).

Table 7.2.
The new and old classification of railway lines according to the UIC in various groups in relation to the daily traffic load T_{th} of the line, (143)

New classification		Old classification	
Daily traffic T_{th} (thousands of tonnes/day)	Group UIC	Daily traffic T_{th} (thousands of tonnes/day)	Group UIC
$T_{th} > 130$	1	$T_{th} > 120$	1
$80 < T_{th} \le 130$	2	$85 < T_{th} \le 120$	2
$40 < T_{th} \le 80$	3	$50 < T_{th} \le 85$	3
$20 < T_{th} \le 40$	4	$28 < T_{th} \le 50$	4
$5 < T_{th} \le 20$	5	$14 < T_{th} \le 28$	5
$T_{th} \le 5$	6	$7 < T_{th} \le 14$	6
		$3.5 < T_{th} \le 7$	7
		$1.5 < T_{th} \le 3.5$	8
		$T_{th} \le 1.5$	9

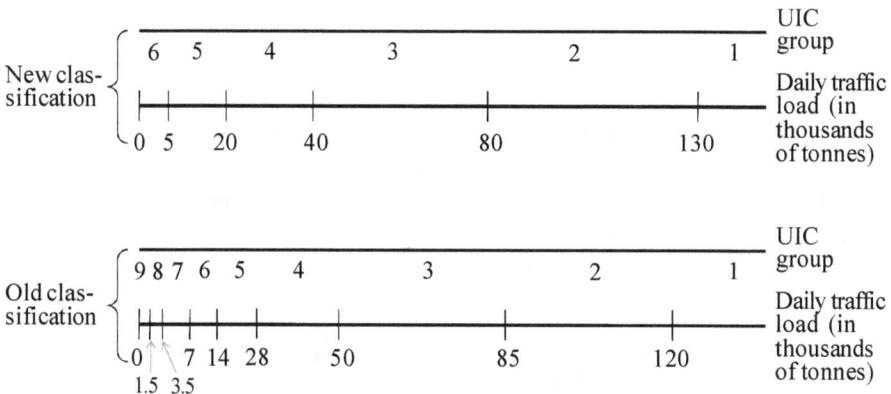

Fig. 7.6. Classification (new and old) of railway lines into UIC groups in relation to the daily traffic load T_{th}, (143)

7.6. Sleeper spacing

The study of track behavior has shown that the closer the sleepers are spaced, the better the load distribution and the smaller the stresses developed. As sleeper spacing is made smaller, however, track maintenance becomes more difficult. A compromise should therefore be found between the above two requirements.

Sleeper *spacing* is defined as the distance between the axes of consecutive sleepers and its optimum value for standard gauge tracks is 0.60 m, which can be reduced to 0.55 m in cases of subgrade inadequacy and small radius of curvature. Accepted tolerances of sleeper spacing during construction of the track are ± 0.02 m. Occasionally the number of sleepers per kilometer is used as a parameter, with 1,666 sleepers per kilometer of track as the average value. In railways with higher values of axle load (e.g. the USA), sleeper spacing may be reduced to 0.50 m. On lightweight railways, sleeper spacing may be increased, but rail fatigue must be carefully considered. For fishplated tracks, (see section 10.12), that is for tracks with joints in the rails, sleepers before and after the joint are spaced half the ordinary distance between them.

Distribution of train loads in the track is different for the various types of sleepers. Effects of sleeper spacing on the stresses developed within the track can be calculated with the help of models of mechanical behavior of the track, (see section 8.4), in relation to values of axle load, speed, type and strength of rail, type and strength of ballast, type and strength of sleeper, and the accepted limit values of stress for the soil of the subgrade. In order to keep similar values of stresses along the track for the various types of sleepers, some railways apply greater spacing to concrete sleepers than to timber ones.

7.7. The wheel-rail contact

A fundamental characteristic of rail vehicles is that the wheel movement is guided by the two rails. Wheel-rail contact, (Fig. 7.7), has an elliptical form, (Fig. 7.8). The rail axis inclination to the vertical is termed conical tread γ and has the value 1/20 (e.g. French railways) or 1/40 (e.g. German railways, Japanese high-speed tracks), (149).

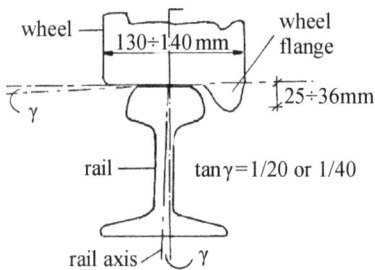

Fig. 7.7. The wheel-rail contact

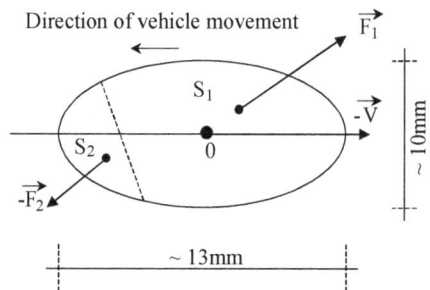

Fig. 7.8. Detail of the wheel-rail contact surface

Wheel movement on the rail gives rise to the creep effect. Indeed, the wheel-rail contact surface can be divided into two areas, S_1 and S_2, the sizes of which depend on the vehicle speed, (147). Thus, the vehicle rolling resistance consists of two components, F_1 and F_2, corresponding to areas S_1 and S_2 respectively and of opposite direction. Force F_1 is generated by vehicle movement, (i.e. it is of kinematic origin), while force F_2 is generated by elastic deformation of the S_2 surface, (i.e. it is of elastic origin).

As speed increases, S_1 becomes larger and S_2 smaller. At high speeds, S_2 almost decreases to zero.

A better approximation of the physical phenomena between wheel and rail considers that the elliptical contact surface may be divided into two sections, (154):

♦ The first section of the contact surface undergoes creeping and each point of the first section transmits to the second section of the contact surface a transverse force given by Coulomb's equation.
♦ The second section of the contact surface transmits to the first one a force with a value lower than that given by Coulomb's equation.

More accurate and analytical methods, such as the finite element method, permit to study more in detail phenomena in the wheel-rail contact surface, (133), (155).

Railways use almost exclusively metal wheels. Rubber wheels started being used after 1970 in metropolitan railways and tramways to reduce vibrations transmitted to the environment and increase permitted values of acceleration and deceleration. Rubber wheels do not permit increased speeds and are subject to deterioration under bad weather conditions. For this reason, they are used principally in metro vehicles.

7.8. Transverse wheel oscillations along the rail

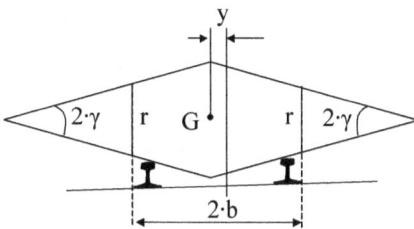

Fig. 7.9. Simulation of a rail vehicle by a solid composed of two cones

A rail vehicle can be simulated by a solid composed of two cones connected at their base and supported by the two rails, (Fig. 7.9). The wheel conical tread γ has a value of 1/20 or 1/40.

Due to the conical tread, the wheel follows a sinuous path along the rail, (Fig. 7.10).

The gap between the rail head and the wheel allows the latter to move transversely, a fact that causes the sinuous movement of the rail vehicle. Transverse wheel movements are opposed by creep forces.

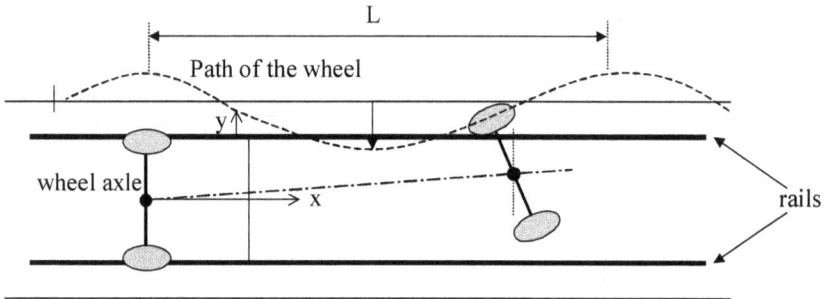

Fig. 7.10. Path of the wheels along the track

Analysis of transverse movements of a rail vehicle can be done by assuming a sinusoidal transverse movement with no attenuation. Let, (Fig. 7.11), (147):

y : the transverse movement from the equilibrium position,
v : the train speed,
s : the track gauge,
γ : the wheel conical tread,
R : the radius of curvature of the sinusoidal movement,
r : the wheel radius,
x : the abscissa.

From Figure 7.11 and the similar triangle relationship, it follows that:

$$\frac{r + \gamma y}{r - \gamma y} = \frac{R + s/2}{R - s/2} \tag{7.7}$$

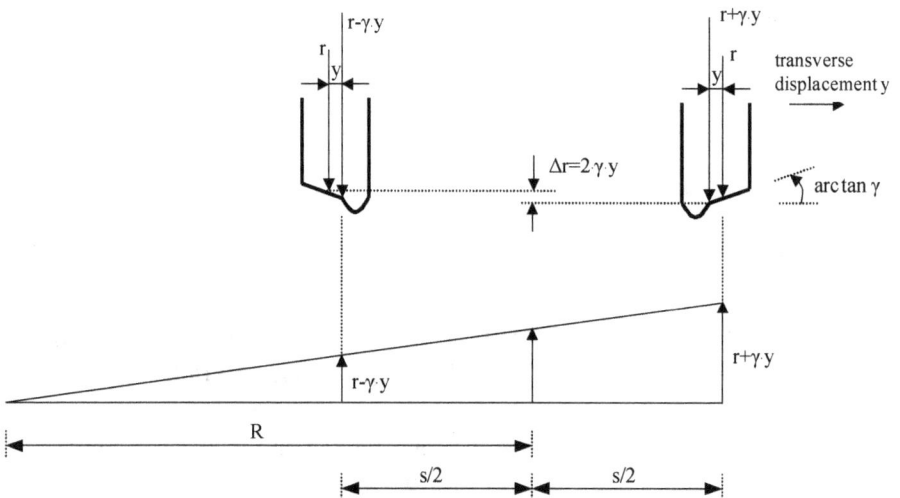

Fig. 7.11. Analysis of transverse wheel movement

The relationship between y and R is deduced from kinematics as follows:

$$\frac{1}{R} = -\frac{d^2y}{dx^2}$$ (7.8)

From equations (7.7) and (7.8) we deduce the differential equation for the sinusoidal wheel movement:

$$\frac{d^2y}{dx^2} + \frac{2\gamma}{rs}y = 0$$ (7.9)

Given the limit condition

$$y(0) = 0$$ (7.10)

the solution for the differential equation becomes:

$$y = y_0 \sin 2\pi\frac{x}{L}$$ (7.11)

with y_0 the amplitude and L the wavelength,

$$L = 2\pi\sqrt{\frac{rs}{2\gamma}}$$ (7.12)

The maximum value of the transverse acceleration is:

$$\gamma_{max} = \frac{d^2y_{max}}{dx^2} = 4\pi^2 y_0 \frac{v^2}{L^2}$$ (7.13)

As a numerical example, let r=0.45m, s=1.435m, γ=1/20, in which case L=15.96m. If, however, γ=1/40, then L=22.57m.

The frequency of the sinusoidal wheel movement can be found from the equation:

$$f = \frac{v}{L}$$ (7.14)

When frequency f is the same as the frequency at which the rolling stock resonates, then wheel movement becomes instable. The transverse acceleration, which is a measure of the forces exerted, shows the opposing effects generated by increasing the speed and decreasing the transverse movement wavelength L. A conical tread of 1/40 instead of 1/20 is therefore more advantageous concerning wheel movement at the same speed. Conversely, as the wheels gradually wear off, conical tread increases and as a result wavelength L of transverse movement decreases.

However, in modern rail vehicles the rolling stock body is not supported directly by the wheel axles but by bogies, which are in turn supported by the axles. Therefore, the movement of rolling stock on bogies is clearly more complex than described above. The related analysis is given in section 19.4.

7.9. Rail inclination on sleeper

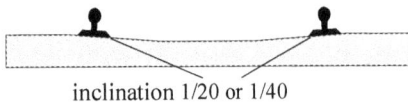

inclination 1/20 or 1/40

Fig. 7.12. Rail inclination on sleeper

Due to the conical tread, rails are mounted on sleepers at an inclination. As explained previously, the conical tread is given in some railways the value 1/20. A reduction of the value of the conical tread has been suggested, however, especially at high speeds. Several railways are already mounting the rails on the sleepers at an inclination of 1/40, (149). According to the TSI, rail inclination on the sleeper should be in the range of 1/20÷1/40, (134).

7.10. Loading gauge

7.10.1. Static and dynamic loading gauge

The *loading gauge* is defined as the minimum external border required to remain free around the rolling stock. The loading gauge is distinguished in:

♦ *static* loading gauge, which is the minimum external border required to remain free from obstacles while the train is not moving. It should take into account all obstacle structures, such as power supply and signaling equipment along the track,

♦ *dynamic* (called also *kinematic*) loading gauge, which is the minimum external border required to remain free from obstacles, while the train is moving. The boundary enclosing the clear spaces required around the dynamic loading gauge is the *structure* gauge. The difference between the structure gauge and the loading gauge is called the *clearance* and depends on the speed of the train and whether the track is on a straight line or on a curve.

The loading gauge mainly depends on two parameters:
- the rolling stock width (usually between 2.60÷3.30 m),
- the spacing b between the axes of the two tracks (usually between 3.60÷4.80 m).

7.10.2. European, British, and American loading gauge

The International Union of Railways has specified the loading gauge, which is required to ensure that trains from one network can run on the tracks of another network without any problems, (Fig. 7.13). The distance b between the axes of the two tracks varies for speeds V<200 km/h between 3.57 m and 3.67 m for French railways and between 3.75 m and 4 m for German railways, (142). Even with the UIC standardization, however, significant differences in the loading gauge are observed for standard gauge tracks, mainly in the United Kingdom, (Fig. 7.14), where the loading gauge has smaller dimensions than in continental Europe, (141). American loading gauge, (Fig. 7.15), also has significant geometrical differences compared to the European ones, (136). Table 7.3 illustrates the required minimum distance b between the axes of tracks in relation to train speed and track gauge (standard gauge and broad gauge tracks) according to the TSI, (134).

Table 7.3.
Distance b(m) between the axes of tracks in relation to train speed and track gauge according to the TSI, (134)

Train speed V (km/h)	Distance b(m) between the axes of tracks for		
	Standard gauge (1.435 m)	Broad gauge (1.520 m)	Broad gauge (1.668 m)
V ≤ 160		4.10	
160 < V ≤ 200	3.80	4.30	3.92
200 < V ≤ 250	4.00	4.50	4.00
250 < V ≤ 300	4.20	4.70	4.30
V > 300	4.50	4.70	4.50

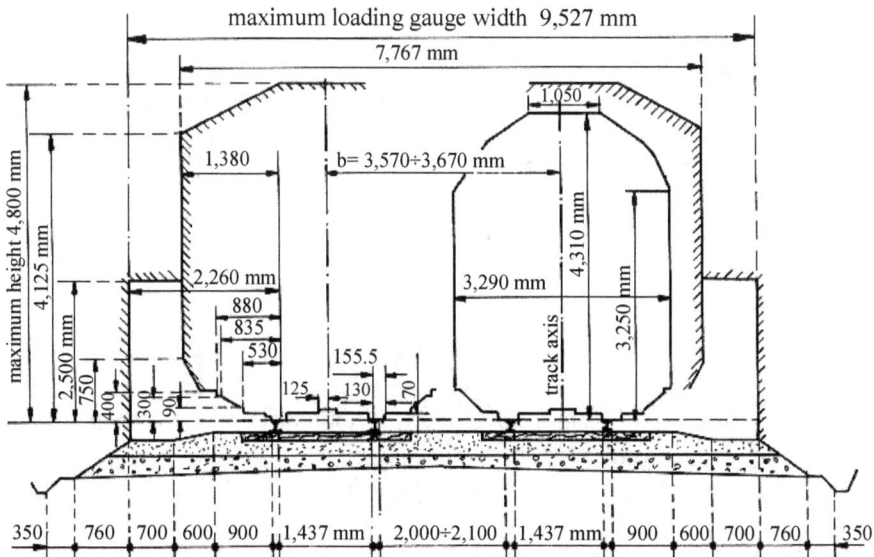

Fig. 7.13. Medium- and low-speed loading gauge, (136)

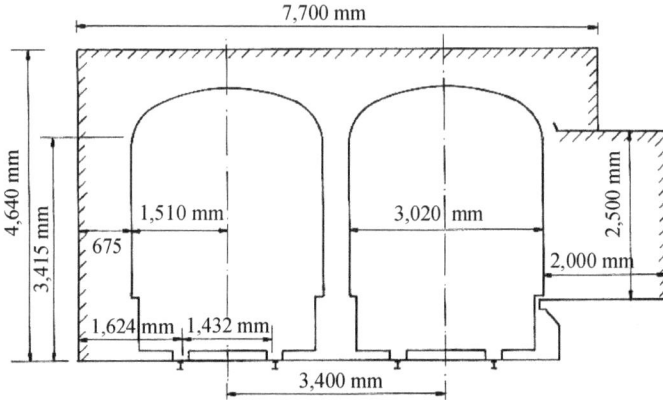

Fig. 7.14. British loading gauge, (141)

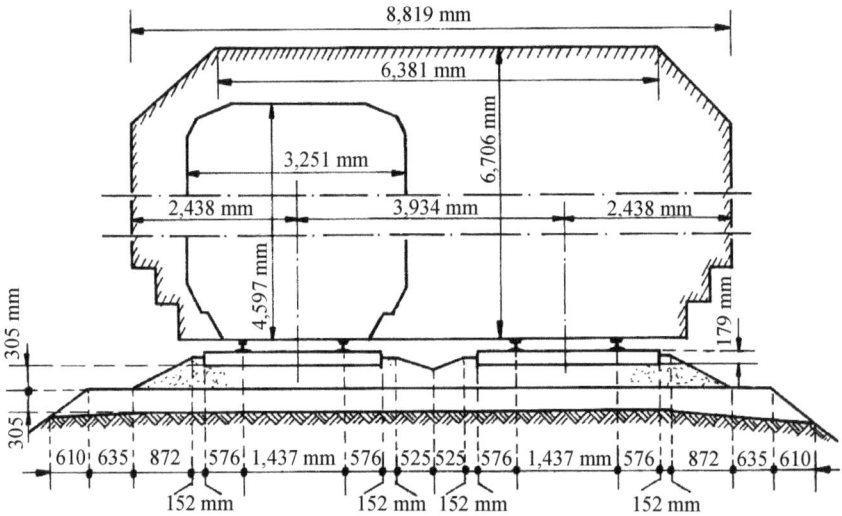

Fig. 7.15. American loading gauge, (147)

7.10.3. Loading gauge for high-speed tracks

The loading gauge is different for high-speed tracks, mainly because of the large spacing b necessary between the axes of the two tracks as well as the larger lateral distances (during movement) in both sides of the rolling stock. Thus, for high-speed tracks, the distance b is:

- b=4.20 m in the case of French railways, with V_{max}: 300 km/h (line Paris– Lyon),
- b=4.70 m in the case of German railways, with V_{max}: 300 km/h. A reason of this greater value of b in German railways is the existence of many tunnels,

218

- b=4.30 m in the case of Japanese railways, with V_{max}: 320 km/h,
- b=4.80 m in the case of French railways, with V_{max}: 350 km/h (line Lyon–Marseille),
- b=4.00 m in the case of Italian railways, with V_{max}: 250 km/h,
- b=5.00 m in the case of Chinese railways, with $V_{max} > 300$ km/h,
- b=5.30 m in the case of Indian railways, with V_{max}: 320 km/h.

According to the TSI, the minimum distance between the axes of tracks either specifically built or upgraded for high speeds should be 4.00 m for speeds between 200÷250 km/h, 4.20 m for speeds between 250÷300 km/h, and 4.50 m for speeds greater than 300 km/h, (134).

7.10.4. Loading gauge for metro systems

The dynamic loading gauge requires special attention when trains are running through tunnels, as well as in the case of metropolitan railways, (Fig. 7.16). Each railway and metro authority must have its own loading and structure gauge requirements, which must be followed in each specific case.

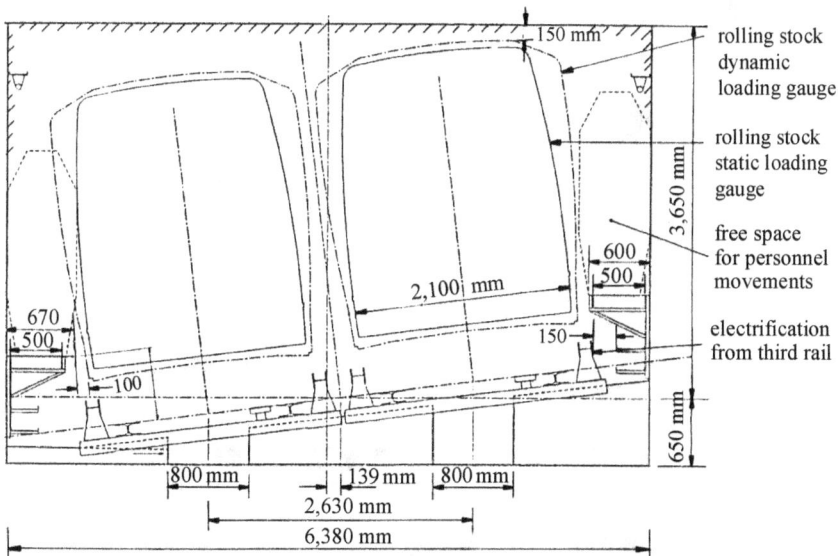

Fig. 7.16. Dynamic and static loading gauge of a metro (with narrow rolling stock) on curved track, (144)

7.10.5. Loading gauge for metric gauge tracks

Figure 7.17 illustrates the rolling stock outlines for some metric gauge railways, (140). It is recommended that when fixed structures are being installed, they should be 250 mm outside the outermost of all of the rolling stock outlines illustrated in Figure 7.17.

Concerning dynamic loading gauge, it should allow a lateral movement of the vehicle of ±43 mm and a rotation of the vehicle of ±2.00 degrees around a roll center that is situated 330 mm above the rail level, (138), (140).

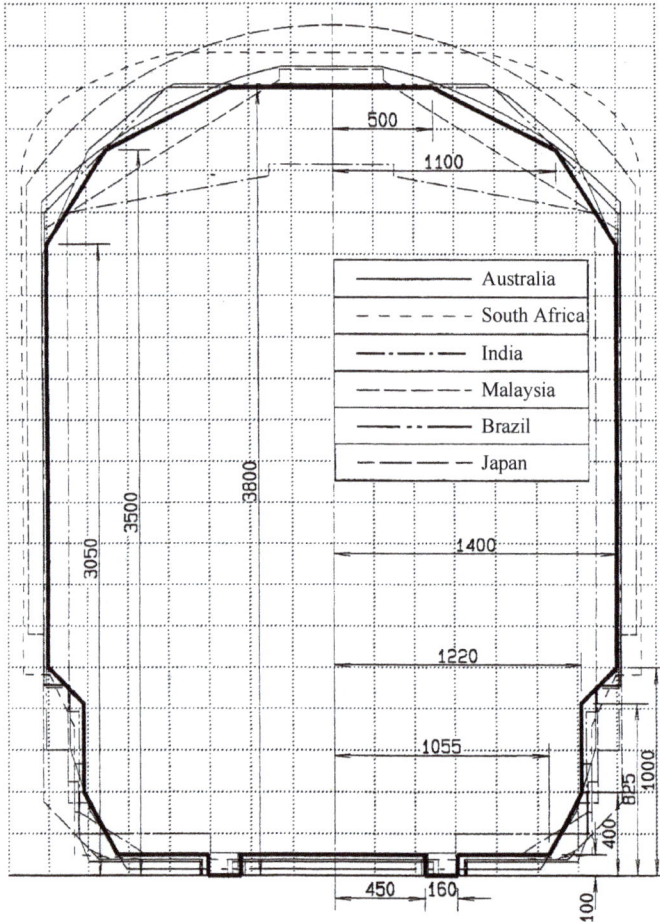

Fig. 7.17. Rolling stock outline for various metric gauge railways all over the world, (140)

7.11. Forces generated by the movement of a rail vehicle – Static and dynamic analysis

7.11.1. Forces generated

Forces exerted on the track during the running of a rail vehicle may be classified, depending on their direction, as follows:

– *Vertical* forces, which are the principal cause of the mechanical stresses in the track. When subjected to vertical forces, the behavior of certain parts of the track (rails, sleepers) is elastic, while that of the ballast and the subgrade is elastoplastic, (148). Vertical forces are critical to the dimensioning of the various components of the track system.

– *Transverse* (lateral) forces, which influence train safety and may, under certain conditions, cause train derailment. The effects of transverse forces are analyzed in chapter 13.

– *Longitudinal* forces, which may have as origin:
  ♦ braking or acceleration of the rail vehicle. According to the TSI, the track must be designed to withstand longitudinal forces equivalent to the force arising from the extreme braking of a value of 2.5 m/sec², (134),
  ♦ changes in the length of continuous welded rails, due to temperature changes. The problem is discussed in detail in section 10.13,
  ♦ creep of the track, (see section 11.9.5).

Although an accurate analysis of the various phenomena has shown a nonlinear behavior, the inaccuracy introduced by the omission of the nonlinearity is often smaller than the inaccuracy introduced by other parameters, e.g. the values of the mechanical characteristics, (148), (152). It is common practice in railway engineering to analyze separately the effects of vertical, transverse, and longitudinal phenomena, generated during train motion, and then sum up the values of stresses and settlements calculated separately. Such an approach is called superposition, which however implies that the phenomena studied are assumed to be linear. It is an approximation, which the engineer must be aware of in the analysis of the various effects. The principle of superposition can be written as:

$$f(a+b)=f(a)+f(b) \qquad (7.15)$$

where f is the effect and a, b are the external forces.

7.11.2. Static and dynamic analysis – Track defects and additional dynamic loads

A frequent assumption in railway engineering is that both the wheel and the rail are free of defects and that metal-to-metal contact of wheel to rail is smooth. Measurements of the stresses have furthermore shown that the influence of time may be considered as negligible in most cases. In such conditions, a static analysis of the various effects is adequate, (137).

In both the wheel and the rail, however, defects do occur, (see section 16.4), causing additional dynamic loads to the wheel-rail system. These additional dynamic loads increase rapidly as train speed increases. Force measurements have shown that for wheel loads of 10 tonnes and speeds of 200 km/h, the additional dynamic loads may attain values up to 4÷6 tonnes, (152). Therefore, if at low speeds the additional dynamic loads can be neglected, this is not so at medium speeds and even less so at high speeds, (see also sections 8.5, 8.6, 8.7, Fig. 8.15).

Due to their random nature, an accurate analysis of the additional dynamic loads is possible by spectral analysis, which consists in decomposing a sequence of signals into oscillations of different lengths, (135). With this method it was found that additional dynamic loads can be classified into two groups:

- Additional dynamic loads caused by *sprung masses (rolling stock)* and influenced by the type and the characteristics of the rolling stock, (Fig. 7.18). Oscillations of sprung masses increase with train speed, but at a lower rate. The increase of the oscillations of the sprung masses is a function of their vertical oscillation resonance frequency, (152).

- Additional dynamic loads caused by *unsprung masses (wheels, rails, sleepers)*, which are proportional to: speed, the magnitude of track defects, the square root of the unsprung masses, and the square root of the vertical stiffness of the track. The standard deviation of the additional dynamic loads ΔQ caused by the unsprung masses may be expressed by the relation, (152):

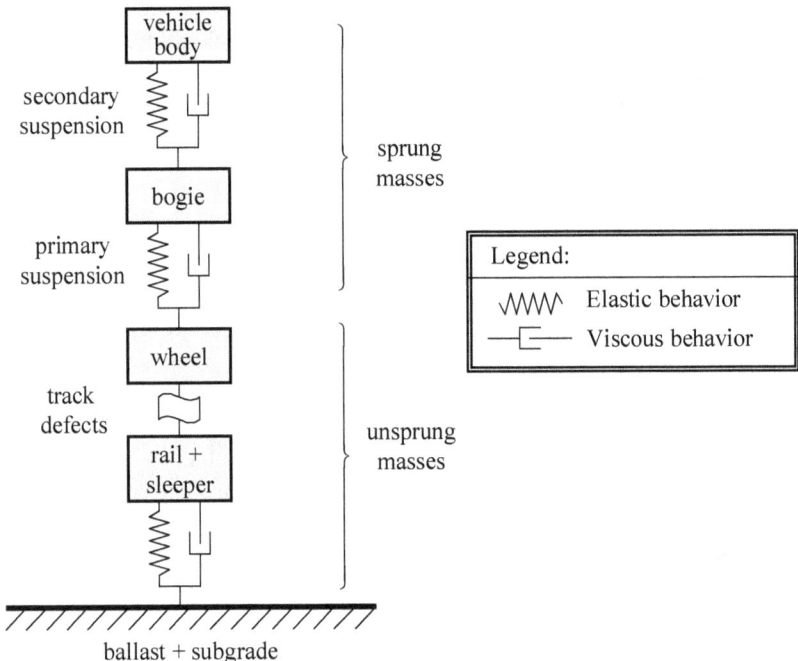

Fig. 7.18. Sprung and unsprung masses in a rail system

$$sd_{\Delta Q} = V \sqrt{\frac{A \cdot m \cdot h}{2 \cdot a}} \qquad (7.15)$$

where: $sd_{\Delta Q}$: standard deviation of ΔQ,

     V   : speed of rail vehicle,

     m   : unsprung mass per wheel,

     h   : vertical stiffness of the track which, as explained in section 8.2.2, is defined as $h = Q/z$, with Q the wheel load and z the vertical settlement at the rail level,

     a   : damping factor,

     A   : empirical coefficient depending on track maintenance conditions.

7.12. Influence of forces on passenger comfort

Passenger comfort is affected both by the values of vertical and transverse accelerations exerted on the human body, but also by the frequency of vibration. It was found that comfort is minimum at frequencies in the order of 5 Hz and that the human body supports more easily vibrations corresponding to frequencies 5÷20 Hz, (132), (147).

8 Mechanical Behavior of Track

8.1. A variety of methods adjusted to the nature of the problem under study

The accurate knowledge of the mechanical behavior of track (stress, strain, moments, etc.) is essential for a rational dimensioning of the various components of the track system, which should satisfy the requirements for both safety and economy.

There is a variety of methods which can be adjusted to the nature of the problem under study. Some methods are based on Boussinesq's analysis (multilayer system with elastic behavior). Other methods consider the track system as one-dimensional problem. A more modern method, finite element analysis, permits to take into account the real geometry of the track and the real stress-strain relation of the various materials. For some problems, boundary element methods may also be used. The problems occurring in contact surfaces (rail-sleeper, sleeper-ballast, etc.) may be approached by unilateral contact theories.

In most cases, satisfactory results can be drawn from a static analysis, that is without considering time as a variable. However, there are problems, such as the analysis of ground vibrations from rail traffic, for which a dynamic analysis, taking into account time among the other variables of the problem, is necessary.

8.2. Track coefficients and Bousinesq's analysis

8.2.1. Definitions – Symbols

We will first examine a static approach of the mechanical behavior of the track. Let, (Fig. 8.1):

Q: wheel load,

z : vertical settlement at the rail level,

r : wheel load uniformly distributed along the rail,

R: vertical reaction between sleeper and rail,

ℓ : sleeper spacing,

S : sleeper seating area,

p : average pressure between sleeper and ballast.

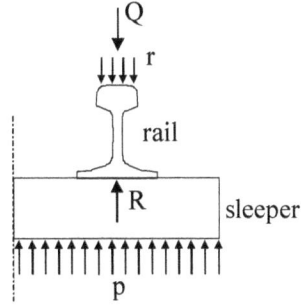

Fig. 8.1. Simplified approach of the track system

8.2.2. Track coefficients

We define the following track coefficients:

$$\text{Track index} \qquad k = \frac{r}{z} \tag{8.1}$$

$$\text{Track stiffness} \qquad h = \frac{Q}{z} \tag{8.2}$$

$$\text{Sleeper reaction coefficient} \qquad \rho = \frac{R}{z} \tag{8.3}$$

Substituting equation (8.1) into (8.3), we obtain

$$\rho = R \cdot \frac{k}{r} \tag{8.4}$$

and since $R = \ell \cdot r$ (equilibrium's equation), then

$$\rho = \ell \cdot r \cdot \frac{k}{r} = k \cdot \ell \tag{8.5}$$

The *ballast coefficient* is defined as $\quad C = \dfrac{\rho}{S}$ (8.6)

Substituting equation (8.3) into (8.6), we obtain

$$C = \frac{R}{z \cdot S} \tag{8.7}$$

and since $\dfrac{R}{S} = p$, we will have $\quad C = \dfrac{p}{z}$ (8.8)

In a more general way, the *reaction coefficient* of a component of the track system is defined as

$$\rho_n = \frac{R}{z_n} \tag{8.9}$$

where z_n is the vertical settlement at the level of the component under consideration.

Hence,

$$\sum z_n = z \Rightarrow \sum \frac{R}{\rho_n} = R \cdot \sum \frac{1}{\rho_n} \Rightarrow \frac{1}{\rho} = \sum \frac{1}{\rho_n} \qquad (8.10)$$

Equation (8.10) gives the total reaction coefficient of the track – subgrade multilayer system.

Below are given values of the reaction coefficient ρ for the various track components, (152):

Rail	5,000÷10,000 t/mm
Timber sleeper	50÷80 t/mm
Concrete sleeper	1,200÷1,500 t/mm
Ballast	10÷30 t/mm
Rubber pad	10÷20 t/mm

Track elasticity depends on the elastic characteristics and the thickness of the ballast, the subgrade, and the elastic pads (placed between the rail and the sleeper). It has been found that along existing tracks with only a ballast layer (i.e. with no subballast layer) the total reaction coefficient ranges between 1.5 and 10 t/mm, with 3.0 t/mm as an average value, (152).

Subgrade elasticity depends on soil quality, with the following values for the reaction coefficient, (152):

Silty subgrade	0.5÷1.5 t/mm
Clay subgrade	1.5÷2.0 t/mm
Gravel or rocky subgrade	2÷8 t/mm
Frozen subgrade	8÷10 t/mm

In civil engineering structures (bridges, slab track etc.), the values of the reaction coefficient range from 10 to 15 t/mm and therefore elasticity is far lower than in a track on subgrade. The rubber pads used in these cases are significantly thicker.

8.2.3. Track coefficients and Bousinesq's analysis

An increase in the thickness of ballast layer will result in lower stresses in the subgrade and increased elasticity of track. Let:

$$\lambda = \frac{\text{stress at the subgrade surface}}{\text{stress under the sleeper}}$$

e : thickness of the ballast layer
ρ_o : track reaction coefficient for e=0

By applying Boussinesq's analysis (multi-layer system with elastic behavior), the values illustrated in Table 8.1 can be derived, (152).

A detailed analysis of the influence of ballast thickness on track and sub-grade stress and strain is given in section 8.4.7.

Table 8.1.
Influence of ballast thickness on track elasticity and on the reduction of subgrade stress according to Bousinesq's analysis, (152)

e(cm)	0	15	20	30	40	50
λ	1	0.70	0.50	0.35	0.25	0.20
p/p_0	1	1.40	2.00	2.85	4.00	5.00

8.3. Approximate one-dimensional elastic analysis of track

8.3.1. Assumptions and equations

The approximate one-dimensional elastic analysis of track is often named after Zimmermann, (178), and is based on Winkler's approach of a beam of infinite length resting on a semi-infinite elastic subsoil with forces between beam-subsoil proportional to settlements. According to this analysis, the track is considered as an one-dimensional system and the rail as an infinitely long beam resting on an elastic support (sleepers, track bed, subgrade), simulated by successive linear springs. Thus, all effects of layers under the rail are represented by a single parameter, the track index (equation (8.1), analyzed previously). The wheel load is simulated as a concentrated load Q. Though this method does not consider the specific structural features and peculiarities of the various materials under the rail, it can be used for a preliminary approach and a simplistic analysis. Let:

      M : bending moment,
      T : shear force,
      k : track index,
      E : modulus of elasticity of rail,
      I : moment of inertia of rail.

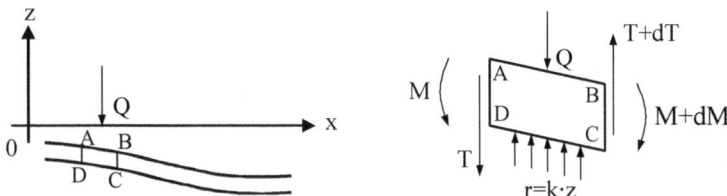

Fig. 8.2. One-dimensional simulation of track (beam of infinite length (rail) on elastic layer) and moments in an elementary section ABCD of the beam

We will start with equations of strength of materials:

$$\frac{dM}{dx} = T \tag{8.11}$$

$$\frac{dT}{dx} = k \cdot z + Q \cdot \delta(x) \tag{8.12}$$

where $\delta(x)$ is the Dirac function, the Fourier transform of which is equal to 1:

$$\delta(x) = 0, \ x \neq 0 \ , \quad \delta(0) = \infty \tag{8.13}$$

The equation of the elastic line of a beam is:

$$\frac{d^2 z}{dx^2} = -\frac{M}{E \cdot I} \tag{8.14}$$

By substituting equations (8.11) and (8.12) into (8.14), it can be derived that:

$$E \cdot I \cdot \frac{d^4 z}{dx^4} + k \cdot z = -Q \cdot \delta(x) \tag{8.15}$$

Let $Z(\omega)$ be the Fourier transform[*] of z, and let

$$\frac{k}{E \cdot I} = w^4 \tag{8.16}$$

Equation (8.15) is transformed as

$$\omega^4 \cdot Z + w^4 \cdot Z = -\frac{Q}{E \cdot I} \tag{8.17}$$

and $\quad Z = -\dfrac{Q}{E \cdot I \cdot (\omega^4 + w^4)} \tag{8.18}$

Applying the inverse Fourier transform, it is derived that:

for $\quad x \geq 0, \ z = z_0 \cdot \sqrt{2} \cdot e^{\left(-\frac{\omega \cdot x}{\sqrt{2}}\right)} \cdot \cos\left(\frac{w \cdot x}{\sqrt{2}} - \frac{\pi}{4}\right) \tag{8.19}$

$\quad x < 0, \ z = z_0 \cdot \sqrt{2} \cdot e^{\left(-\frac{\omega \cdot x}{\sqrt{2}}\right)} \cdot \cos\left(\frac{w \cdot x}{\sqrt{2}} + \frac{\pi}{4}\right) \tag{8.20}$

and $\quad z_{max} = z_0 = \dfrac{Q}{2 \cdot \sqrt{2} \cdot \sqrt[4]{E \cdot I \cdot k^3}} \tag{8.21}$

---

[*] The Fourier transform F_f of a function $f(x)$ is defined by the following formula:

$$F_f = \int_{-\infty}^{+\infty} f(x) \cdot e^{-2 \cdot i \cdot \pi \cdot \omega \cdot x} \cdot dx, \ i = \sqrt{-1}$$

Therefore, the analytical expressions of the bending moment M, the shear force T, and the reaction r between rail and its elastic support will be:

$$M = \frac{Q}{2 \cdot w} \cdot e^{\left(-\frac{w \cdot x}{\sqrt{2}}\right)} \cdot \cos\left(\frac{w \cdot x}{\sqrt{2}} + \frac{\pi}{4}\right) \tag{8.22}$$

$$T = -\frac{Q}{2} \cdot e^{\left(-\frac{w \cdot x}{\sqrt{2}}\right)} \cdot \cos\frac{w \cdot x}{\sqrt{2}} \tag{8.23}$$

$$r = k \cdot z = \frac{Q}{2} \cdot w \cdot e^{\left(-\frac{w \cdot x}{\sqrt{2}}\right)} \cdot \cos\left(\frac{w \cdot x}{\sqrt{2}} - \frac{\pi}{4}\right) \tag{8.24}$$

8.3.2. Results of the one-dimensional analysis

Graphical representations of bending moment M, shear force T, and vertical settlement z, according to previous formulas of the approximate one-dimensional analysis, are provided in Figure 8.3. All three curves representing M, T, and z are oscillating in relation to the parameter λ, where:

$$\lambda = 2 \cdot \sqrt{2} \cdot \frac{\pi}{w}$$

Indeed, moment becomes zero for $x = \lambda/8$, $x = 5 \cdot (\lambda/8)$, $x = 9 \cdot (\lambda/8)$. Shear force becomes zero for $x = 2 \cdot (\lambda/8)$, $x = 6 \cdot (\lambda/8)$. Vertical displacement becomes zero for $x = 3 \cdot (\lambda/8)$, $x = 7 \cdot (\lambda/8)$. From Figure 8.3 it can be deduced that according to the one-dimensional analysis the bending moment M and the shear force T become practically zero beyond a distance of $x = 5 \cdot (\lambda/8)$ from the point of application of the load Q.

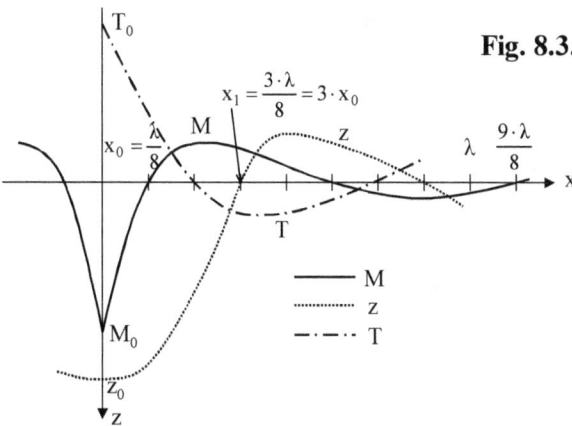

Fig. 8.3. Bending moment, shear force, and settlement of the track system in relation to the distance from the point of application of the wheel load, according to one-dimensional elastic analysis, (178)

229

The maximum values of M, T, z, h will result from the above equations for x=0. Thus:

$$M_{max} = M_0 = \frac{Q}{2 \cdot \sqrt{2}} \cdot \sqrt[4]{\frac{E \cdot I \cdot \ell}{\rho}} \qquad (8.26)$$

$$T_{max} = T_0 = \frac{Q}{2 \cdot \sqrt{2}} \cdot \sqrt[4]{\frac{\rho \cdot \ell^3}{E \cdot I}} \qquad (8.27)$$

$$z_{max} = z_0 = \frac{Q}{2 \cdot \sqrt{2}} \cdot \sqrt[4]{\frac{\ell^3}{E \cdot I \cdot \rho^3}} \qquad (8.28)$$

$$h = \frac{Q}{z_0} = 2 \cdot \sqrt{2} \cdot \sqrt[4]{\frac{E \cdot I \cdot \rho^3}{\ell^3}} \qquad (8.29)$$

Equations (8.26) to (8.29) show that if the sleeper reaction coefficient ρ increases, M_0 and z_0 decrease, and T_0 increases. The vertical settlement z_0, however, which is proportional to $1/\rho^{3/4}$ decreases much faster than the bending moment M, which is proportional to $1/\rho^{1/4}$. Therefore, a high value of the sleeper reaction coefficient is beneficial for track geometry. It should be noted that the sleeper reaction coefficient is mainly affected by the quality of the subgrade, where most of the total vertical settlement occurs.

An increase of sleeper spacing ℓ results in an increase of M_0, R_0, and z_0. However, vertical settlement and sleeper reaction increase faster than moment, since they are proportional to $\ell^{3/4}$, while the moment is proportional to $\ell^{1/4}$. Consequently, a reduction of sleeper spacing affects track geometry more and rail mechanical behavior less.

When rail stiffness E·I increases, M_0 increases, while z_0 and T_0 decrease. Rail stiffness increases mainly as a result of an increase of rail weight per unit length.

From strength of materials, rail bending stresses can be calculated from the equation:

$$\sigma = M \cdot \frac{y}{I} \qquad (8.30)$$

and considering the value y_{max}, we obtain:

$$\sigma_{max} = \frac{Q \cdot y_0}{2 \cdot \sqrt{2}} \cdot \sqrt[4]{\frac{E \cdot \ell}{I^3 \cdot \rho}} \qquad (8.31)$$

where y_0 is the maximum distance from the center of gravity of the rail.

Therefore, an increase of the moment of inertia of the rail influences significantly the stresses generated within the rail and to a lesser degree the track geometry. This is why the increase of the axle load in recent years has led to a proportionally higher increase of the rail cross-section.

8.4.   Accurate analysis of the mechanical behavior of track – Finite element method and elastoplastic analysis

8.4.1. A short description of the finite element method and applications for track problems

Simplified methods (Zimmermann's method, Boussinesq's multi-layer method, etc.) permit an approximate calculation of stress and strain quite easily. However, a comparison of the results of simplified methods with actual values, as measured by on-site measurements, may reach differences as much as 100%, (148). Such a gap between calculated and measured values may not be acceptable in most analyses of track behavior. It is therefore necessary for the mechanical behavior of the track – subgrade system (mainly calculations of strain and stresses, on which the dimensioning of the various layers is based) to be analyzed by more accurate methods. This has become easy for some time now, with the help of numerical methods and powerful computers. An accurate analysis of the mechanical behavior of track can be achieved with applications of the *finite element method* (FEM). In this method, instead of the physical system, (Fig. 8.4.a), a system resulting from dividing the physical system into discrete parts (finite elements) is analyzed, (Fig. 8.4.b), (168), (173).

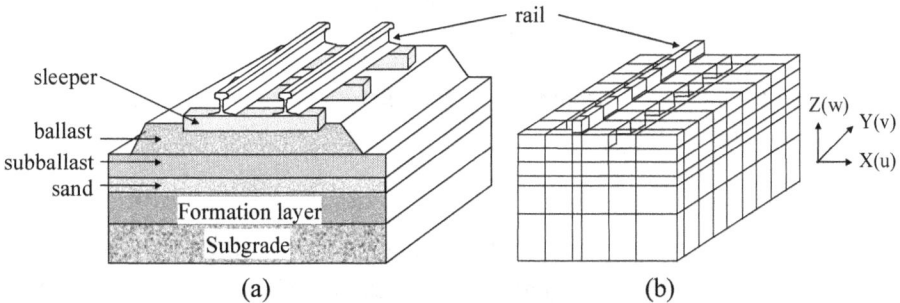

Fig. 8.4. The railway system (a) and the mesh (constituted of finite elements) of the model (b), (164), (165)

Figure 8.5 illustrates the various stages (which are analyzed in detail in the following paragraphs) for application of the finite element method in railway problems.

```
┌─────────────────────────────────────┐
│        Construction of the mesh of   │
│      the railway system under study  │
├─────────────────────────────────────┤
│       Definition of limit conditions │
└─────────────────────────────────────┘
┌─────────────────────────────────────────┐
│              Decision whether:            │
│   • Analysis will be static or dynamic    │
│   • Mechanical behavior of materials will be │
│      elastic or elastoplastic or viscoelastic │
└─────────────────────────────────────────┘
┌─────────────────────────────────────┐
│         Choice of the appropriate    │
│        finite element software *     │
├─────────────────────────────────────┤
│        Values of external forces     │
├─────────────────────────────────────┤
│         Values of mechanical         │
│        properties of materials       │
├─────────────────────────────────────┤
│         Numerical calculations       │
├─────────────────────────────────────┤
│          Results (stresses,          │
│      displacements, moments, etc.)   │
└─────────────────────────────────────┘
┌───────────────────────────────────────────┐
│  Check of convergence of the method and of  │
│  results - Comparison with measurements or  │
│         with results of other methods       │
└───────────────────────────────────────────┘
```

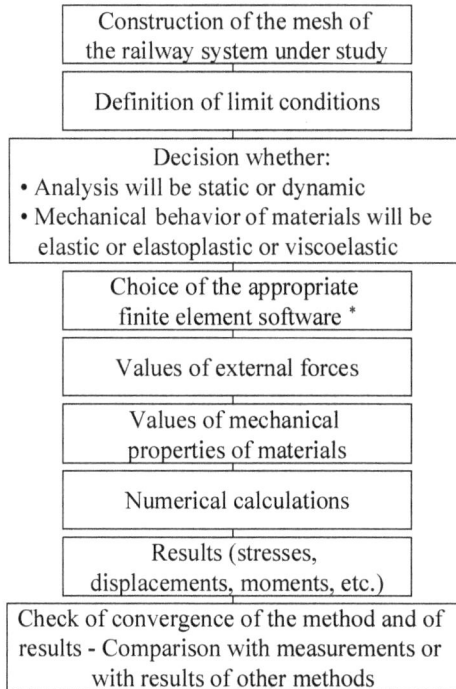

Fig. 8.5. Successive stages for the application of the finite element method in railway problems

The finite element method permits to study the actual physical system, without extreme simplifications, by taking into account the accurate limit conditions (i.e. the conditions imparting to stresses or strain specific values at limit positions, for instance, in the supports displacement is zero) and the accurate constitutive law of behavior (i.e. the relation between stress and strain for every material), (164), (174).

8.4.2. Construction of the mesh of the model

For reasons of symmetry (along the longitudinal and the transverse axes), the study of the problem can be limited to ¼ of the initial system, (Fig. 8.4.b). The construction of the mesh of the model is an essential part of the method and the resulting finite elements must be homogeneous (i.e. of about the same size), otherwise the method may not converge, (168). The rail is simulated with a rectangular cross-section which has the same moment of inertia (I) with the actual rail.

---

* Among the various finite element software we can mention Sofistik, Adina, Abacus, Ansys, etc.

8.4.3. Limit conditions

The limit conditions that were considered are as follows:
♦ conditions of symmetry, i.e. transverse displacement at any plane of symmetry is zero,
♦ conditions at the most distant points of the problem, where displacements are set to zero.

Limit conditions must be set in such a way that the finite element model will have a similar behavior to the physical system investigated.

8.4.4. Stress-strain relation

The constitutive law of behavior (stress-strain relation) must express the real mechanical behavior of the materials. Concerning ballast and subgrade, it was found that the deformation caused by train loads is composed of two components:
– an elastic component which disappears after the passage of the train,
– a plastic component which remains after the train has passed, (146).

8.4.4.1. Case of ballast and subgrade

The behavior of ballast, subballast, and subgrade, as tested by in-situ experiments, (166), is found to be elastoplastic and is given by the following equations, (165):

$$\varepsilon_{ij}^{total} = \varepsilon_{ij}^{elastic} + \varepsilon_{ij}^{plastic} \quad \text{(total deformation is the sum of} \qquad (8.32)$$
$$\text{an elastic and a plastic component)}$$

$$\varepsilon_{ij}^{elastic} = \frac{1+v}{E} \cdot \sigma_{ij} - \frac{v}{E} \cdot I_1 \cdot \delta_{ij} \text{ (linear elasticity)} \qquad (8.33)$$

$$\varepsilon_{ij}^{plastic} = \lambda \cdot \frac{\partial f}{\partial \sigma_{ij}} \quad \text{(law of plasticity, known as Hencky's law}^*) \quad (8.34)$$

where: ε_{ij}^{total} : total deformation,

$\varepsilon_{ij}^{elastic}$: elastic deformation,

$\varepsilon_{ij}^{plastic}$: plastic deformation,

---

* The accurate plasticity law is written as $\dot{\varepsilon}_{ij}^{plastic} = \mu \frac{\partial f}{\partial \sigma_{ij}}$ (law of Hill), (167).

If deformations are small, this law can be simplified as $\varepsilon_{ij}^{plastic} = \lambda \frac{\partial f}{\partial \sigma_{ij}}$ (Hencky's law)

      E   : modulus of elasticity,

      v   : Poisson's ratio,

      I_1 $= \sigma_{11} + \sigma_{22} + \sigma_{33}$,

      δ_{ij} : Kronecker's delta, $\delta_{ij} = 1$ for i=j, $\delta_{ij} = 0$ for i≠j,

      f   : plasticity criterion, with a different formula for each material,

      λ   : a scalar quantity.

The indices i, j take the values 1, 2, 3.

It has been suggested that the plasticity criterion best suited for *soil materials* and *ballast* is the *Drucker-Prager* criterion, defined by the following equation, (167), (169):

$$f(\sigma) = \alpha \cdot I_1 + J_2 - k \tag{8.35}$$

where:

$$J_2 = \frac{1}{6} \cdot [(\sigma_1 - \sigma_2)^2 + (\sigma_2 - \sigma_3)^2 + (\sigma_1 - \sigma_3)^2] \tag{8.36}$$

$\sigma_1, \sigma_2, \sigma_3$: principal stresses

$$\alpha = \frac{\tan \varphi}{(9 + 12 \cdot \tan^2 \varphi)^{1/2}}, \quad \varphi : \text{friction angle} \tag{8.37}$$

$$k = \frac{3 \cdot c}{(9 + 12 \cdot \tan^2 \varphi)^{1/2}}, \quad c : \text{cohesion} \tag{8.38}$$

If the track bed is a *concrete slab*, the plasticity criterion best suited is the *parabolic criterion*, expressed by the formula, (169):

$$f(\sigma) = J_2 + \frac{1}{3} \cdot (R_c - R_T) \cdot I_1 - \frac{1}{3} \cdot R_c \cdot R_T \tag{8.39}$$

where R_c : compressive strength,

           R_T : tensile strength.

8.4.4.2. Case of rail and sleeper

In contrast to ballast and subgrade, rails and sleepers have an almost elastic behavior, therefore plastic deformations are negligible and there is no need to be taken into account. Whenever plasticity effects have to be considered, however, the parabolic criterion should be used as the plasticity criterion for concrete sleepers. For *rails*, the *von Mises* plasticity criterion should be used, which is given by the equation, (169):

$$f(\sigma) = \sqrt{\frac{1}{6} \cdot [(\sigma_1 - \sigma_2)^2 + (\sigma_2 - \sigma_3)^2 + (\sigma_1 - \sigma_3)^2]} - q \qquad (8.40)$$

where   q : shear yield stress of rail.

8.4.5. Numerical calculations

The various finite element models can be classified into three broad categories, (169), (177):

− *Strain* (or kinematic) models, in which the limit conditions concerning strain (deformations) are introduced as given data, while equilibrium equations as well as limit conditions concerning stresses are the object of successive numerical calculations. Strain models are more suitable to construct and implement.
− *Stress* (or static) models, in which equilibrium equations and limit conditions concerning stresses are introduced as known data, while deformations are calculated through successive steps. Stress models are less suitable to construct than strain models.
− *Hybrid* models, in which a strain model is applied in a geometric part of the model and a stress model in another part.

In *strain models,* in particular, static finite element analysis leads to the solution of the matrix equation:

$$[K] [q] = [F] \qquad (8.41)$$

where: [K] : the model's stiffness matrix,
          [q] : the displacement vector of the model's nodes,
          [F] : the vector of the forces exerted on model's nodes.

The quantities [K], [q], [F] for the entire model are the result of the assembly of the elementary quantities $[K_e]$, $[q_e]$, $[F_e]$ corresponding to each finite element, (168), (177).

The elastoplastic constitutive law, correlating stress and strain, may be implemented numerically by two methods, (169), (177):

a. The initial stress method, which is slower in convergence but easier to use, (Fig. 8.6.a).
b. The variable stiffness method, which is faster in convergence but has the disadvantage that the stiffness matrix changes at each successive step, (Fig. 8.6.b).

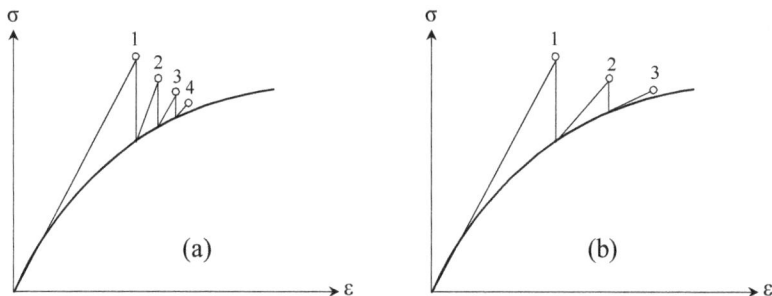

Fig. 8.6. The initial stress (a) and variable stiffness (b) methods to implement the elastoplastic stress-strain relation, (169), (177)

8.4.6. Determination of the mechanical characteristics of the various materials

The subgrade can be of different classes (S_1 (poor), S_2 (medium), S_3 (good), R (rock)) as they are classified by the UIC, (see section 9.5). Table 8.2 gives the average values of the mechanical characteristics of track materials, as determined by a series of tests conducted within the framework of the UIC, (166), (175). In other finite element method analyses of the track system, conducted by other researchers, similar values of the mechanical characteristics of track materials were introduced, (157), (172).

Table 8.2.
Values of the mechanical characteristics of materials of track and subgrade, (166), (175)

Material	Modulus of elasticity E (kp^*/cm^2)	Poisson's ratio v	Cohesion c (kp/cm^2)	Friction angle $\varphi(°)$
Poor quality subgrade (S_1)	125	0.4	0.15	10
Medium quality subgrade (S_2)	250	0.3	0.10	20
Good quality subgrade (S_3)	800	0.3	0	35
Rock subgrade (R)	$3 \cdot 10^4$	0.2	15	20
Ballast	1,300	0.2	0	45
Gravel subballast	2,000	0.3	0	35
Sand	1,000	0.3	0	30
			Tensile strength R_T (kp/cm^2)	Compressive strength R_c (kp/cm^2)
Reinforced-concrete sleeper	$30 \cdot 10^4$	0.25	30	300
Prestressed-concrete sleeper	$50 \cdot 10^4$	0.25	60	90
Tropical timber sleeper	$25 \cdot 10^4$	0.25	100	1,000
Rail (steel)	$2.1 \cdot 10^6$	0.30	$7 \cdot 10^3$	$6 \cdot 10^3$

* The unit kp is the force exerted by 1 kilogram of mass in standard earth gravity.

8.4.7. Stress and strain in the track–subgrade system

Finite element analysis allows all parameters of the track–subgrade system to be taken into consideration, (146), (164), (165):
♦ subgrade soil quality (S_1, S_2, S_3, R), (see section 9.5),
♦ sleeper type, (see sections 11.3, 11.5, 11.6):
  – twin-block reinforced-concrete sleeper,
  – monoblock prestressed-concrete sleeper,
  – timber sleeper,
♦ track bed thickness e (= ballast + subballast).

Figures 8.7, 8.8, 8.9 illustrate the vertical stresses at the subgrade level, as well as the vertical settlements at the rail, sleeper, and subgrade level, according to the elastoplastic analysis with the use of the finite element method, (146), (164). It can be deduced that the values of stresses are primarily affected by the subgrade soil quality and to a lesser degree by the track bed thickness. Indeed, the better the subgrade soil quality is, the lesser the influence of track bed thickness. In particular and with all other parameters unchanged, an improvement of subgrade quality from one class to the next ($S_1 \rightarrow S_2$, $S_2 \rightarrow S_3$, $S_3 \rightarrow R$) will result in an increase of the stresses developed in the subgrade by about 50%.

With respect to the influence of the sleeper type, it can be deduced (except in the case of a rocky subgrade) that timber sleepers and monoblock prestressed-concrete sleepers have a better load distribution, i.e. they result in smaller values of stresses in the subgrade. In any case, the influence of sleeper type is smaller than the influence of subgrade quality.

8.4.8. Distribution of wheel load along successive sleepers

Railway engineers have been accustomed, on the basis of simplified considerations, to the assumption that when a wheel load is applied above a sleeper, then the sleeper below the load supports 50% of the wheel load and each of the neighboring sleepers supports another 25%. Stress measurements and finite element analysis applications, however, have shown that wheel load distribution along successive sleepers is as follows, (Fig. 8.10), (146):
– sleeper under wheel load: 40%,
– first neighboring sleeper: 23%,
– second neighboring sleeper :7%.

The above finding has already been accepted by a number of researchers, (158).

$\sigma_z \ (kg/cm^2)$

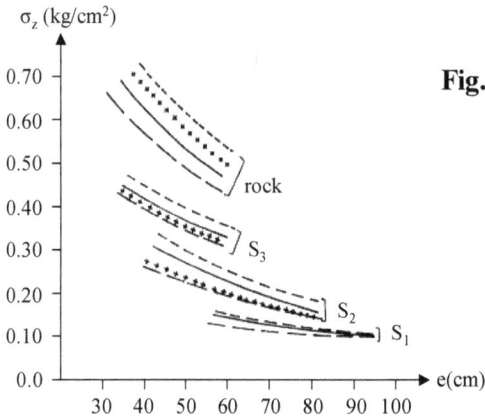

Fig. 8.7. Vertical stresses at the subgrade level for various subgrade and sleeper types, as a function of track bed thickness e (=ballast + subballast). Elastoplastic finite element analysis, (146), (164)

w (mm)

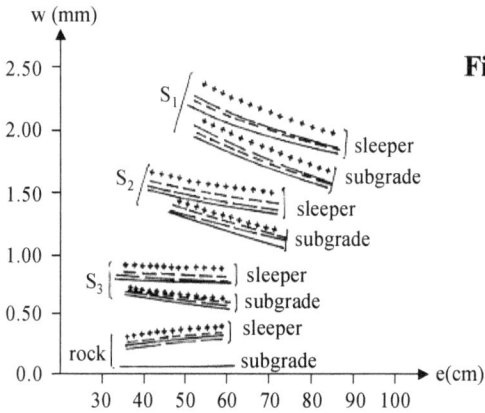

Fig. 8.8. Vertical settlements at the subgrade and sleeper level for various subgrade and sleeper types, as a function of track bed thickness e (=ballast+subballast). Elastoplastic finite element analysis, (146), (164)

w (mm)

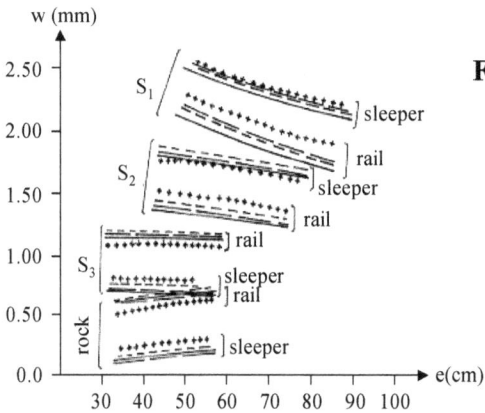

Fig. 8.9. Vertical settlements at the sleeper and rail level for various subgrade and sleeper types, as a function of track bed thickness e (= ballast+subballast). Elastoplastic finite element analysis, (146), (164)

+++++++ Timber sleeper, l=2.60m	S_1 : subgrade of poor quality
— — Monoblock sleeper, l=2.50m	S_2 : subgrade of medium quality
—— Twin-block sleeper, l=2.415m	S_3 : subgrade of good quality
- - - - - Twin-block sleeper, l=2.245m	rock : rocky subgrade

238

Therefore, when a wheel load is applied over a sleeper, its effect is negligible beyond the second successive sleeper. The above load distribution, in conjunction with the value of the wheel load, affects sleeper dimensioning.

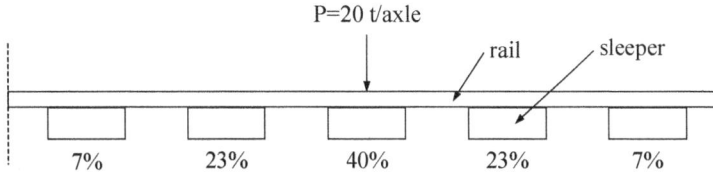

Fig. 8.10. Wheel load distribution along successive sleepers, (146)

8.4.9. Elastic line of sleeper

The elastic line is an essential part of the mechanical behavior of the railway system. Figure 8.11 illustrates a comparison of the elastic line for timber and monoblock prestressed-concrete sleepers. Figure 8.12 illustrates the elastic line of a timber sleeper for various qualities of subgrade, (146). The significant role of the subgrade is again confirmed.

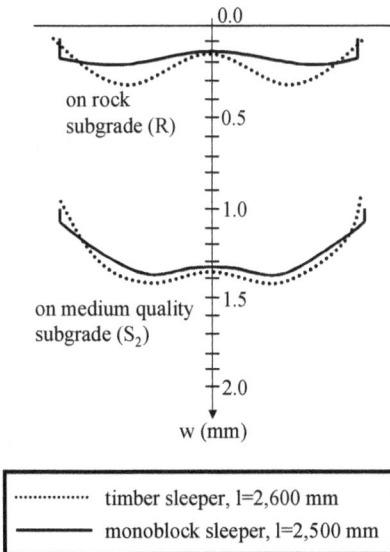

Fig. 8.11. Comparative elastic line for timber sleeper and monoblock prestressed-concrete sleeper, (146)

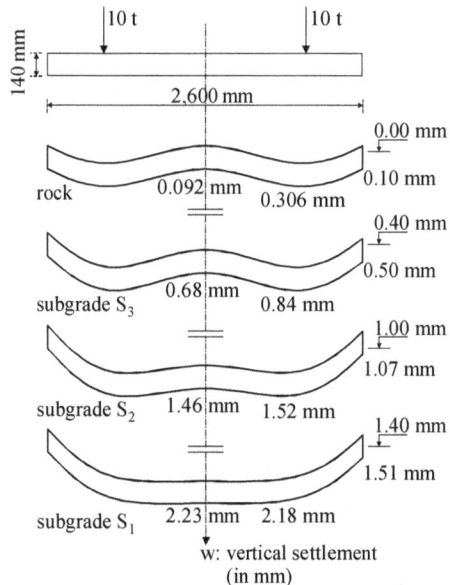

Fig. 8.12. Elastic line of timber sleeper for various subgrade qualities, (146)

239

8.5. Dynamic analysis of the track–subgrade system

As discussed in section 7.11.2, an adequate calculation of the stress and strain of the track–subgrade system may be obtained by static analysis, which however leaves out dynamic effects. A comparison of the results of finite element static analyses with stress and strain measurements has shown deviations not exceeding 20%, thus confirming that the static approach can be considered as satisfactory for stress and strain analysis, (165).

There are phenomena, however, which cannot be adequately simulated by the static approach. These include the problem of the transmission of vibrations from the trains to the environment, the problem of the motion and the suspension of the various rolling stock components, etc., (158), (170).

A satisfactory simulation of dynamic effects can be realized by a viscoelastic constitutive law and is illustrated in Figure 8.13, where:

♦ the symbol ⎯〰〰⎯ represents elastic behavior,
♦ the symbol ⎯⊏⎯ represents viscous behavior,
♦ the symbol ⎡〰〰⎤ represents viscoelastic behavior (Kelvin-Voigt model),
♦ rail vehicles and bogies are modeled as non-deformable solids,
♦ wheels and sleepers are modeled as discrete masses,
♦ the ballast and the various subgrade layers are modeled as horizontal layers,
♦ the various system components are interconnected by a viscoelastic stress-strain relation.

In the dynamic analysis, the relation between external forces and strain (displacements) is given by the following matrix equation:

$$[M] \, [\ddot{q}] + [C] \, [\dot{q}] + [K] \, [q] = [F] + [R] \qquad (8.43)$$

where: $[M]$: the mass matrix,
$[C]$: the viscosity (damping) matrix,
$[K]$: the stiffness matrix,
$[q]$: the displacement vector,
$[\dot{q}]$: the velocity vector,
$[\ddot{q}]$: the acceleration vector,
$[F]$: the external forces vector,
$[R]$: the vector of the reactions exerted by the sleepers on the ballast.

In the dynamic analysis, the calculations are more complex compared to the static one and therefore take a longer time. For this reason, dynamic analysis should be applied to phenomena which cannot be adequately simulated by static analysis.

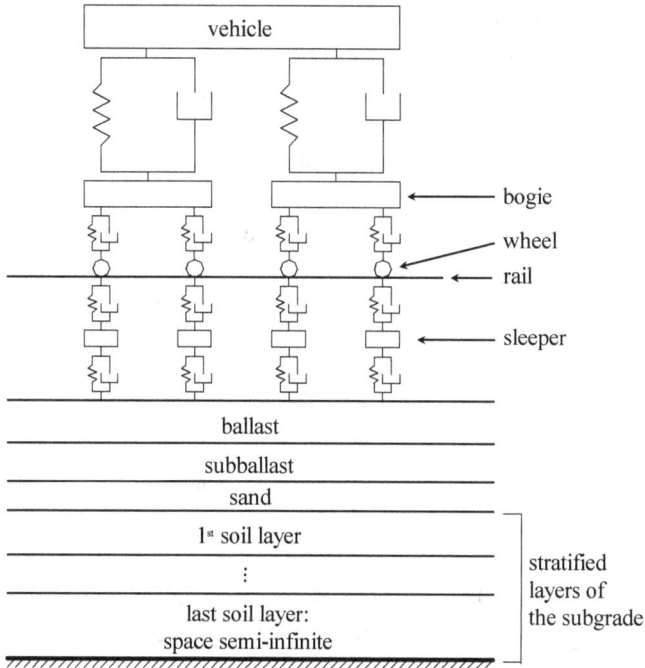

Fig. 8.13. Modeling of the vehicle – track – subgrade system for a dynamic analysis, (148)

8.6. Track defects and additional dynamic loads

Analyses of the mechanical behavior of the rail system have until now been based on the assumption that both rails and wheels are smooth and free of defects. However, this is not the case, and as explained in section 7.11.2, defects that appear stimulate the system and cause additional dynamic loads Q_{dyn}, which may reach values of up to 50% of the wheel load Q_{stat}.

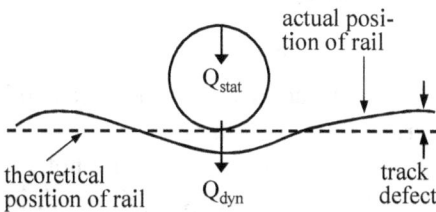

Fig. 8.14. Track defects and additional dynamic loads

The mechanical analysis of the track – subgrade system should therefore be considered not on the basis of the static wheel load Q_{stat}, but by taking into account the total load, (Fig. 8.14):

$$Q_{tot} = Q_{stat} + Q_{dyn} \tag{8.44}$$

Additional dynamic loads may be divided into three categories according to the respective vibration frequency:

– Loads resulting from vibrations with frequencies in the range $0.5\,\text{Hz}<v<15\,\text{Hz}$. These correspond to the movement of sprung masses (rolling stock), (see section 7.11.2), and depend principally on the characteristics and peculiarities of the rolling stock.

– Loads resulting from vibrations with frequencies in the range $20\,\text{Hz}<v<100\,\text{Hz}$. These correspond to the movement of unsprung masses (wheels, rails, sleepers), (see section 7.11.2), and depend mainly on track quality and stiffness.

– Loads resulting from vibrations with frequencies in the range $100\,\text{Hz}<v<2{,}000\,\text{Hz}$. These correspond to short- and long-pitch corrugations of the rail surface, (see section 10.9.4.4).

If we assume a linear behavior, then it is possible to separate each class of additional dynamic loads (corresponding to a specific range of frequencies) from others. In order to correlate accurately and causally track defects and the resulting dynamic loads Q_{dyn}, spectral analysis is used, since track defects may be recorded accurately and in detail by special recording vehicles. This analysis is again based on the dynamic equation (8.43).

8.7. Dynamic impact factor coefficient

The design of track components is usually conducted on the basis of results with the help of static analysis. The question arises, however, what is the dynamic impact factor η by which the static load should be multiplied in order to take into account in the static analysis the dynamic effects. Figure 8.15 summarizes the results of various theories.

Figure 8.15 illustrates for the dynamic impact factor η: the measured values (curve 6), the calculated values for an ideal track without defects (curve 7), and the values suggested by various empirical formulas (curves 1÷5). However, curves 1÷3 are deduced from old rolling stock characteristics and are not valid for modern rolling stock. More close to reality are curves 4, 5, 6, which illustrate that for the speed 200 km/h the dynamic impact factor η varies from 1.35 to 1.6. Thus, for speeds approaching 200 km/h a dynamic impact factor of 1.5 is suggested. For speeds greater than 200 km/h an analytical survey should be conducted on the basis of specific experimental data.

8.8. Design of the track–subgrade system

The design of the track–subgrade system should take into account the following two principles, (146), (164):

dynamic impact factor η

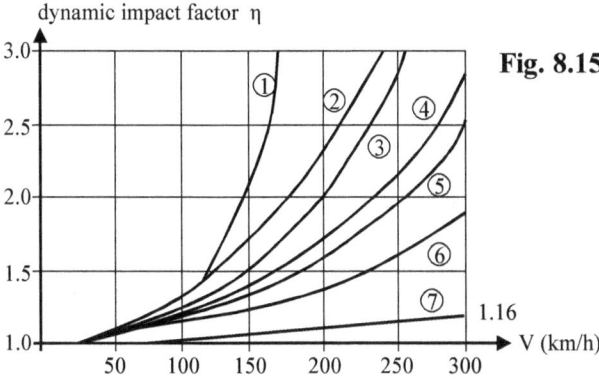

Fig. 8.15. Results for the dynamic impact factor η according to various theories, (163)

Legend:

① Winker (1871), $\eta = \dfrac{1}{1 - \dfrac{M_o \cdot V}{E \cdot I \cdot g}}$,

$M_o = 0.1188 \cdot P(t) \cdot \ell$ (cm), P(t): axle load, ℓ (cm): sleeper spacing, V(km/h): speed I (cm^4): moment of inertia.

② Formula of central European railways (1936), $\eta = 1 + \dfrac{V^2}{30,000}$

③ $\eta = \dfrac{1}{[1 - (9.1 \cdot 10^{-6} \cdot V - 2.957 \cdot 10^{-4})] \cdot \dfrac{P \cdot \ell}{I}}$, for V<65 km/h

$\eta = \dfrac{1}{(1 \div 7) \cdot 10^{-8} \cdot V^2 \cdot \dfrac{P \cdot \ell}{I}}$, for V>65 km/h

④ Schramm's formula (1955), $\eta = 1 + \dfrac{4.5 \cdot V^2}{100,000} - \dfrac{1.5 \cdot V^2}{10,000}$

⑤ Birman's formula (1966), $\eta = 1 + \alpha + \beta + \gamma, a = 0.04 \cdot \left(\dfrac{V}{200}\right)^3, \beta = 0.2, \gamma = \gamma_o \cdot \alpha \cdot \beta$

$\gamma_o = 0.1 + 0.017 \cdot \left(\dfrac{V}{100}\right)^3$

⑥ Values measured on vehicles of the French high-speed TGV 001 (1981)
⑦ Values of theoretical calculation for ideal track and vehicle contact surfaces without any irregularities

– loads must be properly distributed to the various layers, so that stresses developed in the materials of the various components of track must be less than the values causing failure,

$$\sigma_{\text{material of component i}}^{\text{under load}} < \sigma_{\text{material of component i}}^{\text{causing failure}} \qquad (8.45)$$

– adequate flexibility of the track should be ensured, i.e. track stiffness should not be excessive. Track stiffness is mainly determined by subgrade soil quality and track bed thickness. Rocky subgrades (with no problem as regards proper distribution of train loads) have a stiffness more than triple that of clay subgrades. Accordingly, rocky subgrades, although free of load distribution problems, must always have a layer composed of ballast and subballast, (155), (158).

Figure 8.16 illustrates the average contribution of each component of the track system to the total elasticity of the track in the cases of timber and concrete sleepers, (158).

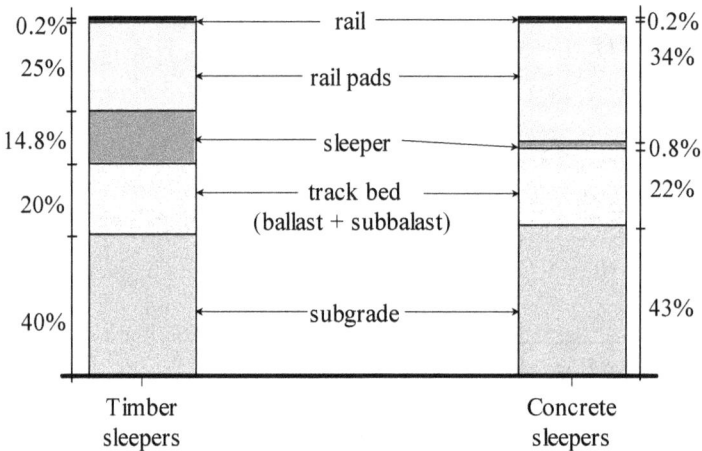

Figure 8.16. Contribution of each component of the track system to the total elasticity of track, (158)

8.9. Vibrations and noise from rail traffic

8.9.1. Origins of rail vibrations

A rail vibrating source produces three types of waves, (160), (171):

♦ *compression* waves (7% of the energy transmitted), which are longitudinal waves with particle motion being an oscillation in the direction of propagation,

♦ *shear* waves (26% of the energy transmitted), with particle motion being an oscillation in a plane normal to the direction of propagation,

♦ *Rayleigh* waves (67% of the energy transmitted), which are surface waves, with an elliptical oscillation in a vertical plane through the direction of propagation.

Rail vibrations have in *low* speeds two principal origins:
– *engines of rolling stock,*
– *wheel-rail interaction.*

In *electrified* lines, a third origin should be added, the catenary noise, caused by friction from the sliding contact of the pantograph along the trolley wire. A fourth noise origin is of *aerodynamic* nature and can be considered as minor for low and medium speeds (V<200 km/h), as important for high speeds (200<V<300 km/h), and as prevailing in very high speeds (V>300 km/h).

8.9.2. Relation of rail noise level to speed

From various analyses, it has been found that there is a logarithmic relation between the level of rail noise L (in dB(A)[*]) and train speed V, of the form, (160), (163):

$$L\,(dB(A)) = a + b \cdot \log V \tag{8.46}$$

with coefficients a, b depending on rolling stock and track characteristics, type of traffic, soil characteristics, etc.

8.9.3. Damping of rail noise in relation to distance

Figure 8.17 illustrates the noise level (in dB(A)) in various distances (100 m, 300 m, 400 m) from the track and for speeds from 130 km/h to 200 km/h. We can note that:

– the noise level does not decrease linearly for each doubling of distance as would be expected, probably due to ground impedance,
– noise levels are influenced more by distance than by changes in speed,
– noise levels are correlated with the logarithm of speed.

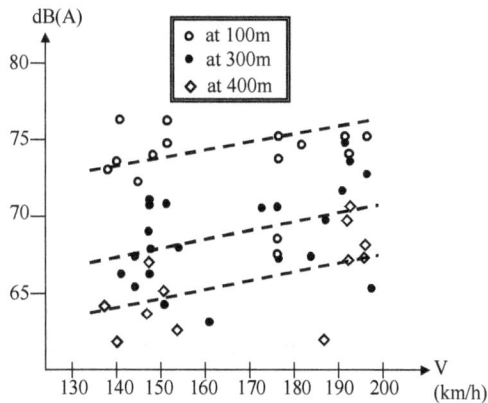

Fig. 8.17. Rail noise level in relation to distance and speed, (163)

---

[*] Decibel (dB) is a unit to measure the level of noise and refers to the pressure received by the human ear. Among the various methods of simulation of rail noise (which is composed of sounds of many frequencies and intensities), the most commonly used is method A, which focuses on frequencies around 2,000 Hz and the resulting monitoring of noise is expressed as dB(A).

8.9.4. Noise level in relation to infrastructure type

Measurements of noise level at 25m from the track centerline have been conducted at a speed of 200 km/h in the Japanese Shinkansen high-speed train (with 12÷16 vehicles) for various infrastructure types: bridge, viaduct, embankment, and cut, (Fig. 8.18).

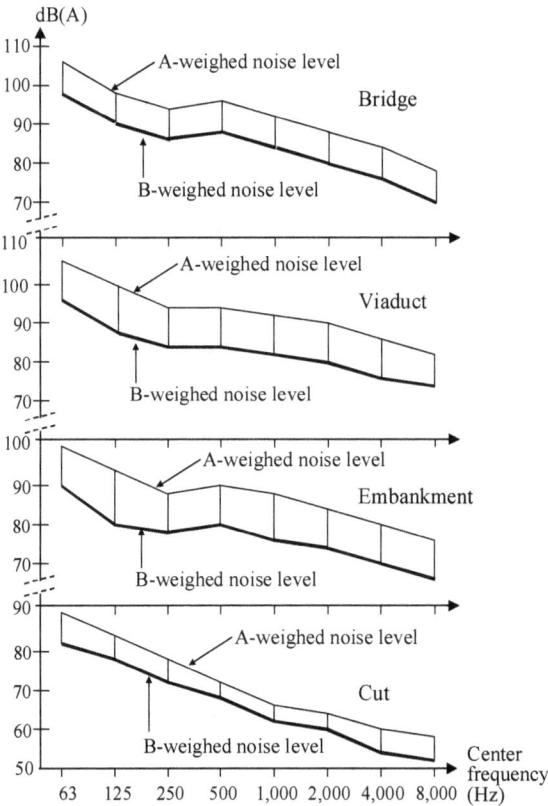

Fig. 8.18. Noise level in relation to infrastructure type, (163)

The noise levels in cut subgrades show the effectiveness of this solution in reducing the noise from rail traffic. Consequently, geometrical design and choice, wherever possible, of cut sections in layout can be used as a way to reduce the impact and disturbances from rail vibrations and noise.

8.9.5. Noise level in high speeds

A major concern in high-speed trains is to reduce the noise levels emitted. Thus, a noise level of 97 dB(A) is reported for the French TGV at 25 m from the track and a speed of 272 km/h. For the German ICE, noise levels of 86 and 93 dB(A) have been reported at a distance of 25 m from the track for speeds of 200 and 300 km/h respectively, (162). Table 8.3 illustrates noise levels in relation to the type of the train and distance.

8.9.6. Noise level standards

If noise level cannot be reduced otherwise (e.g. by the appropriate design of rolling stock and track), the usual means (in order to comply with noise level standards) is to construct noise barriers along the track, so as to protect neighboring sensitive human activities, (see also section 23.3).

Table 8.3.
Noise levels in dB(A) in relation to the type of train and distance, (162)

Type of train	Speed (km/h)	Distance D between track axis and noise reception point		
		D=0.75m	D=15m	D=25m
Short distance train, suburban railway, metro	60	79	75	72
Interurban train	140	97	94	92
High-speed train	272	104	100	97
Freight train	80	93	89	86
	100	96	92	89

In recent years, national and international specifications require studies of environmental effects in cases of important projects, such as new railway lines. Standards for noise level differ from one country to another.

8.10. Analysis of the accurate mechanical behavior of rail

A model for the analysis of the mechanical behavior of rail has been suggested as follows, (Fig. 8.18), (159):
- rail is simulated by the so-called beam of Timoshenko (linear beam submitted to vertical and transverse bending and torsion),
- support of rail to sleeper is modeled by springs,
- ballast and subgrade are represented by three-dimensional finite elements,
- loads of two wheels are applied symmetrically, (Fig. 8.19).

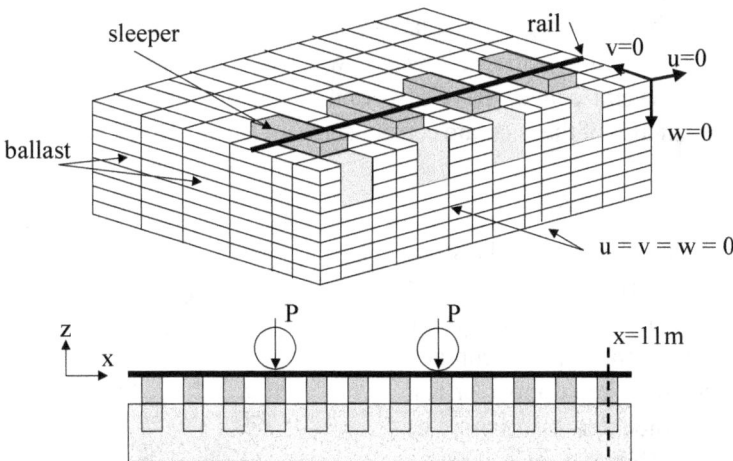

Fig. 8.19. A model for the analysis of the mechanical behavior of rail, (159)

Figure 8.20 illustrates the results of the model concerning vertical settlements of the rail along the longitudinal axis and Figure 8.21 illustrates settlements of the rail in relation to time (t=0.0, application of wheel load).

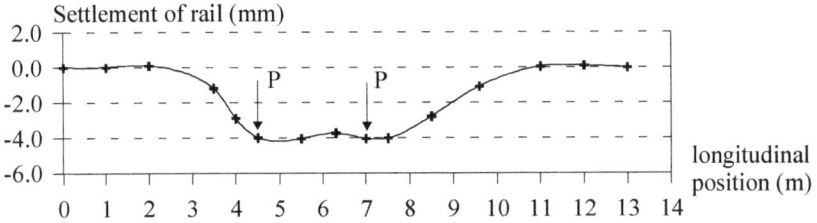

Fig. 8.20. Vertical settlements of the rail along the longitudinal axis, (159)

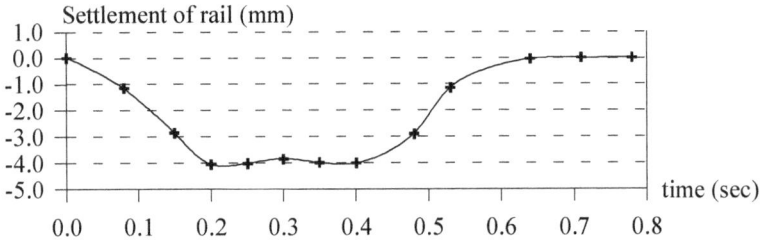

Fig. 8.21. Vertical settlements of the rail in relation to time, (159)

8.11. Application of unilateral contact theories in railway problems

8.11.1. Transmission of forces through contact surfaces

The railway system is based on the transmission of forces through contact surfaces: wheel-rail, rail-sleeper, sleeper-ballast or sleeper-concrete slab, ballast-subballast, subballast-subgrade. Contact surfaces are usually supposed to be continuous and perfect, a hypothesis, however, which is contradicted by physical observations. Unilateral contact theories permit calculation not only of stress and strain fields but also of the accurate contact surfaces between two solids, (137).

8.11.2. Unilateral contact theories

Let us consider the contact between rail and sleeper, (Fig. 8.22), which is perfect in a part Γ_0, whereas in another part Γ_2 there is no contact. We assume that the support of rail to sleeper is modeled by springs of a rigidity k. Calculation of stress, strain, surface Γ_0, surface Γ_2 is based

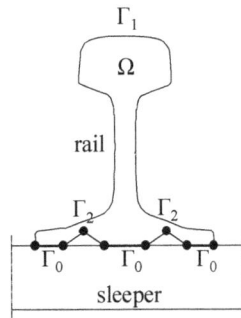

Fig. 8.22. Rail-sleeper contact

on the assumption of Signorini: where contact is perfect, work of external forces is zero; where there is no contact, work of external forces is negative, (161).

8.11.3. Equations of the unilateral contact problem

We assume a mechanical behavior which is static and elastic. Then:

$$\frac{\partial \sigma_{ij}}{\partial x_j} + f_i = 0 \qquad \text{(in volume } \Omega \text{, equation of equilibrium),} \\ \text{(} f_i : \text{mass forces, i.e. gravity)} \tag{8.47}$$

$$\sigma_{ij} \cdot n_j = g_i \qquad \text{(where external forces (wheel load) are given)} \tag{8.48}$$

$$\sigma_{ij} = \lambda \cdot \text{tr}\varepsilon \cdot \delta_{ij} + 2 \cdot \mu \cdot \varepsilon_{ij} \qquad \begin{array}{l}\text{(linear elasticity),} \\ (\lambda, \mu: \text{Lamé coefficients})\end{array} \tag{8.49}$$

$$u_{\text{normal}} + \frac{1}{k} \cdot \sigma_{\text{normal}} \leq 0 \qquad \begin{array}{l}\text{(in surface } \Gamma_2 \text{, no interpenetration} \\ \text{of rail and sleeper)}\end{array} \tag{8.50}$$

$$\sigma_{\text{normal}} \leq 0 \qquad \begin{array}{l}\text{(in surface } \Gamma_2 \text{, reaction of sleeper} \\ \text{is a compression force)}\end{array} \tag{8.51}$$

$$\sigma_{\text{tangential}} = 0 \qquad \text{(in surface } \Gamma_2) \tag{8.52}$$

$$u_i + \frac{1}{k} \cdot \sigma_{ij} \cdot n_j = 0 \qquad \begin{array}{l}\text{(in surface } \Gamma_0 \text{, displacement of slee-} \\ \text{per is equal to displacement of rail)}\end{array} \tag{8.53}$$

$$(u_{\text{normal}} + \frac{1}{k} \cdot \sigma_{\text{normal}}) \cdot \sigma_{\text{normal}} = 0 \quad \text{(in surface } \Gamma_0 \cup \Gamma_2) \tag{8.54}$$

8.11.4. Numerical calculations

The problem described by equations (8.47)÷(8.54) can be transformed to a problem of minimization of the dynamic energy of a field kinematically acceptable, (161). In practice, this means that during successive iterations, if a spring of the modeled system is subjected to tension, it is removed and thus conditions of no contact are created. If a spring is subjected to compression, then there is a perfect contact. Thus, the unilateral contact theories, described above, permit the accurate calculation of contact surfaces of railway (and more generally engineering) problems and the resulting stress and strain. Unilateral contact methods can be combined with applications of finite element methods.

8.12. The boundary element method

The finite element method can provide accurate results for stress and strain but requires a discretization of the whole structure and usually takes time for implementation and application. If we are not interested in what happens in the whole structure but only in its boundaries, then we can use the *boundary* element method, which requires a discretization and construction of the mesh only in the boundary and can provide results at selected points. The boundary element method is flexible, easy to use, and fast to compute, as it requires less computer memory. It can reduce three-dimensional problems to two-dimensional ones and two-dimensional problems to one-dimensional ones. It can be used either separately or in combination with the finite element method.

Figure 8.23 illustrates how the same problem is analyzed by the two methods, (156):

- the finite element method, (Fig. 8.23.a), needs a mesh of 144 elements and 90 nodes,
- the boundary element method, (Fig. 8.23.b), needs a boundary discretization of 30 elements (2-noded) and 30 nodes on the boundary, whereas the interior discretization has 30 linear elements and 30 nodes.

The boundary element method has been used for the calculation of rail problems such as: ground vibrations and noise from rail traffic, stresses in a specific part of the contour of a tunnel, conductance problems to determine the unknown electromagnetic quantities at the interface between materials, etc.

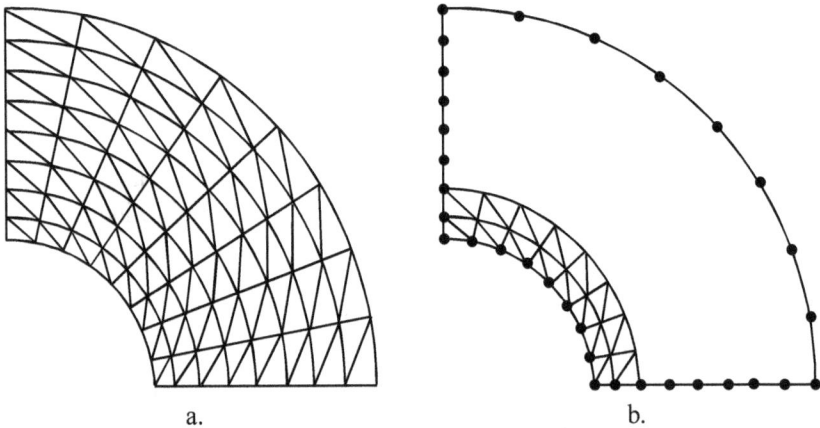

a.   b.

Fig.8.23. Mesh to analyze the mechanical behavior of a tunnel with (a) the finite element method and (b) the boundary element method, (156)

9 Subgrade–Geotechnical and Hydrogeological Analysis

9.1. The importance of the railway subgrade on track quality and its functions

Railway subgrade is particularly important in ensuring that track quality reaches the standard necessary for the safe and comfortable operation of trains. Railway authorities make serious efforts to improve passenger comfort. These efforts, however, concentrate usually on track superstructure (rails, sleepers, ballast, subballast), (see Figure 7.1), and often disregard the fact that many problems appearing at the track superstructure level are traceable to the subgrade, rather than to the superstructure.

It should be stressed that, in the past, studies concerning the railway subgrade were influenced by ideas prevailing in highway engineering. This had the advantage of using the technical experiences acquired with highways, but the disadvantage, when highway design specifications were applied literally, that the techniques implemented were not compatible with the peculiarities of the railway system.

The railway subgrade problem arises in different ways in new and existing track layouts. Accordingly, in new layouts the subgrade design is a function of track loading (axle load and track tonnage), sleeper type, and ballast thickness and strength. A rational consideration of the problem requires that the various parameters describing the subgrade be taken into account: soil type, hydrogeological conditions, and mechanical strengths.

On the other hand, in existing layouts, the problem is encountered in a different way. The policy of the railway authorities for higher speeds and higher axle loads leads to increased subgrade stresses. Since in existing layouts, the lower surface of the subballast and the upper surface of the subgrade have formed a compact zone, which should be disturbed as little as possible, there is only limited possibility of intervention in the subgrade. However, any intervention in the subgrade should be limited to areas where particular problems have arisen and should be scheduled as much as possible to be performed during periodic track maintenance. The decision between improving the subgrade or increasing the ballast layer thickness should be the subject of a technical and economic study and is therefore difficult to make in advance, (191).

The railway subgrade should fulfill the following functions:
♦ enable passenger and freight trains to run safely at the specified speed,
♦ support axle loads of freight and passenger trains,
♦ minimize future track maintenance costs.

These functions can be achieved by:
• limiting settlements of the original ground and of the embankment filling,
• providing stable mechanical behavior under train loads and earthworks,
• facilitating a quick evacuation of rain and ground water,
• ensuring that the condition of the subgrade does not deteriorate during its working life.

9.2. Analytical geotechnical study

9.2.1. Targets of a geotechnical study and soil investigation

Before constructing a new railway line, a geotechnical investigation should be conducted. No new line can be either designed correctly or constructed economically, unless both the nature of soils encountered and the hydrogeology of the route are known in detail.

A geotechnical investigation should indicate:
– whether material for embankment construction is available on site or will have to be transported,
– the appropriate slopes for embankments and cuts,
– whether loosening or densification of soil may take place,
– where weak ground requires treatment before filling can commence,
– where ground water levels may cause problems,
– the measures necessary to ensure the stability of earthwork slopes in the long term,
– where cut sections require particular drainage or protective measures,
– the appropriate type of plant to be used on cut or embankment slopes.

As a geotechnical investigation is costly, it should be conducted in successive stages with the use of the most appropriate techniques.

9.2.2. Preliminary studies

The first stage of a geotechnical investigation is the study of available documents, such as: topographic maps, geological maps, hydrogeological data, aerial photographs, historical investigation records related to the area, etc., (186). Site reconnaissance should also be included at this stage.

The preliminary geotechnical analysis should permit a general understanding of the geotechnical problems likely to be encountered and provide a basis for planning the main geotechnical study.

9.2.3. Techniques and methods of exploration used in a geotechnical study

A geotechnical study is a complex procedure that uses many techniques, such as, (186):

— Geophysical methods (seismic, magnetic, gravimetric, resistivity),
— Physical methods (boreholes, trial pits),
— Mechanical methods (pressuremeter or penetrometer, laboratory tests),
— Hydrogeological methods (piezometers, etc.).

The most widely used method of ground investigation is boring holes into the ground, from which samples may be collected for either visual inspection or laboratory testing. Several procedures are commonly used to drill the holes and to obtain the soil samples.

Table 9.1 lists the wide variety of in-situ tests currently available, (188). Prior to 1960 this list would have included only standard penetration test, mechanical cone test, vane shear test, and plate load test. From the list presented in Table 9.1, several choices are provided in making an in-situ determination of any of the necessary engineering parameters, (184).

9.2.4. Planning the exploration program

The purpose of the exploration program is to determine the stratification and engineering properties of the soils underneath the site where a railway track will be constructed. The main topics of study are strength, deformation, and hydraulic characteristics. The program should be planned so that the maximum amount of information can be obtained at the minimum cost.

The planning of a ground exploration program includes some or all of the following steps:

♦ Assembly of all available information.
♦ Reconnaissance of the area, which includes the following:
  – geological maps,
  – topographic maps,

Table 9.1.
In-situ soil test methods and their applicability, (188)

	Soil identification	Establish vertical profile	Relative density	Friction angle	Undrained shear strength	Pore pressure	Stress history	Modulus of elasticity	Compressibility	Consolidation	Permeability	Stress-strain curve	Liquefaction resistance
Acoustic probe	C	B	B	C	C	-	C	C	-	-	-	-	C
Borehole permeability	C	-	-	-	-	A	-	-	-	B	A	-	-
Cone													
Dynamic	C	A	B	C	C	-	C	-	-	-	-	-	C
Electrical friction	B	A	B	C	B	-	C	B	C	-	-	-	B
Electrical piezo	A	A	B	B	B	A	A	B	B	A	B	B	A
Electrical piezo/friction	A	A	A	B	B	A	A	B	B	A	B	B	A
Impact	C	B	C	C	C	-	C	C	C	-	-	-	C
Mechanical	B	A	B	C	B	-	C	B	C	-	-	-	B
Seismic cone penetration test downhole	C	C	C	-	-	-	-	A	-	-	-	B	B
Dilatometer	B	A	B	C	B	-	B	B	C	-	-	C	B
Hydraulic fracture	-	-	-	-	-	B	B	-	-	C	C	-	-
Nuclear tests	-	-	A	B	-	-	-	C	-	-	-	-	C
Plate load tests	C	C	B	B	C	-	B	A	B	C	C	B	B
Pressuremeter													
Menard	B	B	C	B	B	-	C	B	B	-	-	C	C
Self-boring	B	B	A	A	A	A	A	A	A	A	B	A	A
Screw plate	C	C	B	C	B	-	B	A	B	C	C	B	B
Seismic													
Crosshole	C	C	B	-	-	-	-	A	-	-	-	B	B
Downhole	C	C	C	-	-	-	-	A	-	-	-	B	B
Surface refraction	C	C	-	-	-	-	-	B	-	-	-	-	B
Shear													
Borehole	C	C	-	B	B	-	C	C	-	-	-	C	-
Vane	B	C	-	-	A	-	B	-	-	-	-	-	-
Standard penetration test	B	B	B	C	C	-	-	-	C	-	-	-	A

Legend	A: most applicable
	B: may be used
	C: least applicable

- aerial photographs,
- water and / or oil well logs,
- hydrological data,
- soil data by state authorities.
♦ A preliminary site investigation. In this phase a few borings are made or a test pit is opened to establish in a general manner the stratification, the types of soils to be expected, and the location of the ground water table.
♦ A detailed site investigation.

9.2.5. Geotechnical report and longitudinal section

The results of geotechnical investigations are summarized in the geotechnical report and the longitudinal section. Figure 9.1 illustrates the geotechnical characteristics along the Channel Tunnel (connecting France and the United Kingdom), which is constructed along a layer of blue chalk that was proven resistant to water penetration.

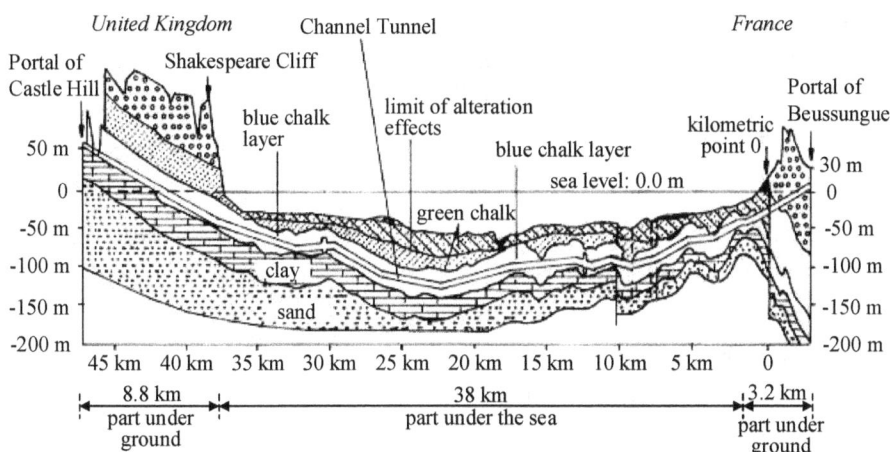

Fig. 9.1. Geotechnical characteristics along the Channel Tunnel

The geotechnical report should give clear and accurate description and recommendations on the following issues, (186):
♦ geotechnical data of each layer,
♦ hydrogeological data: maximum and minimum piezometric levels, drainage requirements,
♦ methods of construction, height of earthworks, suitability of soils for reuse,
♦ values of recommended slopes, embankment design, eventual special techniques such as reinforced soil, etc.,

♦ calculation of mechanical characteristics of soils and of the bearing capacity of the subgrade.

9.3. Geotechnical classifications of soils

In existing railway lines, which were constructed many decades ago, an analytical geotechnical survey is not necessary. Nevertheless, a general knowledge of the basic parameters of the mechanical behavior of the subgrade is essential. The various geotechnical classifications, adopted mainly for highway engineering projects, are a helpful tool for this purpose. These classifications are based principally on the characteristics of granulometric grading and Atterberg limits: liquid limit (from liquid to plastic state), plastic limit (from plastic to semi-solid state), shrinkage limit (from semi-solid to solid state). Occasionally, mechanical parameters are also taken into consideration, such as the California Bearing Ratio (CBR[*]), etc.

Various railway networks within Europe classified soils in the past in a different manner, as illustrated in the cases of the following countries, (191), (199):

♦ the United Kingdom, France, Germany, Switzerland, and others use the Unified soil classification system (USCS), also known as Casagrande classification,

♦ Scandinavian countries mainly rely on granulometric grading of the materials,

♦ Italy, Greece, and others use the AASHO (American association of state highway officials) classification.

Of these, the Unified soil classification is the most generally applicable and most widely used. It was developed from a system proposed by Casagrande. Coarse-grained soils (sands and gravels) are classified according to their grading, whereas fine-grained soils (silts and clays) and organic soils are classified according to their plasticity. Classification is carried out by using particle size distribution data and values of the liquid limit and plasticity index.

The American association for testing and materials (ASTM) has adopted the USCS as a basis for its soil classification, entitled 'Standard test method for classification of soils for engineering purposes'. While the ASTM classification is somewhat different from the USCS, the method followed by both is almost identical. The main difference consists in that the ASTM requires classification tests to be performed, while the USCS allows a tentative classifica-

---

[*] CBR (California Bearing Ratio) is the ratio of the value of load required to achieve settlement of 0.1 inch (2.54 cm) of a sample of the material under study to the value of load resulting in the same settlement of a similar sample of a reference material.

tion based on visual inspection only; however, the ASTM provides a further subdivision of soil classes.

The British standard classification system (BS 5930) is also based on the USCS, but the definitions of sand and gravel are different.

The German classification (DIN 4022) is more analytical and proceeds for silt, sand, and clay at further subdivisions (fine, medium, coarse).

Soils composed of mixtures of two or more groups of fine-grain sizes are usually considered separately. An accurate classification of such soils with similar granulometric composition requires that plasticity characteristics (Casagrande diagram) be also taken into consideration.

Despite the small differences of the various methods, the following classification is commonly acceptable in soil mechanics, (Fig. 9.2):

- Rock: low-, medium-, or high-variability rock, depending on the decay-disintegration it has undergone.
- Gravel (2÷4.76mm<d<20÷76.2mm): Well- or poorly-graded gravel, silty gravel, clay gravel.
- Sand (0.02÷0.074mm<d<2÷4.76mm): silty sand, clay sand.
- Fine-grained soil (0.0001<d<0.05÷0.074mm): Slightly plastic silt, slightly plastic clay, very plastic silt, very plastic clay.

Classification system	Dimensions (in mm)						
	0.0001 0.001 0.01 0.1 1 10 100						
Bureau of soils, 1890-95	Clay	Silt	Sand		Gravel		
	0.005 0.05 1						
Atterberg 1905	Clay	Silt	fine Sand coarse		Gravel		
	0.002 0.02 0.2 2						
MIT 1931	Clay	Silt	Sand		Gravel		
	0.002 0.06 2						
AASHO 1970	Clay colloid	Silt	Sand		Gravel		
	0.002 0.074 2						
Unified classification of 1953 & ASTM of 1967	Fine soils (Clay, Silt)		Sand		Gravel		Stone
	0.074 4.76 76.2						
German DIN 4022	Clay	Silt fine medium coarse	Sand fine medium coarse		Gravel fine medium coarse		Stone
	0.002 0.006 0.02 0.06 0.2 0.6 2 6.3 20 63						

Fig. 9.2. Various systems of geotechnical classifications of soils

9.4. Hydrogeological conditions

Another fundamental parameter, used in determining the subgrade quality, is hydrogeological conditions.

The various railway authorities have tried to determine the maximum groundwater level beyond which hydrogeological conditions are considered to be bad. Figure 9.3 illustrates the minimum distances of the groundwater level from a certain reference position, for hydrogeological conditions to be considered good, according to the regulations of various railway networks, (186), (199), (200).

Even if the groundwater level is below that shown in Figure 9.3, hydrogeological conditions are not generally considered good if suitable drainage devices are not provided, (Fig. 9.4), or the subballast does not have the required transverse slope (3÷5%), (186), (199).

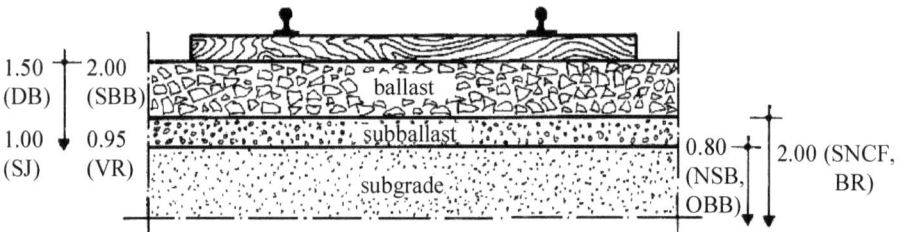

Abbreviations: BR: (former) British railways, DB: German railways, NSB: Norwegian railways, OBB: Austrian railways, SBB: Swiss railways, SJ: Swedish railways, SNCF: French railways, VR: Finnish railways.

Fig. 9.3. Minimum distance (in meters) of the groundwater level from a certain reference position, so as hydrogeological conditions be considered good, according to the regulations of various railway networks

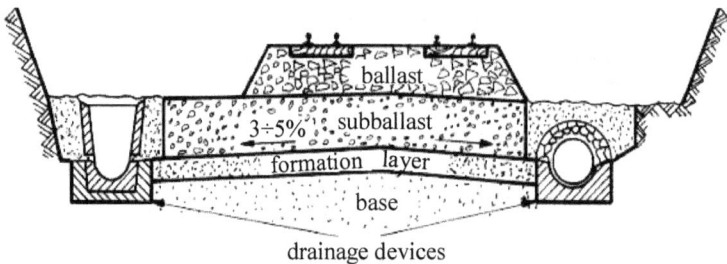

Fig. 9.4. Drainage devices along the railway subgrade

Moreover, areas with large groundwater level fluctuations over time should be the subject of a separate study. In such cases, it is of interest to examine, from a technical and economic point of view, the feasibility of installing a sand filter or a geotextile, (see section 9.15).

For countries experiencing very cold winters, where frost occurs frequently, a third parameter to be taken into account involves the susceptibility of the subgrade to the penetration of frost, (198), (201), (see section 9.11).

9.5. Classification of the railway subgrade

In accordance with the UIC classification, the behavior of the subgrade may macroscopically be described as follows, (186):
- Low settlements and very good support of train loads. This subgrade is hereafter designated as S_3.
- Medium behavior in settlements and in supporting train loads. This subgrade is designated as S_2.
- Large settlements and less satisfactory support of train loads. This subgrade is designated as S_1.
- Extensive settlements and a very inadequate performance in supporting loads. The quality of such a subgrade is designated as S_0.

To the above classes of subgrade should be added the case of a subgrade composed of rock of satisfactory strength. The quality of such subgrade is designated as R. However, more recent UIC classifications include the formerly designated rock subgrade (R) within the subgrade of good quality (S_3).

The criteria for the classification into one of the above categories are geotechnical characteristics of the soil and hydrogeological conditions. Therefore, according to the UIC, (186), the railway subgrade classification is shown in Table 9.2. The reference parameters used in this classification include the percentage of fine grains, plasticity index PI[*], and the Los Angeles coefficient, (see section 12.4.2).

Soils of category S_0 are in principle unsuitable for supporting track properly for the following reasons: they settle extensively, they are inhomogeneous, their characteristics may change over time, and they allow penetration of ballast stones deeply into the subgrade. Such soils should be avoided, whenever possible, when deciding the layout of the track, or replaced by more appropriate soil material. Should this prove impossible and the track has to traverse areas with such unsuitable soils, especially on high earthbanks, the risk of settlements

---

[*] Plasticity index (PI) is the difference between liquid and plastic limits, where liquid limit (LL) is the water content of the soil at the transition between liquid and plastic state and plastic limit (PL) is the water content of the soil at the transition between plastic and semi-solid state. In addition, shrinkage limit is the water content of the soil at the transition between the semi-solid and the solid state. Further decrease in water content beyond the shrinkage limit has no effect in the decrease of volume of the soil. The plastic limit is rather an arbitrary limit between the plastic and semi-solid state of a soil.

should be considered carefully and soil improvement solutions should be examined in combination with an appropriate increase in the ballast and subballast thickness and with an eventual use of geotextiles, (190), (196), (197).

Table 9.2.
Classification of subgrade quality as a function of geotechnical characteristics and hydrogeological conditions, (186)

Geotechnical classification of soils	Hydro-geological conditions	Classification of railway subgrade quality
Rock of very good quality	-	R
Well graded soils containing fine grains[1] < 5% Hard rock with Los Angeles coefficient < 30	-	S_3
Soils containing 5÷15% of fines Sand containing less than 5% of fines Moderately hard rock (e.g. 30 < Los Angeles coefficient < 40)	good bad	S_3 S_2
Soils containing 15÷40% of fines Rocks which are moderately susceptible to weathering Soft rock (e.g. with Los Angeles coefficient > 40) Silty sand and schist with PI > 7	good bad	S_2 S_1
Rocks which are very susceptible to weathering Soils with fine grains > 40% Silty slightly plastic	-	S_1
Organic soils Soils containing more than 15% of fines with a high moisture content Soils containing soluble materials (rock salt, gypsum, etc.) Thixotropic soils (e.g. quick clay) Mixed material/organic soils[2] Contaminated ground (e.g. industrial waste)	-	S_0

[1] Grains are characterized as fine when their dimensions are < 60μ

[2] Certain railways sometimes include these soils in quality class S_1

9.6. Mechanical characteristics of the subgrade

The role of the subgrade is to withstand train loads which have been adequately attenuated by the various track components. In order to withstand loads properly, the subgrade should have the required mechanical properties.

On the basis of a series of tests conducted within the research department (known as ORE[*]) of the UIC, the limits within which the modulus of elasticity ranges were determined for each of the subgrade categories, according to the UIC classification, (Fig. 9.5), (186). For rocky soils, the modulus of elasticity varies in accordance with the nature of the rock material and is in the order of $3 \cdot 10^4$ kp/cm^2, (see section 8.4.6, Table 8.2).

In addition to the modulus of elasticity, classification of subgrade requires the determination of its capacity to withstand train loads. For this purpose, the CBR may be used. Figure 9.5 illustrates values of CBR, which correspond to the various subgrade categories, (186).

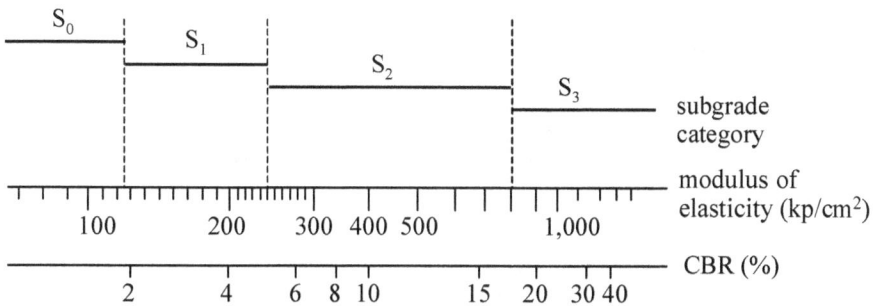

Fig. 9.5. Modulus of elasticity and CBR for various subgrade categories, (186)

9.7. The formation layer

9.7.1. Laying of formation layer in new tracks

If the subgrade subsoil is classified as S_1 or S_2, it is advisable to place on the subsoil an additional top layer composed of a better quality soil material. This layer is often termed the formation layer.

The formation layer should be more compact than the subsoil. Most railways require the formation layer to have a coefficient of 100% by the Standard Proctor Compaction test, while this value is routinely 95% for subsoil layers in the case of embankments, (151).

Use of the formation layer leads to a substantial improvement in the subgrade behavior only if the following two requirements are met, (151):
♦ the subsoil of the subgrade has a low water content, otherwise grains of the subsoil may penetrate the formation layer and deteriorate the transverse slope,
♦ the formation layer should be homogeneous and free of local concentrations of fine-grained material.

---

[*] Initials of the French name of the Research Department ("Organisme des Recherces et d'Essais"), later named ERRI (European Rail Research Institute), of the International Union of Railways (UIC).

The thickness of the formation layer is defined as a function of the sub-grade quality. Values of Table 9.3 were found semi-empirically, (186).

Subgrade quality	Formation layer		**Table 9.3.**
	Quality	Thickness (cm)	**Required thickness of the**
S_1	S_2	30÷55	**formation layer as a func-**
	S_3	20÷40	**tion of the quality of the**
S_2	S_3	20÷30	**subgrade for UIC 1÷4**

Table 9.3.
Required thickness of the formation layer as a function of the quality of the subgrade for UIC 1÷4 group lines, (186)

9.7.2. Improvement of formation layer in existing tracks

Many tracks have been constructed in the past without a formation layer. In some of these old tracks, it is necessary to increase speed and axle load, both of which result in increased stresses in the subgrade. The most practical solution is to in-crease, during maintenance works, the thicknesses of track bed structures, which, however, will be difficult in cases where the height above the track is limited or may not lead to the desired values of stresses in the subgrade. In such cases, it will be necessary to improve the upper surface of the subgrade or install a forma-tion layer in an existing track, which is illustrated in detail in Table 9.4, (189).

9.8. Impact of traffic load on the subgrade

When studying the impact of traffic load (line tonnage) and maintenance con-ditions, Dormon's rule, established for highway engineering, can be used with an accuracy that can be considered sufficient. According to Dormon's rule, the mechanical stresses developed in the subgrade are inversely proportional to the number of the loading cycles, raised to a power λ, (151):

$$\frac{\sigma_1}{\sigma_2} = \left(\frac{N_2}{N_1}\right)^{\lambda} \tag{9.1}$$

where σ_1, σ_2 are the stresses corresponding to N_1, N_2 loading cycles, respec-tively, and λ is an exponent with a mean value of 0.2, (151).

Let P be the axle load and T the daily traffic load (tonnage), (see section 7.5.2). From equation (9.1) it follows that:

$$\frac{\sigma_1}{\sigma_2} = \left(\frac{T_2/P_2}{T_1/P_1}\right)^{\lambda} \tag{9.2}$$

In the case of a constant axle load, $P_1=P_2$, then the equation (9.2) becomes

$$\frac{\sigma_1}{\sigma_2} = \left(\frac{T_2}{T_1}\right)^{\lambda} \tag{9.3}$$

Table 9.4.
Various methods for improving the formation layer, required equipment and machinery, working conditions, and estimated time for execution, (189)

Working methods	Operating conditions							Civil engineering conditions			Man-hours (excluding machine operators), (in hours)	Performance (m/hour)	Time for setting-up and removal (min)
	Possession (h)			Transport of materials		Section of improvement		Excavated depth from top of rail	Speed restrictions (km/h) of improved track	Track removal and replacement			
	5÷8	8÷12	continuous	with | adjacent track	without	up to 100m	over 100m						
Use of earthmoving machinery in short sections (<100m)	+	+	/	+	+	+	-	1.20m***	40÷80**	yes	50÷70** / 60÷80*	152	0
Use of earthmoving machinery over longer sections (>100m)	0	x	+	+	+	-	+	1.20m***	70÷100**	yes	65** / 75*	152	0
Use of tracklaying machinery	0	0	+	/	+	-	+	1.20m***	70÷100**	no / yes*	45** / 55*	20÷40	75
Use of ballast screening machine	+	+	/	/	+	x	+	0.85m	40÷60	no	10 / 15*	90	75
Use of PUSCAL II formation improvement machinery****	0	+	/	/	+	+	x	1.40m	50÷80	yes	252	18	80
Use of PM 200 formation improvement machinery****	0	+	/	/	+	x	+	1.10m	70	no	20	35÷45	120
Laying of thermal insulation sheets by ballast screener SPX****	+	+	/	/	+	x	+	0.85m	40	no	5	100	45

Legend: + = method recommended   / = method not required
x = method possible, decision depends on cost or local circumstances   0 = method excluded for technical reasons

* with laying of geotextile
** cost, performance, and speed restrictions are based on excavated depth of 1.20 m
*** if greater depth excavated, the stability of the adjacent track is critical
**** technical description of this equipment can be found in Technical Specification of UIC under Code 722R, (189)

9.9. Impact of maintenance conditions on the subgrade

9.9.1. The maintenance coefficient

In order to estimate the extent (and therefore the expense) of track mainten-ance works, the maintenance coefficient k is used as a parameter. The entire railway network is divided into sections with approximately the same number of maintenance sessions of track teams along each section; maintenance ses-sions refer to all sessions, with either only manual labor or including the use of mechanical equipment, between two complete renewals of the track. Let I be the annual number of work sessions along a track section and I_m the aver-age number of maintenance sessions along tracks of the same age (i.e. re-newed in the same year), belonging to the same UIC group, and carrying trains with the same axle load. The maintenance coefficient k is defined as:

$$k = \frac{I}{I_m} \tag{9.4}$$

The value k=1 corresponds to an average maintenance level, whereas the value k=0.5 corresponds to a satisfactory maintenance level. It should be noted that when subgrade quality is poor, k may take values up to 10, (Fig. 9.6).

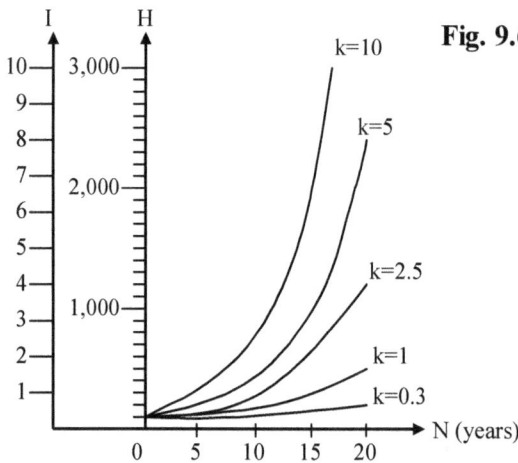

Fig. 9.6. Maintenance expenses for manual work sessions (in man-hours H per km of track) and annual number of work maintenance sessions I (both with manual and me-chanical means) as a func-tion of the maintenance coef-ficient k and the number N of years since the last com-plete renewal. Case of UIC group 1÷3 lines

9.9.2. Impact of the maintenance coefficient on the behavior of track bed and the subgrade

Use of the maintenance coefficient k may contribute to a rational planning of track maintenance works. Figure 9.6 illustrates maintenance expenses (for UIC 1÷3 group lines) as a function of the maintenance coefficient and the number of years elapsed since the last complete renewal. On the basis of the point on the curves beyond which maintenance expenses increase dispropor-tionately, the time for the next complete renewal of the track is rationally de-termined.

When index I, (see Fig. 9.6), exceeds a certain threshold value, track geo-metry standards can no longer be fully ensured. It is then necessary to carry out other methods of track improvement, since maintenance has reached its limits of efficiency, so as to try to reduce the value of maintenance coefficient k. Such a reduction is possible through an increase of the thickness of the track bed layers.

Based on the value of the maintenance coefficient k we can assess whether track bed layers have been properly dimensioned or not, (Table 9.5).

Table 9.5.
Assessment of the proper dimensioning of track bed structures
in relation to the maintenance coefficient k, (186)

	k<1	1<k<2.5	2.5<k<5	k>5
Assessment of dimensioning of track bed structures	Correct dimensioning	Track bed structures slightly under-dimensioned	Track bed structures under-dimensioned. Poor functioning of drainage	Renewal and strengthening of track bed struc-tures is necessary

9.9.3. Impact of the maintenance coefficient on subgrade stresses

Let us now consider two tracks 1 and 2 with different maintenance coeffi-cients k_1 and k_2 respectively. Application of the Dormon rule gives:

$$\frac{\sigma_1}{\sigma_2} = \left(\frac{\tau_2/P_2}{\tau_1/P_1}\right)^{\lambda} \tag{9.5}$$

where τ is the traffic load on each track between two consecutive maintenance sessions. Statistical analysis has shown that τ is proportional to the value of T/k, (151):

$$\frac{\sigma_1}{\sigma_2} = \left(\frac{T_2/k_2/P_2}{T_1/k_1/P_1}\right)^{\lambda} \tag{9.6}$$

Considering the case of two tracks with the same axle load and the same traffic load, equation (9.6) becomes:

$$\frac{\sigma_1}{\sigma_2} = \left(\frac{k_2}{k_1}\right)^{\lambda} \tag{9.7}$$

Equation (9.7) allows calculation of the impact of maintenance conditions on the mechanical stresses of the subgrade.

The use of coefficient k requires the accurate recording of all maintenance problems and expenses.

9.10. Fatigue behavior of the subgrade

Fatigue is defined as the reduction of the mechanical strength of a material under the influence of repeated loads. In the case of metals, it has been found that there is a limit stress σ_0 (called fatigue limit); if this limit stress is exceeded, fatigue effects can occur and may lead to failure without being preceded by any macroscopically large deformations, (see also section 10.8).

However, for the soil materials which constitute the subgrade, fatigue does not involve the development of excessive stresses but of plastic deformations in relation to the loading cycles. Experimental results of the triaxial test under repeated loading conditions show that the following parameter

$$R = \frac{(\sigma_1 - \sigma_3) \text{ of the } 1^{st} \text{ cycle}}{(\sigma_1 - \sigma_3) \text{ of the cycle causing failure}} \tag{9.8}$$

has a limit value in the order of 0.9, beyond which plastic deformations increase very rapidly, as it becomes apparent in Figure 9.7.

For the evolution of plastic deformations ε_N^p as a function of the loading cycles N, the following relation has been suggested, (200):

$$\varepsilon_N^p = a + b \cdot \log N + c \cdot N^\alpha + d \cdot N^\beta + \dots \tag{9.9}$$

where $a < b < \dots$ and the parameters a, b, c, d, α, β are determined experimentally.

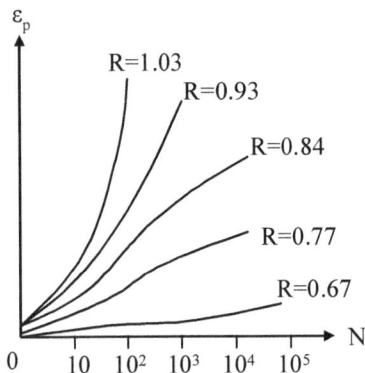

Fig. 9.7. Evolution of plastic deformations ε_p in clay soils as a function of the parameter R and of the number N of loading cycles, (200)

According to equation (9.9), as long as the exponential terms are negligible, plastic deformation proceeds logarithmically and practically stabilizes after a certain number of loading cycles. On the contrary, if the exponential terms of equation (9.9) have a determining influence on total plastic deformation, then the subgrade may show large and dangerously increasing deformations as a function of the loading cycles. Such behavior was observed, under certain conditions, in cases of subgrade quality classified as S_0 or S_1.

9.11. Frost protection of railway subgrades

9.11.1. Frost index

Railway authorities must decide whether the protection of the subgrade against frost should be calculated according to the coldest winter possible, or whether to install a subgrade which would be suitable for average winters, while accepting that frost penetration would occur in extreme conditions.

Frost index is defined as the integral of temperature with respect to time for all periods where the temperature is below zero and is expressed in degrees × hours or in degrees × days. Table 9.6 gives the frost index in relation to the probability of freezing through as well as the expected underratings due to frost penetration in a certain period, (198). Silt is very susceptible to frost, clay is susceptible to frost, but sand and gravel are not susceptible to frost.

Table 9.6.
Frost index, probability of freezing through, and expected number of underratings in a certain period, (198)

Frost index	Probability of freezing through	Expected number of underratings in a certain period
F_2	50%	1 in a 2-year period
F_5	20%	1 in a 5-year period
F_{10}	10%	1 in a 10-year period
F_{100}	1%	1 in a 100-year period

9.11.2. Frost foundation thickness

A layer of material or combination of materials is placed under the ballast layer (or the subballast) in order to protect the subgrade against frost heave. Frost foundation is a term comprising several kinds of frost-heaving prevention materials and measures.

Various materials, such as gravel, cinders, etc., can be used in the frost foundation layer. Figure 9.8 illustrates, in relation to the frost index, the appropriate thickness of the frost foundation layer under the ballast and Figure 9.9 illustrates the appropriate thickness when an insulation layer of foam plastic is used.

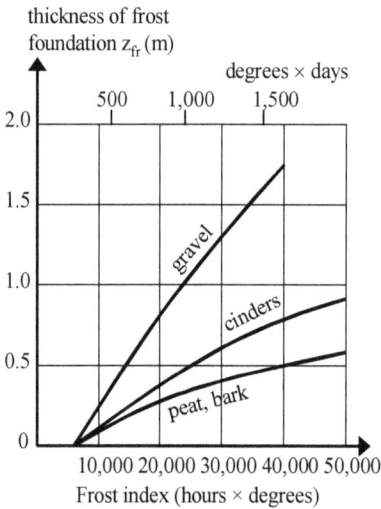

Fig. 9.8. Thickness z_{fr} of frost foundation layer under a ballast layer of 35cm, (198)

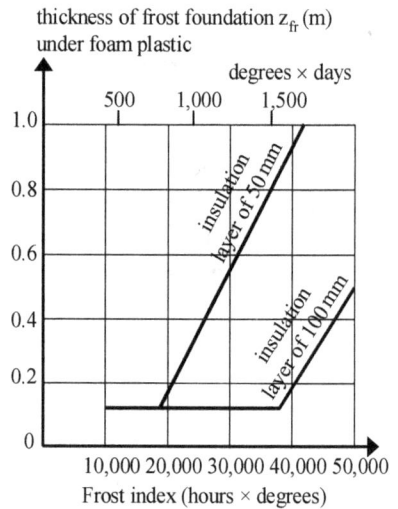

Fig. 9.9. Thickness z_{fr} of frost foundation layer under a ballast layer of 25cm, when an insulation layer of foam plastic is used, (198)

9.11.3. Frost protection methods on existing tracks

Along existing railway tracks which cross areas often freezing in winter, many ways of improving the subgrade (during track renewal) so as to protect against frost have been suggested, (Figures 9.10 to 9.13).

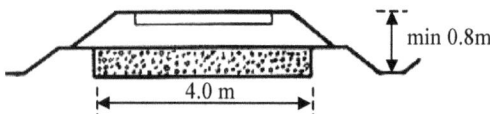

Fig. 9.10. Frost foundation of gravel or cinders

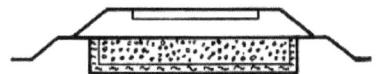

Fig. 9.11. Frost foundation of stone with peat filter

268

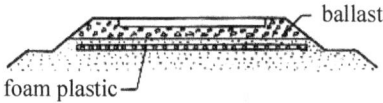

Fig. 9.12. Frost protection with the use of foam plastic

Fig. 9.13. Frost protection with a succession of anti-frost layers

9.12. Track subgrade in cuts and on embankments – Values of slopes

9.12.1. Subgrade in cut sections

Before excavating any cut section, particular attention is paid to studying the geological formations in its path (especially in the case of diaclases), in order to disturb the geological formation equilibrium as little as possible. Parameters to be considered when designing a cut section include safety, cost, and adaptation to the aesthetics of the surrounding environment (and not the other way around).

The slopes of the cut sections are determined according to the results of the geotechnical study, with commonly used values as follows, (184):

$$\text{silt} \frac{1}{2.5 \div 3} \qquad \text{muddy sand} \frac{1}{2 \div 3} \qquad \text{gravel} \frac{1}{1.5 \div 2} \qquad \text{stable clay} \frac{1}{1 \div 1.5}$$

Protection by talus stabilization is usually attained by covering the slopes with shrubs or by planting trees, thus at the same time achieving the merging of the works with the surrounding landscape. Ground drainage is also required along the slopes, to avoid softening.

9.12.2. Subgrade on embankment sections

In the case of an embankment, the quality of geological formations under the embankment should be also considered. Commonly used values of slopes are, (184):

$$\text{usual soil} \frac{1.5 \div 2}{1} \qquad \text{gravel, sand} \frac{2}{1} \qquad \text{erosion-prone soil} \frac{2}{1}$$

If the ground slope is greater than 1:10, it is advisable to secure the embankment base by using a step-like configuration as shown in Figure 9.14.

Due to the subsequent compaction of the embankment, its initial dimensions should be augmented both in width and in height, (Fig. 9.15).

Finally, in the case of very tall embankment sides, a retaining wall or reinforced soil, designed to withstand the soil thrust and train loads, may be used, (Fig. 9.16).

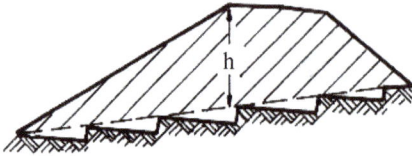

Fig. 9.14. Stepping of the base of the embankment in the case of steep ground

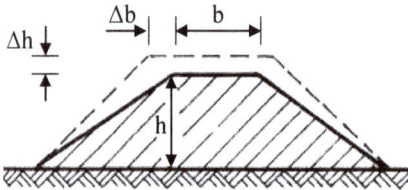

Fig. 9.15. Increase of the initial width and height of an embankment, due to the expected reduction in size by compaction

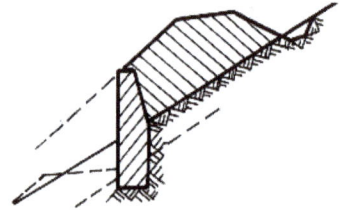

Fig. 9.16. Retaining wall in the case of very tall embankment sides

9.13. The reinforced soil technique

Reinforced soil is a flexible technique which can, in many instances, replace retaining walls. Reinforced soil is an assembly consisting of, (Fig. 9.17):

♦ the embankment edge,
♦ good quality soil material,
♦ metallic bars,
♦ concrete panel.

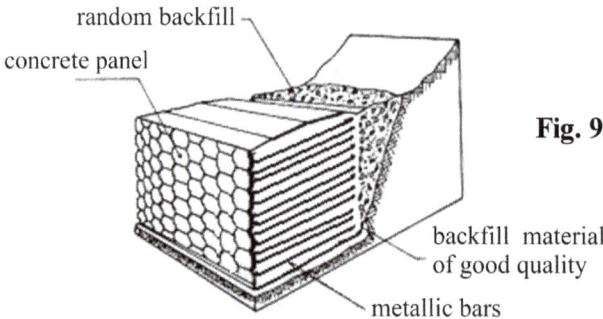

random backfill
concrete panel

Fig. 9.17. The reinforced soil technique, (193)

backfill material of good quality

metallic bars

The reinforced soil technique is especially recommended for medium- and poor-quality subgrades (S_2, S_1) and permits very steep slopes and vertical walls to be safely constructed. Particular attention is required in securing the

270

metallic bars through appropriate anchoring in the soil. A comparative analysis of the construction cost for railway projects in France, (Fig. 9.18), has shown that the reinforced soil solution is both economically and technically advantageous compared to the construction of a retaining wall, especially for heights between 4 m and 12 m, (193). The reinforced soil technique, however, cannot be used in electrified lines, since the electric current return corrodes the metallic bars of the reinforced soil, a situation that may lead to failure.

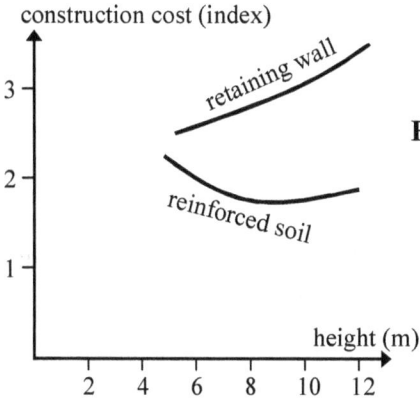

Fig. 9.18. Comparative construction cost of the retaining wall and the reinforced soil technique in railway projects in France, (193)

9.14. Hydraulic analysis and calculation of flows

9.14.1. Level of ground water

The mechanical behavior of the subgrade and the stability of the track are strongly affected by the level of ground water, which should be at least 80 cm lower than the top level of the subgrade, (186). If this is not the case in situ, then the level of ground water must be lowered by using ditches or deep drainage systems, (195).

9.14.2. Semi-empirical formulas for the calculation of run-off flows

Any rainwater likely to penetrate to the subgrade must be quickly channeled away. The top surface of the subgrade should be given the appropriate slope (3÷5%) towards drainage devices, which must be used both transversely and longitudinally along the track, (see Fig. 9.4).

The design of hydraulic devices is based on semi-empirical formulas of hydraulics, which allow to calculate the following two run-off flows during a major storm:

a) run-off flow Q_p coming from rainwater on the surface of the track, which may be in cut or embankment, and can be calculated by the formula:

$$Q_p = k^{\frac{1}{u}} \cdot i^{\frac{v}{u}} \cdot c^{\frac{1}{u} \cdot \frac{w}{u}} \cdot A \tag{9.10}$$

where: i : slope of the longest flow path,

c : run-off coefficient of the subgrade, which is equal to 0.3 or 0.4 for the embankment slope and 0.85 for the track,

A : surface of the catchment area,

k,u,v,w : coefficients depending on the intensity of the storm (10-year, 50-year, 100-year).

b) run-off flow coming from the catchment area, which is calculated in relation to the effective surface A and the average slope i of the catchment area and the average run-off coefficient c of the catchment area. Usual values of c are: around 0.9 for impermeable surfaces, 0.4÷0.8 for cultivated soils, around 0.3 for sandy soils, and around 0.2 for areas with forests, (186).

Among the various methods and formulas, it is worth mentioning the method of the Soil conservation service of the USA, based on Fuller's formula, which calculates the maximum run-off flow Q_{max} as follows:

$$Q_{max} = Q_1 \cdot (1 + 0.8 \cdot \log T) \cdot \left(1 + \frac{2.66}{A^{0.3}} \right) \tag{9.11}$$

where: Q_1 : maximum flow (in m^3/sec) for a return period of T years,

$Q_1 = c \cdot A^{0.8}$, c=1.8,

A : the catchment area (in km^2).

As an example, let us consider a catchment area A of 10 km^2 and a return period T of 10 years. Then: Q_1=11.63 m^3/sec and Q_{max}=47.70 m^3/sec. If A=15 km^2 and T=20 years, then: Q_1=15.71 m^3/sec and Q_{max}=69.90 m^3/sec.

As previous formulas are semi-empirical, they should be checked with actual data, otherwise they can lead to erroneous estimations. For instance, when studying hydraulic aspects of the high-speed line 'TGV Méditerranée' (Valence-Marseille), based on rainfall measurements and analytical formulas, the maximum flow for a 100-year period was calculated at the range of 1÷5 m^3/sec/km^2, whereas the observed extreme values of water flows were around 10 m^3/sec/km^2. For this reason, analytical methods were adapted to actual observations and the minimal diameter of hydraulic devices under the track was 1.0 m, (183).

9.14.3. The rational method for the calculation of run-off flows

An accurate calculation of run-off flows is fundamental to the design of drainage devices and facilities for railway projects. Eventual errors in hydraulic calculations may result in a structure that is either undersized and causes serious drainage problems or oversized and costs more than necessary. The relationship between the amount of precipitation in a drainage basin and the amount of run-off from the basin is complex and becomes even more complex in the changing climate situation of our era. Experience indicates that the design of drainage devices should be based on adequately documented hydrologic analysis. Semi-empirical formulas of the past decades should therefore be complemented with rational analysis, which is possible today with the help of powerful computers, provided that sufficient meteorological data are available.

Although return periods of $50\div100$ years were considered in the past to be satisfactory, the increase of precipitation intensity, as documented by many studies, requires higher return periods on the order of some centuries and considers as a basic run-off flow the probable maximum flood. Various methods have been presented to convert rainfall data into an estimate of peak flow. Each method differs in terms of complexity, data requirements, and reliability of results, as well as in terms of needs and experience of the user of the method. According to the rational method, the maximum run-off flow Q_p can be calculated from the formula:

$$Q_p \ (m^3/sec) = \frac{c \cdot I \cdot A}{3.6} \qquad (9.12)$$

where: c   : run-off coefficient (dimensionless, empirically estimated),
          I   : rainfall intensity,
          A   : the catchment area (in km^2).

The data required to apply formula (9.12) may be obtained with the use of data of the run-off curve or with computer simulations, such as the Hydrological modeling systems developed by the US Army corps of engineers. However, the relation between the variables rainfall and run-off is nonlinear and a large number of hydrological data are nowadays available for these two variables. The method of artificial neural networks (ANN) can provide more accurate forecasts of run-off in relation to rainfall data. Back propagation algorithms of ANN yielded satisfactory results compared to traditional methods based on Eqs. (9.11) and (9.12), (181).

9.15. Geotextiles in railway subgrades

9.15.1. Characteristics, types, and properties of geotextiles

Mechanical and hydrogeological properties of the railway subgrade can be improved with the use of geotextiles, which are permeable geomembranes consisting of synthetic polypropylene or polyester fibres. Geotextiles have a thickness of 0.4÷3 mm and a weight of 70÷350 g/m^2. There are two large geotextile types, (190), (197):

♦ woven geotextiles, composed of two interwoven perpendicular fibre layers. They are strongly anisotropic,
♦ non-woven geotextiles with isotropic behavior; in this type, fibres are laid randomly.

Geotextiles have a large deformability and are used:

• to separate two consecutive layers of granular materials,
• to reinforce a soil layer of insufficient mechanical strength,
• as filters,
• for drainage.

9.15.2. Use and applications of geotextiles in the railway subgrade

Geotextiles are extensively used in railways. They are laid under the subballast (not under the ballast) and their purpose is manifold, (196):

i) To *facilitate proper laying of the track bed structures on the subgrade*. The geotextile laid on top of the subgrade prevents the intrusion of fine-grained elements into the gravel subballast and allows a suitable transverse slope (3÷5%) to be imparted to the subgrade surface. Figure 9.19 illustrates the plasticity characteristics of certain clay soils, in the case of which a strong infiltration of fine-grained soil into the superposed gravel layer was observed, (201).

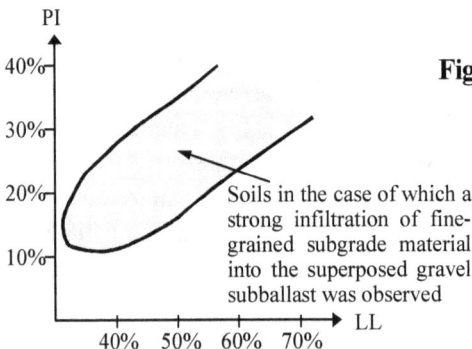

Fig. 9.19. The combination of plasticity index (PI) and liquid limit (LL) at which a strong infiltration of fine-grained soil of the subgrade into the gravel subballast has been observed, (201)

Soils in the case of which a strong infiltration of fine-grained subgrade material into the superposed gravel subballast was observed

ii) To *increase the mechanical resistance of the track bed structures*. Use of geotextiles, however, should not entail an appreciable reduction of the ballast and subballast thickness, because this would result in increased stresses of the subgrade, (196). Geotextiles cannot replace the ballast and gravel in distributing vertical loads. The application, by certain railways, of geotextiles without the laying of subballast in addition to the reduction of the ballast thickness has caused failures (perforation of the geotextile by the ballast, ruining of the transverse slope, etc.). The reinforcing effect of geotextiles may be calculated by numerical methods, such as finite element analysis, (185), (192). Figure 9.20 illustrates the mesh of a finite element model for the assessment of the effects of use of a geotextile on the mechanical stresses of the subgrade. The case of a slab track has been studied and two-dimensional analysis was satisfactory, (185). It has been calculated that use of a geotextile leads to a reduction of stresses on top of the subgrade by around 10%, (185).

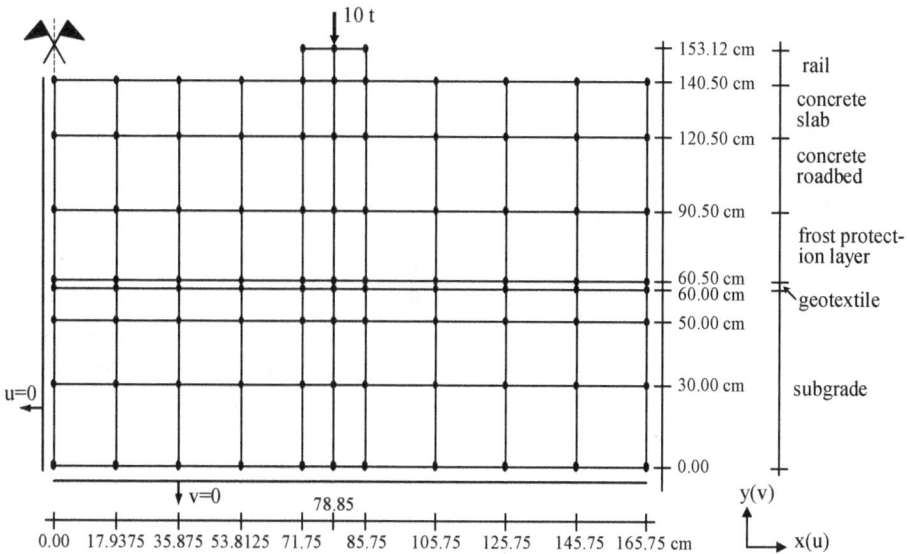

Fig. 9.20. Mesh of a finite element model for the assessment of the reinforcing effects of geotextiles in railway subgrades, (185)

iii) To improve the drainage capability of the subgrade, since geotextiles can operate as *filters* and *drains*. In this case, the geotextile type is selected according to the formulas, (197):

a) for non-cohesive soils:

$$k_g \geq \frac{t_g \cdot k_s}{5 \cdot d_{50}} ,$$

(9.13)

b) for cohesive soils:

$$k_g > 100 \cdot k_s$$

(9.14)

where:

k_g : required geotextile permeability (cm/sec),

t_g : geotextile thickness (mm),

275

k_s : soil permeability (cm/sec),

d_{50} : sieve diameter (mm) allowing passage of 50% of the soil material.

Geotextiles can also protect the subgrade against frost intrusion.

Selection of the appropriate type of geotextile should be based on the required properties such as fracture strength, elongation to failure, perforation strength, compressive strength, water permeability, permeability to avoid infiltration of fine-grained soil in the geotextile, etc. The values of these properties are determined by various tests described in related manuals, (190), (197).

Use of geotextiles along the railway subgrade usually fulfills all the above purposes. Geotextiles, however, are commonly used simply to separate the gravel subballast from the subgrade soil material. Whenever geotextiles have been used, track maintenance expenses have been reduced and thus the geotextile expense is amortized very quickly, (180).

9.16. Vegetation on the subgrade and the ballast

9.16.1. Vegetation on the track and herbicides

As the great majority of railway lines are laid in the countryside, vegetation appears on the subgrade and the ballast. Railways make efforts to control this vegetation by implementing either mechanical or chemical means (herbicides), the latter being the most efficient but with a harmful effect on the environment. Proper drainage, however, of the subballast layer and well-prepared subgrade are important prerequisites for creating conditions which are not favorable to the growth of vegetation.

Spraying of herbicides on both sides of the track is conducted with the use of special rail vehicles and is done either once (on September-October) or twice (in autumn and spring). The most commonly used herbicide is chlorate and staff working on the track should use special and appropriate clothes and undergo yearly medical examinations. Herbicides must have been tested and approved by the relevant authorities. In urban areas and where protection of the water table is necessary, additional restrictions must be set before deciding the use of herbicides. The increasing environmental sensitivity of citizens exercises pressure on railway authorities to minimize the use of herbicides (as far as possible working in daytime, in the absence of wind and rain), (187).

The growth of vegetation along the track can also be reduced by the installation of an asphalt layer under the ballast and on the side paths.

In addition to mechanical and chemical means, other methods such as infrared or electromagnetic or microwave radiation and hot water have emerged recently to control vegetation along the track; these methods, however, are not technically adapted for railway tracks, have a low rate of treatment, disrupt train running, and require 2÷3 treatments per year, (187).

9.16.2. Criteria and dosage for application of herbicides

It is necessary to control vegetation growth, especially along the sides of a railway track, (Fig. 9.21). Chemical control of vegetation growth should be limited to the inspection walkway (D_1), the ballast shoulder (C_1+C_2), the ballast (B), and the inter-track area (A). Each one of the above sections must be treated with different types and quantities of herbicides. However, the ballast (B), the horizontal section of the ballast shoulder (C_1), and the inter-track area (A) should only be treated when absolutely necessary, (187).

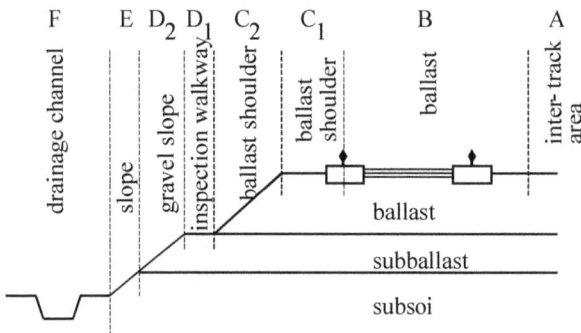

Fig. 9.21. Segmentation of track and use of herbicides, (187)

Herbicides must not be corrosive, combustible, inflammable, or conducting substances. Dosage of herbicides should be adapted to the existing vegetation, (see also section 16.13).

Track sections which are shortly to be renewed must not be chemically treated. New ballast must not be treated for the first few years, when vegetation is sparse. At level crossings, on bridges and in tunnels, as a general rule, no treatment with herbicides should be undertaken.

Environmental awareness puts severe restrictions on the use of herbicides. Their use should be limited and the so-called spread factor* must be greater than 150. Their persistence (i.e. the time required for the herbicides to be transformed) must not exceed 9÷12 months. Their acute toxicity, expressed

---

* Spread factor is defined as: $\dfrac{K_d}{c\,(\%)} \times 100$

  where K_d : the quantity of active substance in an herbicide which is absorbed in μg per g of soil water, balanced with 1 μg of active substance per ml of water,

  c : the percentage of organic carbon content of the soil.

by means of the LD50 index[*], must be generally greater than 500 to avoid oral absorption and greater than 2,000 to avoid skin absorption, for the animals in contact with the herbicides, (187).

Vegetation control along railway lines should not harm the environment. This fact and the increasing pressure to cut costs for vegetation control motivated a number of research projects aiming at balancing environmental protection and vegetation control.

9.17. Earthquakes and the behavior of track and the subgrade

Many areas of the world suffer frequently from earthquakes. It is a crucial problem, which has two aspects, (179):

a) design and dimensioning of structures (bridges, tunnels, buildings), track, and subgrade. Railway bridges and tunnels must be studied in relation to

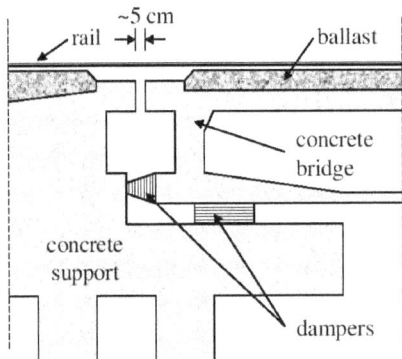

Fig. 9.22. Railway concrete bridge totally isolated, in a region of high seismicity

the maximum seismic ground acceleration (which can attain values up to 4 m/sec^2, but even 6 m/sec^2 for very rare earthquakes), which is decided in each area and country in relation to its seismicity. All structures must have such mechanical strengths, so as to prevent collapse even in the most catastrophic earthquake. In areas with very high seismicity, the most efficient way is to have bridges totally isolated, which is achieved through dampers between the structure and its supports, (Fig. 9.22).

b) Protection of train traffic during an earthquake. This can be achieved by installing a system of seismic sensors along the track, which are connected to an alarm center. Each sensor contains an accelerometer and is installed in a mechanically protected box. If a level of acceleration greater than 0.65 m/sec^2 is registered, traffic must be immediately stopped. For values of acceleration between 0.40 m/sec^2 and 0.65 m/sec^2, a reduction of permitted speeds is suggested. An alarm order is given if three consecutive sensors have sent an alarm message within 5 seconds. All systems must have a high reliability: a false alarm is considered as tolerable only once within a span of 30 years, (182).

---

[*] LD50 (lethal dose in 50% of cases) index measures the acute toxicity of an herbicide and is expressed in μg of active substance per kg of body weight of animals exposed to the herbicide.

10 The Rail

10.1. Rail profiles

Rails are laid on two parallel lines and support and guide the wheels of the train vehicles. Their profile has been the object of continuous improvement since the appearance of railways.

Of the first rail profiles, the only one surviving to this day is the grooved rail, (Fig. 10.1), which is still in use along tracks where the rail top and the pavement surface are at the same level. These include tracks in tramway lines, in level crossings, and in port facilities.

The double-headed rail, (Fig. 10.2), was used in the 19th century, with the expectation that when the upper section was worn out, the rail could be reversed; in this way it was expected that the lower part could be used. Facts did not vindicate this assumption, however, and the double-headed rail was abandoned in many countries at the beginning of the 20th century.

An evolution of the double-headed rail was bull head rail, (Fig.10.3), with a profile of the head of the rail slightly larger compared to the foot of the rail. Few railways (particularly in the United Kingdom) still continue to use bull head rails.

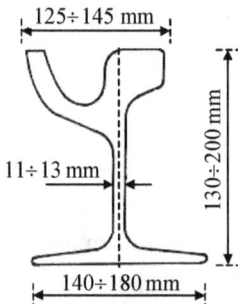

Fig. 10.1. Grooved rail

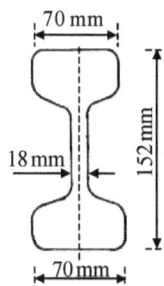

Fig. 10.2. Double-headed rail

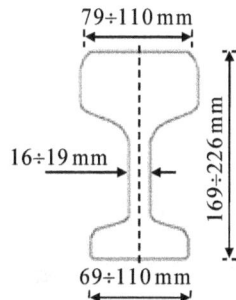

Fig. 10.3. Bull head rail

279

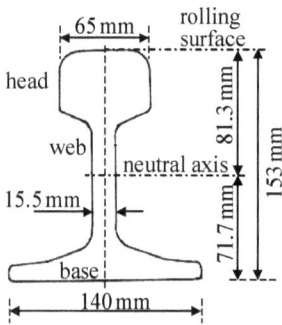

Fig. 10.4. Flat bottom or Vig-noles-type rail; U 36 section (with a weight of 50kg/m)

The rail profile which finally prevailed and is currently widely used is the rail with base, (Fig. 10.4); this profile is known as the flat bottom rail or Vignoles-type rail, named after the Australian engineer who designed it. This rail consists of the head, the web, and the base (foot), (Fig. 10.4). The principal characteristics of its cross-section are the weight w per unit length and the moment of inertia I. A constant goal has been to make any increases of w contingent on a proportionally greater increase of I, to ensure that the I/w ratio increases faster than w. This has led to a constant increase of the height of the rail.

The flat bottom or Vignoles-type rail cross-section was designed on the basis of the need to join rail lengths together, which can be realized with fish-plates, (see section 10.12). The extensive use of continuous welded rails, (see section 10.13), however, is likely to lead in the future to a change in the rail profile.

The increase of axle load and train speed has increased rail loading. The cross-sections of standard gauge rails have been standardized by the UIC, with main types: UIC 50 (weight: 50.46 kg/m), UIC 54 (weight: 54.77 kg/m), UIC 60 (weight: 60.21 kg/m), and UIC 71 (weight: 71.19 kg/m). Figure 10.6, (section 10.4), illustrates cross-sections of rail profiles UIC 50, 54, 60, and 71.

The standardization of UIC has been modified by the European standard EN 13674-1, according to which rail profiles are identified by their weight per meter of length followed by the letter E and a serial number. Thus, similar to UIC 50 rail profile are profiles 50E1÷50E6 of the European standard; similar to UIC 54 rail profile are profiles 54E1÷54E3; and similar to UIC 60 rail profile are profiles 60E1, 60E2 of the European standard, (see also Table 10.4), (203).

10.2. Manufacturing of rail steel

The steel industry manufactures rails following either the oxygen process or the electric arc furnace technique. In the past, Ingot casting has been also used.

The technique of continuous casting, (Fig. 10.5), has been used for some years and can guarantee a rail production more homogeneous than in the past, (159).

Many steel manufacturers have equipment for the continuous quality control of rails, by means of Foucault currents, in order to detect surface and inner defects before delivery.

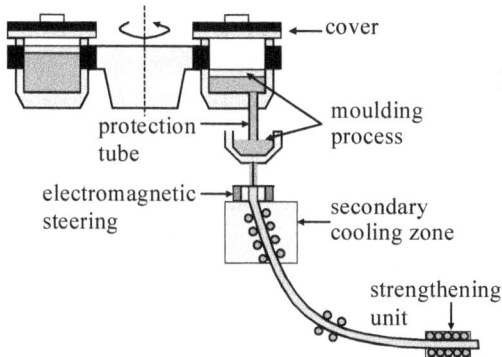

Fig. 10.5. Continuous casting machine for manufacturing rail steel

10.3. Mechanical strength and chemical composition of rail steel

10.3.1. Mechanical strength

The increase of train speed and axle load necessitated the improvement of the mechanical strength of steel used for the fabrication of rails. The greatest tensile strength of rails was 50 kg/mm^2 in 1882, while today it is 70÷130 kg/mm^2. A large increase in rail steel mechanical strength, however, may cause brittle[*] failure and for this reason a further increase of the tensile strength is not desirable.

The quality of the rail steel may be distinguished in two categories:
- normal steel quality, with an ultimate tensile strength of 70÷90 kg/mm^2,
- hard steel quality, used mainly on curves, level crossings, heavy axle loads, etc., with an ultimate tensile strength of 90÷130 kg/mm^2.

10.3.2. Chemical composition

Concerning their chemical composition, rails present a great variety, (202), (214).

10.3.2.1. Carbon

Increased carbon content increases hardness and resistance to wear but at the expense of ductility. Rail steels contain 0.40÷0.80% of carbon.

10.3.2.2. Manganese

All commercial steels contain a small quantity of manganese at a percentage of 0.80÷1.70%. Manganese in excess of this quantity leads to a higher hard-

---

[*] Brittle failure of a material is characterized by a rapid and sudden cracking under stress, while the material does not exhibit any macroscopically visible deformation.

ness. Increasing manganese and reducing carbon can result in rails with high tensile strength and at the same time with high ductility.

10.3.2.3. Chromium and Silicon

Chromium increases hardness and wear resistance. Steels containing 2.0÷2.5% of chromium and 0.30÷0.80% of carbon are very hard and have a high value of tensile strength, of hardness, and of resistance to wear. The content of chromium in rail steel does not exceed usually 1%. Silicon reduces resilience and rails have a varying silicon content between 0.10÷1.30%.

10.3.2.4. Chromium – Manganese

Increasing the content of carbon in the rail steel results in higher hardness but also in lower strength of rail in fatigue. The negative effects of the increase of carbon can be moderated by using more manganese and chromium.

10.3.2.5. Equivalent carbon percentage

The combined effects of carbon, manganese, and chromium can be considered together in a parameter called equivalent carbon percentage, which is given by the formula

$$\text{Equivalent carbon} = C\% + \frac{Mn\%}{3} + \frac{Cr\%}{3} \tag{10.1}$$

It is found that an increase of 0.1% in equivalent carbon results in an increase of tensile strength of rail by 7 kg/mm^2, (147).

As far as the related effects of carbon, manganese, and chromium on wear resistance are concerned, it has been recorded that an increase of 0.1% in equivalent carbon reduces vertical head wear, (see section 10.10), by 4.5÷7.5%, (214).

10.3.3. Rail grades

10.3.3.1. Rail grades according to UIC

The steel industry offers a variety of products for rail profiles, which are classified either according to UIC (based on tensile strength) or according to European standard EN 13674-1 (based on hardness).

Rail grade UIC 700 has a minimum tensile strength of 68 kg/mm^2 and was extensively used until some decades ago. Rail grade UIC 900A (or grade R260 according to European standard) has a minimum tensile strength of

88 kg/mm^2 and a hardness[*] of 300 HB. A variation of this is UIC grade 900B, with a maximum tensile strength of 103 kg/mm^2. There is also rail grade UIC 1100 with a maximum tensile strength of 108 kg/mm^2 and grades 1200, 1200HH, 1400. Table 10.1 gives the chemical composition and mechanical characteristics for the various rail grades according to UIC.

10.3.3.2. Rail grades according to the European standard

Rail grades according to the European standard EN 13674-1 are illustrated in Table 10.2, in which chemical composition, mechanical resistances, and hardness are given, (203).

10.3.3.3. Choice of rail grade

The choice of the appropriate rail grade for a track must take into account the annual traffic load and the radius of curvature of the track. Guidelines, concerning rail grades, of UIC and of various European railways are illustrated in Table 10.3. However, according to the European technical specifications for interoperability, the minimum hardness of rail should be 200 HB, (134), (205), (207).

Choice of rail grade according to the specifications of the various European countries, (Table 10.3), is based on the old version of the European standard and did not take into account the new qualities (R370crHT, R400HT) added in the later amendments of 2011 and 2017 of the European standard. It seems that for very tight curves (R<300 m) rail grade R370crHT could be suitable, for medium curves (300 m<R<700 m) rail grades R350HT and R370crHT could be suitable, and for curves with R>700 m rail grade R260 could be suitable. In any case, the choice of rail grade is not a purely technical issue; aspects such as the lifetime of the rail, maintenance conditions, and the evolution of the wear of the rail, (see section 10.10), should be considered.

American railways use steel qualities with a minimum tensile strength of 90 kg/mm^2 and a hardness of 250 HB.

In any case, for rails supporting heavy axle loads, it is advisable to use rail grades with a higher tensile strength (110÷120 kg/mm^2) and hardness (350÷390 HB). These harder rails are produced following a procedure called thermic hardening, (159).

---

[*] Hardness is the resistance of a material to permanent deformation exerted on it and allows to assess the suitability of the specific material under a given loading. Among the many methods (Brinell, Rockwell, Vickers, named after the scientist who suggested it) for the calculation of hardness, the most commonly used in railway engineering is the Brinell hardness method which indicates the resistance (in kg/mm^2) of a material to the penetration caused by a spherical intender and provides the Brinell hardness, denoted as HB or HBW. Aluminum has a hardness of 15 HB, mild steel of 120 HB, hardened steel of 650÷700 HB, glass of 1,550 HB, and diamond of 8,000 HB.

Table 10.1.
Chemical composition and mechanical characteristics for the various rail grades according to UIC, (207)

Quality	Chemical composition						Tensile strength (kg/mm²)	Elongation after fracture (% of initial length)
	C %	Mn %	Si %	Cr %	P_{max} %	S_{max} %		
Grade 700	0.40÷0.60	0.80÷1.25	0.05÷0.35	-	0.05	0.05	68÷83	≥ 14
Grade 900A	0.60÷0.80	0.80÷1.30	0.10÷0.50	-	0.04	0.04	88 (min)	≥ 10
Grade 900B	0.55÷0.75	1.30÷1.70	0.10÷0.50	-	0.04	0.04	≤ 103	≥10
Grade 1100	0.60÷0.82	0.30÷0.90	0.90÷1.30	0.90÷1.30	0.04	0.03	≥ 108	≥ 9
Grade 1200	0.77	1.1	1.0	0.9	n.a.	n.a.	> 118	n.a.
Grade 1200HH	0.77	0.9	0.3	0.1	n.a.	n.a.	> 117.5	n.a.
Grade 1400	0.3	2.0	1.8	2÷3	n.a.	n.a.	> 140	n.a.

Table 10.2.
Chemical composition and mechanical characteristics for the various rail grades according to the European standard EN 13674-1, (203)

Grade	Quality of steel	Chemical composition						Mechanical characteristics		
		C %	Si %	Mn %	P_{max} %	S %	Cr %	Minimum tensile strength (kg/mm²)	Elongation after fracture (% of initial length)	Hardness (HB)
R200	soft	0.40÷0.60	0.15÷0.58	0.70÷1.20	0.035	0.008÷0.035	≤0.15	68	14	200÷240
R220	standard	0.50÷0.60	0.20÷0.60	1.00÷1.25	0.025	0.008÷0.025	≤0.15	77	12	220÷260
R260	standard	0.62÷0.80	0.15÷0.58	0.70÷1.20	0.025	0.008÷0.025	≤0.15	88	10	260÷300
R260Mn	standard	0.55÷0.75	0.15÷0.60	1.30÷1.70	0.025	0.008÷0.025	≤0.15	88	10	260÷300
R320Cr	intermediate, non heat-treated	0.60÷0.80	0.50÷1.10	0.80÷1.20	0.020	0.008÷0.025	0.80÷1.20	108	9	320÷360
R350HT	hard, heat-treated	0.72÷0.80	0.15÷0.58	0.70÷1.20	0.020	0.008÷0.025	≤0.15	117.5	9	350÷390
R350LHT	hard, heat-treated	0.72÷0.80	0.15÷0.58	0.70÷1.20	0.020	0.008÷0.025	≤0.30	117.5	9	350÷390
R370crHT	hardest, heat-treated	0.70÷0.82	0.40÷1.00	0.70÷1.10	0.020	0.020	0.40÷0.60	128	9	370÷410
R400HT	hardest, heat-treated	0.90÷1.05	0.20÷0.60	1.00÷1.30	0.020	0.020	0.30	128	9	400÷440

Table 10.3.

Guidelines of UIC and of various European railways for the choice of rail grade (in relation to the radius of curvature R) of a track with a maximum axle load of 22.5 t and an annual traffic load of at least 20·10^6 t, (205), (207)

Radius R(m)	≤300	400	500	600	700	800	1200	1500	2500	≤3000	≥3000
UIC	R350HT	R350HT/R260				R260					
Austria	R350HT	R260									
Austria - new single track	R350HT	R260									
Austria- new double track	R350HT										R260
Belgium	R350HT						R260				
Denmark	R350HT						R260				
Germany	R350HT (≥ 30·10^6 gross tonnes)						R260				
Germany- new track	R350HT (≥ 50·10^6 gross tonnes)								R260		
Hungary	R350HT						R260				
Ireland	R260										
Italy	R260										
Luxembourg	R350HT						R260				
Norway	R350HT						R260				
Poland	R350HT						R260				
Romania	R350HT						R260				
Sweden	R350HT	R260									
Switzerland	R350HT		R320Cr/R350LHT				R260				
United Kingdom	R260										

10.4. Choice of rail profile

10.4.1. Standard gauge tracks

The choice of rail profile depends mainly on the traffic load as well as on the expected lifetime of the rail. For a standard gauge track, a general rule is to use a rail with a weight of 54 kg/m for a low-traffic load track and a rail with a weight of 60 kg/m for medium- and heavy-traffic load tracks. Heavier rails (such as UIC 71) were manufactured some years ago, but they have not been used extensively until today, (Fig. 10.6).

The choice of rail profile should take into account the following parameters: speed, axle load, traffic of the track, sleeper spacing, permissible wear, lifetime, and eventual reuse of rail. However, railway authorities have established practical and easy to use guidelines. Thus, it has been customary in a number of European railways, to use for low-traffic load tracks (with a daily traffic load not exceeding 25,000 t) a rail profile with a weight of 54 kg/m. For heavier traffic loads (>35,000 t), a rail profile with a weight of 60 kg/m is suggested. For daily traffic loads from 25,000 t to 35,000 t, if timber sleepers are used, then a rail profile with a weight of 54 kg/m is sufficient; if concrete sleepers are used, then a rail profile with a weight of 60 kg/m is suggested, (147). Some railways suggest the selection of the profile of a weight of 60 kg/m for a higher daily traffic (e.g. 70,000 tonnes/day in Austria for a good quality of the subgrade). However, lower expenditure from the selection of a lower rail profile is marginal and could be surpassed by far if the supposed conditions for ballast and the subgrade quality are not verified in practice.

According to the European technical specifications for interoperability, rail profile should be of the type 60 E2 (which is similar to profile UIC 60, slightly modified), with a weight of 60.18 kg/m, a minimum moment of inertia of 1,600 cm^4, and a minimum hardness of 200 HB, (134).

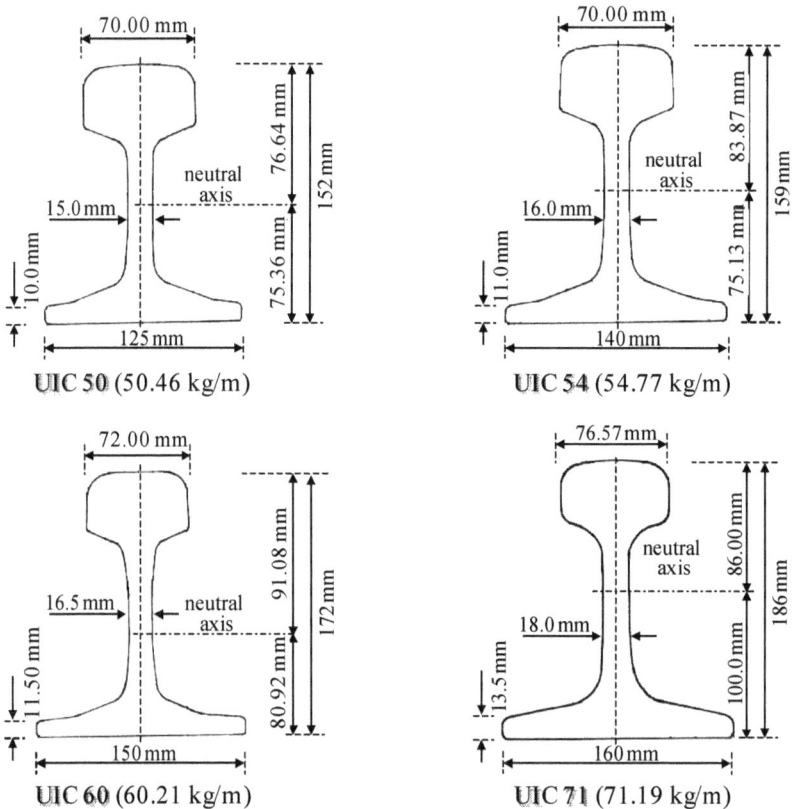

Fig. 10.6. Rail profiles UIC 50 (50 E4*), UIC 54 (54 E1), UIC 60 (60 E1), and UIC 71 (71 E1) for standard gauge tracks, (206)

---

* Equivalent symbol to UIC 50, according to the European standardization.

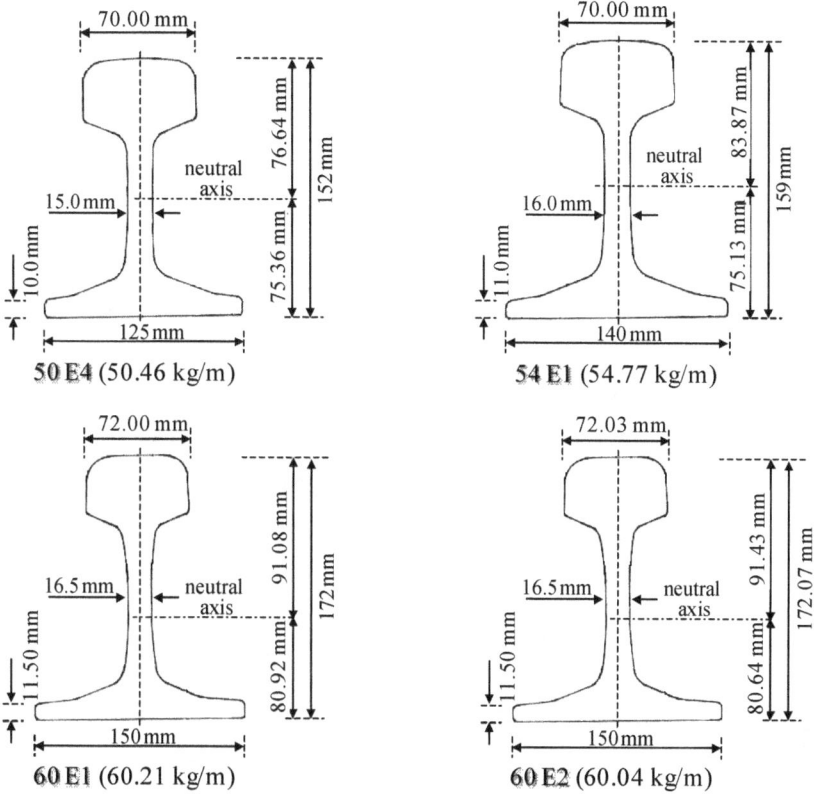

Fig. 10.7. Rail profiles 50E4, 54E1, 60E1, 60E2, according to the European standard 13674-1, (203)

10.4.2. Metric gauge tracks

For metric gauge tracks, there is a variety of rail profiles which have a weight ranging from 30 kg to even 60 kg per meter of length. The most commonly used rail profile for medium- and high-traffic load metric gauge tracks is S49 (weighing 49.05 kg/m of length), (Fig. 10.8), whereas for metric gauge tracks with low traffic, rail profile S33 (weighing 33.47 kg/m of length) can be used.

The choice of the appropriate rail profile for metric gauge tracks is done by taking into account values of speed and axle load. Table 10.4 illustrates recommendations of UIC for suggested rail profiles for metric gauge tracks, (140).

10.4.3. Broad gauge tracks

Broad gauge tracks support greater axle loads compared to standard gauge tracks. For this reason, heavier rails are used in broad gauge tracks. Figure 10.8 illustrates a rail profile extensively used in Russia (with a weight of 65 kg per meter of length).

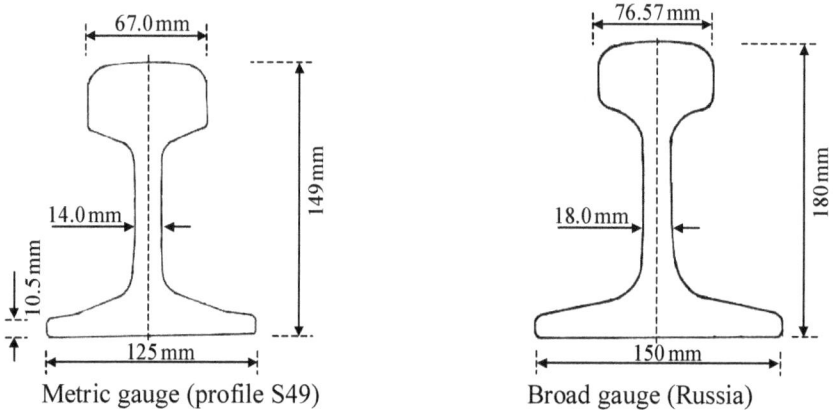

Metric gauge (profile S49) Broad gauge (Russia)

Fig. 10.8. Rail profiles for metric gauge and broad gauge tracks, (206)

Table 10.4.
Choice of rail profile for metric gauge tracks, (140)

V_{max}	160 km/h	120 km/h	80 km/h	100 km/h	60 km/h
Axle load	13t	16t	25÷30t	20t	16t
Traffic	only passenger	mixed	only freight	mixed	principally freight
Suggested rail weight	50÷60 kg/m	50÷60 kg/m	60÷68 kg/m	>40 kg/m	>30 kg/m

10.4.4. Geometrical characteristics of various rail profiles

Tables 10.5 and 10.6 (next pages) present a panorama of rail profiles for standard gauge tracks, according to UIC, the European standard, and national specifications in use by various railway authorities all over the world.

10.5. Transport of rails

The transport of rails from the fabrication area to the worksite should be performed while taking all measures to reduce vertical deflections. Figure 10.9 illustrates for a rail of 36 m long of a weight of 60 kg/m points of suspension, distanced 9 m, during transport. Distances illustrated in Figure 10.9 between suspension points can be reduced to 7m in order to reduce deflection of rail beams during transport.

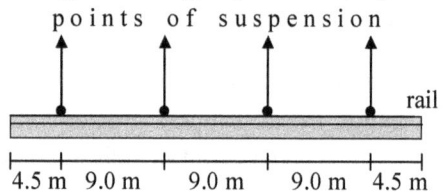

points of suspension

rail

4.5 m 9.0 m   9.0 m   9.0 m   4.5 m

Fig. 10.9. Transport of a rail of 36m long

Table 10.5.
Geometrical characteristics of various profiles of rail according to UIC and national specifications, (206)

	Rail height H (mm)	Base width B (mm)	Head width C (mm)	Web thickness S (mm)	Head height K (mm)	Base thickness D (mm)	Cross-section A (mm^2)	Weight G (kg/m)	Neutral axis X$_H$ (mm)	Moment of inertia J$_X$ (cm^4)
UIC sections										
UIC 50 (50 E4)*	152.00	125.00	70.00	15.00	49.40	10.00	6,428	50.46	75.36	1,934
UIC 54 (54 E1)*	159.00	140.00	70.00	16.00	49.40	11.00	6,977	54.77	75.13	2,338
UIC 54E (54 E2)*	161.00	125.00	67.01	16.00	51.40	12.00	6,856	53.82	76.20	2,307
UIC 60 (60 E1)*	172.00	150.00	72.00	16.50	51.00	11.50	7,670	60.21	80.92	3,083
British sections										
BS 60 R	114.30	109.54	57.20	11.11	35.70	7.60	3,792	29.77	55.70	677
BS 70 A	123.80	111.10	60.30	12.30	39.70	7.90	4,438	34.84	61.30	912
BS 80 A	133.40	117.50	65.40	13.10	42.50	8.70	5,071	39.80	65.60	1,209
BS 90 R	142.90	136.50	66.70	13.90	43.70	9.30	5,684	44.62	68.00	1,600
BS 90 A	142.90	127.00	68.75	13.90	46.00	9.10	5,735	45.02	70.00	1,558
BS 113 A (56 E1)*	158.75	139.70	72.06	20.00	49.21	11.11	7,183	56.39	84.32	2,349
German sections										
SMR 29	115.00	90.00	55.00	12.00	40.00	10.00	3,794	29.78	55.00	624
SMR 32	125.00	70.00	57.58	12.00	50.00	11.00	4,109	32.25	66.40	716
S 30	108.00	108.00	60.30	12.30	31.00	7.00	3,825	30.03	52.14	606
S 33	134.00	105.00	58.00	11.00	39.00	9.50	4,264	33.47	67.33	1,040
S 41, 10R	138.00	125.00	67.00	12.00	43.00	9.50	5,271	41.38	68.23	1,368
S 41, 14R	138.00	125.00	67.00	12.00	43.00	9.50	5,216	40.95	68.23	1,368
S49 (49 E1)*	149.00	125.00	70.00	14.00	51.50	10.50	6,297	49.34	73.30	1,819
S49b (49 E3)*	146.00	125.00	70.00	14.00	48.50	10.50	6,083	47.80	70.95	1,705
S54 (54 E3)*	154.00	125.00	70.00	16.00	55.00	12.00	6,948	54.54	75.00	2,073

* Symbols between parentheses denote codification according to the European standard

Table 10.5 (continued).
Geometrical characteristics of various profiles of rail according to UIC and national specifications, (206)

	Rail height H (mm)	Base width B (mm)	Head width C (mm)	Web thickness S (mm)	Head height K (mm)	Base thickness D (mm)	Cross-section A (mm²)	Weight G (kg/m)	Neutral axis X_H (mm)	Moment of inertia J_X (cm⁴)
American sections										
ASCE 60	107.95	107.95	60.33	12.30	30.96	6.99	3,825	30.03	52.07	606
ASCE 75	122.20	122.20	72.20	13.50	36.10	7.30	4,727	37.11	58.40	952
ASCE 80	127.00	127.00	63.50	13.90	38.10	7.60	5,070	39.80	61.20	1,098
ASCE 90	142.90	130.20	65.10	14.30	37.30	9.10	5,686	44.64	64.50	1,610
ASCE 100 (100 RE)**	152.40	136.50	68.30	14.30	42.10	9.90	6,414	50.35	75.40	2,040
ASCE 115 (115 RE)**	168.27	139.70	69.06	15.87	42.86	11.10	7,236	56.80	75.69	2,730
ASCE 132 (132 RE)**	180.98	152.40	76.20	16.67	44.45	11.11	7,633	65.53	81.28	3,671
ASCE 133 (133 RE)**	179.39	152.40	76.20	17.46	49.21	11.60	8,429	66.17	81.28	3,576
ASCE 136 (136 RE)**	185.74	152.40	74.61	17.46	49.21	11.11	8,606	67.56	85.01	3,949
CB 122	172.21	152.40	74.68	16.51	49.02	11.43	7,743	60.78	80.77	3,080
Various sections										
Australia 141AB	188.91	152.40	77.79	n.a.	n.a.	n.a.	8,901	69.74	n.a.	n.a.
Australia A560	170.00	146.00	70.00	n.a.	n.a.	n.a.	7,725	60.60	n.a.	n.a.
Australia A568	185.70	152.40	74.60	n.a.	n.a.	n.a.	8,602	86.02	n.a.	n.a.
U 33 (46 E2)*	145.00	134.00	64.30	15.00	47.00	10.50	5,898	46.30	67.20	1,588
India IRS 52	156.00	136.00	67.00	15.00	51.00	9.00	6,610	51.89	79.00	2,105
Netherlands SA 42	80.00	80.00	72.20	40.00	46.90	25.00	5,352	42.01	38.20	1,606
Netherlands NP 46 (46 E3)*	142.00	120.00	76.00	14.00	42.50	10.00	5,930	46.55	70.36	1,605
Denmark Form V	141.00	126.00	71.30	13.80	43.00	8.30	5,791	45.46	68.00	1,520
Denmark Form VII	172.00	150.00	74.30	16.50	51.00	11.50	7,687	60.34	80.90	3,055
Switzerland SBB I (46 E1)*	145.00	125.00	65.00	14.00	45.00	9.40	5,880	46.16	69.47	1,631
Turkey 145/46.303	145.00	134.00	64.30	15.00	47.00	10.50	5,898	46.30	67.20	1,588
South Africa SAR 48	150.00	127.00	68.00	14.00	43.00	11.00	6,114	48.00	72.20	1,822
China VRC 43	140.00	114.00	70.00	14.50	42.00	11.00	5,688	44.65	68.80	1,489
China VRC 50	152.00	132.00	70.00	15.50	42.00	10.50	6,562	51.51	70.90	2,037
Russia P50	152.00	132.00	72.00	n.a.	n.a.	n.a.	6,599	65.99	n.a.	n.a.
Russia P65	180.00	180.00	75.00	n.a.	n.a.	n.a.	8,265	82.65	n.a.	n.a.

* Symbols between parentheses denote codification according to the European standard
** Symbols between parentheses denote codification according to AREMA (American Railway Engineering and Maintenance-of-Way Association)

Table 10.6.
Geometrical characteristics of various profiles of rail
according to the European standard, (203)

	Rail height	Base width	Head width	Web thick- ness	Head height	Base thick- ness	Cross- section	Weight	Neutral axis	Moment of inertia
	H	B	C	S	K	D	A	G	X_H	J_X
	(mm)	(mm)	(mm)	(mm)	(mm)	(mm)	(mm^2)	(kg/m)	(mm)	(cm^4)
46 E1	145.00	125.00	65.00	14.00	45.00	4.90	5,882	46.17	69.35	1,641.1
49 E1	149.00	125.00	67.00	14.00	51.50	10.50	6,292	49.39	73.41	1,816.0
50 E1	153.00	134.00	65.00	15.50	49.00	11.50	6,416	50.37	72.44	1,987.8
50 E2	151.00	140.00	72.00	15.00	44.00	11.13	6,365	49.97	70.96	1,988.8
50 E3	155.00	133.00	70.00	14.00	48.00	10.00	6,371	50.02	75.7	2,057.8
50 E4	152.00	125.00	70.00	15.00	49.40	10.00	6,428	50.46	75.36	1,934.0
50 E5	148.00	135.00	67.00	14.00	50.50	10.00	6,362	49.90	71.85	1,844.0
50 E6	153.00	140.00	65.00	15.50	49.00	11.20	6,484	50.90	71.74	2,017.8
54 E1	159.00	140.00	70.00	16.00	49.40	11.00	6,977	54.77	75.13	2,337.9
54 E2	161.00	125.00	67.01	16.00	51.40	12.00	6,856	53.82	77.53	2,307.0
54 E3	154.00	125.00	67.00	16.00	55.00	12.00	6,952	54.57	75.07	2,074.0
55 E1	155.00	134.00	62.00	19.00	53.00	14.00	7,137	56.03	70.74	2,150.4
56 E1	158.75	140.00	69.85	20.00	49.21	11.20	7,169	56.03	74.51	2,321.0
60 E1	172.00	150.00	72.00	16.50	51.00	11.50	7,670	60.21	80.92	3,038.3
60 E2	172.07	150.00	72.03	16.50	51.00	11.50	7,648	60.04	80.64	3,021.0

10.6. Analysis of stresses in the rail

The total stresses developed in the rail are the sum of:
- stresses at the wheel-rail contact (also called Hertz stresses),
- stresses resulting from bending of the whole rail on the ballast,
- stresses resulting from bending of the rail head on the web,
- stresses resulting from thermal effects,
- plastic stresses, remaining in the rail after the removal of external loads.

With the exception of the last category, all other stresses will be calculated with the assumption of an elastic behavior. As discussed in section 8.4.4.2, both theory and experiments show that in most cases rail has an elastic behavior.

10.6.1. Stresses at the wheel-rail contact

The problem of stresses developed at the wheel-rail contact was examined by Dang Van, (219), in accordance with Hertz's assumption that the contact sur-

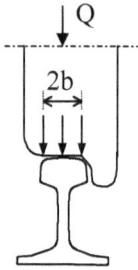

Fig. 10.10. Wheel-rail contact

face between two curved elastic bodies (wheel-rail, Fig. 10.10) is elliptical and the stress distribution along the contact surface is semi-elliptical. Though the problem has a three-dimensional character, measurements have shown that for wheel diameters at the range between 60 cm and 120 cm (covering the majority of cases) the following two-dimensional simplified simulation gives satisfactory results (Eisenmann's theory).

Assuming that all radii of curvature (with the exception of the wheel radius R (in mm)) are infinite and that the wheel load Q (in Newton, Nt) is uniformly distributed, the mean Hertz stress σ_μ is given, according to the Eisenmann analysis, by the formula, (222):

$$\sigma_\mu (Nt / mm^2) = \sqrt{\frac{\pi \cdot E}{64 \cdot (1 - v^2)} \cdot \frac{Q}{R \cdot b}} \qquad (10.2)$$

Substituting the usual values of $E = 2.1 \cdot 10^6$ kp/cm^2, $v = 0.3$, $b = 6$ mm, the following formula is derived:

$$\sigma_\mu (Nt / mm^2) = 1,374 \cdot \sqrt{\frac{Q}{R}} \qquad (10.3)$$

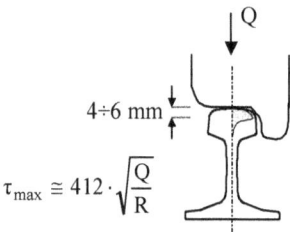

Fig. 10.11. Shear stresses at the wheel-rail contact

The Eisenmann's simplified simulation gives for the shear stress the distribution of Figure 10.11, with a maximum value:

$$\tau_{max} (Nt / mm^2) \cong 412 \cdot \sqrt{\frac{Q}{R}} \qquad (10.4)$$

The maximum shear stress at the wheel-rail contact occurs at a depth of 4÷6 mm from the rolling surface (wheel tread), (222).

10.6.2. Bending stresses of the rail on the ballast

The rail is simulated as a continuous beam on elastic supports, (Fig. 10.12), (209), (211).

The general equation of mechanics:

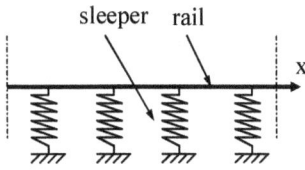

Fig. 10.12. Simulation of rail for the calculation of bending stresses

$$E \cdot I \cdot \frac{d^4u}{dx^4} + k \cdot u = 0 \tag{10.5}$$

gives the following analytical formula for the bending stresses σ_b:

$$\sigma_b = \frac{Q \cdot (h_r - z)}{4 \cdot \gamma_r \cdot I_r} e^{-\gamma_r \cdot x} (\sin \gamma_r \cdot x - \cos \gamma_r \cdot x) \tag{10.6}$$

where:

Q : the wheel load,

I_r : the moment of inertia of the rail in the vertical direction,

h_r : the distance between rolling surface and neutral axis of the rail,

k : the track index, (see section 8.2.2),

u : displacement,

γ_r : $\sqrt[4]{\dfrac{k}{(4 \cdot E \cdot I_r)}}$.

10.6.3. Bending stresses of the rail head on the rail web

The rail head is simulated as a beam lying on an elastic sub-base.

The resulting stresses σ_h are given by the analytical formula, (225):

$$\sigma_h = \frac{Q \cdot (h_c - z)}{4 \cdot \gamma_c \cdot I_c} e^{-\gamma_c \cdot x} (\sin \gamma_c \cdot x - \cos \gamma_c \cdot x) \tag{10.7}$$

where:

h_c : the distance between rolling surface and neutral axis of the rail head,

I_c : the moment of inertia of the rail head,

γ_c : $\sqrt[4]{I_c/4}$.

10.6.4. Stresses caused by temperature changes

Stresses caused by temperature changes are given by the equation:

$$\sigma_{th} = \alpha \cdot E \cdot \Delta\theta \tag{10.8}$$

where: α : the coefficient of thermal expansion for steel,

$\Delta\theta$: the temperature variation.

293

10.6.5. Plastic stresses

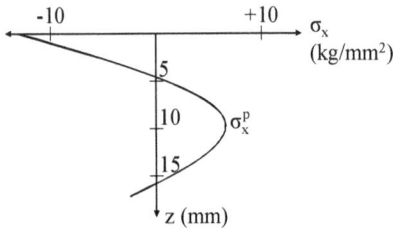

Fig. 10.13. Longitudinal plastic stresses σ_x^p at the plane of symmetry of the rail

Measurements of stresses within the rail have yielded a distribution of plastic stresses as illustrated in Figures 10.13 and 10.14.

Laboratory tests conducted by the Japanese railways on 50T-profile rails (weighing 53 kg/m) have given a distribution of plastic stresses as illustrated in Figure 10.15, (224). Similar results were obtained by the German railways for rail profile S49 (weighing 53 kg/m), (215).

It is worth mentioning that findings of the above tests and measurements are verified to some extent by applications of the FEM for the rail section with an elastoplastic constitutive law for steel. Conditions of contact between the rail and the sleeper can be simulated either with the help of elastic springs or with unilateral contact theories, (204), (161).

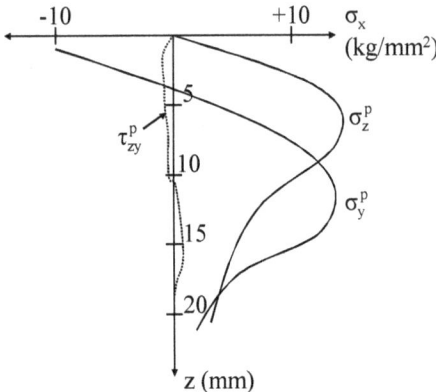

Fig. 10.14. Transverse plastic stresses $\sigma_y^p, \sigma_z^p, \tau_{zy}^p$ at the plane of symmetry of the rail, (224)

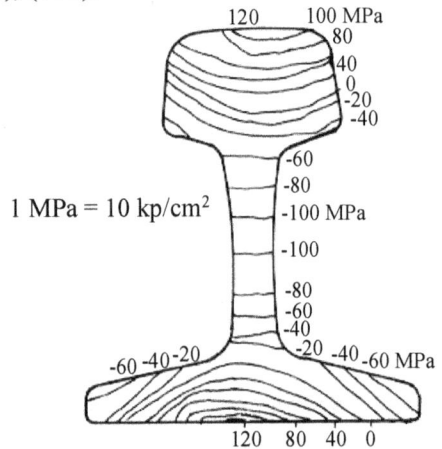

Fig. 10.15. Plastic stresses in rail profile 50T (weighing 53 kg/m), (224)

10.7. Analysis of the mechanical behavior of rail by the finite element and the photoelasticity methods

The mechanical behavior of rail may also be simulated by the finite element method, (Fig. 10.16), (146), (204). In such a simulation, however, and as it was mentioned previously, a delicate problem is to accurately study the limit conditions at the rail-sleeper contact. Therefore, it is often customary to include both the rail and the sleeper in the finite element analysis. Unilateral contact and inequality mechanics theories (presented in section 8.11) can be implemented for the accurate study of the rail-sleeper contact and can be combined with finite element analysis, (137), (161).

Methods of photoelasticity[*] can also be used for the investigation of stress distribution in the rail. Figure 10.17 illustrates curves of equal shear stress based on methods of photoelasticity.

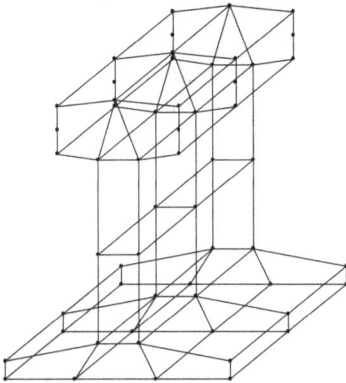

Fig. 10.16. Analysis of the rail with the use of the finite element method, (146)

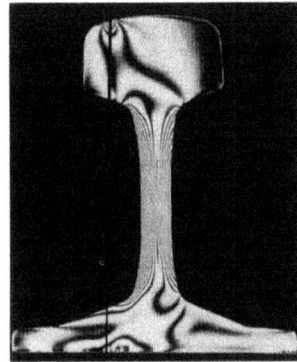

Fig. 10.17. Analysis of stresses in the rail with the use of the photoelasticity method, (147)

10.8. Rail fatigue

10.8.1. Fatigue curve and rail lifetime determination

Fatigue can be defined as the gradual decrease of mechanical strength in a material under the influence of repeated loading, as long as the developed stress exceeds a minimum value σ_o, known as the fatigue (or endurance) limit (Fig. 10.18). For stresses below the fatigue limit ($\sigma < \sigma_o$), fatigue phenomena do not occur. If stresses exceed the fatigue limit ($\sigma > \sigma_o$),

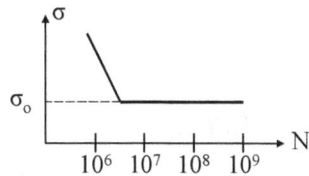

Fig. 10.18. Fatigue curve

then the mechanical strength gradually decreases, leading to failure of the material for stress values lower than values causing fracture during the first loading cycle.

---

[*] Photoelasticity is an optical method to calculate stress and strain in a material.

Theoretical and experimental research of the fatigue phenomenon mainly center on two topics:

a. Determination of the fatigue curve (also known as the Wöhler curve, after the name of the German engineer who first analyzed rail fatigue). The fatigue phenomenon occurs for stresses $\sigma > \sigma_0$, (Fig. 10.18).

b. For a stress history within the fatigue area, determination of the strength reserves of the material. Let σ_1 be a loading history, $\sigma_1 > \sigma_0$, with a lifetime, at the end of which material failure will occur, of N_1 loading cycles. The material is subjected to n_1 loading cycles, and $n_1 < N_1$. Let σ_2 be a second loading history, $\sigma_2 > \sigma_0$, which, in the absence of the σ_1 loading, would have made the lifetime N_2 loading cycles. Unknown is the number n_2 of loading cycles which will lead to failure of the material. The answer is given by the Miner's rule with the approximate formula, (148):

$$\frac{n_1}{N_1} + \frac{n_2}{N_2} \cong 1 \tag{10.9}$$

In the case of more loading histories, the Miner's rule is generalized as follows:

$$\frac{n_1}{N_1} + \frac{n_2}{N_2} + \frac{n_3}{N_3} + \ldots + \frac{n_n}{N_n} \cong 1 \tag{10.10}$$

The origin of the fatigue phenomenon in metals involves internal discontinuities, which are present from the beginning (phase of production of steel). If the developed stresses are sufficiently small, these internal discontinuities do not propagate and thus the state of equilibrium is maintained. However, when stresses exceed the fatigue limit, then internal discontinuities propagate, expand, merge, and may cause fracture of the material under fatigue conditions without any visible macroscopic deformation and for stresses lower than the stresses causing fracture at the first loading cycle.

10.8.2. Rail fatigue criterion

The rail fatigue phenomenon has been extensively researched, both at the experimental, (221), and at the theoretical level, (219), so as to investigate the conditions leading to the commencement of instability due to an internal discontinuity. On the basis of the finding that internal discontinuities tend to propagate towards grains with crystallographic planes less well oriented to resist external loads, and taking into consideration a series of laboratory test results, Dang Van formulated a criterion, named after him, according to which rail fatigue develops in two phases, (219):

1. A first hardening[*] phase, during which stresses develop under the influence of cyclic plastic strains and tend to an equilibrium state. Assuming isotropic hardening of steel, it was deduced that local stresses $\sigma_{ij}(t)$ are related to macroscopic stresses $\Sigma_{ij}(t)$ (those resulting from the continuum mechanics theory) by the equation:

$$\sigma_{ij}(t) = \Sigma_{ij}(t) - \alpha_{ij} \cdot T_o \qquad (10.11)$$

where: α_{ij} : the grain orientation tensor,
  m : the sliding direction,
  n : perpendicular to the sliding plane,
  T_o : the mean shear, defined for the n-cycle as:

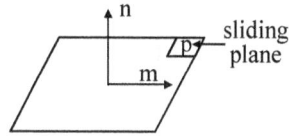

$$T_o^n = \frac{1}{2} \cdot (T_{max}^n + T_{min}^n) \qquad (10.12)$$

2. A second phase during which the propagation of internal discontinuities starts in grains that are already in a plastic state, while surrounding grains are in an elastic state. Since the number of molecules remains constant, the creation of internal voids results in an increase in volume, a fact justifying the investigation of the role of the spherical (or hydrostatic) tensor $(\sigma_{KK}/3)$[**] in the study of the rail fatigue phenomenon,

$$\sigma_{ij} = \frac{\sigma_{KK}}{3} \cdot \delta_{ij} + s_{ij} \qquad (10.13)$$

where: s_{ij} : the deviator tensor,
  δ_{ij} : the Kronecker's delta ($\delta_{ij}=0$ for $i \neq j$ and $\delta_{ij}=1$ for $i=j$).

Macroscopic stresses $\Sigma_{ij}(t)$ result from the continuum mechanics theory, while experimental findings determine n, m, and therefore the tensor α_{ij}.

Local shear $\tau(t)$ in grains with the worst orientation will be:

$$\tau(t) = T(t) - T_o \qquad (10.14)$$

where: T : the macroscopic shear,
  T_o : the mean shear.

---

[*] Hardening is the increase of strength of some materials, when they enter in the plastic area of deformation.

[**] It is worth remembering that a repeating subscript means the sum at all possible values of the subscript (Einstein's notation):

$$\frac{\sigma_{KK}}{3} = \frac{\sigma_{11}}{3} + \frac{\sigma_{22}}{3} + \frac{\sigma_{33}}{3}$$

Analysis of the rail fatigue phenomenon has shown that, (219), (221), (223):

♦ Maximum shear stress develops 10÷15 mm below the rolling surface, (Fig. 10.19). Fatigue analysis of rail (eq. (10.14)) provides global shear stress in the rail, while eq. (10.4) provides shear stress at the wheel-rail contact.

♦ Maximum stresses occur in planes inclined 30° to the vertical.

♦ An increase in wheel diameter causes an increase of internal discontinuities.

♦ Internal discontinuities causing fatigue are proportional to axle load Q raised to a power "a" with a value between 3 and 4 and closer to 4. Thus, rail fatigue is a relation to Q^a.

10.8.3. Evolution of an internal discontinuity

The evolution of an internal discontinuity, elliptical in form, with major axis $2\alpha_c$, is a function of the stress intensity $\Delta\sigma$ exerted on the discontinuity perimeter. For values $\Delta\sigma < \Delta\sigma_{crit}$, discontinuity dimensions remain unaffected by external loading. The region II, (Fig. 10.20), is where the discontinuity presents a large increase, calculated by the equation, (210), (220):

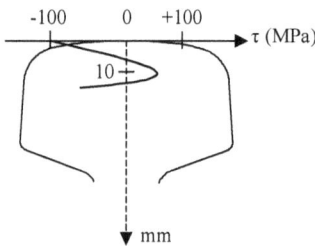

Fig. 10.19. Shear stresses within a rail, (221)

$$\frac{d\alpha_c}{dN} = c \cdot \Delta\sigma^{n_c} \qquad (10.15)$$

where c and n_c are coefficients resulting from laboratory tests. Finite element analysis enables the calculation of the number of loading cycles making an initial discontinuity reach a particular value as a function of cycles of wheel load, (215).

A more empirical relation giving the evolution of an internal discontinuity Y_o as a function of traffic load T of the line has been derived from research conducted within the ORE, (221):

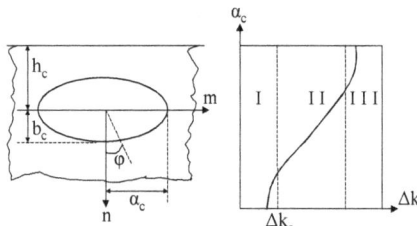

Fig. 10.20. Characteristics of an internal discontinuity

$$Y = Y_o \cdot 2^{T/5} \qquad (10.16)$$

where: Y : value of the discontinuity after passage of traffic load T,
  Y_o : initial value of the discontinuity,
  T : traffic load of the line (in million tonnes per year).

A rail runs a serious risk of fracture when the expanding and merging of internal discontinuities cover more than 55% of the surface of the head of the rail.

10.9. Rail defects

10.9.1. Definition of rail defects

Internal discontinuities of rail which may give rise to rail fatigue are called rail defects. Rail alterations of a mechanical nature occurring under the influence of passing trains are also considered defects. *Rail defects* should be clearly distinguished from *track defects*, the latter being defined as the deviations of actual from theoretical values of the geometrical characteristics of the track. Track defects are exclusively the consequence of train traffic, they are of a macroscopic and geometric nature, and they are usually rectified by track maintenance, (see section 16.4). On the contrary, rail defects are due to initial manufacturing imperfections of the rail, are of a mechanical and microscopic nature, and in most cases are non-reversible.

Defects that are accumulated may render a rail defective in one of the following forms: *broken* (a rail with a gap of more than 50 mm in length and more than 10mm in depth in the running surface), *cracked* (a rail with one or more progressing small gaps of no set pattern, which could lead to breakage), *damaged* (a rail neither broken nor cracked, but with other accumulating defects, generally on the rail surface), (218).

Rail defects may be located at rail ends, away from rail ends, or in welding zones.

10.9.2. Codification of rail defects

Rail defects have been studied, classified, and codified by the International Union of Railways. Thus, broken, cracked, and damaged rails are the object of a code that may comprise as many as four digits, (Table 10.7), (208), (218):

♦ The first digit indicates the point on the rail where a defect appears and can take one of the following four values:
 1. defects in rail ends,
 2. defects away from rail ends,
 3. defects resulting from damage to the rail,
 4. weld and resurfacing defects.
♦ The second digit indicates the location in the rail section, where a defect appears, and more particularly:
 − the place, in the rail section, where the defect originated,
 − the type of welding.

Table 10.7.
Codification of rail defects according to the UIC, (218)

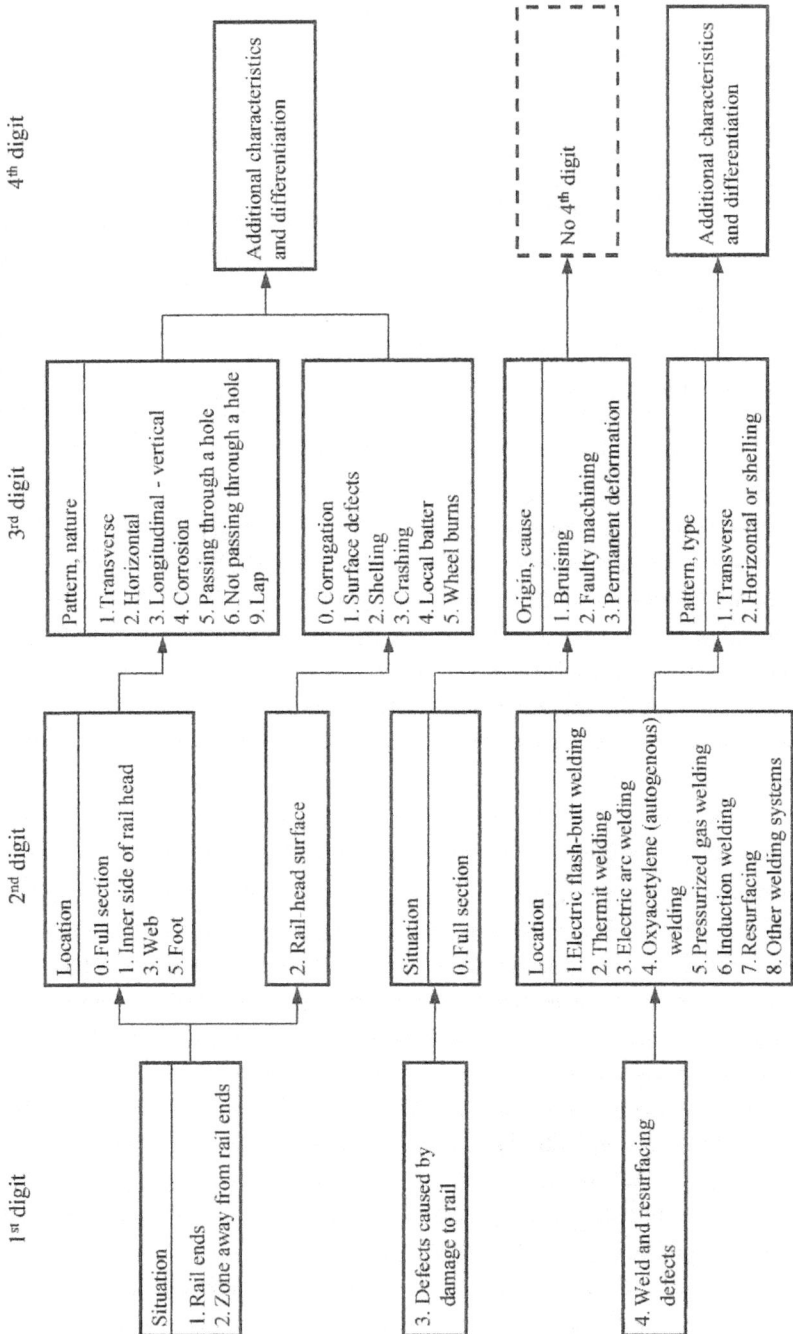

♦ The third digit indicates:
- the pattern of the defect in the case of a broken or cracked rail,
- the nature and the cause of the defect in the case of a damaged rail.
♦ The fourth digit gives the possibility for a further classification in relation to additional characteristics of a specific rail.

The principal rail defects, which are the cause of the most serious risks of rail fatigue and can provoke failure, are described below, (218).

10.9.3. Defects in rail ends

Longitudinal vertical cracking (Rail defect, according to the UIC, 113), causing vertical cracks, which may expand and split the rail head in two. This is a rail manufacture defect. It is detected by ultrasonic equipment, and the affected rail should be immediately replaced.

10.9.4. Defects away from rail ends

10.9.4.1. Tache ovale (Rail defect UIC 211, Fig. 10.21). This corresponds to an initial internal oval discontinuity, caused by thermal effects during rail manufacture. It expands to reach the rail surface. It then becomes visible on the web faces. Breakage of the rail is imminent at this stage. This defect may be the origin of very serious problems and even reach epidemic proportions in rails of the same manufacture. It is detected with the aid of ultrasonic equipment. Most research work on rail fatigue centers on this defect.

Fig.10.21. Tache ovale (expansion of an initial internal discontinuity), (218)

10.9.4.2. Horizontal cracking (Rail defect UIC 212), referring to horizontal cracks of the rolling surface of the rail. It originates at the manufacturing stage (initial internal discontinuities) and may cause locally a separation of the running surface from the head of the rail. It is detected either visually or by ultrasonic equipment.

10.9.4.3. Rolling (running) surface disintegration (Rail defect UIC 221, Fig. 10.22), corresponding to a gradual disintegration of the rolling surface of the rail. Rolling defects are of metallurgical origin and can be detected during track maintenance inspections; affected rails are replaced at scheduled maintenance sessions.

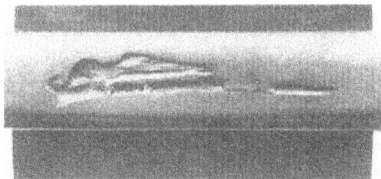

Fig. 10.22. Rolling surface disintegration, (218)

10.9.4.4. *Short-pitch corrugations* (Rail defect UIC 2201, Fig. 10.23). Their cause is train traffic and they consist of corrugations with a wavelength $\lambda=3\div8$ cm. They can provoke many adverse effects: high frequency oscillation of the track leading to higher rail stresses, concrete sleeper fatigue with cracking in the rail seat area, loosening of fastenings, accelerated wear of pads and clips, premature failure of ballast and even of the subgrade, an increase by $5\div15$ dB (A) in noise level. This defect is detected either visually or by appropriate recording equipment. It is repaired by passage of special equipment, which grinds and smooths the rail.

10.9.4.5. Long-pitch corrugations (Rail defect UIC 2202). These have wavelengths $\lambda=8\div30$ cm and occur mainly on the inner rails of curves having a radius of 600 m and smaller. This form of wear is most common on suburban and underground railways carrying a large volume of traffic. Detection and repair are conducted as in the case of short-pitch corrugations.

10.9.4.6. Lateral wear (Rail defect UIC 2203, Fig. 10.24). This affects outer rails in curves and results from rolling stock stresses. It takes a sinusoidal form with a minimum value at the right angle with the fishplated joints. Lateral wear becomes dangerous beyond a certain point, as it affects the track gauge. The various railway authorities specify the permissible value of lateral wear of the rail head, (see section 10.10).

10.9.4.7. Shelling of the running surface (Rail defect UIC 2221). Irregular deformation of the running surface is observed prior to the formation of shells, several millimeters deep in the metal. The cross-section of these shells is extremely variable. Shelling is not an isolated defect and it occurs over a wide area. Detection is done either visually or by ultrasonic testing.

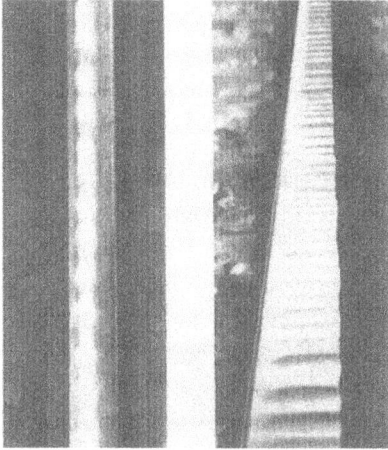

Fig. 10.23. Short-pitch corrugations   **Fig. 10.24. Lateral wear, (218)**

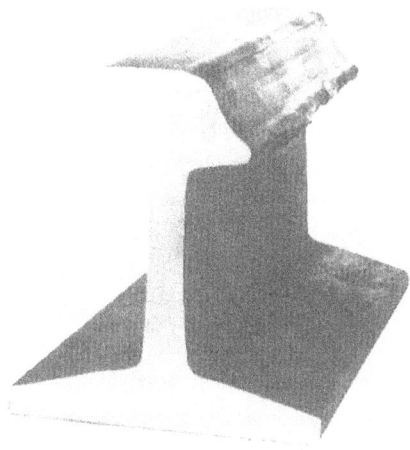

10.9.4.8. Gauge-corner shelling (Rail defect UIC 2222, Fig.10.25). The rails first show long dark spots randomly spaced out over the gauge corner of the rail head. These spots are early signs of metal disintegration which, after a period of evolution, are characterized by the formation of lips on the side face, of cracks, and lastly of shelling in the gauge corner, which can sometimes be quite extensive. This form of shelling usually occurs along the outer rails in curves lubricated to avoid lateral wear. It can be easily detected by visual inspection.

Fig. 10.25. Gauge-corner shelling, (218)

10.9.5. Defects caused by rail damage

10.9.5.1. Bruising (Rail defect UIC 301). This defect is due to traffic load and may be the result of various causes: derailments, dragging parts, damaged tires, handling operation, arcing, or improper use of tools. Cracked and broken rails must be replaced at the earliest opportunity.

10.9.5.2. Faulty machining (Rail defect UIC 302). This is due to traffic load and may have as origin the following: improper in-track drilling of foot or web of rail, faulty cutting, etc. It is inspected visually and may lead to cracking and breakage of the rail, which should be replaced soon after the problem has occurred.

10.9.6. Welding and resurfacing defects

10.9.6.1. Electric flash-butt welding (transverse and horizontal cracking defects). Defects coming from electric flash-butt welding may be either transverse cracking of profile (defect UIC 411) or horizontal cracking of web (defect UIC 412). Transverse cracking may lead either to an internal defect of the head or to a defect located in the foot of the rail. Horizontal cracking develops in a curved shape in the web. Both transverse and horizontal cracking are inspected visually (with a confirmation by ultrasonic equipment) and may cause complete breakage of the affected rail. Fishplating should be urgently carried out and defective rail should be replaced with a new one.

10.9.6.2. Thermit welding (transverse and horizontal cracking) and electric arc welding (transverse and horizontal) defects. Defects due to thermit welding (transverse cracking (defect UIC 421) and horizontal cracking of web (defect UIC 422)) are similar, both in behavior and in treatment, to defects occurring in the case of electric flash-butt welding. Defects due to electric arc welding (transverse cracking (defect UIC 431) and horizontal cracking of rail web (defect UIC 432)) are also similar to other welding defects.

10.10. Permissible rail wear

10.10.1. Vertical wear

The maximum permissible vertical wear of the rail is a function of the maximum train speed and of traffic load. Tables 10.8 and 10.9 give for medium-speed tracks the maximum permissible wear values of the rail head according to British and German regulations, (214). For a rail with a weight of 60 kg/m and a height of 172 mm the maximum permissible vertical wear is, according to German regulations, 14mm on main lines and 22 mm on secondary lines, (158). The evolution of a vertical wear in relation to traffic load of the line has an average rate of 0.7÷1.00 mm per 100 million tonnes, (158).

Table 10.8.
Maximum permissible vertical wear of rail (159 mm high)
according to British regulations, (214)

Maximum speed (km/h)	Maximum permissible vertical wear of the rail head (mm)
> 160	9
120÷160	12
80÷120	15
< 80	18

Table 10.9.
Maximum permissible vertical wear of rail (154 mm high)
according to German regulations, (214)

Category of line	Maximum permissible vertical wear of the rail head (mm)
Lines with an annual traffic load exceeding 19 million tonnes or with a daily load exceeding 25,000 tonnes or speeds exceeding 140 km/h or more than 120 trains per day	12
Lines with an annual traffic load exceeding 7.5 million tonnes or with a daily load between 20,000 and 25,000 tonnes	20
Lines with an annual traffic load exceeding 1.75 million tonnes	26

It should be noted that rail wear caused by locomotive wheels is about 6 times greater than that caused by the wheels of hauled rolling stock.

For broad gauge tracks (such as in India and elsewhere), maximum vertical wear, measured at the central vertical line of the rail, is 13 mm for a rail of a weight of 60 kg/m and 8 mm for a rail of a weight of 52 kg/m. For metric gauge tracks, maximum vertical wear is 4.5 mm for a rail of a weight of 37 kg/m and 3 mm for a rail of a weight of 30 kg/m.

10.10.2. Lateral wear

The maximum permissible lateral wear according to British regulations is defined in relation to a reference point located 3 mm above the lowest point of the rail head and at a 26° angle to the rail axis, (Fig. 10.26).

German regulations measure wear at 45° to the rail axis on a line through the center of the rail shoulder of the full cross-section, (Fig. 10.27). On main tracks and for UIC 60 rail profiles, lateral wear measured on the line MB should not exceed 16 mm for a speed 120 km/h (suburban lines), 14 mm for speeds 120÷160 km/h, and 10 mm for speeds higher than 160 km/h (intercity lines). The sum of the vertical and the lateral wear of the rail head, however, should not exceed 25 mm, (214).

For broad gauge tracks, the permissible lateral wear, measured perpendicularly on the vertical axis at a distance 13÷15 mm below the rail table, is 8 mm for a speed of 130 km/h and 10 mm for a speed of 100 km/h; for metric gauge tracks, permissible lateral wear is 6mm for a speed of 130 km/h and 8 mm for a speed of 100 km/h.

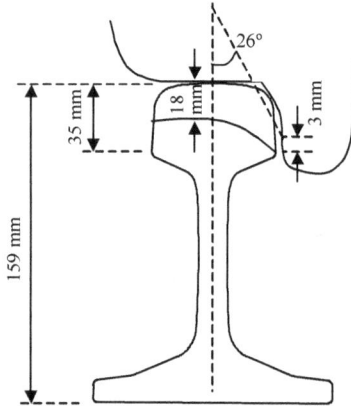

Fig. 10.26. Maximum permissible lateral wear of rail, according to British regulations

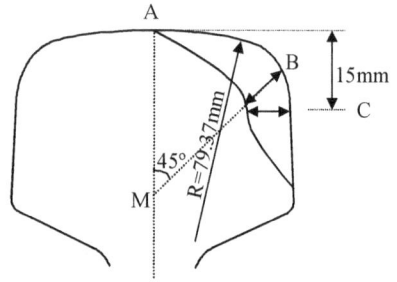

Fig. 10.27. Maximum permissible lateral wear of rail, according to German regulations

10.11. Optimum lifetime of rail

Determining the optimum lifetime of a rail is not a purely technical problem, but should take into account economic aspects. Beyond the service period of the rail, the total cost increases sharply, (Fig. 10.28). It is therefore advisable to replace the rail before all technical strength margins are exhausted. Optimal lifetime of rail is determined by point K, (see Figure 10.28), corresponding to a minimum of total cost (related to the rail), (216). However, a rail removed from a principal line can be used for a certain period on secondary lines.

For a UIC 60 rail profile, the German railways assume a lifetime of around 40 years for principal lines and around 80÷100 years for secondary lines, (142).

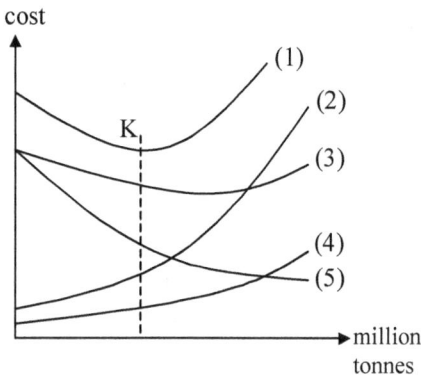

(1) Total cost
(2) Cost of derailments
(3) Total cost excluding derailments
(4) Cost of repairs of wears
(5) Cost of rails

Fig. 10.28. Calculation of the optimum lifetime of rail

306

French railways achieve an average lifetime of 50÷60 years and British railways around 45 years. Once again, the best of the serviceable rail recovered is cropped, welded, and reused for lower category lines, (see also Table 5.4). The average lifetime for rails is around 50 years.

The differences in lifetime among various countries result from different economic considerations and from the fact that some countries (e.g. France) have an increased number of secondary lines, for which evidently rail lifetime has higher values compared to lines with a more dense traffic. As a general rule, a rail profile with a weight of 60 kg/m and a tensile strength of 90 kg/mm^2 is usually replaced after a traffic load of 750 million tonnes.

10.12. Fishplates

Fig. 10.29. Fishplates joining rails

Until about 60 years ago, tracks were laid in all rail networks (and are still laid in some of them) by leaving gaps between consecutive rails, which were joined with fishplates, (Fig. 10.29). The basic purpose of the gaps was to absorb length variations due to temperature variations.

The fishplate technique was detrimental to rail transportation in several ways:

♦ it significantly reduced passenger comfort,
♦ it caused considerable fatigue and wear of both the wheel and the rail,
♦ it greatly increased maintenance expenses, on the one hand due to the necessary inspections to ensure proper condition of all fishplate parts, and on the other because of the height irregularities arising in the fishplate region.

In standard gauge tracks, fishplates are usually installed every 36 or 54 m, after prior welding of the rails of 18 m in groups of two or three. A characteristic of fishplate-joined rails is their capability for contraction-expansion, depending on temperature fluctuations. Every rail profile has a corresponding fishplate type, as well as a particular form of bolt.

10.13. The continuous welded rail

10.13.1. The continuous welding technique

From the time when railways were first introduced, efforts were made to increase the length of rails, the ultimate goal being a continuous track without gaps between rails. The *continuous welded rail (cwr)* is the result of welding together pieces of rail as obtained from the manufacturer in various lengths, commonly 18 m, 24 m, 30 m, or 36m for standard gauge tracks. The usual maximum length for the production of rails is nowadays 36 m (United Kingdom, France, Italy, etc.), but may attain greater values in some countries (60 m in Germany, up to 108 m in Austria, and even up to 120 m elsewhere), (159). In contrast to fishplate-joined rails, cwr are characterized by a rail region where no temperature-induced length change occurs. Continuous welding does away with fishplates, with all the obvious beneficiary consequences this entails.

Although it is a technically simple concept, continuous welding took a long time to be adopted in railway technology. This delay was due to the following reasons, (141), (212):

♦ As aforementioned, a characteristic of the continuous welded rail is the absence of length variation. This is a result of the friction forces between sleeper and ballast as well as between rail and sleeper. These forces, however, cannot be ensured unless the rail-sleeper connection is stable. This was enabled six decades ago by elastic fastenings, (see section 11.9.2.2).

♦ The fatigue behavior of welds, which undergo repeated stresses by the passage of trains, was not adequately known. Research on welds has shed light on this aspect and there are no reservations concerning this matter.

♦ Finally, the risk of buckling was also considered, due to the great length of the cwr. Research on the mechanical resistance of the ballast, which opposes buckling, combined with a track weight increase, has addressed the problem in a satisfactory manner.

In the case of tramway lines, which are fully restrained by being embedded in the road pavement, the problem of longitudinal forces does not occur and longer rails could be implemented.

10.13.2. Mechanical behavior of continuous welded rail

10.13.2.1. Assumptions

In recent years, the development of nonlinear constitutive laws and numerical models as well as the knowledge of fatigue mechanisms have contributed to a more accurate analysis of the mechanical behavior of the cwr, at the price, however, of complex and time consuming calculations, (212). For this reason, railways still use a simplified analysis, which gives a satisfactory qualitative representation of phenomena, in addition to rendering safety-oriented results.

10.13.2.2. Simplified mechanical analysis of continuous welded rail

It is assumed that the behavior of all materials is elastic and that ballast resistance is uniform in space and constant in time.

The continuous welded rail is simulated by a bar of a length L and a cross-sectional area S, (Fig.10.30). Under the influence of a temperature variation $\Delta\theta$, the length change of the bar will be:

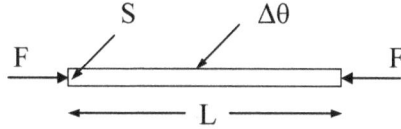

Fig. 10.30. Simplified simulation of the continuous welded rail

$$\Delta\ell_{\Delta\theta} = \alpha \cdot L \cdot \Delta\theta \tag{10.17}$$

where α is the thermal expansion coefficient of steel.

The ballast resists the change of length (resulting from temperature variations) by a force F. The length change due to F will be:

$$\Delta\ell_F = \frac{F \cdot L}{E \cdot S} \quad \text{(Hooke's law)} \tag{10.18}$$

Total stress and strain will result from the superposition of the effects of the two forces (of opposite direction) previously mentioned and therefore:

$$\Delta\ell_{tot} = \alpha \cdot L \cdot \Delta\theta - \frac{F \cdot L}{E \cdot S} \tag{10.19}$$

We are looking for a value of F such that the change in length $\Delta\ell_{tot}$ will be zero. From equation (10.19) it follows that:

$$F = \alpha \cdot E \cdot S \cdot \Delta\theta \tag{10.20}$$

a relation showing that force F is independent of rail length but proportional to the cross-sectional area, therefore depending on the rail profile.

From the aforementioned equation it can be calculated that a force of 1.85 tonnes per degree centigrade ($\Delta\theta$) is generated in the case of UIC 60 profile and of 1.60 tonnes for UIC 54 profile.

10.13.2.3. Distribution of forces along a continuous welded rail

Forces generated along a cwr by temperature variations are transmitted through the fastenings and sleepers to the ballast. Let r denote the ballast

resistance, with values ranging from 0.5 to 1.0 tonnes per
meter of track. This resistance
is obviously zero at the end of
the cwr and, cumulatively
increasing over a length ℓ_A,
(Fig. 10.31), it generates a
force equal to F. Therefore,
and taking into account equation (10.20), it will be:

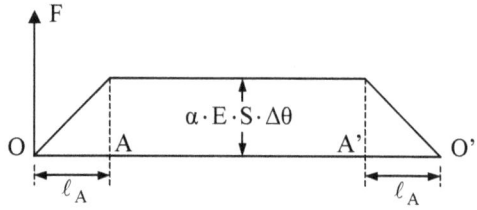

Fig. 10.31. Diagram of forces developed within cwr (O=cwr left end; O'=cwr right end)

$$r \cdot \ell_A = F = \alpha \cdot E \cdot S \cdot \Delta\theta \qquad (10.21)$$

Equation (10.21) gives: $\quad \ell_A = \dfrac{\alpha \cdot E \cdot S \cdot \Delta\theta}{r} \qquad (10.22)$

The length ℓ_A corresponds to what is often referred to as the expansion
zone. Beyond this length, the force F due to the ballast resistance completely
balances out the force developed by temperature variations along the cwr.
Therefore, beyond the length ℓ_A no length change takes place.

Considering an average value r=0.75 t/m for the ballast resistance and the
case of UIC 60 rail (S=76.70 cm^2), we will have for $\Delta\theta$=35 °C:

$$\ell_A = 85.1 \text{ m} \qquad (10.23)$$

which even in extreme cases of temperature variations cannot exceed a limit
value in the order of 150 m.

Since the length of the cwr cannot be smaller than $2 \cdot \ell_A$ (because if it
were, no point of the cwr would remain immobile during temperature variations), it follows from the foregoing that the minimum length of a cwr is
2·150 m = 300 m.

10.13.2.4. Length changes in the expansion zone

The cwr undergoes a change in length due to temperature variations only in
the expansion zone ℓ_A, beyond which every cwr point remains immobile. The
displacement of the point O, (see Fig 10.31), caused by the superposition of
strain generated by temperature variation and ballast resistance, is calculated
as follows:

i. Due to a temperature variation $\Delta\theta$, a length change $\Delta\ell_{\Delta\theta}^{\ell_A}$ will occur:

$$\Delta\ell_{\Delta\theta}^{\ell_A} = \alpha \cdot \ell_A \cdot \Delta\theta = \alpha \cdot \Delta\theta \cdot \frac{\alpha \cdot E \cdot S \cdot \Delta\theta}{r} = \frac{\alpha^2 \cdot E \cdot S \cdot \Delta\theta^2}{r} \qquad (10.24)$$

ii. Due to ballast resistance, the value of which is zero at point O and $r \cdot l_A$ at point A, a length change $\Delta\ell_r^{\ell_A}$ will occur. Assuming a linear evolution between O and A, there will be a resultant force equal to $r \cdot \ell_A / 2$, producing a displacement

$$\Delta\ell_r^{\ell_A} = \frac{r \cdot \ell_A}{2} \cdot \frac{\ell_A}{E \cdot S} = \frac{r}{2 \cdot E \cdot S} \ell_A^2 = \frac{r}{2 \cdot E \cdot S}\left(\frac{\alpha \cdot E \cdot S \cdot \Delta\theta}{r}\right)^2 = \frac{\alpha^2 \cdot E \cdot S \cdot \Delta\theta^2}{2 \cdot r} \quad (10.25)$$

iii. If we combine equations (10.24) and (10.25), we obtain:

$$\Delta\ell_{tot}^{\ell_A} = \left(\Delta\ell_{\Delta\theta}^{\ell_A}\right) + \left(\Delta\ell_r^{\ell_A}\right) = \left(\frac{\alpha^2 \cdot E \cdot S \cdot \Delta\theta^2}{r}\right) + \left(-\frac{\alpha^2 \cdot E \cdot S \cdot \Delta\theta^2}{2 \cdot r}\right) =$$

$$= \frac{\alpha^2 \cdot E \cdot S \cdot \Delta\theta^2}{r} - \frac{\alpha^2 \cdot E \cdot S \cdot \Delta\theta^2}{2 \cdot r} \Rightarrow \Delta\ell_{tot}^{\ell_A} = \frac{\alpha^2 \cdot E \cdot S \cdot \Delta\theta^2}{2 \cdot r} = k \cdot \Delta\theta^2 \qquad (10.26)$$

where $k = (\alpha^2 \cdot E \cdot S)/(2 \cdot r)$ is constant for specific types of ballast and rail.

10.13.2.5. Rail welding

Rail welding can be realized following either the flash-butt (electric) welding technique, usually in depots, or the thermit (aluminothermic) welding technique, usually on site. Whatever the welding technique chosen, it is critical to appropriately control quality of welds in order to safeguard the longevity of the cwr.

10.13.2.5.1. Flash-butt (electric) welding

The flash-butt (electric) welding process is a method of joining metals, in which the heat generated, necessary to forge the joint, is created by the resistance of the rails being welded to the passage of an electric current. Unlike the thermit welding process, no additional chemicals or metals are required to make the weld. In flash-butt welding the parent metal is consumed during the welding cycle, thus creating the necessary heat along the rail ends in order to accomplish the merging action and consolidate the joint. A total length of approximately 25÷35 mm, depending upon the rail section, is consumed per weld.

Flash-butt welding is usually carried out in depots-workshops and the time required per weld is around 3 minutes. It can be carried out also in mobile depots and rarely on the track.

10.13.2.5.2. Thermit (aluminothermic) welding

The thermit (or aluminothermic) welding process is based on the reduction of heavy metals from their oxides with the aid of aluminium. This reaction is strongly exothermic, since a very large quantity of heat is generated, and is brought about by the strong affinity which aluminium exhibits towards oxygen. Thermit welding is usually realized on site (not on depots) and the time required per weld is around 20 minutes. Many European railways use the German thermit process called SKV, a welding process with short preheating, (142).

In every welding procedure, the appropriate control of welds is critical for the longevity of the cwr, (203).

10.13.2.6. Distressing of a continuous welded rail

It is desirable that cwr welding and laying be carried out at a temperature ranging between the upper and lower extremes, so as to minimize cwr stresses.

Regardless of the cwr laying temperature, however, the reduction of stresses resulting from temperature variations is sought. This is achieved by distressing the cwr and creating free expansion (or contraction) conditions. Distressing is done after an elapse of time from the cwr laying, depending upon the traffic load necessary to stabilize the track. This load is usually 100,000 tonnes in the case of timber sleepers and 20,000 tonnes for concrete sleepers.

Distressing should be done successively along 800÷1,000 m and exceptionally on 1,200 m track lengths. The following procedure can be implemented:

i.   If the cwr is longer than 1,200 m, distressing is done in sections. The rails are cut at the end of each section and the ends are diverted to enable free rail change of length.

ii. Fastenings are loosened.

iii. Rails are placed on rollers (of a diameter of 20 mm (Φ20), every 10÷20 sleepers), so as to reduce friction as much as possible.

iv. Further reduction of friction is achieved by lateral blows along the rail by wooden or plastic sledge hammers.

v.   If at the time of distressing, rail temperature is less than the mean temperature of the area, the rail is heated (by propane heaters) to reach the optimum mean temperature, in order to minimize stresses at extreme temperatures. Obviously, if the rail temperature exceeds the mean temperature, no additional heating is required.

vi.   The rollers are removed and the fastenings are tightened.

vii. Distressing should take place on both rails. On each track section, distressing works should be performed during traffic-free intervals.

10.13.3. Expansion devices

The length change at the end of a cwr was calculated in section 10.13.2.4. In order to ensure that this length change will not be accompanied by excessive stresses at certain sensitive parts along the track (e.g. the ends of steel bridges, station entrances-exits, etc.), expansion devices are installed at these points.

Figure 10.32 illustrates the details of an expansion device for UIC 54 rail. There is a great variety of such expansion device types in relation to the selected rail profile.

Expansion devices should not be used in the following cases:

• on transition curves between straight and curved track,

• on curves with small radius of curvature (less than 800 m),

• on large bridges without ballast:

    – if the bridge is more than 30 m long, expansion devices are required at each end of the bridge,

    – if the bridge is less than 30 m long, cwr may be laid with no expansion devices.

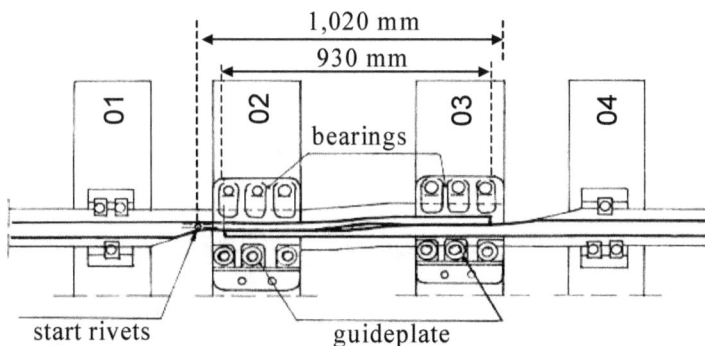

Fig. 10.32. Expansion device for UIC 54 rail

10.13.4. Advantages of the continuous welded rail

Although the cost of installing cwr is higher than that of fishplated track, an adequate return of capital for the initial investment is provided by the reduced maintenance cost of the track, improved track stability, higher running speeds, lower power consumption, and improved passenger comfort. In particular:

♦ cwr offer a much higher comfort level,

♦ evolution of track defects is much slower with cwr,

♦ fatigue of the various components of the track is smaller,

♦ stresses developed in the wheels and the rolling stock are generally much lower.

11 Sleepers – Fastenings

11.1. The various types of sleepers and their functions

Sleepers (which are called *ties* in North America and elsewhere) are the track components positioned between the rails and the ballast. The rails of the first railway lines were mounted on blocks placed directly on the ground. The need for better load distribution led to the addition of sleepers and ballast.

Sleepers must ensure the following functions:

- appropriate transfer and distribution of loads from the rails to the ballast,
- constant rail spacing, as specified by the track gauge,
- mounting of the rails on the sleepers at an inclination from 1/20 to 1/40,
- adequate mechanical strength both in the vertical and in the horizontal direction.

Along electrified lines, sleepers should moreover ensure (either by themselves or with added accessories) the electrical insulation of each rail from the other.

The first material used for sleepers was wood. Scarcity and sensitivity of wooden (*timber*) sleepers led to the introduction of *steel* sleepers around 1880, which were widely used for a long time. Since 1950, advances in concrete technology have led to the use of *concrete* sleepers, which can be:

- twin-block reinforced-concrete sleepers,
- monoblock prestressed-concrete sleepers.

Nowadays, sleepers installed along new tracks or overhauled old ones are mostly made of concrete. However, timber sleepers are used in several countries. The use of steel sleepers is diminishing and concrete or timber sleepers are usually used to replace steel sleepers at track renewals. As an alternative to timber and concrete sleepers, plastic and composite sleepers have been recently introduced, with, however, a limited field of application up until now.

The choice of the most appropriate type of sleeper should be made for each track by a feasibility analysis, which includes an evaluation and assessment of the following economic and technical factors:

economic
- ◆ cost of manufacturing or of purchase of sleeper,
- ◆ cost of purchase of fastenings,
- ◆ lifetime of sleeper,
- ◆ maintenance cost,
- ◆ probable salvage value of sleeper at the end of its lifetime,

technical
- – track characteristics (track gauge, speed, axle load, sleeper spacing),
- – weight of sleeper, resulting in high or low transverse track resistance and thus affecting values of speeds and axle loads that can be safely supported by the track,
- – distribution of train loads,
- – possibility to provide, without additional insulating techniques, electrical isolation of the one track from the other.

11.2. Steel sleepers

11.2.1. Form and properties

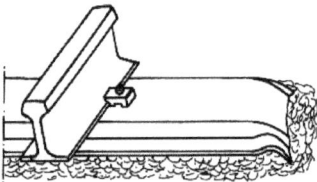

Fig. 11.1. Steel sleeper

The steel sleeper is an industrial product of simple construction. It consists of a profile in the form of ∩. Its ends are forged to provide anchoring in the ballast, so as to ensure transverse track stability, (Fig. 11.1).

The rail is mounted on to the steel sleeper by rail spikes fixed by bolts in holes drilled onto the sleeper top. Elastic fastenings may also be used.

11.2.2. Dimensions, weight, and chemical composition

Steel sleepers are made from low carbon steel of an ultimate tensile strength of 40÷60 kg/mm^2. Generally, sophisticated steels have not been used, and therefore the yield strength is near 50% of the ultimate strength. The chemical composition of steel sleepers is, according to British and Australian specifications: 0.15%÷ 0.25% C, 0.55%÷1.50% Mn, 0.20%÷0.50% Si, 0÷0.04% S, 0÷0.04% P, (238).

The finite element method and computer software have helped in recent years to optimize the cross-section of steel sleeper and its moment of inertia.

Figure 11.2 illustrates the geometrical characteristics of a steel sleeper (used in tracks for low speeds, V<120 km/h), weighing 70÷80 kg. In the area of rail joints, where a greater steel resistance is needed, a twin-type steel sleeper can be used, weighing 130÷140 kg.

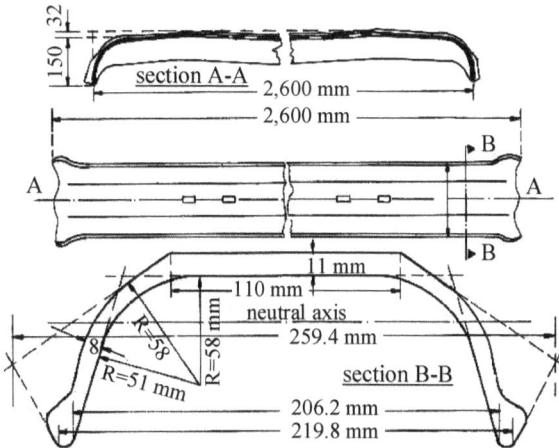

In spite of their decreasing use, there is still a variety of steel sleepers. Usual dimensions of steel sleepers for standard gauge tracks are: length 2,500÷2,600 mm, section width 220÷280 mm, thickness of lateral sides of cross-section: 7÷9 mm, thickness of upper side of cross-section: 11 mm. For metric gauge tracks, steel sleepers have a length of 2,000 mm. Ballast thickness beneath a steel sleeper should be 25÷30 cm (in addition to a gravel subballast of a thickness of 15 cm).

Fig. 11.2. Geometrical characteristics of a steel sleeper

11.2.3. Advantages and disadvantages

Steel sleepers are easily manufactured, installed, and maintained. They keep the track gauge adequately constant for a long time. Their lifetime is relatively long (usually 50 years) and after replacement they have still a certain value as scrap iron.

However, steel sleepers have many disadvantages. They have a low transverse resistance, (see section 13.5), a fact precluding increased speeds on tracks with steel sleepers. Their form makes longitudinal and transverse track positioning difficult. Steel sleepers are noisy, more costly than other sleeper types, they require special insulating devices for signaling, and their maintenance is difficult. Furthermore, steel sleepers are sensitive to chemical attacks and particularly vulnerable in lines close to industrial and coastal areas. All the above disadvantages have led to the economic obsolescence and to the gradual withdrawal of steel sleepers, particularly in Europe. Thus, and according to UIC, they represent in 2013 6% of the total number of sleepers in Germany, 4% in France, and 2.3% in Poland, (227).

11.2.4. Lifetime

Steel sleeper lifetime ranges from 40 to 60 years with an average value of 50 years.

11.3. Timber sleepers

11.3.1. Form, properties, and timber types

Engineers have traditionally tried to make the utmost use of any raw materials near the work in progress. The obvious choice for sleepers was timber and for over 100 years timber was (with steel) the principal rail support used throughout the world.

Timber sleepers achieve a better distribution of train loads than other sleeper types. They are accordingly recommended for tracks laid on fair- or poor-quality subgrade, where concrete sleepers would require a comparatively greater thickness for the ballast layer, (see section 12.5.1). Because of their higher cost and shorter lifetime, they are less used in Europe, compared to concrete sleepers. They represent in 2013, according to UIC, 11% of the total number of sleepers in Germany, 28% in France, 38.4% in Poland, 36.5 % in Austria. However, they are still extensively used in North America and elsewhere. Thus, 14÷16 million timber sleepers are laid per year in the USA (a 93% of the market of sleepers in the country).

The kinds of wood presently used for timber sleepers include beech and oak from European trees and azobé from tropical ones. Pine tree timber has also been used in the past. Timber sleepers in use by the various railways today are mostly of azobé tropical timber, which is stronger and more durable. In underground tunnels, Australian jarrah hardwood sleepers have been used extensively. Due to the great variety of the quality of wood used for timber sleepers, a variation of elasticity modulus is observed in the range $(10 \div 28) \cdot 10^4 \, kg/cm^2$.

Timber sleepers suffer from the effects of the following:

♦ chemical and physical disintegration of wood through exposure to alternate wet and dry conditions, heat, and dust,

♦ attacks by fungi and insects.

There are several methods for the treatment of timber sleepers, the most common of which involves impregnation under pressure. The substances mainly used are:

– 100% creosote,

– creosote/furnace oil, mixed in various proportions,

– a number of other chemicals, alone or in combination.

However, as creosote contains the substance benzopyrene, which has a carcinogenic effect, it is suggested to limit the content of this substance at 50 ppm. Thus, though creosote extends the lifetime of timber sleepers, it harms the environmental friendly image of the railway sector.

In order to prevent the timber sleeper from splintering or slipping on the ballast, it is possible to contain the wood fibres within the ballast. This is achieved by suitable configuration of the sleeper ends, which are either braced with a steel strap surrounding the sleeper end or have special metal plates driven into the vertical section of the sleeper ends.

In an effort to improve and homogenize mechanical (such as the elasticity modulus) characteristics of timber sleepers, the industry fabricated composite sleepers on the basis of wood with the addition of reinforcing fibres (glass fibres and other). Even though composite timber sleepers sustain better track loads, they are more costly, have low stiffness, and are vulnerable to brittle fracture.

Timber sleepers are particularly sensitive and their strength decreases with time as a result of:

- deterioration of their mechanical characteristics,
- influences of a chemical nature,
- influences of a biological nature.

11.3.2. Geometrical characteristics

The geometrical characteristics of timber sleepers are specified by UIC, (239). Timber sleepers in *standard gauge* tracks have typical dimensions, illustrated in Figure 11.3, as follows: length 2,600 mm, width 250 mm, height 150 mm, (239). The following tolerances are allowed to the dimensions illustrated in Figure 11.3:

Length: +40 mm, -30 mm,   Width: -10 mm,   Height: -5 mm

In the USA, recommended dimensions of timber sleepers are 2,600 mm (length), 230 mm (width), 180 mm (height).

In *metric gauge* tracks, timber sleepers have typical dimensions, illustrated in Figure 11.4, as follows: length 1,800 mm, width 220 mm, height 130 mm, (239). The allowed tolerances are proportionate to those for standard gauge tracks.

In *broad* gauge tracks, timber sleepers usually have a length of 2,750 mm, a width of 250 mm, and a height of 130÷200 mm.

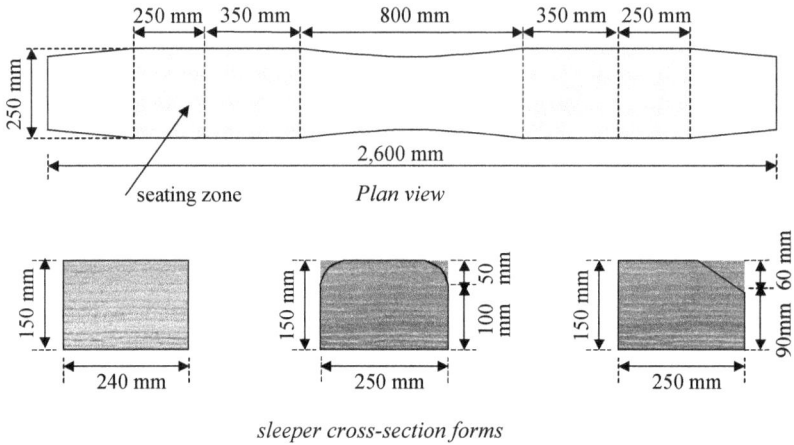

sleeper cross-section forms

Fig. 11.3. Geometrical characteristics (dimensions in mm) of timber sleepers for *standard gauge* tracks according to UIC, (239)

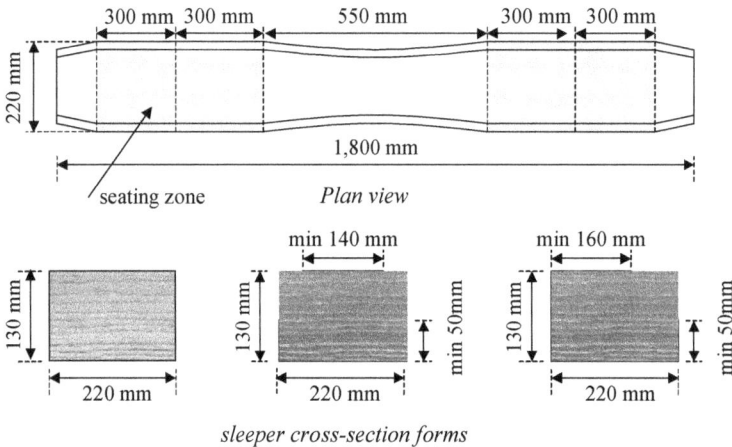

sleeper cross-section forms

Fig. 11.4. Geometrical characteristics (dimensions in mm) of timber sleepers for *metric gauge* tracks according to UIC, (239)

11.3.3. Advantages and disadvantages

The principal advantage of timber sleepers is flexibility and the resulting better load distribution. Timber sleepers are accordingly recommended in the case of poor-quality subgrades (classified as S_1). Timber sleepers moreover provide good insulation and do away with the need for special devices for signaling and electric traction. Finally, compared to concrete sleepers, timber sleepers are shorter in height.

The disadvantages of timber sleepers include their relatively short lifetime, their comparatively higher cost in Europe (though the situation is different in other parts of the world), and their low transverse resistance (a result of their low weight), thus precluding high speeds on their tracks. Timber sleepers are not recommended for tracks run at speeds higher than 160 km/h.

11.3.4. Lifetime

The lifetime of timber sleepers depends on the timber type used and is:
- 25 years for oak timber (impregnated),
- 30 years for beech timber (impregnated),
- 40 years for azobé tropical timber (non-impregnated),
- 45 years for azobé tropical timber (impregnated),
- 50 years for jarrah or similar hardwood used in tunnels.

However, non-impregnated oak and beech sleepers have a short lifetime in the range of 7÷12 years.

11.3.5. Deformability of timber sleepers

Finite element analysis provides an accurate and detailed calculation of deformability of timber sleeper for various subgrade qualities and has already been presented in section 8.4.9, Figure 8.12. It can be observed that the poorer the subgrade soil quality, the more uniform the timber sleeper settlement, (228).

11.4. Concrete sleepers

11.4.1. Inherent weaknesses of concrete sleepers

Monoblock reinforced-concrete sleepers, when first introduced, presented the following serious intrinsic weaknesses:
- a propensity for brittle fracture under the influence of dynamic train loads and for extensive cracking, leading to failure,
- very little resistance to fatigue, resulting in high tensile stresses in the central part of the sleeper, which, if exceeding the limit tensile strength, can lead to slippage of the reinforcing bars.

Overcoming these two weaknesses required:
- laying the rails so that they do not have direct contact with the sleepers, by interposing an absorbing material to blunt load impact. Such material includes rubber pads, which in turn necessitate the use of elastic fastenings,
- using reinforcing bars with the same lifetime as concrete.

11.4.2. The two types of concrete sleepers

In tandem with the reinforced-concrete and the prestressed-concrete technologies, two concrete sleeper types were developed:

- the twin-block reinforced-concrete sleeper, consisting of two trapezoidal reinforced-concrete blocks (each beneath each rail) joined by a connecting bar, (Fig. 11.5, section 11.5.1),
- the monoblock prestressed-concrete sleeper, which can be pre-tensioned or post-tensioned, (Fig. 11.7 and Fig. 11.8, section 11.6.1). Pre-tensioned monoblock sleepers are manufactured by using the long line method, with $25 \div 75$ molds in the line. The concrete has a final strength of 715 km/cm^2. In post-tensioned concrete sleepers, tendons are inserted after molding. Post-tensioned sleepers are more expensive than pre-tensioned ones, but they are more suitable for production in small volumes.

Stress distribution under the sleeper, (see section 11.8), has shown that stresses in the central section are very small, therefore less material can be safely used in this part of the sleeper. As a result, in the central part of the twin-block sleeper, the concrete was replaced by a connecting bar (which principally serves to maintain the track gauge), while in the monoblock sleeper (where the above solution cannot be applied) the cross-section at the central part of the sleeper was reduced.

Concrete sleepers (twin-block and monoblock) represent in 2013, according to UIC, 81% of the total number of sleepers in Germany, 68% in France, 60% in Poland, 80% in Sweden (in spite of abundance of wood in this country).

The twin-block sleeper was developed in France and has been used in: Algeria, Belgium, Brazil, Denmark, Greece, Mexico, Netherlands, Portugal, Spain, and Tunisia.

The monoblock pre-tensioned sleeper was developed in the United Kingdom and has been used in: Australia, Canada, Hungary, Iraq, Japan, Norway, Poland, South Africa, Sweden, USA, and Russia.

The monoblock post-tensioned sleeper was developed in Germany and has been used in: Austria, Finland, India, Italy, Greece, Mexico, and Turkey.

Of all new concrete sleepers, twin-block account for little less than 20% and monoblock for little more than 80%, (233).

The use of concrete sleepers on curved tracks is a controversial issue. In their metric gauge tracks, South African railways do not use concrete sleepers in curves of radius less than 300m. On the other hand, Canadian railways, which experience very extreme temperatures (- 40°C to +30°C), install concrete sleepers in all curves of radius less than 870 m, including many curves of radius less than 200 m, with continuous welded rails and no gauge widening. To some extent the different approaches may arise from shortcomings in certain fastenings, (233).

11.5. The twin-block reinforced-concrete sleeper

11.5.1. Geometrical characteristics and mechanical strength

Figure 11.5 illustrates the geometrical characteristics of the twin-block reinforced-concrete sleeper U41 of the French railways, which weighs 260 kg and has been used for four decades at the TGV tracks, run at a speed of 300 km/h and some of them designed for a speed of 350 km/h, (232). The U41 sleeper is slightly modified and is used today under the name B450, (232). Sleeper type U41 is suitable for tracks with a high traffic load (UIC 1, 2 groups) and high speeds. The connecting bar has a Y or L shaped cross-section. For medium load tracks (UIC 3, 4 groups) and speeds up to 200 km/h, a shorter type of sleeper (named formerly U31 and today B440) with a length of 2.245 m and a weight of 180 kg can be used, (Fig. 11.6.a).

Twin-block sleepers require ballast thickness and strength greater than that required by timber sleepers. Whenever this requirement is met, the track laid on twin-block sleepers has a satisfactory behavior.

Particular care should be taken when the subgrade is of poor quality. In this case the ballast thickness should be further increased.

Because of the flexible connecting bar, twin-block sleepers require extra maintenance, so as to ensure that the two blocks do not tilt differentially and do not loosen.

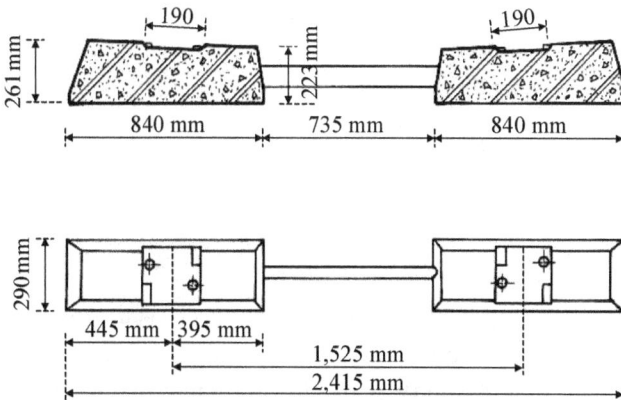

Fig. 11.5. Twin-block reinforced-concrete sleeper U41 (B450) of French railways (for groups UIC 1 and 2 and speeds up to 300÷350 km/h), (232)

According to the European standard for twin-block reinforced-concrete sleepers, the steel connecting bar must fulfill the following requirements, (229):
- chemical composition: 0.28%<C<0.80%, 0.45%<Mn<1.40%, P<0.08%, S<0.08%, Si<0.50%,
- mechanical characteristics: Tensile strength must range between 550÷1,030 MPa[*]. For steel yield strength ≥400 MPa, minimum elongation can be ≥8%, whereas for yield strength between 350÷400 MPa, minimum elongation can be ≥14%,
- Brinell hardness must be 160÷300 HB.

11.5.2. Advantages and disadvantages

Due to its great weight, the twin-block sleeper provides very satisfactory transverse track resistance and allows for high speeds. It keeps track gauge very satisfactorily within strict tolerances and has a long lifetime. It can be manufactured in any country and is less expensive than the timber sleeper in many countries.

The mechanical behavior of twin-block sleeper is less satisfactory when the ballast does not have the suitable thickness and mechanical characteristics. In addition, load distribution and flexibility are less satisfactory with twin-block than with timber or monoblock sleepers. Twin-block sleepers require elastic fastenings and, due to their great weight, handling is difficult. The twin-block sleeper (in contrast to the timber sleeper) requires special accessories, so as to ensure the necessary insulation for signaling and electric traction. Special attention should be given to the behavior of the connecting bar. If the latter is not appropriately placed and anchored, it may produce a maintenance hazard to staff working on the track.

11.5.3. Lifetime

The twin-block sleeper has a lifetime of 50 years.

11.5.4. Deformability of twin-block sleepers

Figure 11.6 illustrates the deformability of the U31 (B440) and U41 (B450) twin-block sleepers for various qualities of the subgrade (S_1, S_2, S_3, R), (228). It is observed that deformability is much lower than that of timber sleepers. Accordingly, in the case of a poor quality subgrade, the use of twin-block

---

[*] 1 MPa = 10.1972 Kg/cm^2

sleepers must be accompanied by an increase of the thickness of ballast, which should have adequate mechanical strength.

11.5.5. Twin block sleepers in high-speed tracks

The twin-block sleeper U41 (B450), (Fig. 11.5), has been used in most of the high-speed tracks of French railways (with V_{max}: 300÷350 km/h). However, in the Paris–Strasbourg high-speed track (with V_{max}: 350 km/h), French railways used monoblock sleepers.

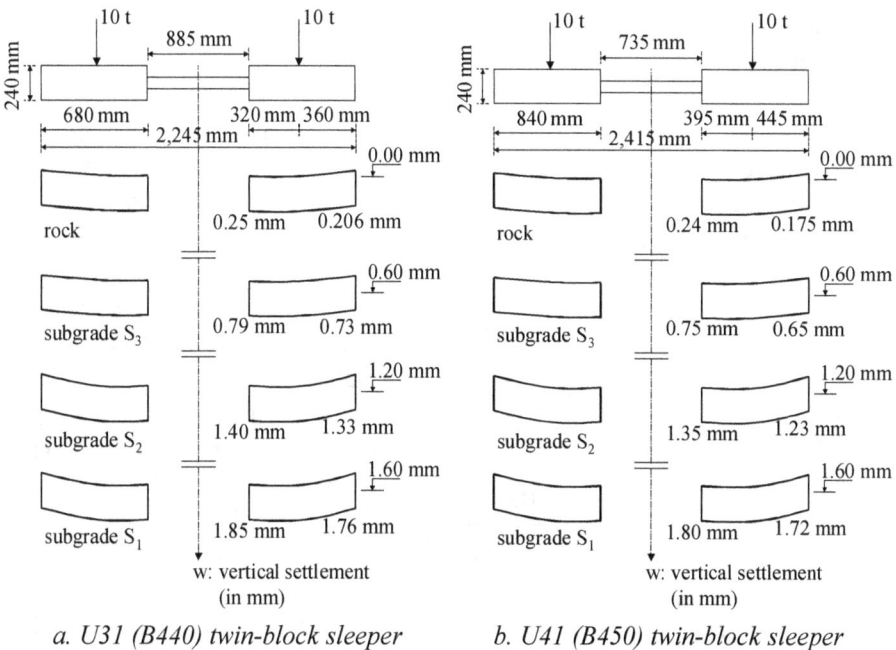

a. U31 (B440) twin-block sleeper b. U41 (B450) twin-block sleeper

Fig. 11.6. Deformability of two types of twin-block sleeper for various subgrade qualities, (228)

11.6. The monoblock prestressed-concrete sleeper

11.6.1. Geometrical characteristics and mechanical strength

The monoblock sleeper has the following characteristics, (240):
- withstands alternating stresses better, since the stress on the concrete is always compressive,
- offers a reduced sleeper height at the central part, since the steel bars do not have to be located, as in reinforced-concrete, as far away from the neutral axis as possible,

325

- allows reduction of the steel used, in comparison to the twin-block sleeper,
- is generally lighter, compared to the twin-block sleeper; this is a fact, however, which also reduces transverse resistance.

Monoblock sleepers come in a large variety of geometrical configurations. All, however, are characterized by a reduction of the cross-section at the central part. Figure 11.7 illustrates the geometrical characteristics of the pretensioned monoblock sleeper of the British railways (with initial prestressing force 38.9 t and residual prestressing force 32.1 t) and Figure 11.8 of the post-tensioned sleeper B70 of German railways (with a weight of 280 kg, maximum approved speed of 250 km/h, initial prestressing force 32.5 t and residual prestressing force 27.0 t), (236). Table 11.1 gives the geometrical characteristics of monoblock sleepers, which have been used in conventional tracks by several railways all over the world and Table 11.2 presents the mechanical characteristics of the monoblock sleepers used in the past decades by various railways, (233).

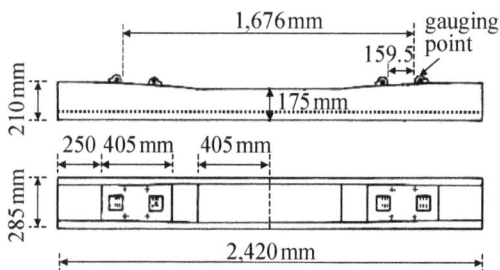

Fig. 11.7. Monoblock sleeper of the British railways, (236)

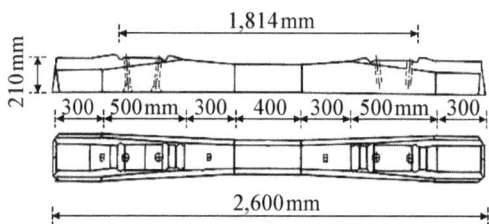

Fig. 11.8. Monoblock sleeper of the German railways, (236)

A critical element in monoblock sleeper design is the ratio λ of the critical moment M_{cr}, which the sleeper can withstand, to the maximum moment M_{max} developing in the sleeper (with an average value for M_{max} of 1.60 tm for axle loads 22.5÷25t). The factor in question takes values between 1.8 and 0.7. Variation in the value of ratio λ reflects differences in the demands from various railways, which in turn are dependent on the various conditions of the track and the rolling stock, together with the general philosophy of safety in various countries. Values of λ should be in principle higher than 1.0. Values of $\lambda < 1.0$ reflect that there are strength reserves not taken into account by theoretical calculations for M_{max}.

Table 11.1.

Geometrical characteristics of monoblock prestressed-concrete sleepers used by various railways, (233)

Country	Track gauge (mm)	Length of the sleeper (mm)	Sectional dimesions (mm)					
			at rail seat			mid-span		
			H	W_B	W_T	H	W_B	W_T
Australia	1,435	2,500	212	250	200	165	250	200
Canada	1,435	2,542	203	264	216	159	264	226
China	1,435	2,500	203	280	170	203	250	161
Germany	1,435	2,600	214	300	170	175	220	150
United Kingdom	1,432	2,515	203	264	216	165	264	230
Italy	1,435	2,300	172	284	222	150	240	190
Japan	1,435	2,400	220	310	190	195	236	180
Sweden	1,435	2,500	220	294	164	185	230	150
USA	1,435	2,591	241	279	241	178	279	250
South Africa	1,065	2,057	221	245	140	197	203	140
India	1,673	2,750	210	250	variable	180	220	variable
Russia	1,520	2,700	193	274	177	135	245	182

Table 11.2.

Mechanical characteristics of monoblock prestressed-concrete sleepers, used by various railways, (233)

Country	Sleeper spacing (mm)	Rail type	Maximum train speed (km/h)	Minimum radius of curvature (m)	Maximum axle load (tn)	Maximum moment developing in the sleeper M_{max} (tm)	Permiss. stress in concrete (kg/cm^2)	Critical permiss. moment M_{cr} (tm)	Coefficient $\lambda = \dfrac{M_{cr}}{M_{max}}$
Australia	550÷600	53/60 kg/m	160	200	24.5	1.62	23	2.38	1.5
Canada	610	132RE/136RE	130	194	29.2	2.01	33	3.06	1.5
China	550	50 kg/m	120	350	24.5	1.62	26	1.34	0.8
Germany	600÷650	S54/UIC 60	250	100	22.1	1.60	30	1.84	1.2
U.K.	650, 700	BS113A	200	400	24.5	1.65	45	2.50	1.5
Italy	600	UIC 60	180	485	22.1	1.19	47	1.50	1.3
Japan	590	50.4/60.8 kg/m	210	1,200	16.4	0.96	n.a.	1.73	1.8
Sweden	600, 650	SJ50	130	300	22.2	1.47	30	1.50	1.0
U.S.A.	610	65/69 kg/m	200	610	32.1	2.33	50	4.24	1.8
South Africa	700	48/47 kg/m	160	150	22.1	1.38	28	1.12	0.8
India	650	UIC 60	130	550	22.0	1.49	20	2.43	1.6
Russia	500÷643	R50/R65/R70	200	350	26.5	1.95	20	1.35	0.7

According to the European standard, concrete of the monoblock sleeper must have a minimum compressive strength[*] C 50/60 MPa, a minimum cement content 300 kg/m^3, and a water / cement ratio less than 0.45. The minimum concrete cover for the prestressing tendons must be 30 mm from the bottom surface and 20 mm from the other surfaces.

Steel must have at the minimum a tensile strength of 1,600 MPa, a leakage limit of 1,400 MPa, and a relaxation less than 3% within 1,000 h, (229).

Normally the bending moment capacity of monoblock sleepers is calculated considering the prestressing force after all losses (20÷25%, due to elastic shortening, shrinkage, creep, and relaxation). The allowable concrete tensile stress is 20.4÷30.6 kg/cm^2 and the allowable concrete compressive stress is 204÷306 kg/cm^2.

Design of prestressed concrete sleeper is based on theories of either permissible stress or limit state, the latter making utilization of full strength of concrete.

As sleepers are subjected to cyclic loads, special care must be taken to ensure resistance in fatigue of all materials involved, including prestressed steel. Engineers should aim at crack-free sleepers, since cracks in the concrete caused by bending moments lead to a large increase in stress values of the prestressing steel, which could cause failure due to fatigue. As a high quality prestressing wire or strand is apt to withstand a stress variation of only 5÷10% of its ultimate strength, most railways are conservative in taking into account concrete tensile stresses as the basis for the moment capacities, and a few of them even exclude any tensile stresses, (233).

11.6.2. Advantages and disadvantages

Monoblock sleepers have a behavior similar to that of the twin-block sleepers. They maintain the track gauge in a satisfactory manner and have a long lifetime. They require elastic fastenings and special accessories for signaling.

However, monoblock sleepers distribute loads better than twin-block sleepers, but not as well as timber ones. Their transverse resistance is lower compared to twin-block sleepers, but higher compared to timber sleepers; monoblock sleepers provide also a good surface for the staff in charge of inspection and maintenance.

11.6.3. Lifetime

The lifetime of monoblock sleepers is 50 years.

---

[*] The symbol C 50/60 demotes a concrete grade with a cylinder strength of 50 MPa and a cube strength of 60 MPa (1 MPa=10.1972 Kg/cm^2)

11.6.4. Deformability of monoblock sleepers

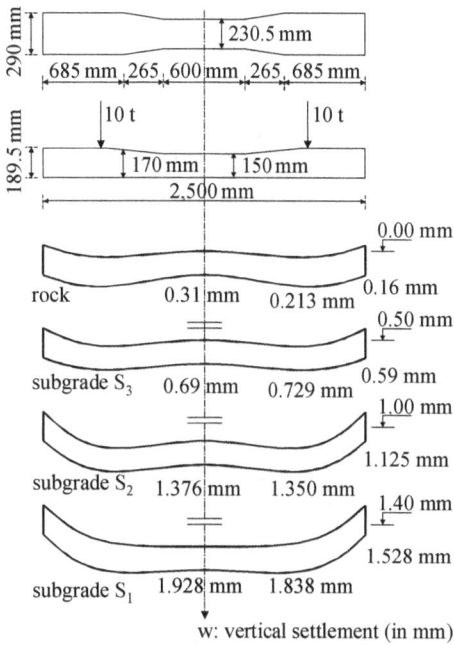

Fig. 11.9. Deformability of monoblock sleeper for various qualities of the subgrade, (228)

Figure 11.9 illustrates the deformability of a monoblock sleeper for various subgrade qualities (S_1, S_2, S_3, R), (228). It is observed that the monoblock sleeper has a deformability similar to that of the timber sleeper, but less flexibility. Monoblock sleepers should therefore be laid on ballast of suitable thickness and mechanical strength.

11.6.5. Monoblock sleepers in high-speed tracks

The monoblock sleeper B70 illustrated in Figure 11.8 has been used in high-speed tracks in Germany, with V_{max}: 250 km/h. Variations of this type of sleeper can withstand higher speeds (up to 300 km/h for the sleeper type B90 of a weight of 332 kg and a concrete grade C 50/60, up to 350 km/h for the sleeper types AI-99, AI-04[*] used in Spain) and higher axle loads (25 tonnes for sleeper B90).

However, many high-speed lines in Germany, Japan, Taiwan, China, and elsewhere are laid on a slab track, (see chapter 17).

11.6.6. Manufacturing, quality control, and testing of concrete sleepers

The manufacturing of both twin-block and monoblock sleepers has special requirements, (233):

– demanding tolerances, typically ±3 mm for overall dimensions and reinforcement location and ±0.8 mm for the position of cast-in fastening components,

---

[*] The sleeper AI-04 is used in Spain and has the following characteristics: maximum speed 350 km/h, maximum axle load 25 tonnes, length 2,600 mm, width 300 mm, maximum height 267 mm, height at the sleeper center 210 mm, weight 325 kg, concrete grade C 50/60 MPa.

– for pre-tensioned sleepers, the development of high strength of 35÷40 MPa in early ages (14÷15 h),
– concrete of high durability.

Manufacturing methods can be classified into three categories, (233):
♦ long line, for pre-tensioned, full bonded monoblock sleepers,
♦ short line, for pre-tensioned, full bonded and end-anchored monoblock sleepers,
♦ instant demolding, for twin-block reinforced concrete and post-tensioned monoblock sleepers.

Cement should be of a high quality and aggregates well-graded with a proven durability.

An inherent problem of any construction is to ensure that it is made following the qualities and strengths specified. In concrete sleepers, this requires inspection and testing procedures from the selection and control of materials, during the manufacturing process, and until the point of delivery, (229).

Testing of concrete materials includes three steps to confirm acceptability: basic design, materials, and finished product.

The European standard describes in detail the steps for testing concrete sleepers: test arrangements and procedures, acceptance criteria, design approval tests, and routine tests.

The manufacturing rules for *monoblock* concrete sleepers include, (229):
– water/cement ratio (and tolerances),
– weight of each component (and tolerances),
– grading curves for each aggregate,
– characteristic compressive and tensile strength of concrete samples after 7 and 28 days,
– maximum relaxation for prestressing tendons after 1,000 hours,
– description of the prestressing system, including prestressing force and tolerances on each tendon,
– methods of concrete vibration,
– curing time and temperature cycle,
– method used for releasing prestressing force,
– stocking and stacking rules after manufacturing,
– minimum concrete compressive strength before releasing prestressing tendons,
– the position of the centroid of the prestressing tendons should be within 3mm of the theoretical position relative to the rail seat and ±6mm for each

individual prestressing tendon. Concerning the tolerances of the prestressing force, they should be within 5% of the specified force.

The manufacturing rules for *twin-block* concrete sleepers include, (229):
- water/cement ratio (and tolerances),
- weight of each aggregate of the concrete (and tolerances),
- grading curves for each aggregate,
- characteristic compressive and tensile concrete strength,
- methods of concrete vibration,
- methods of demolding and curing,
- stocking and stacking rules after manufacturing.

11.7. Plastic and composite sleepers

11.7.1. Definition and distinction of plastic from composite sleepers

No material used for the fabrication of sleepers can be considered that it meets satisfactorily the required mechanical characteristics for a modern track, which provides increased speeds and axle loads, high levels of safety, reasonable construction costs, low levels of maintenance costs, and quasi-elimination of unpredicted failures. Timber sleepers, particularly, if they are not impregnated, they have a short lifetime and are vulnerable; if they are impregnated with creosote they become a toxic product, harmful to human health and the environment, whereas their disposal must be treated as hazardous waste. Thus, plastic and composite sleepers appeared in recent years as an alternative solution, (especially) for the replacement of timber sleepers.

Plastic sleepers are composed of some plastic material (usually hard polyurethane foam); in composite sleepers some form of reinforcing fibres have been added to the original material (usually plastic, but in some cases timber). Some researchers consider plastic sleepers as a specific category of composite sleepers.

11.7.2. Categories and mechanical strength

Plastic and composite sleepers can be classified in three broad categories, (230):
- plastic sleepers with no or short fibre reinforcement. They consist of recycled plastic to which small quantities of sand, gravel, recycled glass or short glass fibres can be added. They have a modulus of elasticity $E = (15 \div 18) \cdot 10^3$ kg/cm^2, a modulus of rupture $170 \div 200$ kg/cm^2, and a shear strength of 40 kg/cm^2,
- composite sleepers with long glass fibre reinforcement in the longitudinal direction and no or very random fibres in the transverse direction. They

have a modulus of elasticity $E=80 \cdot 10^3$ kg/cm², a modulus of rupture of 1,420 kg/cm², and a shear strength of 100 kg/cm². It is clear that this category of composite sleepers has much better mechanical strength than simple plastic sleepers,

– composite sleepers with fibre reinforcement both in the longitudinal and transverse direction. They have a modulus of elasticity $E=(50 \div 80) \cdot 10^3$ kg/cm², a modulus of rupture of $700 \div 1,400$ kg/cm², and a shear strength of $150 \div 200$ kg/cm².

11.7.3. Advantages and disadvantages

Plastic and composite sleepers have a high durability, damping capacity, and electrical resistance, no vulnerability to insect attacks, resistance to moisture and corrosion, and are easy for handling and installation. However, they are anisotropic with low stiffness and strength, they experience plastic deformations, and have low weight of $45 \div 75$ kg, against $60 \div 70$ kg for timber sleepers, $70 \div 80$ kg for steal sleepers and $260 \div 320$ kg for concrete sleepers. The low weight of plastic and composite sleepers results in low transverse resistance and does not allow high speeds, unless the axle load is reduced, as is the case of some Japanese high-speed tracks, laid on composite sleepers, with axle loads of 16 t.

In order to lessen the handicap of low transverse resistance, some plastic and composite sleepers have textured the bottom and vertical sides of sleeper. It must be noted that until 2021 there were no recognized standards for plastic and composite sleepers, while long-term effects of UV radiation upon plastic sleepers still remain unknown.

11.7.4. Lifetime, cost, and applications

The lifetime of plastic and composite sleepers is estimated to be around 60 years. As there is a variety of products, there are diverging costs and prices. It is reported that composite sleepers containing fibres of high strength have costs many times higher compared to timber sleepers, (230). Thus, until 2021 composite sleepers have a limited application and have been used in countries with high labor costs, such as Japan (with 1,300 km of tracks on composite sleepers), Germany, Australia, the Netherlands, the USA, Taiwan, and elsewhere.

11.8. Stresses beneath the sleeper

The stresses developing beneath the sleeper may be studied by the simplified simulation of Figure 11.10, where:

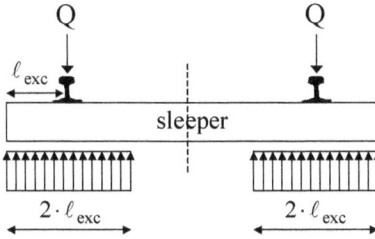

Fig. 11.10. Simplified model of sleeper

- the sleeper is simulated as a beam protruding at both ends,
- wheel load is assumed to be applied at a point,
- stresses between ballast and sleeper are considered uniform over a length $2 \cdot \ell_{exc}$ below each rail.

On-site stress measurements under the sleeper have yielded the stress distribution illustrated in Figure 11.11, with a maximum stress σ, given by the empirical formula, (241), (242):

$$\sigma = \frac{P}{\alpha \cdot \left(\dfrac{L}{2} + \dfrac{3 \cdot \ell_{exc}}{2} \right)} \qquad (11.1)$$

where:

α : width of sleeper,
L : length of sleeper,
ℓ_{exc}: distance between sleeper end – wheel load application point,
P : axle load, $P = 2 \cdot Q$.

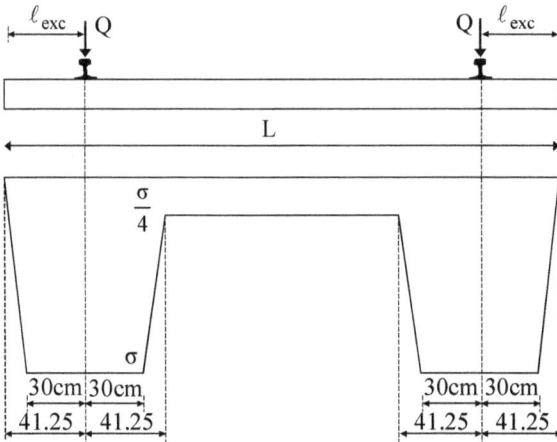

Fig. 11.11. Stress distribution under the sleeper, (241)

11.9. Fastenings

11.9.1. Functional characteristics

The set of parts and materials making the rail-sleeper connection are termed fastenings and they should ensure the following properties:
- keep track gauge as close as possible to its nominal value,

- keep the inclination of the rail on the sleeper constant,
- transfer loads from the rail to the sleeper,
- attenuate and dampen vibrations caused by train traffic.

Fastening should provide in addition:
- electrical insulation,
- resilience and adequate deflection,
- avoidance of abrasion between components of the track and of over-stressing,
- adequate resistance to corrosion,
- reasonable cost,
- lifetime compatible to that of the sleeper,
- easy installation and maintenance,
- resistance to vandalism.

11.9.2. Types of fastenings

Fastenings are distinguished into *rigid* and *elastic* fastenings.

11.9.2.1. Rigid fastenings

Rigid fastenings are used only with timber or steel sleepers. In rigid fastenings, the rail is connected to the sleeper with bolts or nails. During train passage the rail compresses the sleeper and part of the strain is plastic (i.e. it does not disappear when train loads leave the track), resulting in the creation of a gap between nail head and rail. With successive train passages the gaps grow, causing a gradual slackening of the fastening, which affects safety and may be the origin of a derailment. In addition to plastic strain, high frequency vibrations may also contribute to the widening of the gaps and the slackening of the fastening.

Rigid fastenings may be installed either with or without a seating plate (Fig. 11.12, 11.13), the latter being the preferable solution.

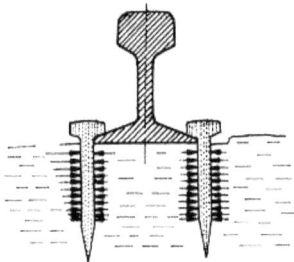

Fig. 11.12. Rigid fastening without a seating plate

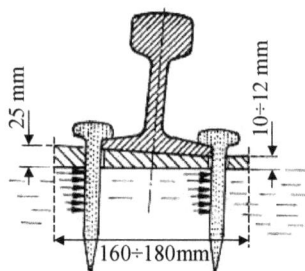

Fig. 11.13. Rigid fastening with a seating plate

11.9.2.2. Elastic fastenings

The use of elastic fastenings is mandatory with concrete sleepers and optional with timber and steel sleepers. Elastic fastenings are distinguished in screw-type and spring-type ones:

– *Screw-type* elastic fastenings, (Fig. 11.14). These have the advantage of great fastening strength and easy maintenance and replacement. They have

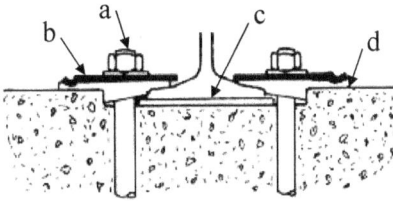

Fig. 11.14. Screw-type elastic fastening

the disadvantage that correct installation is affected by local conditions. Screw-types are RN, Vossloh, Nabla, and other fastenings, (Fig. 11.15 and Fig. 11.16), (226). The common elements in all these designs are, (Fig. 11.14):

• a threaded element (a), which is used to apply a force to a spring steel element, this threaded element being removable from the sleeper,
• the spring steel element (b), which can be a bar or a plate,
• a pad (c) between rail and sleeper to absorb vibrations, to provide a suitable layer between rail and sleeper and also electric insulation,
• insulating elements (d) to electrically isolate the rail from any metallic path into the sleeper.

Fig. 11.15. Vossloh fastening, (226)

Fig. 11.16. Nabla fastening

– *Spring-type* elastic fastenings, (Fig. 11.17). These are less adaptable than screw-type fastenings but less affected by installation conditions, thus any error is easily located visually. Pandrol, (Fig. 11.18), Lineloc, etc., are fas-

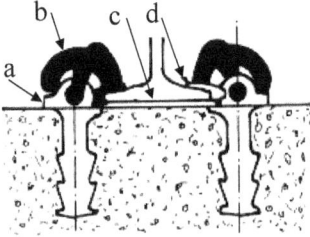

Fig. 11.17. Spring-type elastic fastening

Fig. 11.18. Pandrol fastening, (226)

tenings of the spring-type. The common elements in spring-type fastenings (which in principle do not require any maintenance after installation) are, (Fig 11.17):

- some form of anchorage (a) in the sleeper, generally at the time the sleeper is manufactured,
- a spring steel element (b) to generate clamping forces on the rail foot,
- a rail pad (c) between rail and sleeper to attenuate forces and stresses and to provide electrical insulation, which is necessary for the signaling and electrification systems,
- insulators or a layer of insulating materials (d), to provide electrical insulation between the rail and any metallic path, such as via (a) and (b), to the sleeper.

11.9.2.3. Types of elastic fastenings

There is a great variety of elastic fastenings, Nabla, Vossloh, and Pandrol being only some of them. Some types of elastic fastenings have a seating plate, whereas others do not. Thus, elastic fastenings can be categorized as follows:

- fastenings with direct mounting without a seating plate,
- fastenings with indirect mounting without a seating plate,
- fastenings with direct mounting with a seating plate,
- fastenings with indirect mounting with a seating plate.

11.9.2.4. Operating principles of elastic fastenings

During operation, elastic fastenings should ensure the following principles, (235):

- The rail-sleeper fastening force should be sufficient to make the rail-sleeper slippage resistance much greater than the resistance to longitudinal motion of the sleeper on completely stabilized ballast.

- The fastening resonance frequency should be distinctly higher than the rail resonance frequency.
- Fastenings should retain sufficient clamping force over the years.
- Fastening tightness should be easily checked on the track without disassembly.
- Fastenings should retain their elastic characteristics for a long time after installation.
- The ratio of the force applied on the rail base (foot) to the force transmitted by the fastenings to the sleeper should be as high as possible.

11.9.3. Forces and stresses in rigid and in elastic fastenings

The difference between rigid and elastic fastenings becomes apparent mainly in the diagram of the tensile force developed in the fastening as a function of time, (Fig. 11.19). The better behavior of elastic fastenings is thereby confirmed. Figures 11.20 and 11.21 illustrate the force-elongation curves for screw-type and spring-type fastenings, respectively.

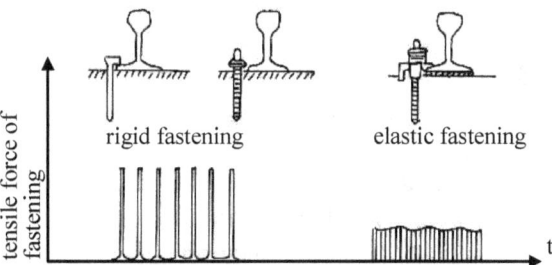

Fig. 11.19. Tensile force developed in rigid and in elastic fastenings as a function of time

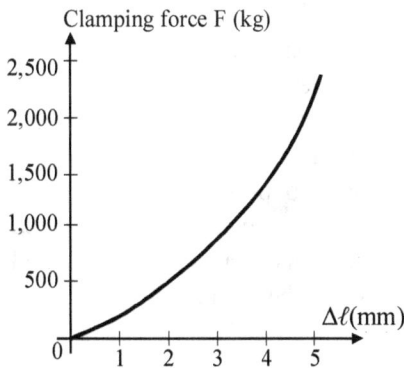

Fig. 11.20. Force-elongation curve for screw-type fastenings, (226)

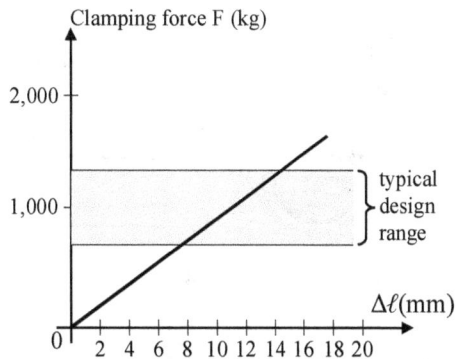

Fig. 11.21. Force-elongation curve for spring-type fastenings, (226)

11.9.4. Design criteria, anchorage, and insulation of a fastening

Clamping forces and the resulting elongation are different for screw-type and spring-type fastening systems, (Fig. 11.20 and Fig. 11.21). Most spring-type fastening systems provide a clamping force within the range 750÷1,250 kg for corresponding elongations of 5÷15mm. Spring-type fastenings have a greater elongation (for the same force) than screw-type ones. However, it is important for the spring to have a large load capacity beyond its working range, as this increases the life expectancy of the fastening. The rail clamping force requirements are calculated in relation to rail profile, permitted speed, vehicle weight, stiffness of the track, radius of curvature, outside temperature, etc.

Specific care should be taken with the appropriate anchorage of the fastening. For screw-types, anchors are made of nylon or polypropylene plastic. For spring-types, anchors are made of cast iron or forged steel. In addition to vertical forces, the anchorage should be designed to transmit safely transverse forces on the sleeper, (235).

Where track circuiting is used for signaling, insulation is an important requirement for the fastening. The insulation requirements of the track depend on the characteristics of the signaling system used. A dry assembly should have an infinite resistance and a wet one a resistance not significantly more than 20,000 Ω per assembly. The insulator should be resistant to wear, to degradation by ultraviolet light, and to attacks from track chemicals, (235).

11.9.5. Rail creep and anti-creep anchors

Along fishplated tracks (i.e., not continuous welded), it has been observed that the rails (or even the entire track) are subjected to longitudinal creep. Creep

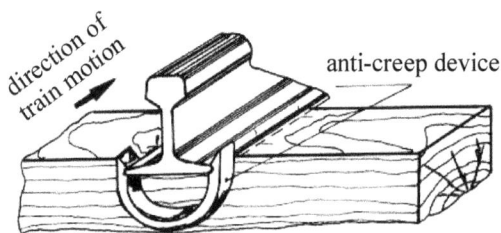

Fig. 11.22. Rail anti-creep device

usually occurs in the train's running direction. On high-gradient tracks, however, rails tend to move downwards, regardless of the direction of traffic. To prevent slippage, special devices, called anti-creep devices or anchors, are installed along the track, (Fig. 11.22).

11.10. Resilient pads

11.10.1. Pads with or without a baseplate

As explained in section 7.2, Figure 7.2, resilient pads are used between rail and sleeper (for ballasted tracks) or between rail and concrete slab (for non-

338

ballasted tracks). When a baseplate is used (both in ballasted and non-ballasted tracks), pads are used between rail-baseplate as well as between baseplate-sleeper (or concrete slab).

11.10.2. Functions and properties of pads

Pads must fulfill a number of functions and properties, (231), (237):
– *load distribution*. The pad should provide load distribution between the rail foot and the sleeper, so as to accommodate irregularities on both the rail and the sleeper,
– *vibration attenuation*. The pad should attenuate the transmitted vibrations, created by wheel loads and track irregularities,
– *resilience*. The pad should be designed to provide optimum deflection compatible with the fastening system, so that the fastening is able to ensure the necessary resistance to the longitudinal and lateral rail forces at all times,
– *resistance to creep*. The pad, together with the rail fastening system, should provide adequate creep and torsonial resistance, which should not change significantly with respect to pad's age or tonnage transported,
– *electrical insulation*. The pad should have good electrical insulation properties, so as to isolate the rails from the sleeper, thus enabling track circuits[*] to be used for signaling and control purposes,
– *durability*. The pad should have a lifetime of at least as long as the rail. The ideal condition is to install pads during rail replacement. Furthermore, pads should have properties which resist contamination by dirt, water, oil, and chemicals, and be able to perform with similar characteristics regardless of ambient temperatures and weather conditions. The Japanese railways have experienced after 10 years of operation of their Shinkansen high-speed train an increase in the pad stiffness of 66%, (237).

11.10.3. Dimensions, materials, and design

The thickness of the pad (which varies from 5 mm to 10 mm) is chosen to suit the particular installations and depends on several factors:
• the width of the flat-bottomed rail foot,
• the type of elastic fastening used,
• the size of the sleeper and baseplate, if any,
• the type of traffic, e.g. slow-speed heavy freight traffic or high-speed passenger traffic.

---

[*] Track circuit is an electrical system that detects the presence or absence of a train on a section of track, (see section 21.3.2).

Three main types of materials have been used for pads:
– rubber (both natural and synthetic),
– plastic,
– rubber bonded cork.

Thus, French railways use rubber pads, whilst German railways use a harder plastic pad. However, certain pads are provided with a rough surface to absorb more efficiently the dynamic and vibration effects of train loads.

In relation to their stiffness, pads can be considered as soft for a stiffness of 8 tonnes/mm, medium for a stiffness of 8÷15 tonnes/mm, and hard for a stiffness higher than 15 tonnes/mm.

11.10.4. Force-elongation curves

Figure 11.23 illustrates the force-elongation curve for a foamed polyurethane pad of a thickness of 7 mm.

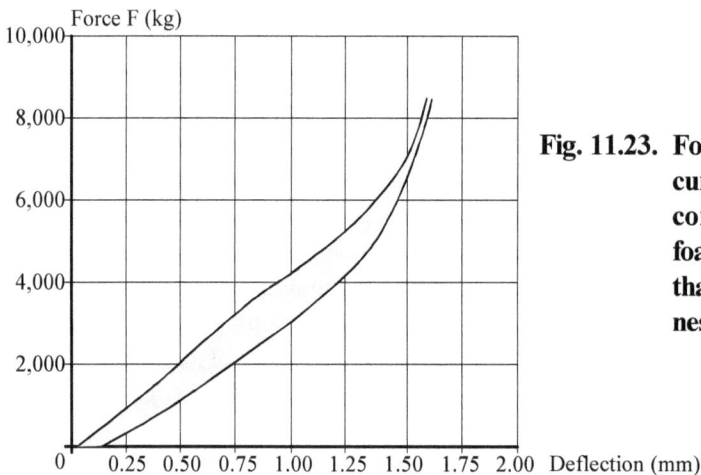

Fig. 11.23. Force-elongation curve for a pad constituted of foamed polyurethane of a thickness of 7mm

11.11. Requirements of the European specifications for the sleeper-fastening system

Sleepers and fastenings (together with rail pads) constitute a discrete sub-system which transfers and distributes loads from the rail to the ballast or to the concrete slab. According to the European technical specifications for interoperability, (134):

♦ sleepers should ensure track gauge, equivalent conicity, and transverse resistance of the track so as to be safely run at the specific speed,

♦ the longitudinal force required to cause rail slip should be at least 7 kN, and for speeds of more than 250 km/h it should be at least 9 kN,

♦ fastenings should resist application of a number of 3,000,000 cycles of axle loads in sharp curves, so that the clamping force and longitudinal restraint are not degraded by more than 20% and the vertical stiffness by more than 25%,

♦ for fastenings on concrete sleepers, the dynamic stiffness of the rail pad should not exceed 600 MN/m,

♦ the minimum electrical resistance should be 5 kΩ; it is permissible, however, to require higher values for the electrical resistance, in cases of particular control-command and signaling systems.

11.12. Numerical application for the design of the various track components

A standard gauge continuous welded railway track has a daily traffic load of 30,000 tonnes, a maximum axle load of 20 tonnes, a maximum speed of 140 km/h, and is laid on medium-quality subgrade (S_2). We will study and choose:

– the most suitable rail type,
– the most suitable sleeper type. We will examine the cases of timber, twin-block, and monoblock sleeper,
– the most suitable type of fastening,
– the stress distribution under the sleeper.

We will have:

a. The rail cross-section will be chosen on the basis of the average daily traffic load of 30,000 tonnes, according to analysis of section 10.4.1. For timber sleepers we choose UIC 54 rail, while for twin-block or mono-block concrete sleepers we choose UIC 60 rail.

b. Since we have a standard gauge track, run at a speed lower than 160 km/h, both timber or concrete sleepers can be used. If timber sleepers are chosen, they will have the geometrical dimensions of Figure 11.3. If twin-block concrete sleepers are chosen, given that this is a UIC 4 medium-load track with a speed V<200 km/h, we will choose the sleeper type with the geometrical characteristics of Figure 11.6.a. Were it a track with a higher traffic load (UIC group 1, 2) and higher speed (V>200 km/h), however, we would have selected twin-block sleepers with the geometrical characteristics of Figure 11.5. In the case of monoblock sleepers, a choice of geometrical characteristics can be made based on Tables 11.1, 11.2 and Figures 11.7 and 11.8.

c. Stress distribution under the sleeper is illustrated in Figure 11.11. We will calculate the maximum stress in the case, for instance, of timber sleepers 2.60 m long and 0.25 m wide, (Fig. 11.3). In order to take into account the dynamic effects, (see section 8.7), the nominal static axle load will be multiplied by a dynamic impact factor of 1.3, derived from Figure 8.15 for V=140 km/h. Given that the sleeper under loading supports only 40% of the axle load (section 8.4.8), the actual total load exerted on the sleeper will be: 20 t · 1.3 · 0.4 = 10.4 t

The formula (11.1), (section 11.8), gives:

$$\sigma = \frac{10.4 \text{ t}}{0.25 \cdot \left(\dfrac{2.60}{2} + \dfrac{3 \cdot (2.60 - 1.50)/2}{2} \right) \text{m}} = 2.25 \text{ kg/cm}^2$$

The order of magnitude of stress σ is also confirmed by values given in Figure 7.3 (section 7.3).

d. In the case of timber sleepers, rigid or elastic fastenings will be chosen, while in the case of twin-block and monoblock sleepers, elastic fastenings are mandatory. Each sleeper type has usually its appropriate type of fastening. Thus, for twin-block sleepers, Nabla fastenings will be selected, whereas for monoblock sleepers, Vossloh or Pandrol fastenings, or another type compatible with the characteristics of the sleeper will be selected.

12 Ballast

12.1. Functions of ballast and subballast

12.1.1. Functions of ballast

The term ballast denotes in railway engineering the layer of crushed stone (and only in exceptional cases of gravel), on which the sleepers rest. Furthermore, the ballast fills the space between sleepers as well as at some distance (called ballast shoulder) beyond the sleeper ends.

The railway ballast, (see also section 7.2, Figure 7.1), performs several functions, such as the following:
♦ further distributing stresses transmitted by the sleepers,
♦ attenuating the greatest part of train vibrations,
♦ resisting track shifting (transverse and longitudinal),
♦ facilitating rainwater drainage,
♦ allowing track geometry to be restored and correcting track defects (with the use of track maintenance equipment, see section 16.8).

The above functions are clearly contradictory in some aspects, thus the ballast cannot completely fulfill all of them. It could be argued that for good load bearing characteristics and added track stability, the ballast needs to be well graded and compact, which, in turn, makes dispersal of water more difficult, together with associated maintenance. A balance, therefore, among the various functions of ballast is aimed at.

12.1.2. Functions of subballast

The gravel subballast is laid under the ballast and has the following functions:
- protection of the upper surface of the subgrade from the intrusion of ballast stones,
- further distribution of stresses,
- quick runoff of rainwater,
- impart a transverse slope (commonly 3÷5%) to the upper surface of the subgrade for proper runoff.

343

The usual thickness of the gravel subballast layer is 15 cm. However, some railways do not use a subballast layer; they simply use a greater thickness of the formation layer, which is placed on top of the subgrade. In this case and if a geotextile is placed on the upper surface of the formation layer without a gravel subballast above it, there is a high risk of perforation of the geotextile from the ballast stones in touch.

12.2. Geometrical characteristics of ballast

12.2.1. Granulometric composition

To fulfill the above functions, the ballast must be of good hard stone, angular in shape (cubic or polyhedral), with hard corners; it must also have the various dimensions of stones nearly equal and be clean and free from dust.

The ballast consists of a mixture of sizes, expressed as percentage by weight, which should be evenly graded.

Figure 12.1 illustrates a typical granulometric composition of ballast according to French regulations. Pieces larger than 63 mm and smaller than 16 mm are acceptable up to 3% above and 2% below the limit values. The granulometric composition of ballast according to British[*] and American regulations is given in Table 12.1.

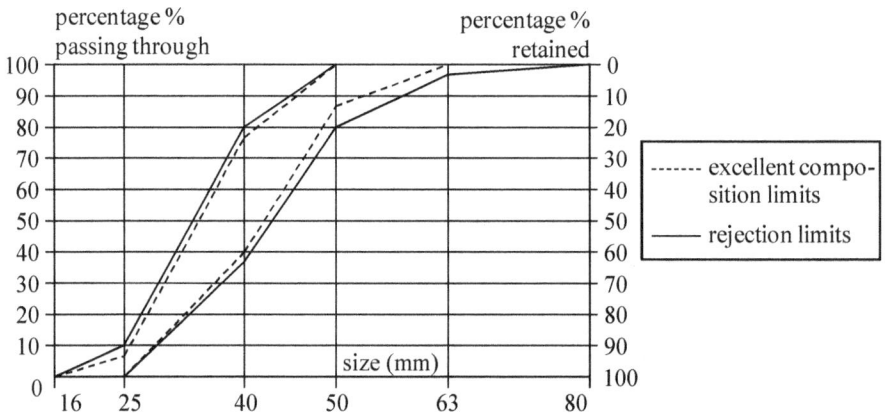

Fig. 12.1. A typical diagram of the granulometric composition of ballast according to French regulations, (252)

---

[*] British railways operated as a unified railway enterprise responsible for both infrastructure and operation until the mid-1990s. Since that time responsibility for infrastructure has been given to Railtrack and later to Network Rail. Whenever the term British regulations is used in this book, it includes regulations not only of former British railways but of Railtrack and Network Rail also.

Table 12.1.
Ballast size according to British and American regulations, (250)

British regulation		American regulation (AREMA)	
Sieve size D (mm)	Percentage to pass	Sieve size D (mm)	Percentage to pass
63 mm	100%	< 76 mm	100%
50 mm	97% ÷ 100%	< 60.96 mm	90% ÷ 100%
28 mm	0% ÷ 20%	<45.72 mm	60% ÷ 90%
14 mm	0% ÷ 2%	< 30.48 mm	10% ÷ 35%
1.18 mm	0% ÷ 0.8%	< 11.43 mm	0% ÷ 3%

Figure 12.2 illustrates the granulometric composition of ballast according to German regulations[*]. Pieces smaller than 22.5 mm must not exceed 3% of weight and pieces smaller than 31.5 mm 25% of weight, (158).

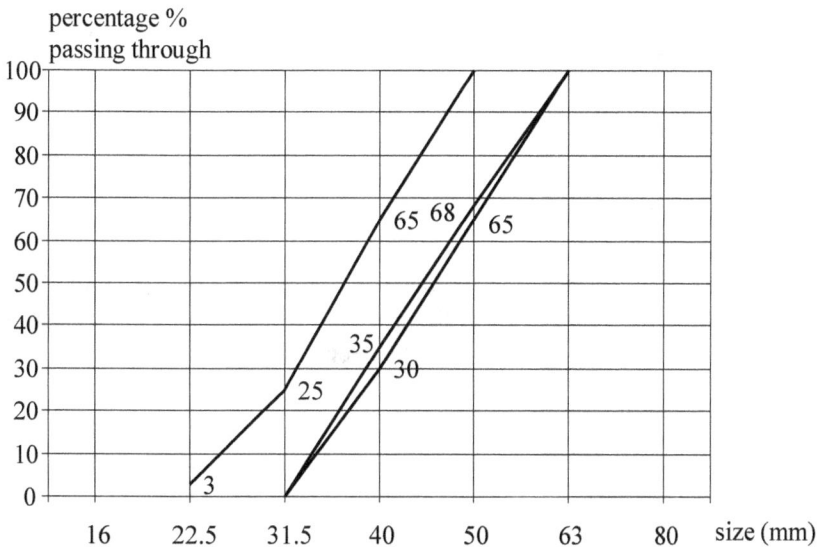

Fig. 12.2. Granulometric composition of ballast according to German regulations, (158)

[*] To avoid any confusion, it is worth remembering the difference between regulations, specifications, codes, standards, and guidelines. Regulations and specifications are rules issued by governmental or inter-governmental bodies that impose specific instructions or methods. Codes are a form of legislation which defines a procedure or performance to be followed. Standards are uniform criteria and methods developed by a national or international regulatory body and represent suggested (but not obligatory) requirements. Guidelines are non-mandatory suggestions and recommendations.

According to the European standard, railway ballast is designated by a pair of sieve sizes, with 31.5 mm being the lower limit and 50 mm or 63 mm the upper limit, (Table 12.2), (246).

Table 12.2.
Granulometric composition and categories of ballast according to the European standard, (246)

Sieve size (mm)	Railway ballast size 31.5mm ÷ 50mm			Railway ballast size 31.5mm ÷ 63mm		
	Percentage passing by weight					
	Grading category					
	A	B	C	D	E	F
80	100	100	100	100	100	100
63	100	97 ÷ 100	95 ÷ 100	97 ÷ 99	95 ÷ 99	93 ÷ 99
50	70 ÷ 99	70 ÷ 99	70 ÷ 99	65 ÷ 99	55 ÷ 99	45 ÷ 70
40	30 ÷ 65	30 ÷ 70	25 ÷ 70	30 ÷ 65	25 ÷ 75	15 ÷ 45
31.5	1 ÷ 25	1 ÷ 25	1 ÷ 25	1 ÷ 25	1 ÷ 25	0 ÷ 7
22.4	0 ÷ 3	0 ÷ 3	0 ÷ 3	0 ÷ 3	0 ÷ 3	0 ÷ 7
31.5 ÷ 50	≥ 50	≥ 50	≥ 50	-	-	-
31.5 ÷ 63	-	-	-	≥ 50	≥ 50	≥ 85

12.2.2. Fine particles

According to the European standard, fine particles are defined as the ballast grains passing from a sieve size of 0.5 mm. Based on the content of fine particles, various categories of ballast can be declared, (Table 12.3), (246).

Table 12.3.
Categories of ballast in relation to fine particles content according to the European standard, (246)

Maximum percentage of ballast (by weight) passing from sieve size of 0.5 mm			
Category of fine particles			
A	B	Declared	C
0.6%	1.0%	>1.0%	No requirement

12.2.3. Fines

According to the European standard, fines are defined as the ballast grains passing from a sieve size of 0.063 mm. Based on the content of fines, various categories of ballast can be declared, (Table 12.4), (246).

Table 12.4.
Categories of ballast in relation to fines content
according to the European standard, (246)

Maximum percentage of ballast (by weight) passing from sieve size of 0.063 mm				
Category of fine particles				
A	B	C	Declared	D
0.5%	1.0%	1.5%	>1.5%	No requirement

12.2.4. Particle shape

12.2.4.1. Flakiness index

The shape of railway ballast is determined in relation to the flakiness index, which is defined as the percentage (by weight) of particles, whose least dimension is less than 3/5 of their mean dimension. Based on the value of the flakiness index, various categories of ballast can be declared, (Table 12.5), (246).

12.2.4.2. Shape index

The shape of railway ballast is determined in relation to the shape index, which is defined for a surface (with L its longest axis) to be equal to $1.27^4 \cdot L^2$. Based on the value of the shape index, various categories of ballast can be declared, (Table 12.6), (246).

Table 12.5.
Categories of ballast in relation to the flakiness index
according to the European standard, (246)

Flakiness index	Category of railway ballast
≤ 15	Fl_{15}
≤ 20	Fl_{20}
≤ 35	Fl_{35}
> 35	$Fl_{declared}$

Table 12.6.
Categories of ballast in relation to the shape index
according to the European standard, (246)

Shape index	Category of railway ballast
≤ 10	SI_{10}
≤ 20	SI_{20}
≤ 30	SI_{30}
5 ÷ 35	$SI_{5 \div 30}$
> 30	$SI_{declared}$

12.2.4.3. Particle length

Based on the value of particle length, various categories of ballast can be declared, (Table 12.7), (246).

Table 12.7.
Categories of ballast in relation to the value of the particle length according to the European standard, (246)

Percentage by weight with length ≥ 100 mm in a sample greater than 40 kg					
Particle length category					
A	B	C	D	Declared	E
4%	6%	8%	12%	>12%	No requirement

12.3. Mechanical behavior of ballast and subballast

12.3.1. Elastoplastic behavior

On-site measurements of settlements and stresses at the time of the passage of train loads have shown that the mechanical behavior of ballast and subballast is elastoplastic, with as most suitable criterion of plasticity the criterion of Drucker-Prager, (see also section 8.4.4.1), (148), (175).

12.3.2. Fatigue behavior

12.3.2.1. Ballast

Both laboratory tests and on-site measurements have shown that on initial loading, the ballast undergoes a considerable permanent (plastic) deformation. In view of its peculiar granulometric composition, the probable cause of this phenomenon is the rearrangement of the ballast stones to attain a state of equilibrium, (244), (249). In subsequent loadings, the contribution of the plastic component to the total deformation is smaller. Triaxial tests have shown that the plastic deformation ε_N^p of ballast at the n-th loading cycle may be expressed as a function of the plastic deformation at the first loading cycle ε_1^p by the following formula, (253), (256):

$$\varepsilon_N^p = \varepsilon_1^p \cdot (1 + c \cdot \log N) \tag{12.1}$$

Research in ORE and the British railways has suggested for c the value 0.2, (256), (257). However, research conducted by American railways has suggested for c values between 0.25 and 0.40, (251), (254).

Most of the laboratory results fit with the linear form of equation (12.1). Nevertheless, in a very small number of tests, data have revealed a nonlinear character for the evolution of plastic deformation of ballast, (251), (255).

According to equation (12.1) and taking into account the aforementioned values of c, it would take 100,000÷300,000 loading cycles to double the plastic deformation caused in the first loading cycle.

Laboratory tests under constant stress conducted by the British railways have yielded the following semi-empirical formula for the plastic deformation ε_N^p of ballast after N loading cycles, (257):

$$\varepsilon_N^p = 0.082 \cdot (100 \cdot n - 38.2) \cdot (\sigma_1 - \sigma_3)^\alpha \cdot (1 + 0.2 \log N) \qquad (12.2)$$

where:

n : ballast porosity,

a : coefficient depending on the level of the stress applied. It takes values between 1 and 2 for low stresses, but may reach the value 3 for high stresses.

12.3.2.2. Subballast

For gravel subballast the following formula has been suggested for the total deformation after N loading cycles,

$$\varepsilon_N^{tot} = \varepsilon_1^{tot} \cdot N^\alpha \qquad (12.3)$$

where:

ε_N^{tot} : deformation at the N-th loading cycle,

ε_1^{tot} : deformation at the first loading cycle,

α : a parameter depending on the characteristics of gravel.

12.3.3. Modulus of elasticity

12.3.3.1. Ballast

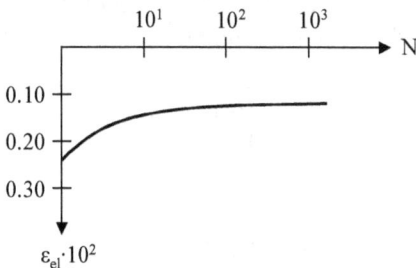

Fig. 12.3. Evolution of elastic deformation of ballast in relation to the number of loading cycles, (249)

With respect to the modulus of elasticity, triaxial tests have shown it to change during the first 1,000 loading cycles and thereafter to remain about constant, (Fig. 12.3). This is similarly explained as with the appearance of important plastic deformations during the first loading cycle. The modulus of elasticity at the one-thousandth loading cycle was found to be about double that at the first cycle, $E_{1,000} \cong 2 \cdot E_1$, (249), (255).

12.3.3.2. Subballast

Concerning gravel subballast, a series of tests have suggested that the modulus of elasticity does not change in relation to the number of loading cycles, (256).

12.4. Ballast hardness

Ballast must have adequate hardness, otherwise it disintegrates and cannot fulfill its functions. Ballast hardness can be assessed according to the Deval, the Los Angeles, and the Microdeval laboratory tests.

12.4.1. The Deval test

This is the oldest of the tests still in use. It was designed in 1896, at a time when road traffic was composed of carriages with wheels surrounded by steel hoops (tires).

The weight of the test sample (as close to cubic shape as possible) is 5 kg. In the case of the Deval standard test, known also as Deval dry test, the sample pieces are washed and dried before being weighed. They are thereafter placed in the cylinders of the Deval machine, which have an internal diameter of 20 cm, an internal length of 34 cm, are inclined by 30°, and are connected to a horizontal axle, (Fig. 12.3). The machine is then started (2,000 revolutions per hour) and the entire test takes about 5 hours (a total of 10,000 revolutions).

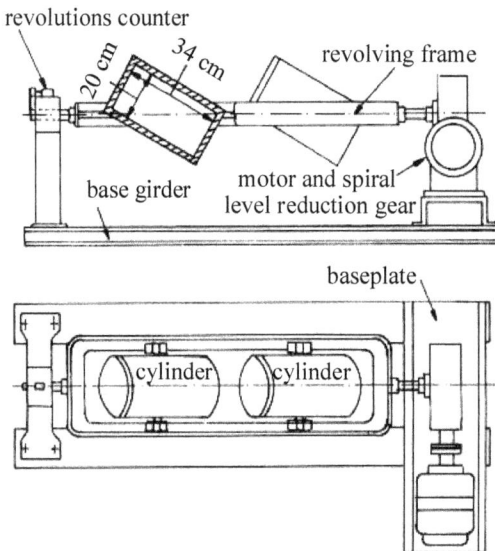

Fig. 12.4. Deval attrition machine

Let A be the initial weight of the sample and B the weight of the sample material retained after the test by a sieve of a diameter d (mm). The value of d is according to French regulations 1.6 mm and according to British regulations 2.36 mm, (250), (252). Hence, the percentage of attrition will be:

$$W_D = \frac{A - B}{A} \cdot 100 \qquad (12.4)$$

The Deval coefficient Q is derived from the formula:

$$Q = \frac{40}{W_D} \cdot 100 \qquad (12.5)$$

Some regulations specify that the ballast should have a Deval coefficient greater than 14 in the case of hard rock and greater than 12 in the case of limestone rock. However, other regulations require a greater hardness for ballast and they specify that the Deval coefficient be grater than 16. If this value is taken into account, ballast from limestone rock may prove inappropriate and railway authorities should look for ballast coming from granite rock.

The attrition action during the course of the Deval test (the sample completes 10,000 revolutions at the end of the test) is much stronger than the vibrating action. Therefore, only very soft rock is broken to a considerable extent. Pieces with sharp corners in particular are rounded off.

A variation of the Deval standard test is to carry out the whole procedure in the presence of water, in which case the result is termed as the wet Deval coefficient.

12.4.2. The Los Angeles test

In the Los Angeles test (designed in 1926), the test equipment consists of a steel cylinder with an internal diameter of 71.1 cm and an internal length of 50.8 cm. A 5 kg sample is placed inside the cylinder together with 12 steel balls, each one weighing 420 gr. The cylinder is then set in rotation (30÷33 rounds per minute) until 500 revolutions are completed (duration of the test is about 15 minutes).

Let A be the initial weight of the sample and B the weight of the sample material retained after the test by a sieve of a diameter d (mm) (d=1.6mm according to French regulations and d=2.36 mm according to British regulations). The percentage of attrition is called the Los Angeles coefficient and is:

$$W_{LA} = \frac{A - B}{A} \cdot 100 \tag{12.6}$$

Many regulations specify that the ballast must have a Los Angeles coefficient smaller than 25.

The Los Angeles test has the following characteristics:
- action on the inert material is sufficiently strong to bring out any weaknesses,
- it is equally suitable for testing inert materials, crushed rock, and gravel,
- the time required to complete the test is short,
- the results of the test agree to a satisfactory degree with the behavior of the crushed and inert materials in various construction projects.

Many current technical regulations are based on the Los Angeles test. Several variations of the Los Angeles test are in use.

According to the European standard for ballast, the Los Angeles test method should be the reference test for the determination of ballast hardness.

Thus, based on the value of the Los Angeles coefficient, various categories of ballast can be declared, (Table 12.8), (246).

However, it should be emphasized that according to the European standard, the sample in the Los Angeles test is 10 kg and the machine is rotated for 1,000 revolutions at a speed of $31 \div 33$ rounds per minute.

Table 12.8.
Categories of ballast in relation to the Los Angeles coefficient according to the European standard, (246)

Los Angeles coefficient	Category of ballast LA_{bal}
≤ 12	$LA_{bal}\ 12$
≤ 14	$LA_{bal}\ 14$
≤ 16	$LA_{bal}\ 16$
≤ 20	$LA_{bal}\ 20$
≤ 24	$LA_{bal}\ 24$
> 24	LA_{bal} declared

12.4.3. The Microdeval test

The Microdeval test is used principally to determine the hardness of gravel subballast. The test equipment consists of a cylinder of a length of 154mm with an internal diameter of 200 mm. The sample consists of 500 gr of gravel with grains ranging between sieve diameters 10 mm and 14 mm. A steel ball weighing 5 kg and 2.5 liters of water are put in the cylinder together with the sample. The cylinder performs 12,000 revolutions at a speed of 100 revolutions/minute. Let be w the weight (in gr), after the test, of grains smaller than the 1.6mm sieve. The Microdeval coefficient MDE is defined as

$$MDE = 100 \cdot \frac{w(gr)}{500} \qquad (12.7)$$

The European standard suggests use of the Microdeval test when required. Based on the value of the Microdeval test, various categories of ballast can be declared, (Table 12.9), (246).

However, it should be emphasized that according to the European standard, the sample in the Microdeval test is 10 kg and the machine is rotated for 14,000 revolutions.

12.4.4. Required strength and hardness of ballast

The required strength and hardness of ballast depend upon the line traffic, the frequency of renewal of ballast (usually every $15 \div 20$ years), the material of the crushed stone, etc. French regulations mandate that the Los Angeles and Deval coefficients intersect at a point lying within the band specified in Figure 12.5, (252).

Table 12.9.
Categories of ballast in relation to the Microdeval coefficient according to the European standard, (246)

Microdeval coefficient	Category of ballast MDE_{bal}
≤ 5	MDE_{bal} 5
≤ 7	MDE_{bal} 7
≤ 11	MDE_{bal} 11
≤ 15	MDE_{bal} 15
> 15	MDE_{bal} declared

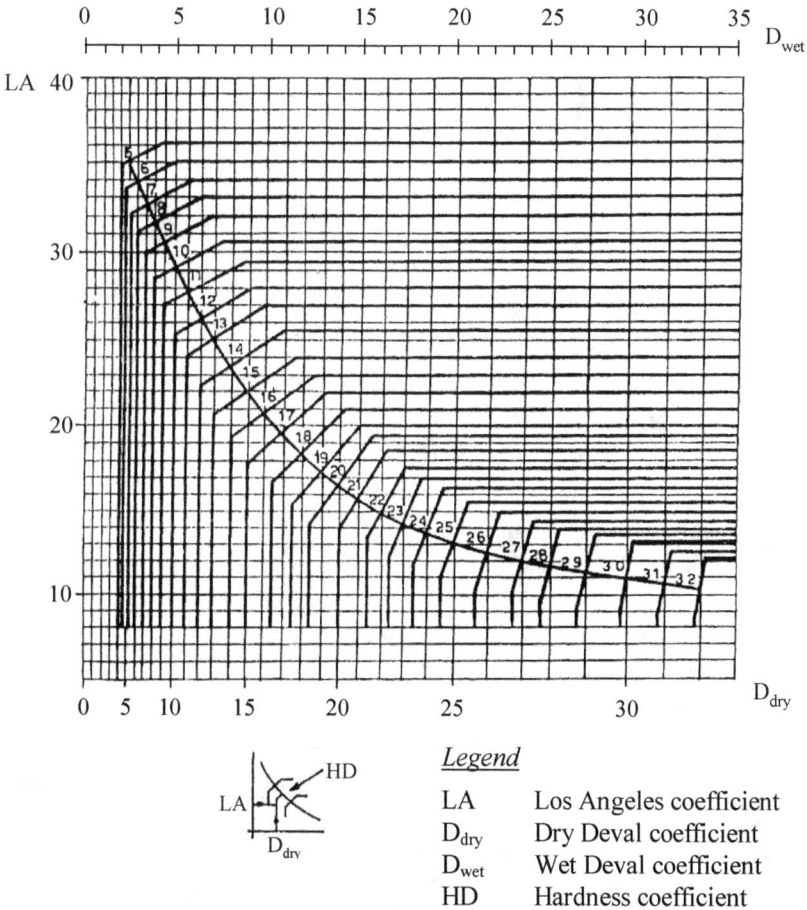

Legend

LA	Los Angeles coefficient
D_{dry}	Dry Deval coefficient
D_{wet}	Wet Deval coefficient
HD	Hardness coefficient

Fig. 12.5. Combination of the Los Angeles and Deval coefficients for ballast according to French regulations, (252)

12.5. Determination of the appropriate thickness of ballast

12.5.1. Determination of the appropriate thickness of track bed

Until the mid-1980s, ballast thickness was calculated based on the Boussinesq equations. However, a more accurate analysis, with application of the finite element method, has allowed all railway parameters concerning ballast dimensioning to be taken into account:

- quality of soil and bearing capacity of the subgrade,
- type of sleeper,
- traffic characteristics (traffic load and axle load),
- volume of maintenance works,
- train speed,
- use or not of a geotextile.

Thickness e of track bed structures (e = ballast + subballast) will be derived from stress analysis given in Figure 8.7 of section 8.4.7 and is calculated by the following formula, which is suggested by the UIC, (186):

$$e(m)=N(m)+a(m)+b(m)+c(m)+d(m)+f(m) \tag{12.8}$$

where:

♦ N (parameter depending on subgrade quality):
- 0.70 m for subgrade of poor quality (S_1),
- 0.55 m for subgrade of average quality (S_2),
- 0.45 m for subgrade of good quality (S_3).

♦ a (parameter depending on the daily traffic load of the line):
- 0 for UIC groups[*] 1 to 4,
- −0.05 m for UIC groups 5 and 6,

♦ b (parameter depending on sleeper type):
- 0 for timber sleeper (with a length L=2.60m),
- (2.50−L(m))/2 for concrete sleepers of length L (b may be negative for L>2.50 m),

♦ c (parameter depending on maintenance conditions):
- 0 for normal conditions,
- −0.05 m for difficult maintenance conditions on tracks with only freight traffic and a daily traffic load < 7 thousand tonnes,
- −0.10 m for difficult maintenance conditions on tracks with passenger and freight traffic and a daily traffic load < 7 thousand tonnes,

♦ d (parameter depending on the value of axle load Q):
- 0 for Q=20.0 tonnes,

---

[*] UIC has classified railway lines in various groups according to their daily traffic load. See section 7.5.2, Fig.7.6

- – 0.05 m for Q=22.5 tonnes,
- – 0.12 m for Q=25.0 tonnes,
- – 0.25 m for Q=30.0 tonnes,
- ♦ f (parameter depending on train speed and subgrade quality):
  - – 0 for V≤160 km/h (whatever the subgrade quality) and for high-speed tracks on good subgrade quality (S_3),
  - – 0.05 for high-speed tracks on average subgrade quality (S_2),
  - – 0.05 for high-speed tracks on poor subgrade quality (S_1).

Use of a geotextile beneath the gravel subballast layer can increase both the mechanical and the hydraulic behavior of the track. According to UIC, the use of a geotextile is obligatory for subgrades of poor quality (S_1) and average quality (S_2) and optional for subgrades of good quality (S_3). Even in the last case (subgrade S3), however, the use of geotextile is recommended.

12.5.2. Required thickness of track bed (ballast + subballast) to avoid frost penetration

Penetration of frost in the subgrade may cause serious problems in railway tracks. A number of theoretical and experimental studies have permitted to calculate the thickness of ballast + subballast, in relation to the frost index, (see section 9.11), so as to avoid frost penetration in the subgrade, (Fig. 12.6). The shaded area of Figure 12.6 represents conditions encountered in northern Europe and America or in areas with difficult and lasting winters.

12.5.3. Thickness of ballast and subballast

Gravel subballast usually has a thickness of 15 cm; it should be well-graded, with the following required mechanical characteristics:
- ♦ Microdeval coefficient < 15 or 20,
- ♦ Los Angeles coefficient < 20 or 25,

depending on traffic load and maintenance conditions of the line.

However, some railways require, particularly on new lines, the gravel sub-ballast to contain at least 30% of crushed stone, (186).

The thickness of ballast is calculated by subtracting the thickness of the subballast (usually 15 cm) from the thickness of the track bed, as calculated previously.

12.5.4. Calculation of thickness of ballast according to British regulations

The thickness of ballast is calculated according to British regulations in relation to the speed and tonnage of the line, as shown in Table 12.10.

Required thickness h of ballast+subballast
for protection against frost (m)

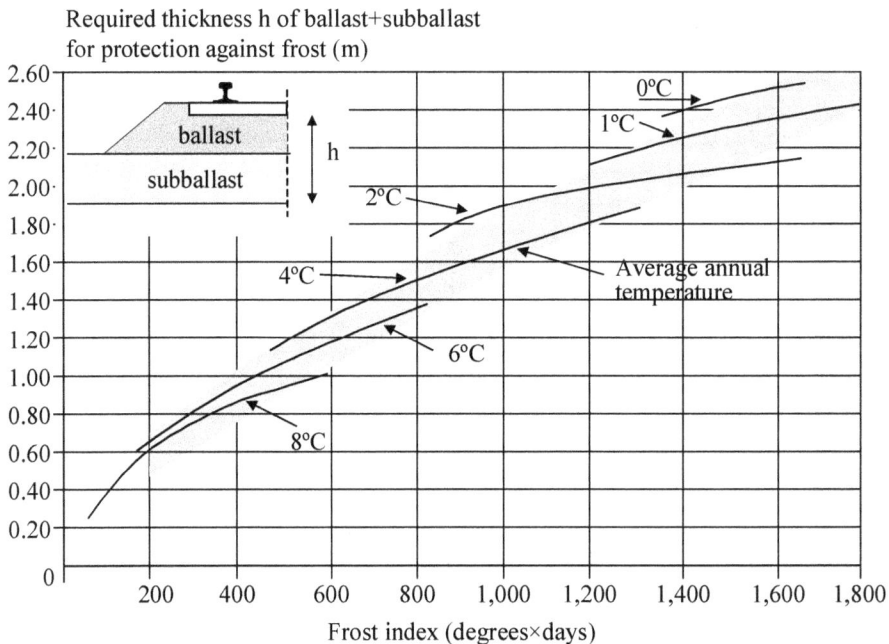

Fig. 12.6. **Required thickness of ballast + subballast to avoid frost penetration in the subgrade, (186), (198)**

Table 12.10.
Thickness of ballast according to British regulations, (141), (250)

Line speed (km/h)	Yearly line tonnage (million tonnes)	Ballast thickness (m)
160÷200	All	0.38
120÷160	>12 million	0.38
120÷160	2÷12 million	0.30
120÷160	<2 million	0.23
80÷120	>12 million	0.30
80÷120	<12 million	0.23
< 80	>2 million	0.23
< 80	<2 million (concrete sleepers)	0.20
< 80	<2 million (timber sleepers)	0.15

356

12.5.5. Numerical application

Let us consider a track with V_{max} =160 km/h and a daily traffic load of 40 thousand tonnes (UIC group 4), which is representative of a great number of tracks all over the world. The track is laid on monoblock prestressed-concrete sleepers with a length L=2.60 m. Subgrade soil is of average quality (S_2), axle load is 22.5 tonnes, and the working conditions (for maintenance) are normal.

Thickness e of track bed structures will be calculated according to formula (12.8),

$$e(m)=N(m)+a(m)+b(m)+c(m)+d(m)+f(m)$$

with the various parameters taking the following values:

N= 0.55 m (subgrade quality: S_2),

a = 0 (track of UIC group 1 to 4),

b = (2.50 m–2.60 m)/2=–0.05 m (concrete sleepers with a length L=2.60 m),

c = 0 (normal maintenance conditions),

d = 0.05 (axle load: 22.5 tonnes),

f =0 (track with V=160 km/h).

Substituting these values in the above formula, we calculate the thickness e of the track bed (ballast + subballast):

$$e(m) = 0.55 + 0 – 0.05 + 0 + 0.05 + 0= 0.55 m$$

Subballast will be given the usual thickness of 0.15 m, thus the thickness of ballast will be: e–0.15 m=0.40 m.

It is to note that this value of ballast thickness (calculated according to formula (12.8)) is very close to the value suggested by British regulations (=0.38 m).

If the track is laid in areas with cold whether, the thickness of the track bed should be calculated to ensure frost protection of the subgrade. Suppose an average annual temperature of 6°C and that for 100 days per year negative temperature is –3°C. Thus, the frost index will be: 100×3=300 degrees×days. From Figure 12.6 we deduce that, in order to avoid frost penetration, the thickness e (ballast + subballast) shall be 0.80 m, far greater than the value calculated according to the mechanical requirements of the track. It is obvious that in this case the greater value of ballast + subballast will finally be taken into account for the dimensioning of the track bed.

12.5.6. Appropriate thickness of ballast for metric gauge tracks

For metric gauge tracks, recommended values according to UIC for the appropriate thickness of the ballast are given in Table 12.11, (140).

Table 12.11.
Appropriate thickness of ballast for metric gauge tracks, (140)

V_{max}	160 km/h	120 km/h	100 km/h	80 km/h	60 km/h
Axle load	13t	16t	20t	25÷30t	16t
Traffic	only passenger	mixed	mixed	only freight	principally freight
Appropriate thickness of ballast	25÷30 cm	20÷25 cm	20÷25 cm	25÷30 cm	10÷15 cm

12.6. Track cross-sections

In the present and previous chapters we analyzed how dimensioning and mechanical characteristics of the various components of the track should be calculated. All these analyses are usually reflected in summary in the track cross-section, which illustrates dimensions of all components of the track system.

Design of a track cross-section depends on the following:
– whether track is single or double,
– the distance b between the two tracks, which depends on train speed. As explained in section 7.10.3, the distance b ranges between 3.60 m÷4.00 m for V<200 km/h, between 4.00 m÷4.70 m for V=200÷300 km/h, and between 4.50 m÷5.00 m for V>300 km/h. In figures presented below, (Fig. 12.7 to Fig. 12.15), we can remark that for the same value of b, various railways apply different values of the permitted maximum speed. This is principally related to whether or not there is a great number of tunnels in the specific track,
– the length of the ballast shoulder,
– whether a superelevation of the ballast shoulder is given or not,
– whether the track is electrified or not,
– whether wires of signaling systems are placed within or outside the track,
– the width of rolling stock.

Some representative cross-sections of tracks are illustrated below:
• single track with steel or timber sleepers, (Fig. 12.7),
• single track with twin-block reinforced-concrete sleepers, (Fig. 12.8),
• double track with monoblock prestressed-concrete sleepers and a distance between tracks b=4.20 m (track appropriate for speeds up to 250 km/h). Cases of straight track, (Fig. 12.9), and curved track, (Fig. 12.10), are given,

- high-speed Paris-Lyon track of French railways (V_{max}=300 km/h), (Fig. 12.11),
- high-speed Paris-Marseille track of French railways (V_{max}=350 km/h), (Fig. 12.12),
- high-speed track of German railways (V_{max}=300 km/h), (Fig. 12.13),
- high-speed track of Italian railways (V_{max}=250 km/h), (Fig. 12.14),
- high-speed track of Japanese railways (V_{max}=300 km/h), (Fig. 12.15).

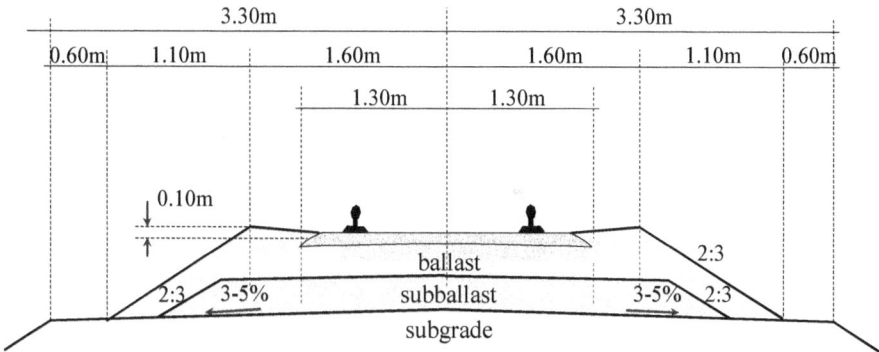

Fig. 12.7. Cross-section of a single track with steel sleepers

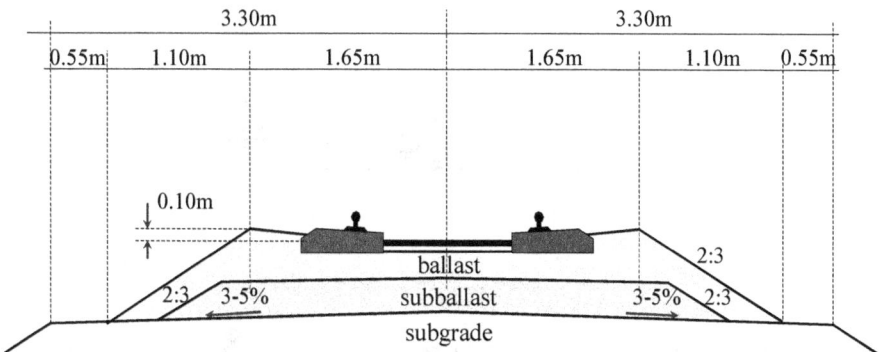

Fig. 12.8. Cross-section of a single track with twin-block sleepers

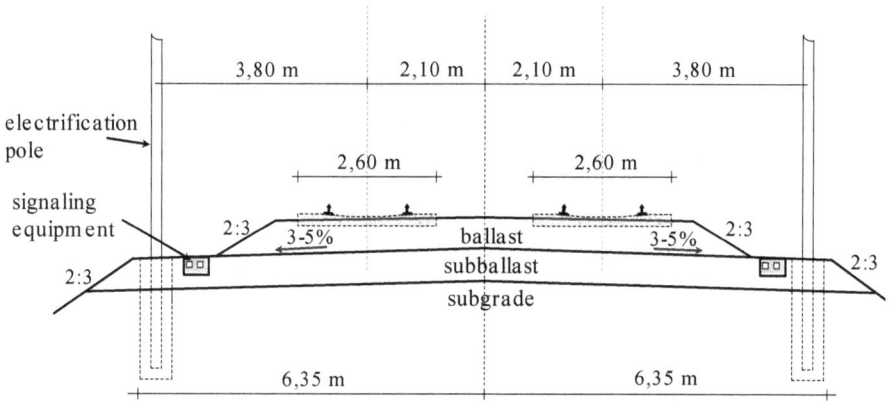

Fig. 12.9. Cross-section of a double track with monoblock prestressed-concrete sleepers (V_{max}=250 km/h), (straight track)

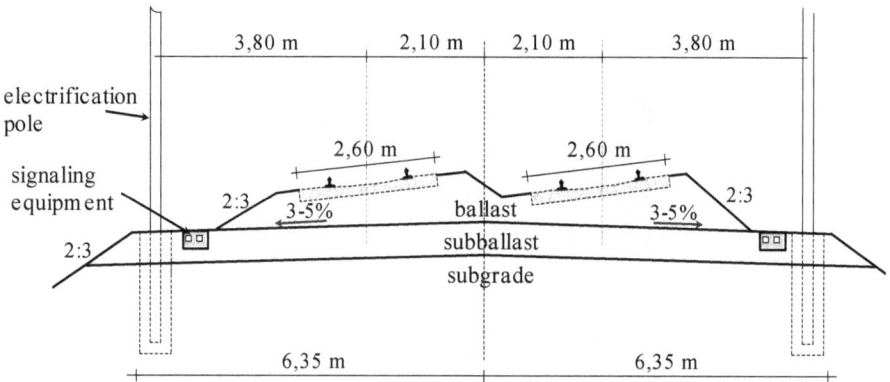

Fig. 12.10. Cross-section of a double track with monoblock prestressed-concrete sleepers (V_{max}=250 km/h), (curved track)

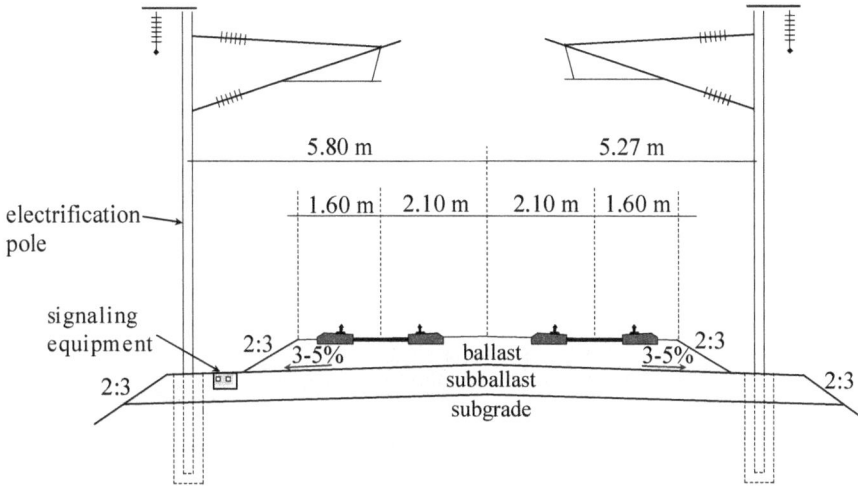

Fig. 12.11. Cross-section of the high-speed Paris-Lyon track of French railways (V_{max}=300 km/h)

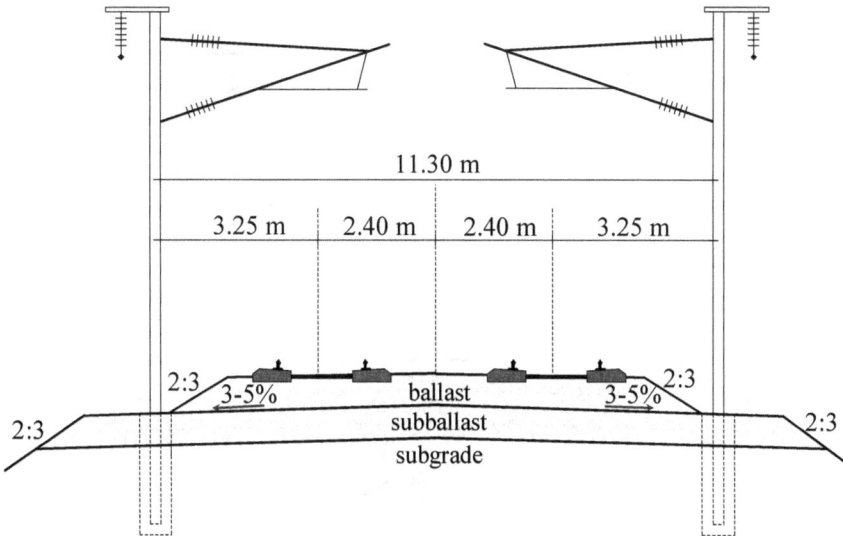

Fig. 12.12. Cross-section of the high-speed Paris-Marseille track of French railways (V_{max}=350 km/h)

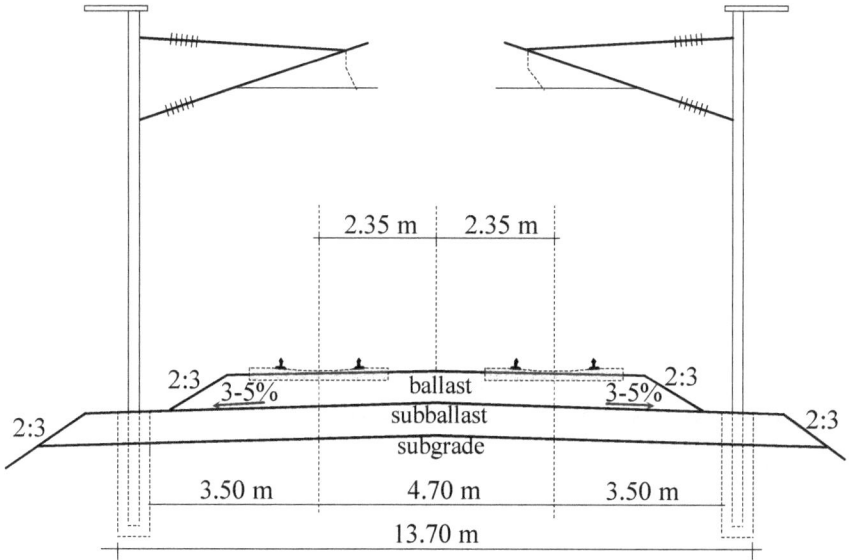

Fig. 12.13. Cross-section of a high-speed track of German railways
   ($V_{max}= 300$ km/h)

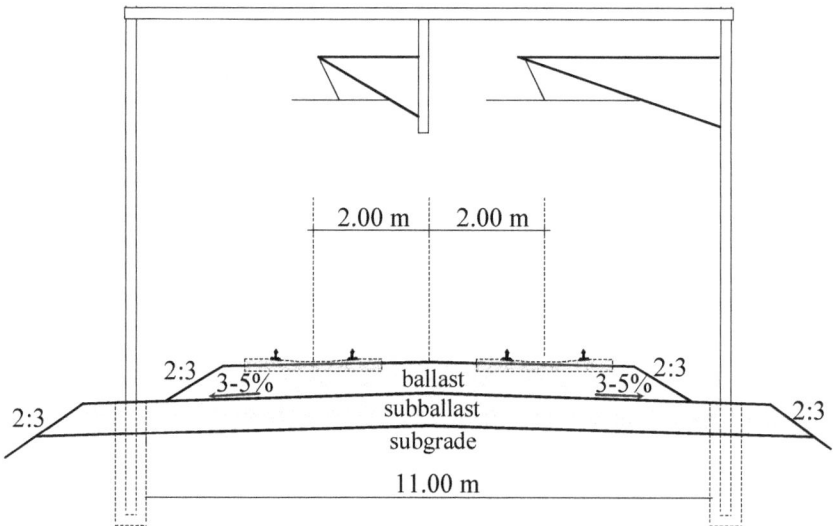

Fig. 12.14. Cross-section of a high-speed track of Italian railways
   ($V_{max}= 250$ km/h)

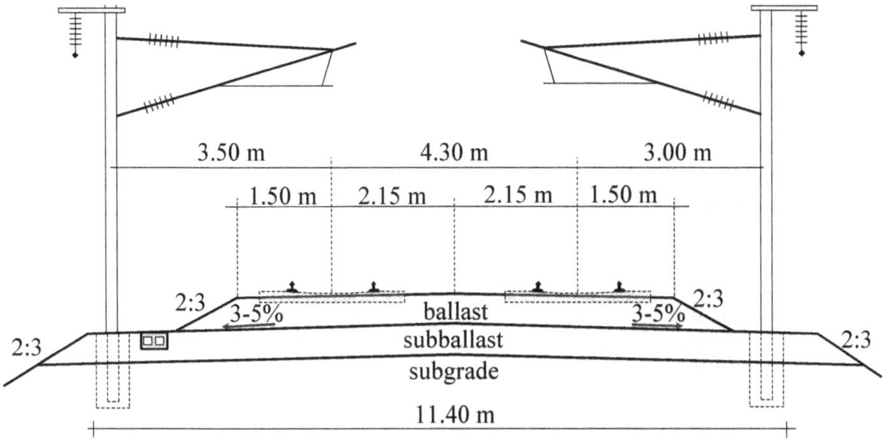

Fig. 12.15. Cross-section of a high-speed track of Japanese railways (V$_{max}$= 320 km/h)

12.7. Lifetime and reuse of ballast

It is suggested that the lifetime of ballast, sleepers, and rails should be combined, so that they are all replaced during the same renewal of the track. In high-speed tracks, ballast renewal is done once per 15÷20 years, whereas concrete sleepers and rails have a lifetime of approximately 50 years.

However, and due to fatigue, ballast stones can support only a limited number of maintenance sessions, during which ballast receives forces of high intensity.

The lifetime of ballast can be increased by using stones of greater hardness, by making the width of ballast layer greater, and by carrying out a more homogeneous compacting of the ballast layer. Such measures have resulted in Switzerland in an increase of the time between successive maintenance sessions from 4 to 7 years and in a decrease of maintenance expenses of 40% within 10 years, (247).

Ballast taken from a track during maintenance is a polluted material particularly in the case of timber sleepers (which are impregnated with special fluids), with a weight of 1.7 t/m^3, instead of a weight of 1.5 t/m^3 for a new ballast. If this ballast has sufficient remaining mechanical resistances, then it can be washed and reused as subballast or as formation layer during the renewal of secondary lines.

If the remaining resistances of ballast are even higher, then after a mechanical treatment and washing, it can be used as ballast for secondary lines (V<140 km/h). A great part of ballast (900,000 tonnes of ballast per year are taken out of tracks during maintenance in France) is in this way reused in France, Germany, and elsewhere, (247).

Reuse of ballast is not a purely technical problem; the environmental and economic aspects should also be taken into account. The cost of the reuse of ballast should be compared to the cost of new ballast and to the cost of transport of new or reused ballast to its final area of use. However, costs of disposal of the used ballast may render the reuse of old ballast prohibitive.

12.8. Monitoring of ballast characteristics with the use of radar systems

Any engineering material undergoes changes in its initial mechanical characteristics during the course of time. In order to follow this irreversible process, engineers have devised methods which can be classified into destructive (a part of the material is taken for testing) and non-destructive (the material is not affected) ones. Among the materials of a railway track that are used for construction purposes, ballast is the most vulnerable to traffic and weather conditions. Under repeated loading from successive wheel loads, railway ballast adapts to changing conditions and shows changes in its grading, decay, pollution of the air voids between stones by fine-graded material, fouling, and eventual differential settlement. The accurate knowledge of the condition of ballast is essential to safeguard safety and quality of service and can be performed by visual inspection of the track by railway staff (costly and inaccurate), destructive methods (a sample of ballast is taken for testing), and non-destructive methods.

An efficient non-destructive method for the monitoring and tracing of ballast characteristics is based on the electro-magnetic response of ballast to radiation from ground penetrating radar systems. A source within such systems emits electromagnetic fields that are partially back reflected and partially transmitted through the voids between ballast stones. The information attained is then processed and can provide an accurate image of the situation of ballast and more particularly of its dimensions, degree of fragmentation, decay, fouling, and pollution. By introducing data of successive monitorings in artificial intelligence algorithms, it is possible to calculate the rate of decay of ballast, areas with more problems locally, and finally to schedule maintenance sessions and the time of replacing old ballast with new one by exhausting all existing mechanical reserves of the material, (see also section 16.12).

13 Transverse Effects – Derailment

13.1. Transverse effects

When a rail vehicle runs on the track, vertical, transverse, and longitudinal forces are developing on the railway system, (see section 7.11.1). Up to this chapter, we have focused on the effects of vertical forces, which determine the dimensioning of the various components of the railway track and the subgrade. Transverse forces affect both passenger comfort and train safety. Exceeding the limits of transverse track resistance may cause track shifting and eventual derailment. Derailment may also be the result of either wheel climbing on the rail or of vehicle overturning, (261). Speed increases have mandated additional measures to increase and ensure safety. It should be stressed that compared to other means of transportation, railways are the safest, (see sections 1.2.3 and 22.4.6).

13.2. Transverse track forces

Let us first investigate what transverse forces are exerted during train motion on the track as a whole. Transverse forces are composed of one static and one dynamic component.

13.2.1. Transverse static force

This is defined as the force due to the non-compensated centrifugal acceleration and to driving forces on curves. Transverse static force $H_s(t)$ will be calculated by the following semi-empirical formula, (269):

$$H_s(t) = \frac{P(t) \cdot NT(mm)}{1,500} \tag{13.1}$$

where: P : axle load,

NT: transverse defect, (see section 16.4.2), if the train is on a straight track, or cant deficiency $h_{d\,max}$, (see section 14.2.2), if the train is on a curve.

13.2.2. Transverse dynamic force

This is defined as the force caused by the various forms of track defects and by rolling stock defects. The transverse dynamic force $H_d(t)$ will be calculated by the following semi-empirical formula, (269):

$$H_d(t) = \frac{P(t) \cdot V(km/h)}{1,000} \tag{13.2}$$

where: P : axle load,
          V : train speed.

13.3. Transverse track resistance

Transverse resistance of the track develops by 65% on the ballast and the ballast-sleeper interaction, by 25% on the rail, and by 10% on the fastening system between rail and sleeper. As rail and fastenings have standardized characteristics in modern tracks, we will principally consider the effects of ballast and sleeper characteristics as well as whether maintenance works have been recently realized on the track. We will consider the worst case, i.e. a track immediately after maintenance works, which destabilize the track. Under the influence of rail traffic, the ballast is compacted, thus resulting in an increase of the transverse resistance.

On tracks with timber sleepers, for which maintenance is performed by *non-mechanical* (manual) means, the transverse track resistance is calculated by the formula, (267):

$$L(t) = 0.85 \cdot \left(1 + \frac{P(t)}{3}\right) \tag{13.3}$$

On tracks with timber sleepers, for which maintenance is performed *mechanically*, the transverse track resistance is calculated by the formula:

$$L(t) = 1 + \frac{P(t)}{3} \tag{13.4}$$

On tracks with twin-block reinforced-concrete sleepers, for which maintenance by mechanical means is mandatory, the transverse track resistance is:

$$L(t) = 1.5 + \frac{P(t)}{3} \tag{13.5}$$

For tracks with monoblock prestressed-concrete sleepers, such an analytical formula is not available; however, tests have shown that the constant term of equations (13.4) and (13.5) has in the case of monoblock prestressed-concrete sleepers values between 1.0 and 1.5, (269).

The above formulas are semi-empirical and are the result of a series of tests conducted by the French and German railways, (267), (269). Most railway authorities are currently using them and no objections or reservations have been expressed.

Research on the effects of speed on transverse track resistance has shown that the latter is not affected by an increase of speed, (267).

Previous formulas are applicable provided that additional dynamic loads, (see section 8.6), are not greater than 20% of nominal static load. If, however, the additional dynamic loads exceed 20% of static load, the above formulas should be multiplied by a correction factor in the order of 0.9. This applies also to tracks of average or poor quality, (265).

13.4. Influence of ballast characteristics on transverse track resistance

13.4.1. Influence of the geometrical characteristics of the ballast cross-section

Transverse track resistance in the area of ballast and sleeper is the resultant of the following three components:

♦ A component generated by friction on the lower surface of the sleeper, proportional to sleeper weight.

♦ A component resulting from friction between the sleeper sides and the ballast filling the space between consecutive sleepers. This component depends on the degree to which the spaces between sleepers are filled, (Fig. 13.1), as well as on the degree of ballast compacting. This lateral component amounts to about 40÷50% of the total resistance in the case of timber sleepers, 15÷25% in the case of twin-block reinforced-concrete sleepers, and 30% in the case of monoblock prestressed-concrete sleepers, (268).

♦ A component developed at the two ends of the sleeper and depending both on the width of ballast shoulder c and whether the ballast is superelevated or not, (Fig. 13.2).

Figure 13.3 illustrates the increase of transverse track resistance caused by an increase of ballast width beyond sleeper ends as well as by a superelevation of the ballast section. Therefore, an increase of the ballast width combined with a simultaneous superelevation is preferable to a simple increase of width.

The effect of the slope of the ballast cross-section to transverse resistance is secondary, (268).

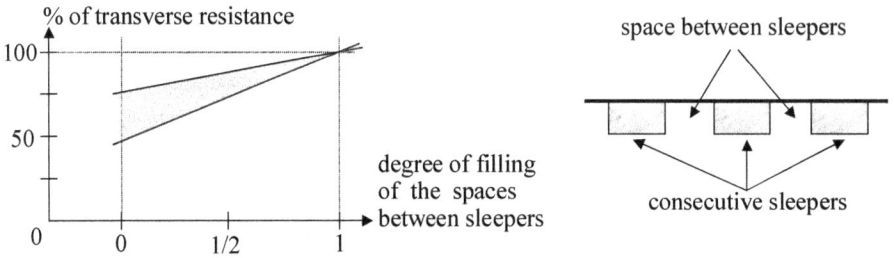

Fig. 13.1. Influence on transverse track resistance of the degree of ballast filling between sleepers, (268)

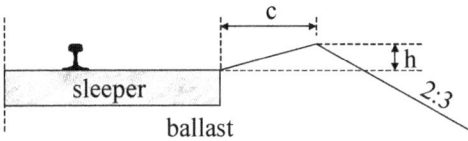

Fig. 13.2. Sleeper end, ballast shoulder width c, and ballast superelevation

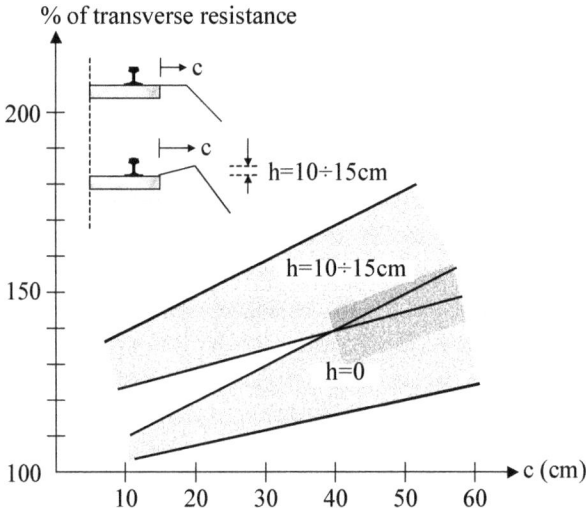

Fig. 13.3. Correlation of transverse track resistance with the geometrical characteristics of the ballast cross-section, (268)

13.4.2. Influence of the granulometric composition of ballast

The shape and size of the ballast stones, their granulometric composition, and the hardness of the material, all have a considerable influence on transverse track resistance, (Fig. 13.4).

368

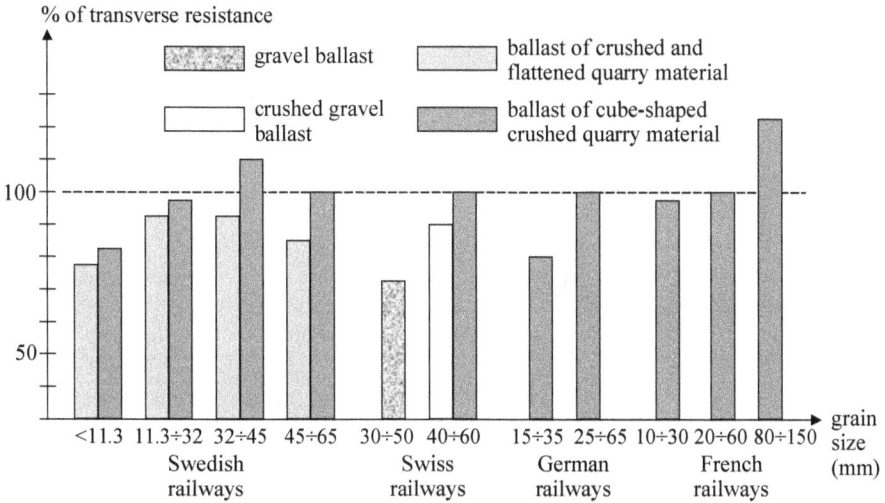

Fig. 13.4. Influence of the granulometric composition of ballast on transverse track resistance, (268)

13.4.3. Influence of the degree of ballast compacting

After track maintenance works[*], the track loses its transverse resistance to a considerable degree, (Fig. 13.5). In order to recover transverse resistance, it is necessary to compact the ballast.

Transverse track resistance is almost fully recovered after the passage of a certain amount of traffic, in particular after the passage of 2 million tonnes, (Fig. 13.6).

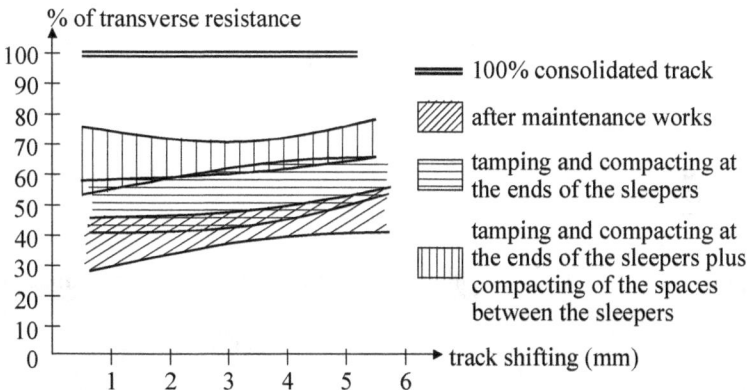

Fig. 13.5. Track stabilization for various forms of compacting, (268)

---

[*] As explained in section 16.8, track maintenance works involve repeatedly raising the track or shifting it horizontally; both cause destabilization of the track.

369

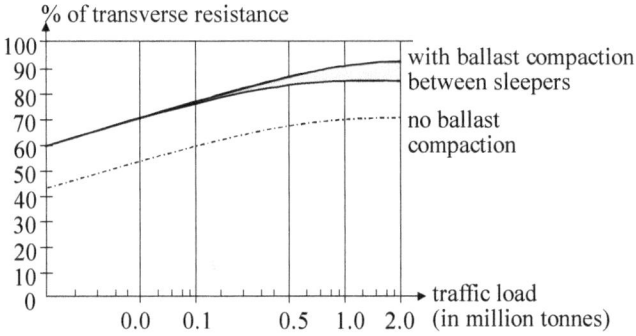

Fig. 13.6. Recovery of transverse track resistance, after maintenance, as a function of traffic load, (261)

13.5. Influence of sleeper type on transverse track resistance

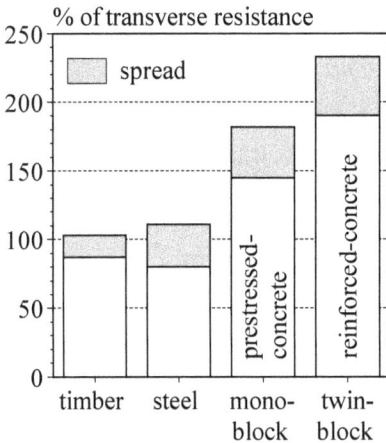

Fig. 13.7. Influence of sleeper type on transverse track resistance, (262)

A series of experimental tests on fully stabilized tracks have shown the unquestionable superiority of concrete sleepers, especially twin-block ones. Figure 13.7 illustrates the transverse track resistance for various sleeper types. The relatively large spread is attributable to manufacturing variations (dimensions, weight, sleeper form, etc.) as well as to ballast quality and properties.

The high transverse resistance of twin-block sleepers, more than double that of timber sleepers, is mainly due to the following two factors, (258):

♦ Due to the greater weight of twin-block sleepers, the transverse resistance component corresponding to the friction between the lower surface of the sleeper and the ballast is greater.

♦ The transverse resistance component generated at the sleeper ends is much greater.

Compared to twin-block sleepers, the transverse resistance of tracks with monoblock sleepers is smaller, but clearly higher than that with timber sleepers. This is due to the greater weight, the greater height, and the larger contact surface of monoblock sleepers (compared to timber and steel sleepers).

The increase of sleeper length from 2.40 m to 2.60 m in the German railways has resulted in an increase of transverse track resistance by 15÷20%, (268).

The transverse resistance of steel sleepers depends to a certain degree upon the sleeper shape (curvature at the ends, ballast contained in the sleeper, etc.). However, the transverse resistance of steel sleepers has values similar to those of timber sleepers, (Fig. 13.7).

Concerning timber sleepers, a comparison between the various qualities of timber leads to the following:

– Differences between sleepers made of hard timber (e.g. oak) and those made of soft timber (e.g. pine) are minor. Sleepers placed a long time ago, with surfaces roughened by the ballast, especially if the latter has been subjected to compaction, present a transverse resistance slightly higher than new (unused) sleepers.

– Sleepers made of tropical timber, due to their great hardness and smooth surfaces, have a transverse resistance reduced by 15%, compared to other qualities of timber, (268).

– A reduction of sleeper spacing leads to a slight reduction in the value of the transverse resistance per sleeper, which, however, is more than offset by the greater number of sleepers per kilometer. Overall, transverse track resistance increases when sleeper spacing decreases, (Fig. 13.8).

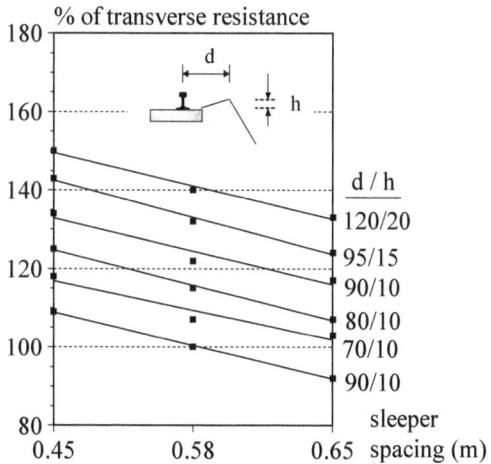

Fig. 13.8. Transverse track resistance as a function of sleeper spacing, (268)

13.6. Additional measures and special equipment used to increase transverse track resistance

In certain cases (e.g. small radius of curvature, turnouts, bridges, etc.), it is necessary to increase locally transverse track resistance by special measures which do not entail a large expense, such as a special sleeper shape, roughened seating surfaces, transverse anchors, etc.

The problem is encountered particularly in certain mountainous railway tracks, which have very small radii of curvature and need a high transverse track resistance to overcome high centrifugal forces and internal stresses in rails. Roughening the side and bottom surfaces of timber sleepers increases transverse resistance only slightly. In contrast, cutting grooves into the seating surface of tropical origin timber sleepers results in an increase of transverse resistance by 20÷25%, (268). The grooves, however, should have sufficient width and depth so as to ensure that the sleepers grip the ballast well, (Fig. 13.9).

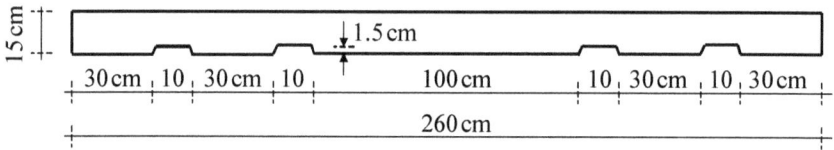

Fig. 13.9. Grooves cut into the seating surface of timber sleepers in order to increase transverse track resistance, (261)

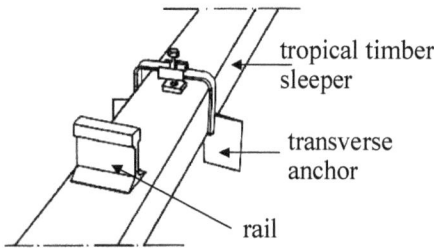

Fig. 13.10. Anchors for increase of transverse track resistance, (261)

A considerable increase of transverse resistance (20÷80%) may be achieved by so-called transverse anchors, (Fig. 13.10), (261). An even greater increase (in the order of 170%) is attained by placing concrete posts against sleeper ends. This is an expensive solution, which in addition interferes with track maintenance conducted with the use of mechanical equipment, (268).

13.7. Derailment

The derailment of a rail vehicle may occur as a result of one of the following, (261), (266):
- track shifting,
- wheel climbing on the rail,
- vehicle overturning.

We will discuss each case separately.

13.7.1. Derailment caused by track shifting

Under the influence of considerable transverse forces, the track shifts as a whole and causes derailment of the vehicle. This form of derailment mainly occurs at high speeds. The condition for derailment by track shifting is that the transverse force H, (Fig. 13.11), which may cause track shifting, exceeds the transverse track resistance L, given by formulas (13.3) to (13.5), (section 13.3):

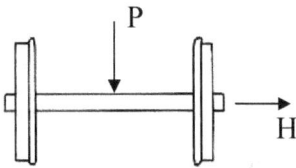

Fig. 13.11. Vertical and transverse forces on a wheel

$$H > L \tag{13.6}$$

$$\text{where} \quad H = H_s + H_d \tag{13.7}$$

13.7.2. Derailment caused by wheel climbing on the rail

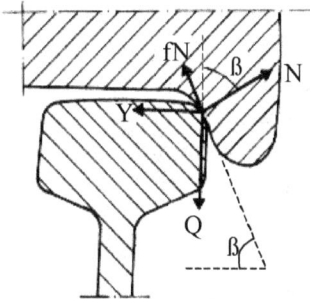

Fig. 13.12. Vertical and transverse forces between wheel and rail

When the transverse force Y developed between wheel and rail exceeds a certain value, then the wheel climbs on the rail and causes derailment. This form of derailment occurs mainly at low speeds and the condition to avoid derailment is given by Nadal's formula, (Fig. 13.12):

$$\frac{Y}{Q} < \frac{\tan\beta - f}{1 + f \cdot \tan\beta} \tag{13.8}$$

where: β : the rail-wheel (flange) angle,

f : the rail-wheel friction coefficient.

Studies of various cases of derailment have shown that equation (13.8) can be simplified as follows, (147), (269):

$$\text{vehicle on axles: } \frac{Y}{Q} < \sim 1.2 \text{ , vehicle on bogies: } \frac{Y}{Q} < \sim 1.3 \tag{13.9}$$

In equations (13.9), Y and Q are the total exerted forces. To the static load Q should therefore be added the dynamic loads, (see section 7.11.2 and section 8.6), which may augment the nominal value of Q (e.g. 10 t/wheel) by up to 50%. With respect to the transverse force Y between wheel and rail, it is of a strongly stochastic[*] nature and no analytical formulation of Y as a function of the rolling stock and track parameters is available. The only way to calculate values of Y is by on-site measurements on the rail, which, however, are difficult, not very reliable, and of course the site of a measurement cannot be expected to coincide with a likely derailment site.

Calculation of the force Y may be obtained by considering forces at both rails. In fact, equation (13.8) applies usually at the outer rail. Nevertheless, forces at the inner rail may be taken into consideration. In this case we will have:

$$Y_1 = Q_1 \cdot \frac{\tan\beta_1 - f}{1 + f \cdot \tan\beta_1} \qquad \text{for the outer rail} \qquad (13.10)$$

$$Y_2 = Q_2 \cdot \frac{\tan\beta_2 + \tan\gamma_2}{1 - \tan\gamma_2 \cdot \tan\beta_2} \qquad \text{for the inner rail} \qquad (13.11)$$

$$Y_1 = Y_2 + H \qquad \text{equilibrium equation} \qquad (13.12)$$

with γ_2 being the conical tread.

Equations (13.10), (13.11), (13.12) give the possibility to calculate transverse forces Y_1, Y_2 at the outer and inner rail.

From a series of derailment accidents, (263), (264), it was derived that there is a high risk of wheel climbing on a rail when the angle β between wheel and rail, (see Fig. 13.12), attains critical values from 58° (case of a wet or lubricated rail, f=0.10÷0.12) to 70° (case of a dry rail, f=0.25÷0.30).

However, the wheel climbing on the rail is most likely to occur when a vehicle is starting from rest on a sharp curve with a high cant, dry rails, and an unlubricated and badly side-worn high rail. The reasons for this are, (260):

♦ the value of Q on the outer rail is minimized by wheel weight transfer, due to cant excess,

♦ transverse force Y is in this case the gauge spreading force, which is related to the wheel weight on the inner rail (maximized by cant excess) and the coefficient of friction across the inner rail (maximized by the dry rail and starting conditions),

♦ the wheel-rail friction coefficient is maximized by the lack of lubrication and the starting condition,

♦ the angle β is reduced by the side-worn rail condition.

---

[*] A process is termed stochastic, if it can only be approximated by statistical measurements (e.g. earthquakes). In deterministic processes, in contrast, correlation of cause and effect is possible in advance. Most known processes in railways, in spite of the observed spread of the results, are considered as deterministic (e.g. elasticity, etc.).

13.7.3. Derailment caused by the overturning of the vehicle

In this case, the vehicle capsizes due to overall unstable equilibrium. It was found that in the worst case for standard gauge tracks (with the center of gravity elevated at 2.25 m above the track), a vehicle would overturn when the transverse acceleration reaches g/3, (269).

As explained in section 14.3, tracks are laid for a maximum value of non-compensated centrifugal acceleration ranging between $0.5 \div 1.0$ m/sec^2 and never exceeding a maximum value of 1.0 m/sec$^2 \cong$ g/10. Therefore, the safety factor against derailment by overturning will have as a lower value

$$\frac{g}{3} / \frac{g}{10} = 3.3.$$

13.7.4. Derailment safety factor – Numerical application

We will investigate the derailment safety factor for a train with vehicles on bogies moving on a curve with a cant deficiency $h_d = 100$ mm, (see section 14.3, Table 14.1), at a speed of 120 km/h. The value of axle load is 20 t; the track is laid on twin-block sleepers and is maintained with mechanical equipment. The wheel-rail friction coefficient is f=0.30 (case of a dry rail).

a. Derailment by track shifting

According to equation (13.6), derailment by track shifting will occur when transverse track forces exceed transverse track resistance, i.e. when H > L. Since $H = H_s + H_d$, from equations (13.1) and (13.2) it follows that:

$$H_s(t) = \frac{P(t) \cdot h_{d\,max}(mm)}{1,500} = \frac{20 \cdot 100}{1,500} = 1.33\,t$$

$$H_d(t) = \frac{P(t) \cdot V(km/h)}{1,000} = \frac{20 \cdot 120}{1,000} = 2.4\,t$$

As we study the case of derailment on a curve, the parameter NT of equation (13.1) has been taken equal to the limit cant deficiency value $h_{d\,max}$ (see section 14.2.2, equation (14.13) and section 14.3, Table 14.1).

Transverse track resistance is calculated by equation (13.5):

$$L(t) = 1.5 + \frac{P(t)}{3} = 1.5 + \frac{20}{3} = 8.16\,t$$

The safety factor against derailment for this particular case will be:

$$v = \frac{L}{H} = \frac{8.16\,t}{1.33\,t + 2.4\,t} = 2.19$$

b. Derailment by wheel climbing on the track

According to equation (13.9), wheel climbing on a rail requires that the ratio Y/Q attains the value 1.3 (vehicle on bogies). As already explained, (section 13.7.2), no analytical expression of Y as a function of track and rolling stock parameters has been formulated. Therefore, the wheel climbing on the rail is considered likely, if certain rolling stock characteristics have values different from those specified during preventive maintenance. This form of derailment is predominant at low speeds, especially in the case of empty rail vehicles.

The critical value of angle β between wheel and rail can be determined by combining equations (13.8) and (13.9). Concerning transverse force Q, additional dynamic loads should also be taken into account from Figure 8.15, from which for V=120 km/h the dynamic impact factor is 1.2, and thus:

$$Q_{tot} = Q_{stat} + Q_{dyn} = Q + 0.2 \cdot Q = 1.2 \cdot Q$$

Therefore, we will have

$$\left. \begin{aligned} \frac{Y}{Q_{tot}} &= 1.3 \\[2em] \frac{Y}{Q_{tot}} &= \frac{\tan\beta - f}{1 + f \cdot \tan\beta} \end{aligned} \right\} \Rightarrow \beta_{critical} = 69°13'$$

and derailment safety factor for this angle is equal to 1.

c Derailment by overturning of the vehicle

This form of derailment can be studied in relation to the geometrical characteristics of the rolling stock. In any case, the safety factor, as discussed in section 13.7.3, has in this case values greater than 3.3, (261).

13.8. Effects of transverse winds

For tracks laid in areas with transverse winds of high intensity, the risk of overturning of the rail vehicle should be carefully evaluated. This risk is greater (by 50%) for metric gauge tracks, (260).

Let us consider a rail vehicle of a mass M on a curve of a radius R with a cant h for a track with a gauge G, (Fig. 13.13). In addition to the transverse forces previously discussed, the vehicle is submitted to a transverse force F_w, due to a wind of a transverse (to the movement of the train) speed u, which causes an additional transverse force ΔQ between the vehicle and the rail.

The static analysis of the phenomenon shows that the overturning of the rail vehicle is a relation of the factor $\Delta Q/Q_o$, which is found to be, (259):

$$\frac{\Delta Q}{Q_o} = \frac{2}{G} \cdot \frac{I}{G} \cdot h_g + \frac{1}{G} \cdot \frac{h_w}{M \cdot g} \cdot \rho \cdot S \cdot c \cdot (V^2 + u^2) \qquad (13.13)$$

where: Q_o : static axle load $[=(M \cdot g)/2]$,

h_g : height of the center of gravity of the vehicle,

h_w : height of the point of application of transverse wind,

ρ : specific mass of the air,

S : surface of transverse cross-section of the vehicle,

c : aerodynamic coefficient at the vertical direction,

V : train speed,

u : transverse (to the movement of the train) wind speed.

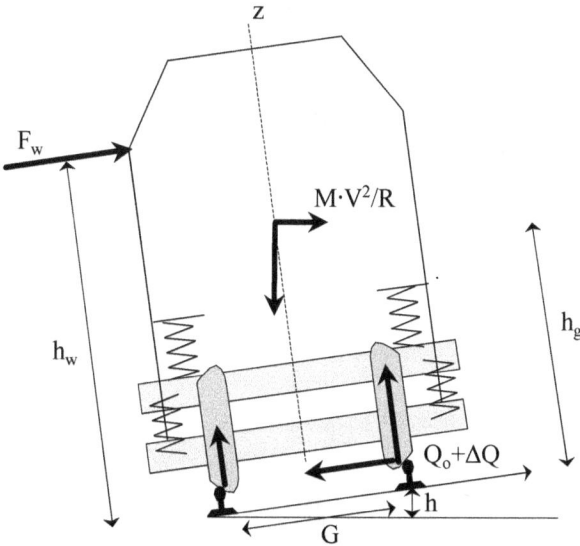

Fig. 13.13. Rail vehicle submitted to a transverse wind force F_w

It was found that the overturning of a vehicle occurs when, (259)

$$\frac{\Delta Q}{Q_o} < 0.9 \qquad (13.14)$$

Formulas 13.13 and 13.14 permit, in relation to the topography of track and wind data, the identification of areas with a high risk of overturning of the rail vehicle, due to transverse winds, as follows:

♦ areas of restriction of speed. Thus, French railways, in their high-speed Paris – Marseille track (operated at a maximum speed of 300 km/h) limit the train speed at 80÷170 km/h in relation to the wind speed (case of winds with a speed 100÷120 km/h), (259),

♦ areas with the highest risk, for which physical or technical fences along the track should be installed for protection against winds.

A series of wind measures (speed and direction) is necessary in order to evaluate any coming risk and take the appropriate measures. These data are introduced in simulation models and forecasts of wind speed (and direction) at any point are transmitted at least 5 minutes before the passing of a train from that specific point with the appropriate instruction for a limitation (or not) of speed.

14 Track Layout

14.1. Rail vehicle running on a curve

14.1.1. Effects during movement of a rail vehicle on a curve

When a rail vehicle runs at a speed V on a curve of radius R, it develops a centrifugal acceleration $\gamma=V^2/R$ and a centrifugal force $F=m\cdot V^2/R$, with the following adverse consequences:

♦ reduction in passenger comfort,
♦ important transverse forces favoring conditions for derailment,
♦ increased transverse loading of both track and rolling stock, resulting in considerable wear,
♦ increased vibrations.

In order to reduce the above unfavorable effects, the following measures are available:

- Using as large a radius of curvature R as possible. Such a measure is not easily implemented, however, due to topographical constraints, which often make large radii conditional on expensive civil engineering projects (bridges, tunnels, high embankments or cuts).
- Transverse superelevation (also called cant) of the outer rail in relation to the inner rail, to offset centrifugal forces. Cant greatly decreases transverse effects, without, however, completely counteracting them in most cases, since it cannot exceed certain values beyond which rolling stock and track wear become prohibitive.
- Reduction in train speed, which constitutes a last resort solution, since the trend is to increase train speed.

14.1.2. Transition curve – Cubic parabola or clothoid

On a straight line, curvature is zero, while on a curve of radius R curvature is 1/R. Therefore, between a straight and a curved track, the curvature changes abruptly from zero to 1/R. Passengers feel this sudden change of curvature as a jolt.

Therefore, a variable-radius transition curve, with zero curvature at the beginning and 1/R curvature at the end, should be interposed for smooth transition from rectilinear to curvilinear motion.

As a transition curve between a straight line and a circular arc, a cubic parabola or a clothoid (as in highway engineering) may be used. Most railway authorities use cubic parabola as transition curve. However, some railways (among them British railways) use the clothoid as a transition curve. Curvature ρ is defined as, (Fig. 14.1):

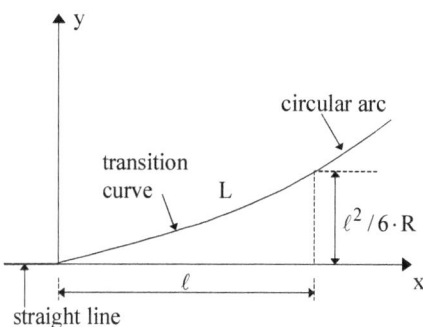

$$\rho = \frac{1}{R} = -\frac{d^2y}{dx^2} \qquad (14.1)$$

In the cubic parabola, curvature ρ is proportional to the projection of the parabolic curve on the x-axis:

$$\frac{1}{R_{cub.bar.}} = k \cdot x \qquad (14.2)$$

Fig. 14.1. Cubic parabola

where k is a coefficient.

In the case of cubic parabola, it may be assumed for small radii of curvature that the length L of the transition curve is equal to its projection ℓ on the x-axis. The approximation introduced by this assumption was found satisfactory in most cases.

In the clothoid, curvature ρ is proportional to the length L of the curve:

$$\frac{1}{R_{clothoid}} = k \cdot L \qquad (14.3)$$

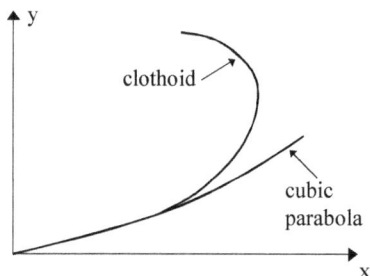

Using the previous assumption, $L = \ell$, it is found that in most cases the use of cubic parabola and of clothoid give similar results.

The critical difference between a clothoid and a cubic parabola is that whereas a clothoid goes round and round, a cubic parabola can never turn through more than a right angle, (Fig. 14.2).

Fig. 14.2. Comparison between clothoid and cubic parabola

Both cubic parabola and clothoid were suggested as transition curves to enable smooth transition between straight and curved segments of a railway line two centuries ago on the basis of mathematical (more precisely geometrical) considerations. In the course of time, other forms of transition curves, which take into account train dynamics, human health considerations, bogie suspension, steering railway vehicles, and tilting vehicles, have been suggested, without however receiving wide acceptance by railway science. Very recently, a new transition curve, called clothoid symmetrically projected, that combines the accuracy of clothoid curve and the simplicity of cubic parabola, was suggested, (277).

14.2. Theoretical and actual values of cant – Permissible values of transverse acceleration

14.2.1. Theoretical value of cant for the complete compensation of centrifugal forces

Let us consider a rail vehicle running at a speed V (km/h) on a curve with a radius R(m). We seek the value of the cant of the outer rail in relation to the inner rail, at which the centrifugal forces are fully compensated. We will designate this as theoretical cant h_{th}(mm). Thus, we have:

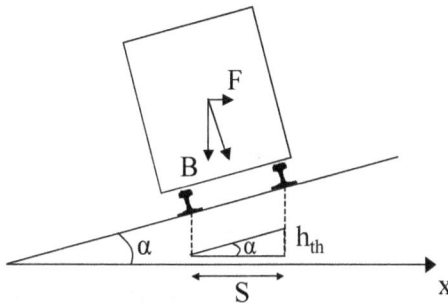

$$B = m \cdot g \qquad (14.4)$$

$$F = \frac{m \cdot V^2}{R} \qquad (14.5)$$

From Figure 14.3 we have:

$$\tan \alpha = \frac{F}{B} \qquad (14.6)$$

as well as:

$$\tan \alpha = \frac{h_{th}}{S} \qquad (14.7)$$

Fig. 14.3. Forces exerted on a rail vehicle when running on a curve and theoretical cant

where   S: the distance of the axes of the two rails, which, in the case
of a standard gauge track, is S = 1,500 mm $\qquad (14.8)$

From equations (14.4)÷(14.8) and after appropriate conversion of units, it is derived for *standard gauge* tracks that:

$$h_{th}(mm) = 11.8 \cdot \frac{V^2(km/h)}{R(m)} \qquad (14.9)$$

In the case of *metric gauge tracks* (with a gauge of 1,000 mm) it will be:

$$h_{th}(mm) = 8.3 \cdot \frac{V^2(km/h)}{R(m)} \tag{14.10}$$

In the case of *broad gauge tracks* (with a gauge of 1,524mm) it will be:

$$h_{th}(mm) = 12.5 \cdot \frac{V^2(km/h)}{R(m)} \tag{14.11}$$

14.2.2. Applied value of cant, cant deficiency, and cant excess

Equations $(14.9) \div (14.11)$ show that the theoretical value of cant for complete compensation of centrifugal forces is proportional to the square of vehicle speed. Assuming that the latter is constant on a curve, a single value h_{th} of theoretical cant can be calculated. This condition, however, is fulfilled only on metropolitan railways or on high-speed lines used only by passenger trains. By contrast, on conventional railway lines, fast (passenger) and slow (freight) trains coexist and the traffic of the line is mixed.

Thus, if the maximum speed of passenger trains is used in equations $(14.9) \div (14.11)$, then passenger comfort is ensured. With freight trains, however, problems arise due to wear of both the wheels and track equipment (specifically of the heads of the inner rails). Furthermore, if a freight train stops on a curve, it will have trouble starting (it will even be unable to do so if the radius of curvature is too small).

If in equations $(14.9) \div (14.11)$ the usual running speed of freight trains is applied, then no problems are encountered in relation to freight trains. Passenger comfort, however, is greatly impaired, and there are greater stresses on the rail placed higher.

A compromise between the two previous conditions should therefore be found by adopting a cant value, which ensures passenger comfort, increases only moderately rolling stock and track stresses, and allows trains to stop on a curve. This intermediate value of cant h is often termed *applied* or *normal* cant (or *standard* cant by some railways). We will have:

$$h_{th}(V_{min}) < h < h_{th}(V_{max}) \tag{14.12}$$

Selecting the applied value of cant results in cant deficiency for fast trains and cant excess for slow trains.

The difference between the theoretical value of cant for the maximum speed and the applied value of cant is termed *cant deficiency* h_d:

$$h_d = h_{th}(V_{max}) - h \tag{14.13}$$

The difference between the applied value of cant and the theoretical value of cant for the minimum speed is termed *cant excess* h_e:

$$h_e = h - h_{th}(V_{min}) \qquad (14.14)$$

The applied value of cant, as explained in section 14.4, will be calculated for standard gauge tracks by the equation:

$$h(mm) = \frac{h_{max}}{h_{max} + h_{d\,max}} \cdot 11.8 \cdot \frac{V^2(km/h)}{R(m)} \qquad (14.15)$$

14.2.3. Cant deficiency and tilting trains

In order to deal with the problem of non-compensated centrifugal acceleration, certain types of rolling stock tilt automatically on small-radius curves.

The so-called tilting trains try (and often fully succeed) to reduce cant deficiency in curves by tilting the vehicle body in relation to the wheel-base (Fig. 14.4). When using tilting trains, speed can be increased for small-radius curves by up to 30%, compared to conventional rolling stock. This technique has been applied in the UK, Spain, Italy, Sweden, Japan, and elsewhere (tilting technology is further analyzed in section 19.9).

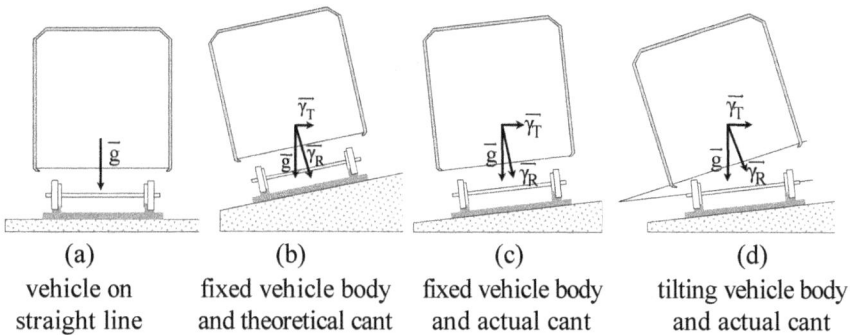

(a)	(b)	(c)	(d)
vehicle on straight line	fixed vehicle body and theoretical cant	fixed vehicle body and actual cant	tilting vehicle body and actual cant

Fig. 14.4. The additional superelevation generated by tilting trains, (338)

14.2.4. Permissible values of transverse acceleration

In section 7.12 we have seen that passenger comfort depends both on the value of the transverse acceleration and on the duration and frequency which are felt by the human body. The direction in which the transverse acceleration is exerted is also critical. It is found that an acceleration of $0.05 \cdot g$ at a frequency of 1.5 Hz can be tolerated by the human body for 5h 30min in the vertical direction and 3h 30min in the horizontal direction, (147).

Considerations of human physiology, therefore, determine the maximum value of transverse acceleration as well as its rate of change. There is a general agreement that maximum transverse acceleration should never exceed g/10, i.e. a value of 1 m/sec², (276).

In track layout, however, a considerable reduction of passenger comfort cannot be tolerated. Consequently, the non-compensated (residual) centrifugal acceleration b should not exceed a percentage of the maximum transverse acceleration γ acceptable by the human body. Many railway authorities set this limit as follows, (279):

$$b \cong \frac{2}{3} \cdot \gamma \qquad (14.16)$$

In metropolitan railways, where the duration of the whole trip is smaller, a higher value of non-compensated centrifugal acceleration up to 0.8 m/sec² can be considered acceptable for some tight curves.

The selected value of b affects the maximum value of cant deficiency.

14.2.5. Variation in time of cant deficiency

The variation of cant deficiency in time is:

$$\dot{h}_d(mm/sec) = \frac{\Delta h_d}{\Delta t} \qquad (14.17)$$

The parameter $\dot{h}_d$ may be expressed as a function of the variation of cant deficiency per unit length:

$$\dot{h}_d(mm/sec) = \frac{\Delta h_d}{\Delta t} = \frac{\Delta h_d}{\Delta \ell} \cdot \frac{\Delta \ell}{\Delta t} = \frac{\Delta h_d}{\Delta \ell} \cdot \frac{V_{max}(km/h)}{3.6} \qquad (14.18)$$

14.3. Limit values of cant, cant deficiency, cant excess, and non-compensated transverse acceleration

14.3.1. Limit values according to UIC

As will be analyzed in the next sections, once values of cant h and non-compensated acceleration b are defined, then for a given value of speed the radius of curvature R can be calculated, (see equation (14.36a), section 14.9).

Limit values of cant and non-compensated acceleration are prescribed by UIC, (276). Lines are classified in 4 classes:

Class I: V_{max}: 80 ÷ 120 km/h,
Class II: V_{max}: 120 ÷ 200 km/h,
Class III: V_{max}: 250 km/h, mixed traffic. Standards of German and Swiss railways are given,

Class IV: V_{max}: 300 km/h, only passenger traffic (case of the French TGV).

For each class, applied, maximum, and exceptional values of cant, cant deficiency, cant excess, and non-compensated transverse acceleration are given in Table 14.1, (276). Exceptional values can be applied only after the running characteristics of the rolling stock have been verified.

Table 14.1.
Limit values of geometrical characteristics of layout
according to UIC, (276)

Traffic class	I			II			III				IV	
Maximum speed V_{max} (km/h)	80÷120			120÷200			250 Germany		250 Switzerland		300 France	
Limit value	appl.	max.	excp.	appl.	max.	excp.	appl.	max.	appl.	max.	appl.	max.
Cant (mm)	150	160	-	120	150	160	65	85	125	-	180	-
Cant deficiency h_d (mm)	80	100	130	100	120	150	40	60	120	-	50	100
Cant excess h_c (mm)	50	70	90	70	90	110	50	70	100	-	-	110
Cant deficiency in time h_d (mm/sec)	25	70	90	25	70	-	13	-	36	-	30	75
Non compensated transverse acceleration b (m/sec²)	0.53	0.67	0.86	0.67	0.80	1.00	0.27	0.40	0.81	-	0.33	0.67

14.3.2. Limit values according to European specifications

According to the European technical specifications for interoperability, (134):

a) The limit value of cant for new high-speed tracks dedicated only to passenger traffic is set to 180 mm and for mixed traffic tracks to 160 mm,

b) cant deficiency for high-speed tracks should be calculated in relation to the value of the non-compensated transverse acceleration b. For tracks with speeds ≤ 200 km/h, the limit value of cant deficiency is set to 130 mm for $b=0.85$ m/sec² and to 150 mm for $b=1.0$ m/sec²,

c) no reference is made concerning track excess,

d) the rate of change of cant as a function of time is set to 70 mm/sec, which under certain condition can be increased to 85 mm/sec,

e) no explicit value is given for the radius of curvature, which is calculated in relation to the value of speed, cant, and non-compensated transverse acceleration.

14.3.3. Geometrical characteristics of layout in some high-speed tracks

Table 14.2 illustrates the geometrical characteristics for the layout of high-speed tracks according to regulations of some countries and specifications of the EU and the UIC.

14.4. Calculation of the transition curve

In section 14.2.2 we have explained that the value of applied cant h must lie between two limits to ensure that no problems are caused to either slow (freight) or fast (passenger) trains. After the limit values given in Tables 14.1 and 14.2, it should be

$$h_{th}(V_{max}) - h_{d\,max} < h(mm) < h_{th}(V_{min}) + h_{e\,max} \qquad (14.19)$$

and in each case

$$h < h_{max} \qquad (14.20)$$

The selection of a value between the two limits of equation (14.19) depends on the relative density of passenger and freight traffic on the particular line. More passenger traffic raises this value towards the upper limit of equation (14.19), while more freight traffic makes it approach the lower limit of equation (14.19).

In all cases, however, the ratio of the maximum cant h_{max} to the maximum theoretical cant $h_{max} + h_{d\,max}$ should remain constant. The theoretical cant will be multiplied by this constant ratio to calculate the applied cant:

$$h(mm) = \frac{h_{max}}{h_{max} + h_{d\,max}} \cdot 11.8 \cdot \frac{V^2(km/h)}{R(m)} \qquad (14.21)$$

The minimum value of cant should not result in a non-compensated centrifugal acceleration greater than b_{max}:

$$h_{min}(mm) = 11.8 \cdot \frac{V^2(km/h)}{R(m)} - 152 \cdot b_{max}(m/sec^2) \qquad (14.22)$$

The cant values found from the foregoing equations are rounded off to multiples of 5 mm.

To ensure smooth train running, the value of cant should vary gradually from zero (at the end of the straight track) to h (at the beginning of the circular arc). *This requires that the superelevation ramp and the transition curve coincide.*

Table 14.2.

Geometrical characteristics of layout of some high-speed and conventional tracks (compiled from data of railway authorities)

Country	Japan	Japan	France	France	Germany	Germany	Italy	Spain	European technical specifications for interoperability		UIC
Line	Tokyo-Osaka	Osaka-Hakata	Paris-Lyon	TGV Atlantique	Hannover-Würzburg	Cologne-Rhein	Milan-Florence	Madrid-Valladolid	only passenger traffic	mixed traffic	mixed traffic
Maximum speed (km/h)	270	300	300	350	250	300	350	350	–	–	120÷200
Minimum radius of curvature (m)	2,500	4,000	4,000	6,250	7,000	3,350	7,000	6,500	TSI do not specify any value for the radius of curvature, which is calculated as a function of the other parameters of the layout		–
Cant (mm)	200	180	180	180	150	160	130	150	180	• 160 for ballasted track • 170 for slab track	120÷160
Cant deficiency (mm)	100	100	85	160	80	150	85	65	in relation to b and R	• 100 for V>300 km/h • 153 for V≤300 km/h	100÷150
Cant excess (mm)	–	–	–	–	50	–	–	–	–	non specified	70÷110
Maximum longitudinal gradient	20 ‰	15 ‰	35 ‰	25 ‰	12.5 ‰	40 ‰	8 ‰ Milan-Bologne, 15 ‰ Bologne-Florence	20 ‰	35 ‰	12.5 ‰	–

If L is the length of the transition curve and ℓ its projection on the extension of the straight section, (Fig. 14.5), then the minimum value of the transition curve can be calculated by the formula, (276), (279):

$$\ell_{min}(m) = \frac{h(mm) \cdot V(km/h)}{144} \quad , \quad \text{for } V \geq 57.5 \text{ km/h} \qquad (14.23a)$$

$$\ell_{min}(m) = \frac{h(mm)}{2.5} \quad , \quad \text{for } V < 57.5 \text{ km/h} \qquad (14.23b)$$

The ordinates of the transition curve, which is usually in railways a cubic parabola, are calculated by the equation, (279):

$$y = \frac{x^3}{6 \cdot R \cdot \ell} \cdot \left[1 + \left(\frac{\ell}{2 \cdot R}\right)^2\right]^{3/2} \qquad (14.24)$$

In the event that the term $\left(\dfrac{\ell}{2 \cdot R}\right)^2$ is much less than ℓ, it can be omitted in equation (14.24), in which case we have a small-length cubic parabola. Its equation, applicable as long as $\ell < \dfrac{R(m)}{3.5}$, is:

$$y = \frac{x^3}{6 \cdot R \cdot \ell} \qquad (14.25)$$

The ordinates of the cubic parabola are commonly calculated every 10m, or, whenever a greater point density is required, every 5m.

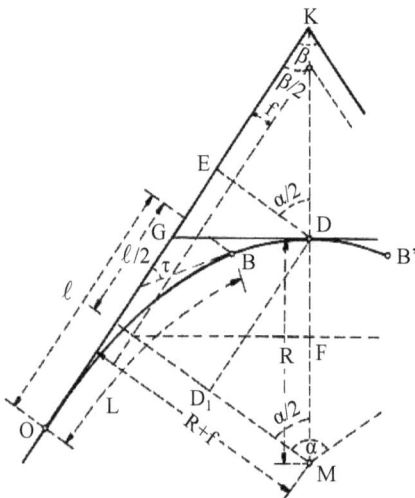

Fig. 14.5. Transition curve (cubic parabola (OB)) and circular arc (BB')

The length L of the cubic parabola and its projection ℓ on the straight line are related by the equation:

$$L = \ell + \frac{\ell}{10} \cdot \left(\frac{\ell}{2 \cdot R}\right)^2 \tag{14.26}$$

Certain railways use parabolic transitions of a higher degree (third or fourth degree parabolas).

Transition curves are not used if:

- the calculated values of cant are practically zero,
- between two adjacent curves (of the same direction), the variation of acceleration has values between 0.2 m/sec^2 and 0.3 m/sec^2.

14.5. Calculation of the circular arc

Let f be the shift produced by the cubic parabola between the straight line and the circular arc, (Fig. 14.5). The characteristics of the circular arc are calculated by the following equations, (142):

$$OK = (R + f) \cdot \tan\frac{\alpha}{2} + \frac{\ell}{2} \tag{14.27}$$

$$KD = (R + f) \cdot (\sec\frac{\alpha}{2} - 1) + f \tag{14.28}$$

$$OBD = \frac{1}{2} \cdot (R \cdot \frac{\pi \cdot \alpha}{200} + \ell] \tag{14.29}$$

where $\sec\dfrac{\alpha}{2} = \dfrac{1}{\cos\dfrac{\alpha}{2}}$ is the secant of the angle $\dfrac{\alpha}{2}$ (angle α expressed in grades).

The shift f is calculated by the equation:

$$f = \frac{\ell^2}{24 \cdot R} \tag{14.30}$$

that is, in most cases the influence of f on the length OK is negligible compared to R.

14.6. Case of consecutive same sense and antisense circular arcs

Between two consecutive circular arcs of the same sense with radii R_1 and R_2, a transition curve is placed adjacent to each circular arc and an intermediate rectilinear section is interposed between the transition curves. For medium-speed tracks (V_{max}=200 km/h), this rectilinear section has a usual value of 30m.

Using the following symbols:

$$\delta = \frac{\ell_2^2}{24 \cdot R_2} - \frac{\ell_1^2}{24 \cdot R_1},$$

$$\rho = \frac{R_1 \cdot R_2}{R_1 - R_2},$$

$$\ell = \sqrt{24 \cdot \rho \cdot \delta},$$

the transition curve adjacent to the circular arc of radius R_1 will be:

$$y = \frac{x^2}{2 \cdot R_1} + \frac{\delta}{2} - \frac{1}{6 \cdot \ell \cdot \rho} \cdot \left[\left(\frac{\ell}{2} \right)^3 - \left(\frac{\ell}{2} - x \right)^3 \right] \qquad (14.31)$$

The transition curve adjacent to the circular arc of radius R_2 will be:

$$y = \frac{x^2}{2 \cdot R_2} + \frac{\delta}{2} - \frac{1}{6 \cdot \ell \cdot \rho} \cdot \left[\left(\frac{\ell}{2} \right)^3 - \left(\frac{\ell}{2} - x \right)^3 \right] \qquad (14.32)$$

If the interposition of an intermediate rectilinear section is not feasible, then, instead of two transition curves, a single transition curve can be used with the following equation:

$$y = \frac{x^3}{6 \cdot \ell_2 \cdot R_2 \cdot \cos^3 \tau_2}, \qquad \text{when } L_2 > \frac{R_2}{3.5} \qquad (14.33)$$

or

$$y = \frac{x^3}{6 \cdot \ell_2 \cdot R_2}, \qquad \text{when } L_2 < \frac{R_2}{3.5} \qquad (14.34)$$

where L_1, L_2 are the required curve lengths for transition between the rectilinear section and the two circular arcs (with radii R_1 and R_2) and τ is the angle between the straight line and the tangent at the beginning of the circular arc, (Fig. 14.5).

Between two consecutive *antisense* circular arcs, one parabolic transition curve adjacent to each circular arc and an intermediate rectilinear section at least 30m long (preferably V(km/h)/2) are interposed. Should the latter not prove feasible, the rectilinear part is omitted and the two transition curves have a common beginning point, a common tangent, and the same curvature variation, (Fig. 14.8).

14.7. Superelevation ramp

As explained in section 14.4, the superelevation ramp and the cubic parabola should coincide. In this case, the following cant variation diagram results, (Fig. 14.6):

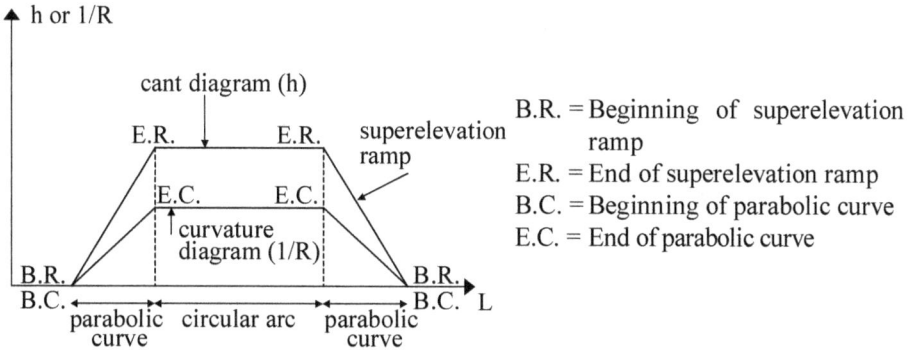

Fig. 14.6. Diagram of variation of cant and curvature between rectilinear section and circular arc

A similar linear variation of cant should be applied between same sense, (Fig. 14.7), or antisense circular curves, (Fig. 14.8).

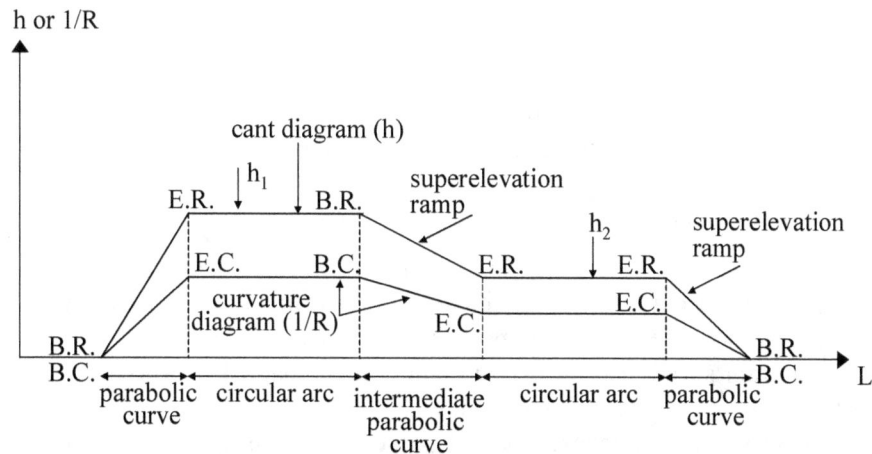

Fig. 14.7. Diagram of variation of cant and curvature between consecutive same sense circular arcs

391

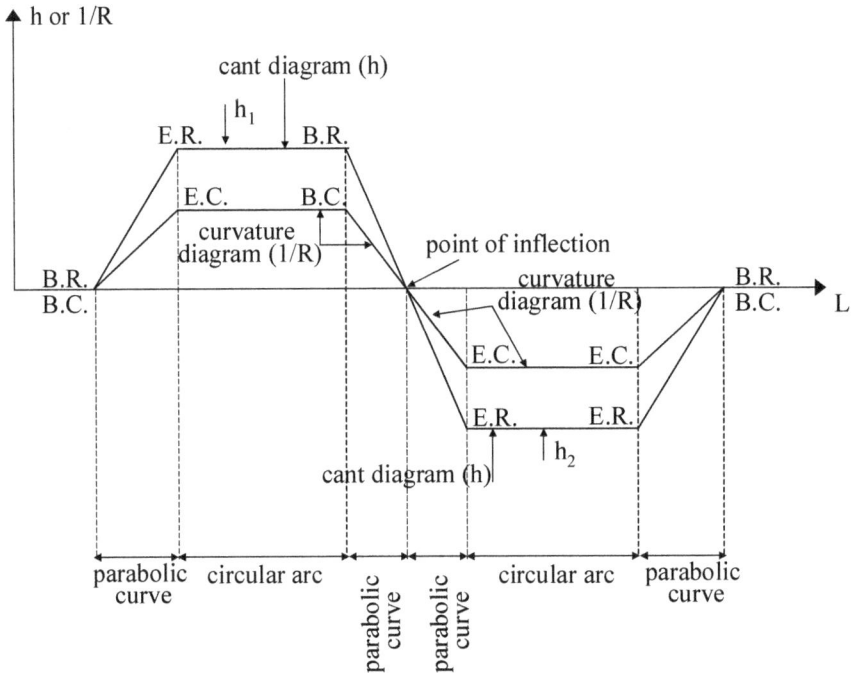

Fig. 14.8. **Diagram of variation of cant and curvature between consecutive antisense circular arcs**

The maximum gradient ω of the superelevation ramp should not exceed the value $144/V_{max}$, i.e.:

$$\omega_{max}\,(mm/m) < \frac{144}{V_{max}\,(km/h)} \tag{14.35}$$

Superelevation ramps should not be located in areas where turnouts or expansion devices are placed. If this is not possible, speed restrictions should be applied.

14.8. Combining maximum and minimum speeds

Equation (14.19), (section 14.4), implies that when maximum and minimum train speeds on a curve differ significantly, it is difficult to find an applied cant value which does not cause problems to freight or passenger trains. A passenger train speed increase is accordingly accompanied by a freight train speed increase, as shown in Table 14.3.

For high speeds, the coexistence of passenger and freight trains is more complicated. For this reason, most railways have specialized their high-speed tracks only for passenger traffic.

Table 14.3.
Maximum and minimum speeds on a layout

$V_{max} < 100$ km/h	$\rightarrow$	$V_{min} \geq 60$ km/h
100 km/h $< V_{max} < 140$ km/h	$\rightarrow$	$V_{min} \geq 70$ km/h
140 km/h $< V_{max} < 200$ km/h	$\rightarrow$	$V_{min} \geq 80$ km/h

14.9. Relationship of train speed with radius of curvature

We shall now calculate the maximum permissible speed on a curve of radius R, or, for a given speed V, the minimum required radius of curvature.

Obviously, for a given radius R, the speed V reaches a maximum when the margins for cant h, cant deficiency h_d, and cant excess h_e are exhausted.

From equations (14.9), (14.15), (14.19), it follows that:

$$\frac{11.8 \cdot V_{max}^2}{R} - h_{d\,max} < 11.8 \cdot \frac{h_{max}}{h_{max} + h_{d\,max}} \cdot \frac{V_{max}^2}{R} < \frac{11.8 \cdot V_{min}^2}{R} + h_{e\,max} \quad (14.36a)$$

Solving equation (14.36a) for V_{max} we obtain the maximum permitted speed for a given radius R, whereas solving for R we obtain the minimum required radius for a given speed V_{max}.

With respect to R_{min}, however, it should be ensured that the maximum cant excess for the minimum speed V_{min} of slow trains can be applied. Equation (14.36a) gives:

$$\frac{11.8 \cdot V_{max}^2}{R} - h_{d\,max} < \frac{11.8 \cdot V_{min}^2}{R} + h_{e\,max} \quad (14.36b)$$

while setting up the maximum values for $h_{d\,max}$, $h_{e\,max}$ and solving for R, we obtain the minimum radius required by slow trains (with V_{min}).

With respect to the minimum speed, therefore, equations (14.36a) and (14.36b) should be simultaneously valid, and the higher value found for R_{min} will be used.

A general empirical formula for a first preliminary calculation of R in relation to V is:

$$V(km / h) = 4.70 \cdot \sqrt{R(m)} \quad (14.37)$$

Whenever possible, railways try to apply the maximum feasible value of R. There are great differences among railways, concerning policy on the lower values of radius, principally due to the mountainous or plane character of the

Table 14.4.
Percentage of curves with a radius smaller than 500 m for various European railways (metro systems are not taken into account), (278)

Country	% of curves with a radius $R \leq 500m$
United Kingdom	3.0%
France	9.0%
Germany	13.0%
Switzerland	15.5%
Austria	21.6%

ground. Table 14.4 gives the percentage of tracks curved at 500 m or less in some European railways.

When the radius of curvature of a track is small, track gauge is increased, resulting in a value higher than in straight track sections. The increase is applied to the inner rail. For radius R<400 m, the track gauge can be increased up to 1.455 m (in the case of timber and steel sleepers) and up to 1.440 m (in the case of concrete sleepers), (see also sections 7.4 and 16.4.4).

14.10. Transition curves in the case of variation of the distance between the axes of two tracks

The distance between the axes of two tracks can change (e.g. at the entrance and exit of stations) from b to c, (Fig. 14.9). The transition between b and c is realized with the use of two antisense circular arcs without any intermediate rectilinear part. The radius of curvature of each circular arc is calculated by the formula:

$$R(m) \cong V^2(km/h) \tag{14.38}$$

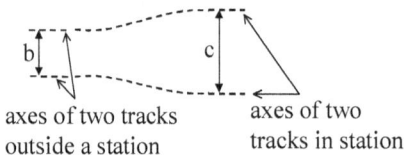

axes of two tracks outside a station

axes of two tracks in station

Fig. 14.9. A case of variation of distance of axes of two tracks

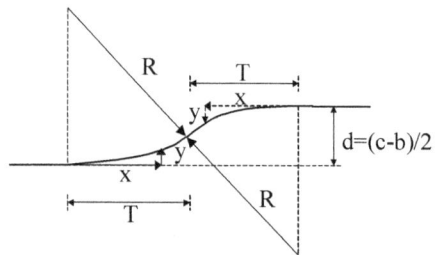

Fig. 14.10. Consecutive antisense curves for transition in the case of variation of distance of axes of two tracks

The tangent T of each circular arc is calculated by the equation, (279):

$$T = \sqrt{d \cdot R} \tag{14.39}$$

and the ordinates of the circular arc are calculated by the equation:

$$y = \frac{x^2}{2 \cdot R} \tag{14.40}$$

14.11. Longitudinal gradients and vertical transition curves

14.11.1. Longitudinal gradients

Wherever possible, the longitudinal profile of a railway line follows the ground profile. Longitudinal gradients of railways are much smaller compared to those of highways. Maximum values of longitudinal gradient mainly depend on the characteristics and power of the rolling stock and usually range between 12‰ ÷ 15‰ for principal lines with mixed traffic and speeds up to 200 km/h. Thus, the maximum longitudinal gradient on the main lines of German railways is 12.5‰, but in the French TGV Paris–Lyon and the German Cologne–Rhein (both with only passenger traffic) gradients are 35‰ and 40‰ respectively, (see also section 2.3.1, Table 2.5 and section 14.3, Table 14.2). For reasons of adhesion between wheel and rail, maximum gradients can hardly exceed the limit value of 40‰. For instance, some lightweight rail systems, which operate vehicles with 50 % of the axles motorized, have gradients up to 40‰, (141). Above 40‰, the use of a rack* railway must be considered.

Figure 14.11 illustrates the effect of a very high value (40‰) of longitudinal gradient and a lower and more usual one (12.5‰) on the number and length of tunnels and bridges of the line.

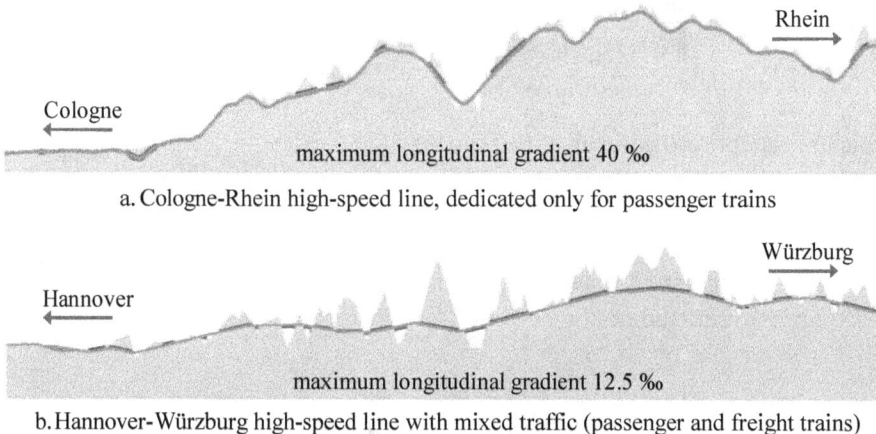

a. Cologne-Rhein high-speed line, dedicated only for passenger trains

b. Hannover-Würzburg high-speed line with mixed traffic (passenger and freight trains)

Fig. 14.11. Longitudinal profiles of two high-speed railway lines with gradients 40‰ and 12.5‰ and effects on the number and length of tunnels and bridges, (38)

---

* A rack (or funicular) railway is used when longitudinal gradients are higher than 40‰. It consists of two running rails (upon which the wheels of the funicular vehicle move) and a third toothed rack rail between the two running rails. A gear wheel or pinion along the toothed rail provides additional traction force at steep grades and protects the vehicle from sliding.

14.11.2. Vertical transition curves

The transition between longitudinal sections with different gradient values is made by interposing a circular arc of radius R_v, whose principal purpose is to limit the vertical acceleration, experienced by passengers, to a comfortable level.

The transition curve is not necessary as long as the difference of the respective gradients (if of the same sense) or their sum (if of opposite sense) is less than 2.5‰, i.e. provided that:

$$\Delta i < 2.5‰$$

The vertical curve radius R_v is calculated for high-speed tracks (with V>200 km/h) by the following formula, (276):

$$R_v = \frac{V^2(km/h)}{12.96 \cdot \alpha_v(m/sec^2)} \tag{14.41}$$

with α_v: vertical acceleration which has a recommended limit value of 0.22 m/sec² and a maximum limit value of 0.44 m/sec².

However, for conventional tracks (V<200 km/h) the vertical transition radius may be calculated by the approximate formula:

$$R_v(m) \cong \frac{V^2(km/h)}{2} \tag{14.42}$$

which in exceptional cases may be reduced to

$$R_{v\,min}(m) \cong \frac{V^2(km/h)}{4} \tag{14.43}$$

Table 14.5 gives the minimum vertical transition radius as a function of speed for conventional tracks.

Table 14.5.
Minimum values of vertical transition radius
as a function of speed for conventional tracks

	Normal value	Exceptional value
V_{max} < 100 km/h	5,000 m	2,500 m
100 km/h < V_{max} < 140 km/h	10,000 m	5,000 m
140 km/h < V_{max} < 200 km/h	20,000 m	10,000 m

The tangent E of the vertical transition circular arc is calculated by the equation:

$$E = \Delta i \cdot \frac{R_v}{2} \qquad (14.44)$$

where Δi is the gradient variation, (Fig.14.12).

The ordinates of the vertical transition arc are calculated by the equation

$$y = \frac{x^2}{2 \cdot R_v} \qquad (14.45)$$

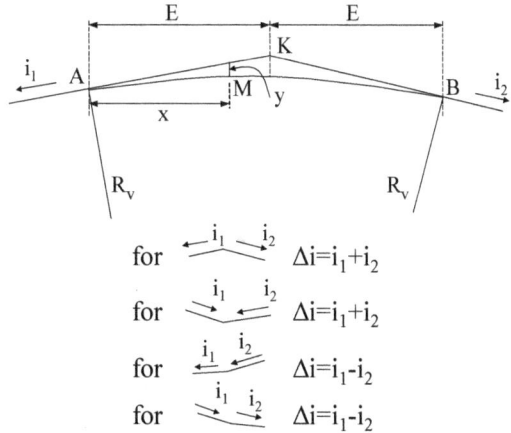

for $\overset{i_1}{\longleftarrow} \overset{i_2}{\longrightarrow}$ $\Delta i = i_1 + i_2$

for $\overset{i_1}{\longrightarrow} \overset{i_2}{\longleftarrow}$ $\Delta i = i_1 + i_2$

for $\overset{i_1}{\longrightarrow} \overset{i_2}{\longrightarrow}$ $\Delta i = i_1 - i_2$

for $\overset{i_1}{\longrightarrow} \overset{i_2}{\longrightarrow}$ $\Delta i = i_1 - i_2$

Fig. 14.12. Vertical transition

No changes of gradient should be made where there are transition curves at the horizontal level and hence superelevation ramps exist. Wherever simultaneous vertical and horizontal transition cannot be avoided, the maximum radius of curvature should be used.

Vertical transitions should terminate at least $5 \div 10$ m from the beginning or the end of switches and crossings. Vertical transitions should moreover be avoided on steel bridges without ballast.

According to the European technical specifications for interoperability, (134):

a) the maximum gradient for high-speed tracks dedicated for only passenger traffic is 35‰. However, this maximum value can only be applied on a continuous length lower than 6km, while the average gradient over a length of 10 km should be lower than 25‰, (134).

b) the maximum gradient for high-speed tracks with mixed traffic is 12.5‰, but for sections up to 3 km the maximum gradient of 20‰ is permitted.

14.12. Some considerations for metric gauge tracks

Previous theoretical analyses focused on standard gauge tracks, but are valid for broad and metric gauge tracks as well. For metric gauge tracks the following considerations should also be taken into account.

The minimum radius of a horizontal curve on main metric gauge tracks should not be less than 100 m. Radius of vertical curves for speeds up to 100 km/h should be $2{,}000 \div 4{,}000$ m. Maximum cant should be $100 \div 110$ mm, maximum cant deficiency $70 \div 90$ mm, and maximum cant excess $45 \div 80$ mm, (140), (274).

14.13. Layout design with the use of tables and computer methods

To facilitate layout design, most of the aforementioned equations can be used in the form of tables. Such tables spare the designer tedious calculations and give values at a glance. Almost all railway authorities established such tables many years ago (before the extensive use of computers).

However, developments in computer hardware and software have revolutionized railway layout design. Several software[*] permit track layout calculation and design, requiring only the topography and the limit values of the layout parameters. Figure 14.13 illustrates the track layout design of a new line using CAD (Computer Aided Design) software. Furthermore, with the help of computer applications, more alternative routes can be easily surveyed, and the solution chosen can be studied in greater detail. Thus, it is possible to easily study many alternative solutions, compare them, and choose the best solution (that is the one which maximizes straight lines and low gradients, while at the same time minimizes earthworks, civil engineering constructions, and costs).

Optimization of the layout and vertical design of a track can be obtained with the use of artificial intelligence and more particularly of Artificial Neural Networks (ANN). A number of existing layout and vertical design cases are compiled and transformed to cropped images that are optimized by an ANN algorithm, (271).

Layout design may be further facilitated with the use of information provided by satellite systems, (270).

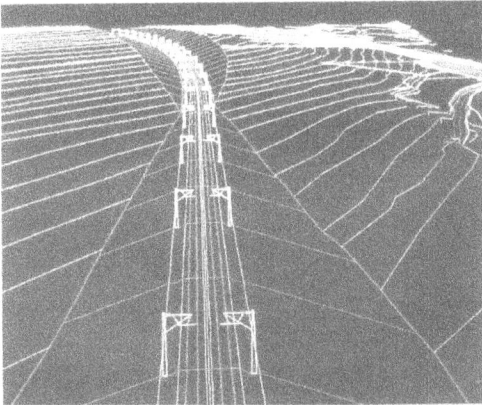

Fig. 14.13. Track layout design using a computer aided design method

---

[*] Among the various software for the design of track layout (horizontal level, vertical level, cross-sections) we mention MXRail, ODOS, INRail, etc.

14.14. Numerical application for the layout and the longitudinal design of a track

14.14.1. Layout design

Advances in computer aided design give nowadays the possibility to conduct all layout design calculations with the help of computers. However, a railway specialist must be able to check the validity of results provided by computer software and to conduct on his/her own, whenever necessary, the appropriate calculations.

We will consider a layout of a standard gauge track between two rectilinear parts of a track intersecting at an angle of 140°, (Fig. 14.14). The track is considered to be plane horizontally and the available length KZ for the insertion of the layout is 600 m on each side. The track has a mixed traffic with a maximum speed of 140 km/h. Referring to Figure 14.5 it will be: $\beta/2 = 140°/2 = 70°$ and $\alpha/2 = 20°$, thus $\alpha = 40°$.

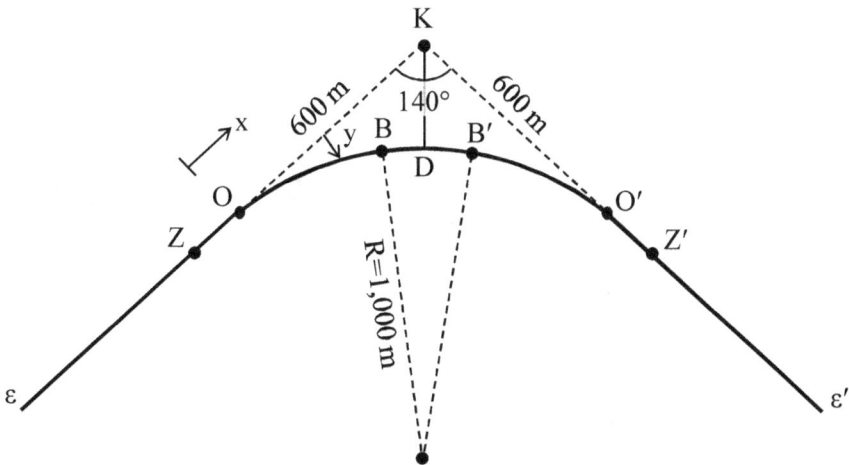

Fig.14.14. Layout design between two rectilinear parts of a track

The layout between the two rectilinear parts εO and $\varepsilon'O'$ of the track will have a circular arc BB' and two transition curves OB and O'B'. The maximum length for KO is KZ=600m. Curvature at the circular arc BB' is 1/R, whereas it is zero at points O and O'. Therefore, we will choose the appropriate transition curve which ensures a smooth change of curvature from zero at points O and O' to 1/R at points B and B'. We select as form of transition curve the cubic parabola and we will calculate the radius of curvature R, the cant h, the cant deficiency h_d, the cant excess h_e, and the ordinates between the two rectilinear parts εO and $\varepsilon' O'$.

The value of radius R is provided by eq. (14.36a) and is a relation of V_{max} and h, h_d, h_e, which in turn are a relation of the non-compensated (residual) value of the centrifugal acceleration b that can be accepted (usually $b<0.65$ m/sec^2). Solving equation (14.36a) for R requires the limit values for h, h_d, h_e, which depend on the regulation of the specific railway authority and sometimes on the conditions encountered on the ground. To overcome differences from one regulation to another, a general empirical rule for the relationship between V_{max} and R_{min} is given by the formula (14.37):

$$V_{max}(km/h) = 4.70 \cdot \sqrt{R_{min}(m)} \tag{14.37}$$

Solving eq. (14.37) for $V_{max} = 140$ km/h we obtain $R_{min} = 890$ m, a radius providing a limit value of the non-compensated centrifugal acceleration b. We choose a better level of comfort for the passengers of the specific layout and we give for R the value R = 1,000 m.

The value of cant h will be obtained with the help of eq. (14.15):

$$h(mm) = \frac{h_{max}}{h_{max} + h_{d\,max}} \cdot 11.8 \cdot \frac{V^2(km/h)}{R(m)}$$

We consider a track regulation with the following limit values: $h_{max}=160$ mm, $h_{d\,max}=105$ mm, $h_{e\,max}=100$ mm, $b_{max}=0.65$ m/sec^2. Substituting in eq. (14.15) we obtain:

$$h(mm) = \frac{160}{160 + 105} \cdot 11.8 \cdot \frac{140^2}{1,000} \Rightarrow h = 139.5 \text{ mm} < h_{max}$$

We give for h the value h=140 mm.

From eq. (14.22) we obtain for h=140 mm the value of the non-compensated centrifugal acceleration:

$$h(mm) = 11.8 \cdot \frac{V^2(km/h)}{R(m)} - 152 \cdot b_{max}(m/sec^2) \Rightarrow b = 0.6 \text{ m/sec}^2 < b_{max}$$

From Table 14.3 it follows that for $V_{max}=140$ km/h the minimum speed of slow (freight) trains will be $V_{min}= 70$ km/h.

We calculate cant deficiency h_d and cant excess h_e with the help of eq. (14.14) and eq. (14.15):

$$h_d(mm) = 11.8 \cdot \frac{V_{max}^2(km/h)}{R(m)} - 140 \Rightarrow h_d = 91.3 \text{ mm} < h_{d\,max}$$

$$h_e(mm) = 140 - 11.8 \cdot \frac{V_{min}^2 (km/h)}{R(m)} \Rightarrow h_e = 82.2 \text{ mm} < h_{e\,max}$$

The length ℓ of the transition curve OB (cubic parabola) will be obtained by eq. (14.23a):

$$\ell_{min}(m) = \frac{h(mm) \cdot V(km/h)}{144} \Rightarrow \ell_{min}(m) = 136.1 \text{ m}$$

We give for ℓ the value $\ell = 140$ m.

We calculate the distance KD, (Fig. 14.14), of the center point of the layout (D) in relation to the intersection point (K) of the two rectilinear parts from eq. (14.28):

$$KD = (R+f) \cdot (\sec\frac{\alpha}{2} - 1) + f \Rightarrow KD = 65.03 \text{ m}$$

with f from eq. (14.30), f= $\ell^2/24R$.

However, for higher values of R, the distance KD can also take even higher values; this often comes to field engineers as a surprise, since point D becomes very distant from point K.

We calculate whether the specific layout can be applied within the limit value for KZ (=600m). The value of OK is calculated by eq. (14.27):

$$OK = (R+f) \cdot \tan\frac{\alpha}{2} + \frac{\ell}{2} \Rightarrow OK = 434.2 \text{ m} < 600 \text{ m}$$

The superelevation ramp, along which curvature changes from 0 to 1/R, will coincide with the transition curve OB. Thus, the gradient ω of the superelevation ramp will be:

$$\omega = \frac{h}{\ell} = \frac{140}{140} = 1 \text{ mm/m}$$

which should be compared with the limit value for ω provided by eq. (14.35):

$$\omega_{max}(mm/m) = \frac{144}{V_{max}(km/h)} = \frac{144}{140} = 1.03 \text{ mm/m}$$

The ordinates y, perpendicular to abscissas x on the segment OK (for point O, x=0) will be calculated by the formula (14.24):

$$y = \frac{x^3}{6 \cdot R \cdot \ell} \cdot \left[1 + \left(\frac{\ell}{2 \cdot R}\right)^2\right]^{3/2}$$

As we have a radius R>400 m, no widening is required to be given to the track gauge (=1.435 m).

14.14.2. Longitudinal design

We will now consider a rectilinear track (a track without any curved parts at the horizontal level) for which the longitudinal gradient changes from $i_1 = 15\ ‰$ to $i_2 = 7‰$, (Fig.14.15). The track is run by trains with a maximum speed of 150 km/h. The available length for the change of gradient is 40 m at each side, $(OC = OD = 40\ m)$.

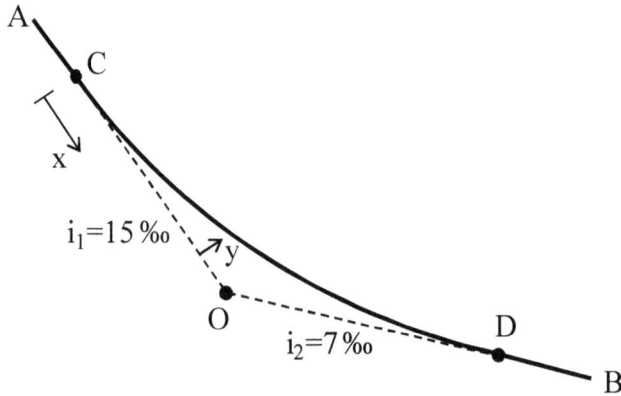

Fig. 14.15. Longitudinal design between sections with different gradients

Between the two consecutive gradients i_1 and i_2, the difference $\Delta i = i_1 - i_2 = 15‰ - 7‰ = 8‰ > 2.5\%$. Therefore, the smooth change of gradient from i_1 to i_2 will be realized by a vertical transition curve R_v given by the formula (14.42). As we have a conventional track with $V_{max} = 150$ km/h < 200 km/h, it will be:

$$R_v(m) \cong \frac{V^2(km/h)}{2} \Rightarrow R_v \cong 11,250\ m$$

We take for R_v the value 12,000 m.

The tangent E (segment OC in Fig. 14.15) will be calculated by formula (14.44):

$$E = OC = \Delta i \cdot \frac{R_v}{2} \Rightarrow E = 8‰ \cdot \frac{12,000}{2} \Rightarrow E = 48\ m$$

We realize that the value for E (=48 m) is greater than the available length $AO = 40$ m. We can select a lower value for R_v up to $R_v = V^2/4$. Thus, for $R_v = 10,000$ m, it will be:

$$E = OC = \Delta i \cdot \frac{R_v}{2} \Rightarrow E = 8‰ \cdot \frac{10,000}{2} \Rightarrow E = 40\ m$$

The ordinates y (perpendicular to abscissas x, situated on the segment CO, for point C, x=0) of the circular arc CD will be calculated by the formula (14.45) as follows:

$$y = \frac{x^2}{2 \cdot R_v}$$

It is evident that for R_v=10,000 m, C≡A.

14.15. Construction of a new railway line

14.15.1. Feasibility study

The decision for realizing a railway line is the outcome of a complex procedure in which politicians, managers, economists, and engineers are involved. Feasibility studies, (see section 6.3), are a powerful tool in rationalizing (economically) the choice of a specific project to be realized after a highly selective procedure.

Once the decision to realize a specific rail project is made, the next step is to conduct the environmental and technical studies (preliminary, outline, and final design).

14.15.2. Preliminary design

Based on the forecasted demand characteristics, the appropriate types of rolling stock for the specific project can be selected. Each rolling stock type is characterized by its power, maximum speed and acceleration, maximum gradient, etc. The analysis at this stage should consider not only current technologies but also future ones to emerge. Rolling stock can change, but track infrastructure is constructed once in a lifetime of a century.

However, the travel times taken into account in the feasibility study determine medium and maximum speeds, which in turn prescribe the maximum radius (for horizontal and vertical transition).

Before beginning the preliminary study, the engineer must collect as much data as possible, which should include the following:
- mapping at a scale of 1/50,000 or 1/25,000,
- any available aerial photography (ideally from satellites),
- land use and town plans, as well as agricultural plans,
- any available geological, hydrological, meteorological, and other information,
- any previous reports on the study area.

At this preliminary stage, all reasonably possible routes (2÷5) should be investigated. For each route, the horizontal and longitudinal sections are studied. The engineer should look for a good vertical profile with as few changes up and down as possible and for a good horizontal profile with as few reverses of curvature as possible. Based on these, major technical projects (bridges, tunnels), public utilities to be displaced, and a first estimation of cost are identified, (278).

14.15.3. Outline design

Completion of the preliminary design should result in defining a route corridor of interest, which may vary from 50 m wide, in reasonably flat terrain, to perhaps 2 km or more in mountainous areas.

The outline design is usually prepared at a scale of 1/5,000, with cross-sections surveyed at 100 m intervals. Two or three alternative routes may be studied at this stage.

During this stage, considerations should cover all aspects including the following, (275):

– maximum speed, acceleration, and deceleration,
– future traffic and number of trains,
– axle load, track design, and loading gauge parameters,
– minimum radius, cant, and other layout characteristics,
– longitudinal gradients,
– subgrade and drainage aspects,
– bridges and tunnels,
– construction planning,
– estimation of the various costs.

The solution chosen at the end of this stage is studied in detail in the final design.

14.15.4. Final design

The final stage of the study is generally carried out at scales of 1/2,000 or 1/1,000 in difficult terrain and 1/1,000 or 1/500 in urban areas. Even at this stage of the study, it may take several attempts to attain the right compromise between speed, curvature, gradient, and soil mechanics considerations.

Engineers should always have in mind future maintenance requirements, which they should try to minimize.

14.15.5. Staking of the track layout

The first step for the implementation of the layout, after completing in office calculations and design, is the staking of the track. Stakes are driven as follows:

♦ on double tracks, in the axis of the double track, both on straight and on curved sections,

♦ on single tracks, on straight sections regardless of the side of the track (right or left), and in curved sections on the side of the outer rail.

Staking the outer side of the track on curves facilitates the precise laying of the outer rail according to the layout. The alignment of the outer rail is crucial, because the outer rail guides fast moving trains. The specific value of track gauge is given by the suitable positioning of the inner rail.

Double staking is usually avoided on double tracks, and stakes are driven in the axis of double tracks. In this case, the outer rail should be laid with great care on curved sections, taking into account the value of the track gauge at the particular point.

On transition curves and circular arcs, stakes are driven every 10 m. Whenever a closer staking is necessary, stakes are driven every 5 m. Along straight sections, stakes are driven every 50 m.

At the one end of the transition curve (cubic parabola), which coincides with the beginning of a straight track, it should be ensured that the extension of the straight line is tangent to the end of the parabola. This is why the staking of a parabolic transition is extended by 4 stakes (spaced 10 m) along the straight section to provide at least two zero-deflection points. A surveying instrument, from a point at least 200 m away, should check the alignment of these 4 stakes.

The specific number of each fixed point and the required cant are marked on each stake. The laying of the track on the horizontal plane follows the staking. This stage consists of placing each rail at the proper position, on the basis of the fixed points, at which the stakes were driven, and the values of the track gauge.

However, satellite systems can greatly facilitate a more accurate stacking of the track to be constructed or renewed, (270), (273).

14.16. Environmental aspects of track layout

Environmental considerations should be carefully considered, right at the beginning of the preliminary stage of track layout. Railway planners, managers, and engineers should be aware that if the environmental considerations are not thoroughly taken into account, there is a high risk of designing a geometrically perfect layout, which however will never be realized, due to environmental restrictions.

A team of specialists, including environmentalists, landscape architects, civil engineers, and agriculturists, should study the environmental aspects of the proposed project. The following steps are necessary for an environmental approach to the track layout, (272):

- avoiding areas of natural beauty in order to minimize any risk coming from environmental reasons,
- trying to minimize disturbances in neighboring areas, due to rail vibrations. The best way is to have the major part of layout in cut sections. In layouts where embankments are necessary, noise barriers should be installed in areas neighboring with villages or cities,
- taking care of the areas where raw material coming from an excavation are deposited, in order to reduce any case of pollution,
- preserving the variety of animals, plants, and birds. As many layouts intersect some areas and thus make difficult the communication of animals from the one side to the other, special transverse passages along the layout should be installed so that frogs, foxes, etc., can easily go from the one side of the track to the other,
- adapting to the aesthetics of the environment. At the final stage of the layout, the quantities of plants and trees that will be planted along the track must be carefully calculated,
- assuring stabilization of soils (both in cuts and embankments) by giving the appropriate slope and by planting the best suited plants,
- taking measures and installing the necessary equipment, so that all plants and trees will have the required moisture,
- installing a system of monitoring, so that an evaluation of the efficiency of the measures taken can be done at regular time intervals.

15 Switches, Tracks in Stations, Marshaling Yards

15.1. Functions of switches

A fundamental characteristic of railways is the one degree of freedom of the movement of the rail vehicle on the track. However, trains must have the possibility to change course from one track to another. This is realized by switches[*], defined as the equipment and parts through which the direction of movement of a rail vehicle can be changed without interrupting its course.

Switches take a great variety of forms. In spite of their apparent complexity, they can be distinguished into two basic forms, and a third, combining the two:

♦ Simple, (Fig. 15.1), or multiple turnouts, allowing a track to be split in two (sometimes three) and the moving rail vehicle to change course.

♦ Crossings, (Fig. 15.2), where two tracks meet at grade with no change of course.

♦ Turnout crossings, combining the functions of turnouts and crossings, (see below section 15.3, Figures 15.10 and 15.11).

Fig. 15.1. Turnout

Fig. 15.2. Crossing

---

[*] Although switches are sometimes referred to as turnouts, the former, strictly speaking, do not include the frogs and check rails, (see below section 15.2 and Figure 15.3), enabling one rail to cross the other, while the latter do.

Thus, the functions of the various forms of switches are to enable rail routes to branch from or to join up with one another; to provide flexibility within a route so that trains may move from one track to another track; and finally to enable vehicles to be sorted out. In order to respond efficiently to these requirements, switches must fulfill certain conditions, which include the following, (288):
- impose the fewest possible speed restrictions,
- be sited exactly where operational exigencies demand,
- provide maximum operational flexibility,
- support the axle load required to be carried,
- be cheap to manufacture, simple to lay, easily worked, robust, and easy to replace,
- resist wear, corrosion and decay, and require minimum maintenance,
- be compatible with signaling requirements.

15.2. Components of a turnout

In a turnout we distinguish, (Fig. 15.3):
- the main track and the turnout (or diverging) track, to which the vehicle can be diverted,
- the mathematical (or intersection) point 0 of the turnout, which is the point where the axes of the two tracks intersect,

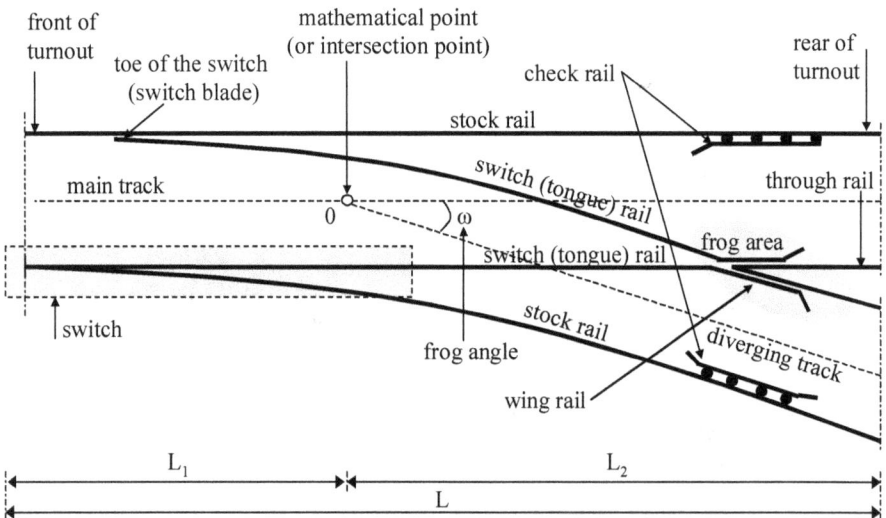

Fig. 15.3. Components of a turnout

- the frog angle, defined by the axes of the two tracks. The frog angle is commonly denoted by its tangent (e.g. 1:9). The frog angle consists of high-grade material (usually manganese steel),
- the stock rail, which is the rail that stays motionless,
- the switch or tongue rail, which is the moving rail that changes the course of the vehicle. A critical parameter is the radius of curvature R of the switch. Depending on their position, switch rails allow rail vehicles to proceed to one or to the other track,
- the check rail, which is a rail (3÷10 m long) placed exactly opposite the frog. Shortly before the frog, a wheel reaches a rail gap and it is necessary to provide the other wheel with a guide bar preventing irregular and uncontrolled movement, which is achieved by installing a check rail. The gap between stock rail and check rail is 38÷46 mm,
- the distances L_1 (from the beginning of the turnout to the mathematical point) and L_2 (from the mathematical point to the end of the turnout),
- the turnout length L ($L = L_1 + L_2$),
- the fouling distance c, which is the distance from the beginning of the turnout to the point beyond which a vehicle may lie on one track of the turnout without interfering with the movement of another vehicle on the other track. This point is specified so that the distance between the axes of the two tracks is at least 3.50 m for standard gauge tracks and 3.00 m for metric gauge tracks.

Values of the switch radius R for conventional tracks usually range between 150÷500 m, permitting speeds at the diverging track of 35÷65 km/h. For low- and medium-speed tracks, the frog angle (tangent of the angle ω) in old turnouts was given values of 1:8 and 1:10, while in more recently installed turnouts it takes usually values of 1:9 or 1:12.

The cross-section of the switch rail takes form gradually, as shown in Figure 15.4.

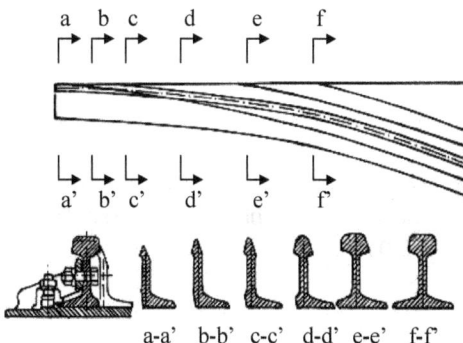

Fig. 15.4. Changing cross-section of the switch rail in relation to the distance from the toe of the switch

15.3. Various forms of turnouts

Turnouts and crossings take a great variety of forms, depending on the intended change of course. The following are the principal ones.

Fig. 15.5. Standard turnout

- Standard turnout, in which one track is split in two and the main track remains rectilinear, (Fig. 15.5).

Fig. 15.6. Simple symmetrical turnout

- Simple symmetrical turnout, with one track split in two and both tracks curving outward, (Fig. 15.6).

Fig. 15.7. One-sided double turnout

- One-sided double turnout, with one track successively split into three tracks on the same side and with the main track remaining rectilinear, (Fig. 15.7)

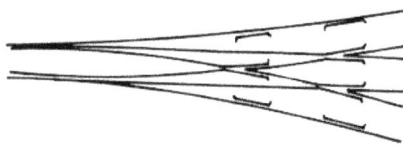

Fig. 15.8. Two-sided double turnout

- Two-sided double turnout, with one track symmetrically split into three tracks: a middle rectilinear track and two symmetrical side tracks, (Fig. 15.8).

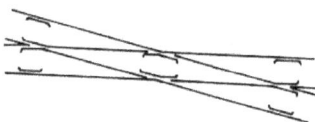

Fig. 15.9. Diamond crossing

- Diamond crossing, where two tracks meet with no change of course, (Fig. 15.9).

Fig. 15.10. Single slip

- Single slip, where two tracks meet and their course can only be changed from one track to the other in one direction, (Fig. 15.10).

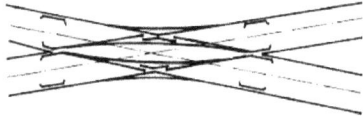

Fig. 15.11. Double slip

- Double slip, where two tracks meet and their course can be changed from one track to the other in both directions, (Fig. 15.11).

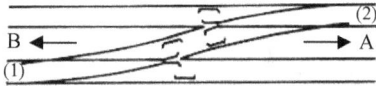

Fig. 15.12. Single crossover between two parallel tracks

- Single crossover between two parallel tracks (1) and (2). Course can be changed from (1) to (2) in the direction A (or from (2) to (1) in the direction B) but not from (2) to (1) in the direction A, (Fig. 15.12).

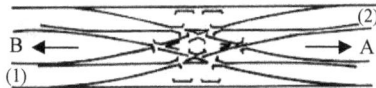

Fig. 15.13. Double crossover between two parallel tracks

- Double crossover (sometimes called 'scissors') between two parallel tracks (1) and (2). Course can be changed both from (1) to (2) and from (2) to (1), (Fig. 15.13).

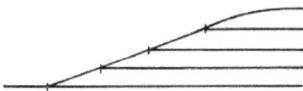

Fig. 15.14. Series of successive turnouts

- Series of successive turnouts, where one track is successively split into several tracks, (Fig. 15.14).

Fig. 15.15. Track 'fan'

- Track 'fan' with successive track splittings, a technique used in depots and marshaling yards, (Fig. 15.15).

411

15.4. Running speed on turnouts

Turnouts differ from regular tracks in that neither cant nor transition curves are used. Therefore, the maximum running speed on a turnout depends on the value of the non-compensated centrifugal acceleration b and the radius of curvature R of the turnout.

The minimum value of cant in relation to the non-compensated centrifugal acceleration is, (see section 14.4, equation (14.22)):

$$h_{min}(mm) = 11.8 \cdot \frac{V^2(km/h)}{R(m)} - 152 \cdot b_{max}(m/sec^2) \tag{15.1}$$

The non-compensated centrifugal acceleration b at turnouts must not be too high for reasons of comfort and wear. Limit values of b_{max} usually range between $0.6 \div 0.7$ m/sec². As the turnout's cant is zero, $h_{min}=0$, from formula (15.1) we obtain:

$$11.8 \cdot \frac{V^2}{R} - 152 \cdot b_{max} = 0 \Rightarrow V = 3.58 \cdot \sqrt{b(m/sec^2)} \cdot \sqrt{R(m)} \tag{15.2}$$

Some railways calculate the running speed on the turnout in relation to cant deficiency h_d. In this case, instead of formula (15.2) the following formula can be used:

$$V(km/h) = 0.29 \cdot \sqrt{h_d(mm)} \cdot \sqrt{R(m)} \tag{15.3}$$

Thus, the minimum radius of curvature of the turnout will be calculated when considering the limit value of the non-compensated centrifugal acceleration b_{max} or of cant deficiency h_d. For many railways $b_{max}=0.7$ m/sec², and by substituting this value* in the formula (15.2), we conclude that the relationship between the running speed on the turnout and the radius of the turnout is:

$$V(km/h) = 3 \cdot \sqrt{R(m)}, \text{ for } b_{max}=0.7 \text{ m/sec}^2 \tag{15.4}$$

Turnouts are designed as cubic parabolas of small length, (see section 14.4, equation (14.25)), in accordance with the equation:

$$y = \frac{x^3}{6 \cdot R \cdot \ell} \tag{15.5}$$

From equation (15.4) it is deduced that in order to have at the diverging track a train speed of V=120 km/h, the turnout radius of curvature should be at least R=1,600 m, while for a speed of V=150 km/h a radius R=2,500 m is required.

Such a layout, however, would clearly be excessively extravagant in space requirements. Furthermore, it assumes the ability to design a switch in which

---

* While b_{max} for track layout takes usually the value 0.65 m/sec², a higher value of 0.7 m/sec² is accepted in switches – turnouts.

the tongue rail can be brought truly tangential to the stock rail. For these reasons, in practical turnout design the switch is made much shorter than previous theoretical considerations would demand. This is realized by cutting the stock rail at a finite angle, known as the switch entry angle, (288).

The main characteristics of a turnout usually include its radius of curvature, the frog angle (tangent of the angle ω, see Fig. 15.3), and the tongue rail.

15.5. Geometrical characteristics of turnouts

The railway industry offers a great variety of turnouts and constructors can often adjust them to conditions in situ. Table 15.1 gives a panorama of some representative turnouts in use for standard gauge and metric gauge tracks. However, minor differences in geometrical characteristics may be observed from one constructor to another.

Table 15.2 gives geometrical characteristics of some representative forms of crossings for standard gauge and metric gauge tracks, (282), (286).

Table 15.1.
Geometrical characteristics of some representative
forms of turnouts, (282), (286)

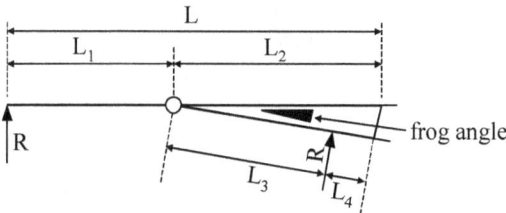

track gauge	frog (intersection) angle	R (m)	L (mm)	L₁ (mm)	L₂ (mm)	L₃ (mm)	L₄ (mm)
	1:5	100	19,804	9,902	9,902	9,902	0
	1:6	100	19,804	8,276	11,528	8,276	3,252
	1:6	140	23,174	11,587	11,587	11,587	0
	1:7	140	23,174	9,950	13,224	9,950	3,274
	1:7	190	27,006	13,503	13,503	13,503	0
	1:9	190	27,006	10,523	16,483	10,523	5,960
standard	1:9	300	33,230	16,615	16,615	16,615	0
	1:12	500	41,594	20,797	20,797	20,797	0
	1:14	760	54,216	27,108	27,108	27,108	0
	1:18.5	1,200	64,818	32,409	32,409	32,409	0
	1:21	1,540	73,292	36,646	36,646	36,646	
	1:26.5	2,500	94,306	47,153	47,153	47,153	
	1:29.74	3,000	100,846	50,423	50,423	50,423	
metric	1:7	100	18,276	7,107	11,169	11,169	0

Table 15.2.
Geometrical characteristics of some representative
forms of crossings, (282), (286)

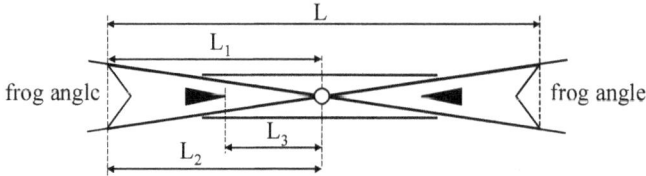

track gauge	frog (inter-section) angle	R (m)	L (mm)	L_1 (mm)	L_2 (mm)	L_3 (mm)
	1:9	190	33,230	16,615	16,615	12,955
standard	1:9	300	45,278	22,639	22,639	9,525
	1:9	500	55,386	27,693	27,693	12,955
metric	1:7	50	22,339	11,169.5	11,169.5	7,035

15.6. Derailment criterion for turnouts and crossings

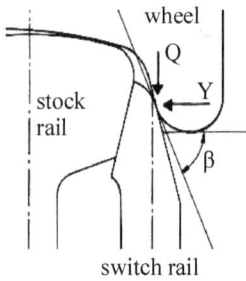

Fig. 15.16. Wheel-rail contact at a turnout

On a turnout or a crossing, a wheel flange may climb a rail and cause derailment. To prevent this event, the ratio Y/Q (where Y is the transverse force between wheel and rail and Q is the wheel load) should not exceed a value given by equation (13.8) of section 13.7.2 (Nadal's formula, known also after the names of Boedecher and Chartet, who presented the same formula at the same time with Nadal):

$$\frac{Y}{Q} < \frac{\tan\beta - f}{1 + f \cdot \tan\beta} \qquad (15.6)$$

where: β : the rail-wheel (flange) angle,
f : the rail-wheel coefficient of friction.

Starting at the lowest Y/Q value found from empirical data and the mean value of f, a value preventing derailment can be calculated for the angle β and therefore the maximum permissible wear of the inner surface of the wheel flange can also be calculated.

In order to prevent derailment on a turnout, it is suggested, in the light of study of many cases of derailments on turnouts, that the necessary condition for derailment is, (284), (287):

$$\frac{Y}{Q} \leq 0.4 \div 0.8 \qquad (15.7)$$

Given the laying and maintenance criteria for turnouts, it is more advisable to take into account for the critical condition of Y/Q a value on the order of 0.8, which is more typical of actual conditions of safety and derailment, (281), (288).

Therefore, for a wheel load of 20 tonnes, an average value f=0.3, and considering the critical value Y/Q=0.8, it can be calculated from formula (15.6) that

$$\beta_{critical} \cong 55° \tag{15.8}$$

15.7. Turnouts on a curved main track

Until now it has been assumed that the main track of a turnout is on a straight line. However, if the main track is on a curve, the speed at which a turnout can be run will be calculated as follows. Let:

R_o : radius of standard turnout out of a straight main track,
R_m : main track radius of curvature,
R_t : desired radius of turnout from a main track curved at R_m.

A turnout on a curved main track can be in contrary or similar flexure. For contrary flexure, curvature of curve R_t will be:

$$\frac{1}{R_t} = \frac{1}{R_o} - \frac{1}{R_m} \Rightarrow R_t = \frac{R_o \cdot R_m}{R_o - R_m} \quad , \tag{15.9}$$

whereas for similar flexure it will be:

$$\frac{1}{R_t} = \frac{1}{R_o} + \frac{1}{R_m} \Rightarrow R_t = \frac{R_o \cdot R_m}{R_o + R_m} \tag{15.10}$$

15.8. Turnouts run with increased speeds

In turnouts run with increased speeds, the frog angle is reduced. Thus the German railways use a turnout with a frog angle of 1:42, in which the diverging track can be run at a speed of 200 km/h (non-compensated lateral acceleration is 0.5 m/sec^2 in this case).

Table 15.3 gives the geometrical characteristics of turnouts that can be run with increased speeds, according to the specifications of UIC, (284). It is to note that in the two last cases of Table 15.3 (R=3,000 m and R=6,720 m) for which the radius of curvature changes from value R (3,000 m or 6,720 m) to R'(∞), the frog (intersection) angle also changes from 1:43.65 to 1:46 (case of radius R=3,000 m) and from 1:61.68 to 1:65 (case of radius R=6,720 m).

For the various turnouts which can be run with increased speeds, (Table 15.3), Table 15.4 illustrates the values of running speed on the turnout in relation to the radius of curvature and the cant deficiency or the non-compensated centrifugal acceleration b, (281).

Table 15.3.

Geometrical characteristics of turnouts that can be run with increased speeds, (284)

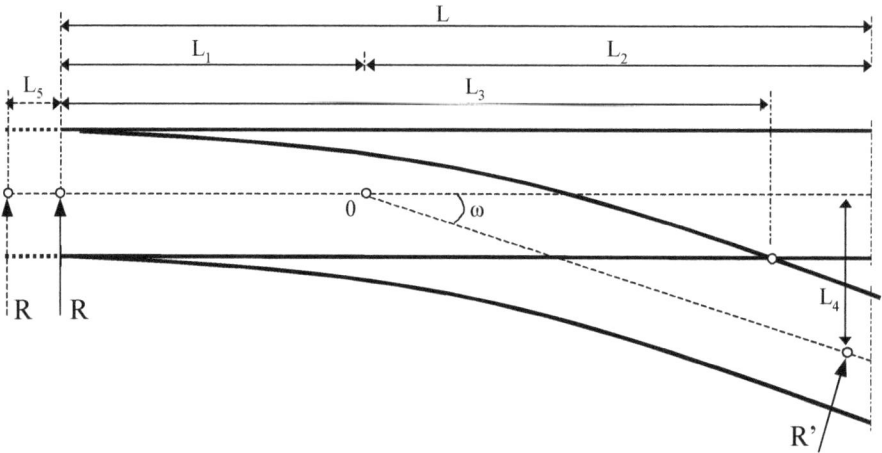

tangent of frog (intersection) angle	R (m)	R' (m)	L (mm)	L₁ (mm)	L₂ (mm)	L₃ (mm)	L₄ (mm)	L₅ (mm)
1:18.5	R = R' = 1,200		64,818	32,409	32,409	58,686	1,749	
1:21	R = R' = 1,540		73,292	36,646	36,646	66,482	1,743	
1:26.85	R = R' = 2,500		93,078	46,539	46,539	84,705	1,732	
1:26.5	R = R' = 2,500		94,306	47,153	47,153	84,705	1,778	
1:29.74	R = R' = 3,000		100,846	50,423	50,423	92,790	1,694	
1:43.65	3,000	∞	136,920	45,260	91,660	107,435	2,100	578
1:61.68	6,720	∞	193,445	63,915	129,530	151,780	2,100	2,004

Table 15.4.

Maximum running speed (km/h) on turnouts run with increased speeds, in relation to the radius of curvature, cant deficiency, and non-compensated acceleration, (284)

cant deficiency h_d	80 mm	85 mm	90 mm	95 mm	100 mm
non-compensated acceleration b	0.523 m/sec²	0.556 m/sec²	0.589 m/sec²	0.621 m/sec²	0.654 m/sec²
radius of curvature (R) — 1,200 m	90 km/h	93 km/h	95 km/h	98 km/h	100 km/h
1,540 m	102 km/h	105 km/h	108 km/h	110 km/h	114 km/h
2,,000 m	116 km/h	120 km/h	123 km/h	127 km/h	130 km/h
2,500 m	130 km/h	134 km/h	138 km/h	142 km/h	145 km/h
3,000 m	142 km/h	147 km/h	151 km/h	155 km/h	160 km/h

15.9. Track layout and positioning of sleepers in turnouts

In the case of track on twin-block sleepers, timber sleepers are used in the turnout area. If the track is laid on other sleeper types (monoblock, timber, steel), then the same sleeper type is used for both the turnout area and the remainder of the track.

Sleepers are laid perpendicular to the axis of the main track up to the edge of the check rail, (Fig. 15.17). Beyond this point, they are laid perpendicular to the bisectrix of the angle of the turnout.

Figure 15.17 illustrates the track and sleeper layout for a turnout type UIC 60, while Figure 15.18 illustrates a turnout according to the American specification.

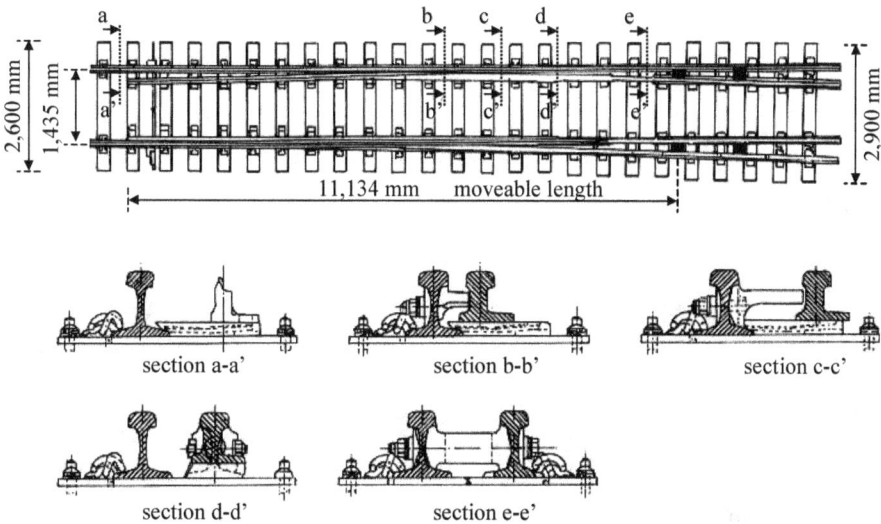

Fig. 15.17. Track layout in the case of a turnout type UIC 60

15.10. Manual and automatic operation of turnouts

A turnout may be operated either manually (by local or remote levers), (Fig. 15.19), or automatically, (Fig. 15.20). Automatic operation is driven by electric activators working on commands from electric control boards, which are operated by personnel either of the station or the operation control center, (285).

A turnout operates as follows: One of the two switch rails, (see Fig. 15.3), stays tangent to the rail adjacent to it, while the other switch rail leaves from its neighboring rail a gap sufficient for passage of the wheel flange, (Fig 15.3 and Fig. 15.21). When the set of the two switch rails changes position, either manually or automatically, the above states are interchanged and the switch rail in contact opens, while the other one closes the gap.

In automatic switch operation, the following controls are performed automatically, (Fig. 15.21):
- control of the gap between the stock and switch rail,
- control of check rail gauge and wear in the frog area.

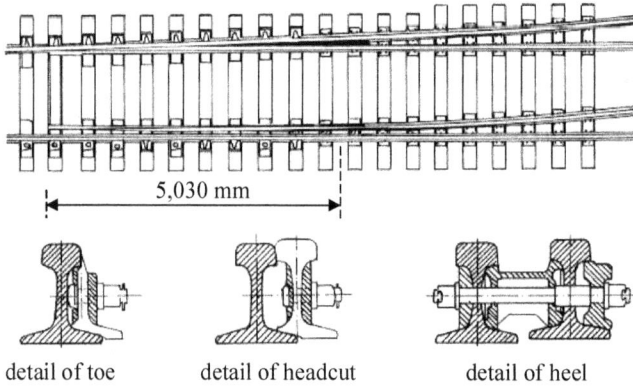

5,030 mm

detail of toe   detail of headcut   detail of heel

Fig. 15.18. Track and sleeper layout in the case of an American turnout according to the American (AREA) specification

Fig. 15.19. Manual operation of a turnout

Fig. 15.20 Automatic operation of a turnout

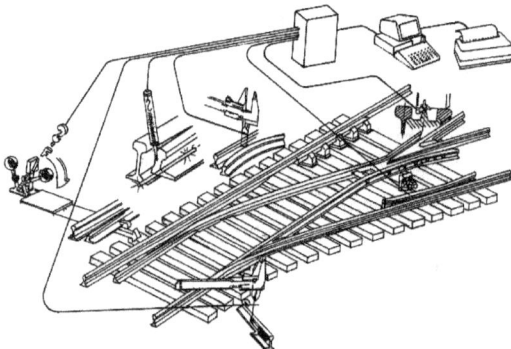

Fig. 15.21. Mode of operation of an automatic turnout

418

15.11. Design principles for turnouts and crossings

In addition to previous analytical methods and formulas, some empirical considerations should also be taken into account when designing turnouts, (282):

♦ the tensile strength of the rails, switch rails, and check rails used in the switches should be at least 8,800 kg/cm². All surfaces must have an industrial finish in a special heat-treatment process, which increases tensile strength to 13,000 kg/cm²,

♦ switches are manufactured in the form of spring switch blades or flexible tongues. The running edges of the diverging track are in the form of a circular arc. Elastic stock rail bracing is used inside in order to fasten the stock rails,

♦ the check rail, which is made of special sectional steel, is fastened to support plates and is thereby connected to the running rail. To account for check rail wear, spacers can be inserted to correct the switch opening and the space between rail faces.

Switches and crossings should not be located in the following areas:

– in tunnels and bridges,
– on sharp horizontal circular curves,
– on horizontal transition curves,
– in cases of high cant.

However, according to the European technical specifications for interoperability, (134):

• the rail in switches and crossings can be designed either to have an inclination (from 1:20 to 1:40) or not,
• all movable parts of switches and crossings should be equipped (for both new and upgraded high-speed tracks) with a means of locking.

Switches and crossings should be inspected in regular intervals in order to check for correctness of the check and wing rail, (see Fig.15.3), flangeways, that all bolts, screws, and fastenings are fitted, and that there is no need for the maintenance of weldings.

15.12. Lifetime and maintenance costs of turnouts

The lifetime of a turnout is a relation of the traffic load and the speed, at which the turnout is run. Turnouts run by high-speed trains on heavy routes are replaced every 25 years, whereas in lower traffic routes of conventional tracks turnouts can be replaced every 60 years. For most railway authorities, the average lifetime of a turnout is 40 years.

Maintenance of turnouts represents an important component of track maintenance, around 12%÷24%. The great difference between the lower and the upper value depends, among others, on the number of turnouts per km of track, which is 0.42 for Germany, 0.43 for France, 0.38 for the UK, 0.32 for Italy, 0.5 for the EU, 1.0 for the USA, 0.32 for Canada, 0.13 for Russia, 0.48 for India.

Efficient operation of trains, respect of time schedules, and safety are greatly dependent on the operation of turnouts. A number of delays (ranging from 5% to 10%) is attributed to turnouts.

15.13. Turnouts and tracks in railway stations

15.13.1. Railway station: a node connecting the railway with life and economy

Railways exist to provide safe, fast, and efficient transport of people and goods at a reasonable cost. Railway stations are the entities of the railway system, where trains can stop safely, embark-disembark passengers, and load-unload freight. A station is the node between the railway system on the one side and real life and the economy on the other. Any station must have some kind of access: on foot, by road (private car, bus, taxi), by metro, or by another railway line. Thus, from an operational point of view, a station is the *node* between two transport processes, the one at least of which is the railway.

A railway station is a complex comprising the following: buildings for passengers (ticketing offices, waiting areas, restaurants-bars, access to trains), buildings for freight, platforms (where embarking-disembarking of passengers and loading-unloading of freight take place), surrounding area and access to road and metro systems (with the required parking facilities), tracks - turnouts - signaling ensuring safe stopping and (eventually) parking (of a train or wagon) or overtaking operations of trains that do not stop at a specific station.

Considering the above, design of buildings and of the surrounding area are clearly tasks in the vast scientific area of architecture and are not analyzed in this book. Some essential characteristics of a station and its environment are: good accessibility, easy localization, safety and security, adaptation to the local architecture, creation of a friendly environment for customers, selection of materials resistant to vandalism, appropriate lighting so as to avoid attracting activities other than those related to railway transport.

15.13.2. Topologies of tracks in stations

Layout of tracks in a station gives the impression of a complex situation, which however is a combination of three basic topologies, (Fig. 15.22), (280):

- running station: a train can overtake another (usually stopping) train. For two principal tracks, overtaking tracks can be designed either on both sides (as in Fig. 15.22.a) or only on one side,
- junction station: a train can continue forward or change its course,
- crossing station: two or more trains can cross, without any overtaking or change of course in the station area.

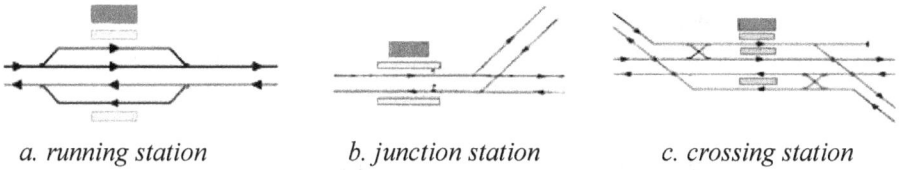

a. running station   b. junction station   c. crossing station

Fig. 15.22. Basic topologies of railway stations, (280)

15.13.3. Layout of turnouts and tracks in a medium-size station

Figure 15.23 illustrates a typical layout of turnouts and tracks for a medium-size station with a population in its catchment area of 10.000÷50.000 inhabitants. The layout is composed of the following, (Fig. 15.23): principal running tracks, overtaking tracks, parking track, waiting track, service track (for maneuvers of pulling rolling stock), freight loading-unloading tracks, dead end tracks (where trains are decomposed and sorted).

As length of a track in a station is considered the length from the mathematical point of the turnout at the entrance till the end of the fouling distance, (see section 15.2), of the turnout at the exit of the track, with usual values at the order of 750 m.

The running speed V on a turnout is calculated in relation to the radius of curvature of the turnout by formulas (15.2) to (15.4).

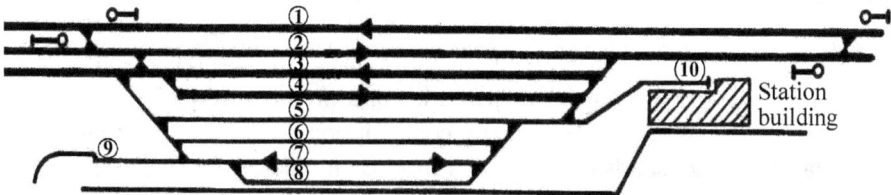

Station building

1, 2 : principal running tracks	7 : service track
3, 4 : overtaking tracks	8, 9 : freight loading-unloading tracks
5 : parking track	10 : dead end track
6 : waiting track	

Fig. 15.23. Typical layout of tracks in a medium-size station

15.13.4. Length, width, and height of platforms

Platform is the area of a station where embarking-disembarking of passengers and loading-unloading of freight take place. The length of a platform is calculated in relation to the maximum length of a train that will stop in the station and varies from 80 m to 400 m. Transverse slope of a platform is given an average value of 2%, which should not exceed 4%. For safety reasons, transverse slope is given by some railways at a direction opposite to the track.

In relation to their height, platforms can be of *low level* when the distance between the level of platform and the upper surface of rail is 30÷36 cm, or of *high level* when the aforementioned distance is 76 cm.

Width of the platform is calculated in relation to the number of passengers that are expected to find themselves on the specific platform. Minimum width for platforms in front of the station building is 7.5 m, for platforms serving two tracks is 9 m, and for platforms serving only one track is 6 m, (Fig.15.24).

Advance signals warning the train driver about the main signal that follows are located 900 m before the entrance turnout, whereas the main signal (red or green) is located 250 m before the entrance turnout.

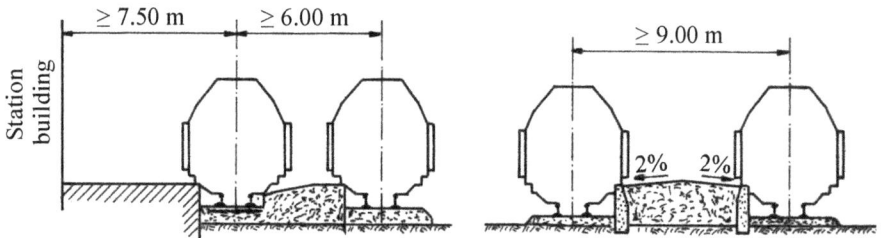

Fig. 15.24. Width of platforms in railway stations

15.14. Turnouts and tracks in marshaling yards

15.14.1. Definition and functions of a marshaling yard

Economic conditions determine the quantities of a product produced at an origin point O that will be transported at a consumption point C. Points O and C may be far apart thousands of kilometers from each other. If the specific product is transported by rail, produced and consumed quantities very rarely justify the formation of a full train from origin point O to destination point C. Thus, railways devised the following procedure: a short-distance train transports the specific product from origin point O_1, (Fig.15.25), to a central station M_1. Another short-distance train transports another product from origin point O_2 to the central station M_1, and so on. All trains arriving at the station M_1 are decomposed and each wagon becomes a unity of a new long-distance through train formation in relation to the destination point, where the specific product is required for final consumption. Such a central station is called *marshaling* (or classification, particularly in the USA) *yard* and it is the place where short-distance freight trains are received, decomposed, sorted out, and recomposed in long-distance trains. Long-distance trains composed in a marshaling yard near the origin point seldom go directly until the destination point D_i. Very often they end up at a marshaling yard M_2 near the destination point, where long-distance trains are decomposed and wagons are recomposed to short-distance trains, which transport freight to the final destination point D_i (Fig.15.25).

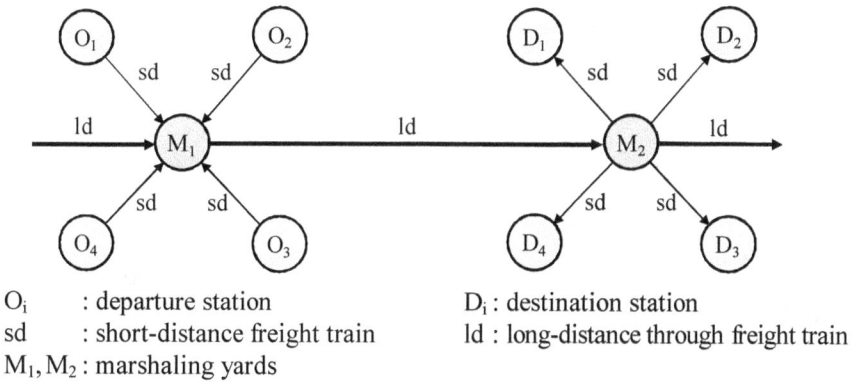

O_i : departure station D_i : destination station
sd   : short-distance freight train   ld : long-distance through freight train
M_1, M_2 : marshaling yards

Fig. 15.25. Functions of marshaling yards M_1, M_2

Track configuration of a marshaling yard is usually characterized by what we called earlier a track "fan", (see Fig. 15.15). Any marshaling yard has three basic components: the receiving yard, the classification yard, and the departure yard, (Fig. 15.26). Marshaling can take place either in one direction (one sided yard, from M_1 to M_2 but not from M_2 to M_1) or in both directions (two-sided yards, from M_1 to M_2 but also from M_2 to M_1), (Fig.15.25). In relation to whether wagons in a marshaling yard move with the help of gravity forces or need a shunting engine, we distinguish three types of marshaling yards: flat, gravity, and hump marshaling yards, (see below section 15.14.2).

The accurate location of a marshaling yard is calculated in relation to the origin-destination pairs of traffic flows, the volume of flows, and the need to minimize waiting times (in marshaling yards) and total origin-destination transport times.

15.14.2. The various types of marshaling yards

15.14.2.1. Flat yards

Flat (also called flat-shunted) yards are located on level ground and can handle, classify, and sort out up to 1,000÷1,200 rail wagons per day. A shunting engine is attached to the train, draws it onto the shunting neck AB, (Fig. 15.26), and pushes it to the classification yard, where each wagon, one after the other, is successively taken out from the train and moves towards a specific track according to its destination. Then the engine draws back to section AB. In spite of its apparent simplicity, flat yards are very slow and costly.

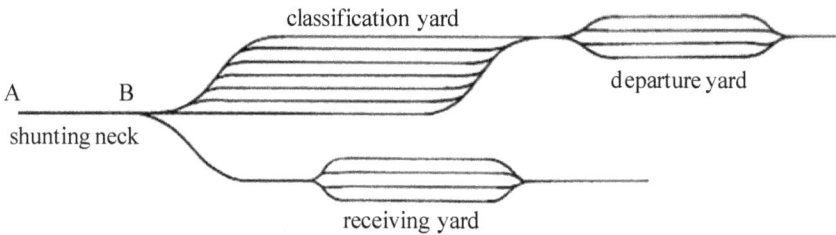

Fig. 15.26. Flat yard

15.14.2.2. Gravity yards

Gravity yards are characterized by gentle gradients and rail wagons move towards the classification yard with the help of gravity forces. Gravity yards strongly depend on local topographical characteristics.

15.14.2.3. Hump yards

In hump yards, rail wagons also move with the help of gravity forces (as in gravity yards), but the whole configuration and the various gradients are the result of technical works, so as to adapt the existing topography to the needs and gradients required. Hump yards are more effective than previous ones and can handle a high number of rail wagons (up to 10,000÷15,000 wagons per day).

A hump yard has the following parts, (Fig.15.27): the receiving yard, the marshaling-sorting out yard (which can have the so called "sick" yard, where light maintenance and repairs of wagons can be brought out), and the departure yard.

The process of marshaling rail wagons in hump yards has the following steps, (280):

– reception lines are joined up into a single line, (Fig.15.28), which has a gentle upward gradient, (Fig. 15.28, section AB),
– a shunting engine pushes the train till the first wagon reaches the top of the hill (point B, Fig. 15.28),

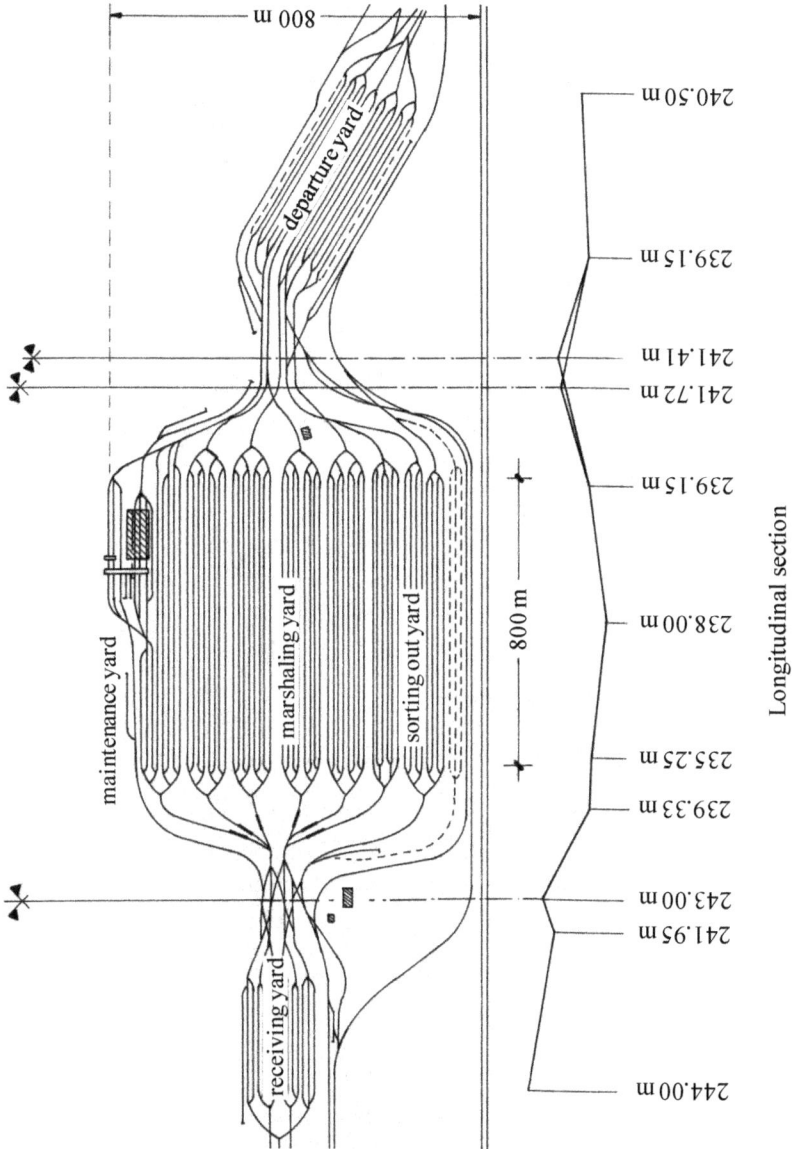

Fig. 15.27. Parts of a hump marshaling yard

- in relation to the destination of the wagon on top of the hill, a computer system arranges the switches, so that the wagon will move on the appropriate track,
- the wagon is cut out from the train and moves towards its track (section BCD),

425

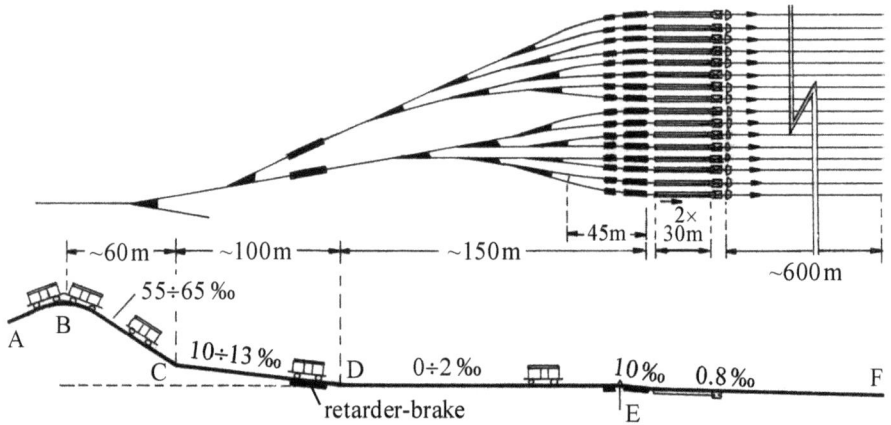

Fig. 15.28. Hump marshaling yard with a gentle gradient

- a system of automatic retarders and brakes ensures that the specific wagon, while moving downwards, stops in time behind the preceding wagon on the specific track of the classification yard,
- the wagon is classified in relation to its destination with the help of a shunting engine (section DE),
- the wagon moves towards the departure tracks (section EF). Wagons with a common destination are queuing the one after the other.

The previous procedure is applied to hump marshaling yards with a gentle gradient. If we want to avoid the use of shunting engine in the section DE of the above procedure, (Fig. 15.28), we can try a hump marshaling yard with a high gradient, (Fig. 15.29). Now, the section DE has a higher gradient, a shunting engine is not required, but in spite of the use of retarders-brakes, there is the risk of collision between successively moving wagons.

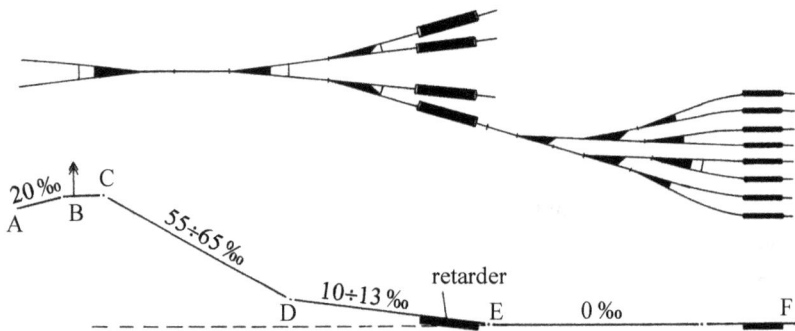

Fig. 15.29. Hump marshaling yard with a high gradient

426

There are many configurations of marshaling yards in relation to the availability of space. Representative types are either in line (receiving, classification, departure) or parallel, or some combination of the above.

15.14.3. Automatic regulation of turnouts in marshaling yards

Even the smallest error in the successive steps of marshaling may be disastrous. Advances in computer applications give the possibility of automatic regulation in turnouts of marshaling yards as follows:
- any new arrival of freight train is recorded by a central computer,
- for each wagon, the necessary data, such as destination, load, weight, number of axles, and length, are provided,
- the computer calculates the speed of each wagon and regulates the appropriate switches automatically,
- every wagon is routed to its preassigned classification track,
- if the time between the movement of successive wagons is not enough for a safe marshaling, retarders are activated to decelerate the last released wagon,
- big data and artificial intelligence can provide an accurate mapping of procedures at any moment and thus absorb any unpredicted events (delays, etc.).

15.14.4. Design of a marshaling yard

Most marshaling yards were designed as hump yards and were constructed many decades ago at the golden era of railways (1880÷1950s) with a high rail share in the freight market. Thus a restructuring of old yards is necessary and should take into account: (i) current and future rail and transport technology, (ii) the huge possibilities of computers to regulate from a central point numerous moves of trains, and (iii) the applications of big data and artificial intelligence for the optimal use of existing rail infrastructure.

The biggest marshaling yard in the world is in Nebraska, USA, where 140 freight trains with 14,000 wagons are passing through on average every day. The area has a length of 13 km, a width of 3 km, 200 separate tracks (in forms of track 'fans'), 985 switches, 766 turnouts. A number of 2,600 people are employed on the site.

The biggest marshaling yard in Europe is situated south of Hamburg and covers an area of 7,000 m long and 700 m wide. It is an over-dimensioned yard with a capacity of 11,000 wagons per day, whilst 8,000 wagons were using the yard per day some years ago and only 4,000 wagons in late 2010s.

In the UK, block trains are made up from origin to destination and thus there is little need for marshaling yards (of only small size).

16 Laying and Maintenance of Track

16.1. Laying of track

16.1.1. Mechanical equipment

The laying of track is carried out, nowadays, with the use of a great variety of mechanical equipment.

Before laying the track, it should be verified that the subgrade has been properly compacted, (see chapter 9), that the transverse slope (3÷5%) is correctly applied, and that the geotextile is well placed on top of the subgrade.

Ballast is transported with special wagons and is unloaded on-site. The ballast bed should be properly leveled, graded, and consolidated. A gantry or a light-type vehicle is used to pull a scarifier for the usual scarifying of the top ballast, to grade the top ballast with a small grading machine, and also to consolidate the ballast bed with a vibrating plate or roller vibrator.

Laying of rails and sleepers is done with the use of more sophisticated machines. Rails are laid continuous welded. They are transported on-site in lengths of 70÷100 m, after having being welded in depots (flash-butt welding) and next are welded on site (thermit welding) in lengths of many hundreds of meters and very frequently of many kilometers. A careful control of the welding procedures is required. Another essential feature of rails is cleanliness, that is freedom from oxide inclusions and minimal hydrogen levels, (299).

Sleepers should be correctly and uniformly spaced. The uniformity of sleeper spacing is just as important as the nominal spacing.

Pads and fastenings should be properly adjusted on the sleepers. The ideal fastening does not require maintenance, but if it does, then it should be easy and at the lowest possible cost.

There are many types of track laying machines. Thus a high-speed laying machine (with a workshift of 6 hours) can achieve an average output of 1.3 km per shift. Peak outputs can reach 500 m/h and 1.5÷2.0 km per shift, (289).

Fig. 16.1. Rail positioning machine

Once the track is laid, rails are positioned with the help of a rail positioning machine, (Fig. 16.1).

Similar mechanical equipment and methods are used both for renewing old tracks and for laying new ones on virgin territory. However, additional mechanical equipment is needed for the removal of the old track.

It should be emphasized that fully mechanized track renewal and laying methods would have to be adapted to the particular conditions of each railway authority (railway network or infrastructure manager).

However, developing countries with limited funds may not find it possible to invest in all the sophisticated material of a fully mechanical laying process. In such cases, there is equipment available which enables countries with a surplus of low-cost labor to install modern track assemblies and some of the smaller items of the physical plant together with hand tools such as: sleeper and rail handling tools, manual rail changers, rail skate roller equipment, rail scooters, small hydraulic fastening installation equipment, hand ballast tamping machines, rail saw, rail drills, jacks, slewing bars, etc, (303).

16.1.2. Sequence of construction of the various track works

In order to save both labor and time and achieve the best use of the available mechanical equipment, the various trackworks must be well scheduled. Optimal scheduling can nowadays be done with the use of specialized software, such as Primavera, Microsoft Project, etc.

In the scheduling of track laying works, the failure of just one operation, at one location, on one day, will disrupt the whole sequence. Such disruptions should be as few as possible and where a disruption is expected, the sequence of works should be rescheduled.

16.2. Track maintenance and parameters influencing it

16.2.1. Preventive, corrective, and condition-based maintenance

In previous chapters we have examined methods for optimizing the design and construction of the track. However, after the various components of a railway track start operating, degradation begins and maintenance becomes necessary within a certain time. Track maintenance affects decisively both train safety and passenger comfort. Track maintenance expenses represent a significant percentage of total railway infrastructure expenses and amount annually for high-speed tracks at 90,000 €/km, (38).

Therefore, track maintenance expenses should be kept as low as possible, while ensuring for a specific train speed, that safety and passenger comfort remain at an acceptable and satisfactory level. With respect to safety, maintenance should be *preventive*; regarding comfort, maintenance should be *corrective*; and as regards the economic aspects of maintenance, an *optimum solution* should be sought, so as to ensure a satisfactory safety margin and prevent a quick degradation of track quality.

Preventive maintenance is based on keeping under surveillance and monitoring each component of the track on regular intervals and on taking into account previous experience of similar problems. Preventive maintenance permits to organize a rational maintenance schedule, without however ensuring that all track components (e.g. ballast) under maintenance or replacement have attained their ultimate strength or operational possibilities. Thus, preventive maintenance may lead to replacement or maintenance of a material, while it still keeps strength reserves.

Corrective maintenance aims at undertaking maintenance or replacement works of a railway component when failure is imminent. It is expensive, not optimal, and in some cases even a risky method, as the time interval between signs of a coming failure and a real failure may be extremely short.

Advanced possibilities to continuously record, monitor, and transmit data related to the situation of any material, combined with forecasting methods for the evolution of a phenomenon, have recently made possible the so-called *condition-based* maintenance: monitoring devices (e.g. measurement of stresses or settlements) which are appropriately placed can monitor continuously all parameters of the real situation of a material and transmit immediately all relevant data to decision centers. By continuously comparing the remaining strength reserves of a material with its actual situation, the optimal time of maintenance or replacement can be accurately calculated.

The various systems of track maintenance are based on scheduled (preventive) maintenance and unscheduled (corrective) repairs. Condition-based maintenance is a combination of preventive and corrective maintenance. It is stemmed of the values of measuring instruments that record, describe, and evaluate continuously the condition of any material and leads to an optimization of lifetime and quality of track, of safety and quality of transport, and of maintenance cost.

16.2.2. Geometrical and mechanical parameters

The previously analyzed objectives of track maintenance depend on two fundamentally different classes of parameters: on the one hand, *geometrical* parameters, the degradation of which is usually reversible; and on the other hand, *mechanical* parameters, which in most cases cannot be restored without parts replacement (rails, fastenings, sleepers, welds, etc.).

Geometrical parameters, however, evolve much faster, about 5÷15 times, than mechanical parameters, (304). Accordingly, in lines with an average traffic load (20,000÷40,000 tonnes/day, UIC group 4), which are representative of the majority of lines all over the world, systematic restoration of geometrical characteristics is done after a traffic load of about 40÷50 million tonnes, while rails are replaced after about 500÷600 million tonnes. This means about four years between scheduled maintenance sessions and rail and concrete sleeper replacement every 40÷50 years (the above figures are indicative of the order of magnitude only).

Deviations between the actual and theoretical values of geometrical characteristics of the track are termed *track defects* and their restoration is accomplished through track maintenance. Track defects should be distinguished from *rail defects*, (see section 10.9.1).

The origin for the creation of track defects should be researched in the situation in the ballast under the sleepers. Indeed, under repeated loadings from successive trains, the upper part of the ballast layer presents loosening characteristics, whereas the lower one remains compacted. Between the loosening and the compacted part of the ballast, a gap is created and results in altering the geometrical characteristics of the track at the longitudinal, transverse, and horizontal level.

16.3. Definitions and parameters associated with track defects

Let $z_i(T,x)$ and $z_o(T,x)$ be the elevation of the inner and the outer rail respectively, corresponding to a traffic load T since the last track maintenance, at a kilometric position x. We define the following parameters, (Fig. 16.2):

Track elevation $z(T,x)$

$$z(T,x) = \frac{z_i(T,x) + z_o(T,x)}{2}$$

Track settlement $e(T,x)$

$$e(T,x) = z(0,x) - z(T,x)$$

431

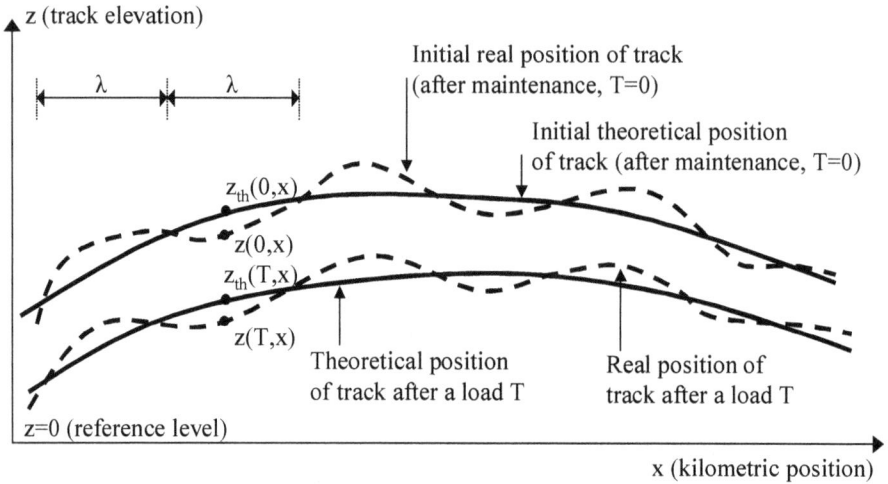

Fig. 16.2. Definition of basic parameters for track maintenance

Mean settlement $m_e(T)$ over a track length L

$$m_e(T) = \frac{1}{L} \cdot \int_{x=0}^{x=L} e(T,x)dx$$

For measurements performed at discrete positions and not continuously, it will be:

$$m_e(T) = \frac{1}{N} \cdot \sum_{i=1}^{N} e(T,x)dx$$

Differential settlement $\Delta e(T,x)$

$$\Delta e(T,x) = e(T,x) - m_e(T)$$

Standard deviation of the settlement, $sd_e(T)$, over a track length L

$$sd_e(T) = \sqrt{\frac{1}{L} \cdot \int_{x=0}^{x=L} \left[e(T,x_i) - m_e(T) \right]^2 dx}$$

and for discrete values

$$sd_e(T) = \sqrt{\frac{1}{N} \cdot \sum_{i=1}^{N} \left[e(T,x_i) - m_e(T) \right]^2}$$

432

Theoretical elevation of the track $z_{th}(T,x)$

The real position $z(T,x)$ of the track oscillates around a theoretical position, which, being unknown, is approximated over a certain length 2λ around position x, (Fig.16.2), by the value $z_{th}(T,x)$ given by the equation:

$$z_{th}(T,x) = \frac{1}{2\lambda} \cdot \int_{x-\lambda}^{x+\lambda} z(T,\xi)d\xi$$

16.4. Track defects

16.4.1. Longitudinal defect

The longitudinal defect LD, (Fig. 16.3.a), is defined as the difference between the theoretical and the real value of track elevation and is given by the equation

$$LD = z_{th}(T,x) - z(T,x) \tag{16.1}$$

The longitudinal defect (called also unevenness defect) is the most reliable in illustrating the effect of the vertical loads on track quality and is the principal factor (together with the transverse defect, see below, which accompanies the longitudinal defect) in determining the magnitude of the track maintenance expenses.

16.4.2. Transverse defect

The transverse defect TD, (Fig. 16.3.b), is defined as the difference between the theoretical and the real value of cant:

$$TD = (z_i - z_o)_{th} - (z_i - z_o)_{act} \tag{16.2}$$

where: z_i: elevation of inner rail,
            z_o: elevation of outer rail.

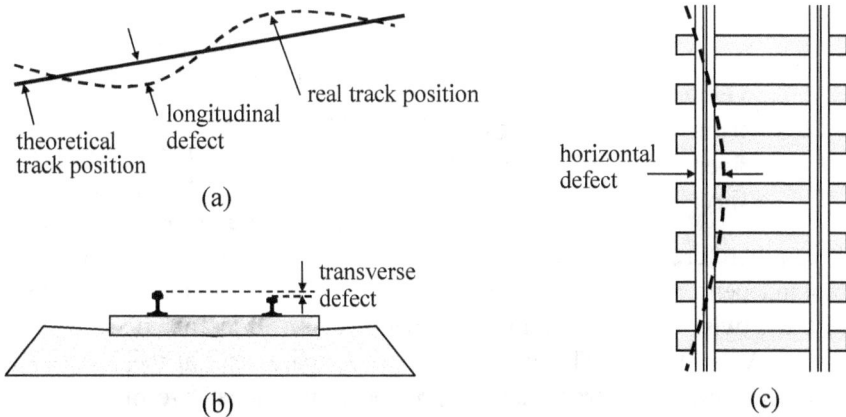

Fig. 16.3. Longitudinal, transverse, and horizontal track defects

For rectilinear parts of the layout, where curvature and cant are zero, the transverse defect is the difference of elevation between inner and outer rail: $z_i - z_o$.

16.4.3. Horizontal defect

The horizontal (or alignment or lateral alignment) defect HD, (Fig. 16.3.c), is defined as the horizontal deviation of the real position of the track from its theoretical position. The horizontal defect depends on the transverse track effects (more than the two previous types of defects) and on the characteristics and particularities of the rolling stock.

16.4.4. Track gauge

Deviations of real values of track gauge from nominal values are affected by the mechanical properties of track materials, the particularities of the rolling stock, and the train speed, as given in Tables 16.2 and 16.5.

16.4.5. Track twist

Along straight and circular sections of the track (where cant is constant), four points of the track lying on two transverse sections (e.g. on two sleepers, as shown in Figure 16.4) must lie in the same plane. Track twist is defined as the deviation of one point from the plane defined by the other three.

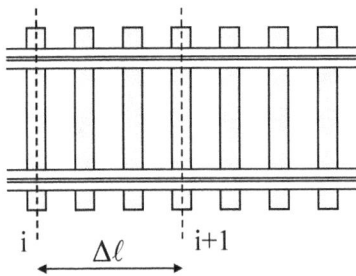

If i and $i+1$ are two successive transverse sections of the track, spaced $\Delta\ell$ apart, (e.g. at the positions of two sleepers), track twist is defined as the variation of the transverse defect per unit length,

$$\text{Track twist} = \frac{TD_{i+1} - TD_i}{\Delta\ell}$$

Fig. 16.4. Track twist: the deviation of one point from the plane defined by the other three

The risk of derailment is reduced when the real value of twist is smaller than its critical value causing derailment, which depends mainly on speed and to a lesser degree on the type of the track equipment and of the rolling stock.

It can therefore be concluded that the track twist and the transverse defect are not independent parameters. However, they are often examined separately, since track twist is one of the most frequent causes of derailment, especially for speeds V<140 km/h. The main critical safety parameter at these speeds is track twist, while other track defects previously mentioned are of lesser importance, (301), (305).

16.5. Recording methods of track defects

The personnel in charge of track maintenance were the ones detecting track defects until the late 1960s either visually (this method, permitting the detection of only large defects, did not prove rational nor free of subjective assessment) or with the help of simple instruments. However, for more than 5 decades now, modern railway technology has been using recording vehicles, (Fig. 16.5), traveling the track at specified intervals (for main routes: 3÷4 times/year, for intermediate routes: 2 times/year, for branch lines: once/year). These vehicles are provided with recording equipment, which measures the values of the various track defects in accordance with a specific basis of measurement (in the order of 10 m for longitudinal, transverse, and horizontal defects and in the order of 2.5÷3 m for track twist). As there exist many types of recording vehicles, whenever values of track defects are given, they should be accompanied with values of the measurement basis. The recording vehicle illustrated in Figure 16.5 corresponds to the chord offset type. There also exists the inertial measuring type recording vehicle, based on technologies similar to those used to measure highway and airport runway irregularities. Figure 16.6 illustrates a recording of longitudinal defects.

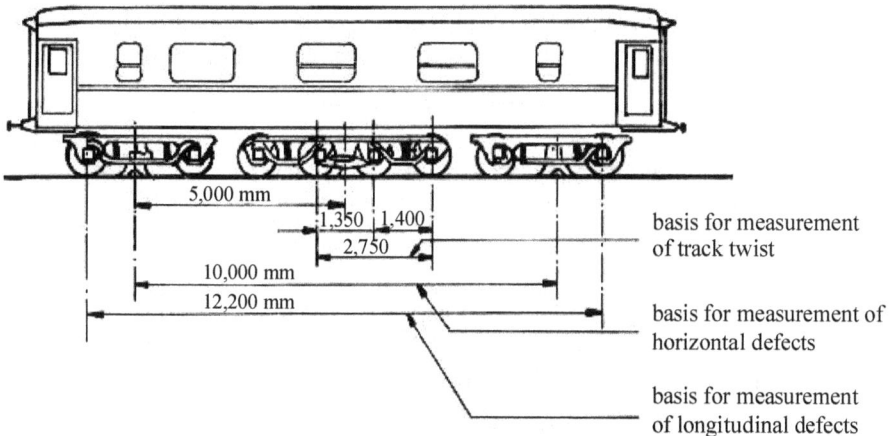

Fig. 16.5. Recording vehicle of track defects

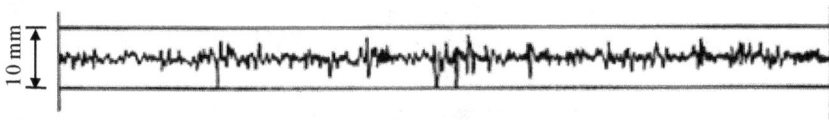

Fig. 16.6. Longitudinal defects as recorded by the recording vehicle

However, evolutions in digitalization and easy transmission of data have given the possibility to use as recording vehicles common commuter trains, upon which the appropriate recording equipment of the track is boarded. Thus, the status of deterioration of the track is monitored continuously and many times on a daily basis thanks to monitoring devices located on the track and immediate transmission of information with the help of laser scanners and color cameras. The recording system can monitor, in addition to track defects, other essential characteristics of the track such as: deterioration of rail and rail defects, clearance limits in tunnels, catenary height, signaling facilities, deficiencies in wheels and the rolling stock. With the help of mobile mapping systems, all these data can be transposed in appropriate figures and designs of the accurate position of the track, which in turn are transmitted to decision centers for track maintenance but also to any agent involved in maintenance.

The distribution of the various types of defects is of a stochastic nature, and can be approximated with the aid of spectral analysis. Thus, for each class of defects, the following can be calculated: their frequency of occurrence, the wavelength to which they correspond, their relation to train speed, etc.

The simplest approach is to calculate the mean (unsigned) values of a defect as well as its maximum values over a particular length. Both the former and the latter will be designated as absolute values of defects and they are used at low and medium speeds as the critical and determining parameters, on which safety depends.

However, for medium-, rapid- and high-speed tracks, the decisive parameters are those determining passenger comfort. At these speeds, ensuring a high level of passenger comfort also ensures traffic safety. Consequently, as indices of track quality at the above speeds, the processed values of the various types of defects are used (obtained from the values recorded by the recording vehicle). Most characteristic of these processed values is the standard deviation of a particular type of defect over a specified length (usually 200÷300m), which reliably simulates variations in the values of the defect in question, (304).

It should be noted that on medium-speed tracks both the absolute and the processed values are used as indices, the former more often.

Recording vehicles are equipped, nowadays, with computer software which, in addition to the results of recording, can provide the following: ranking of data according to severity, comparison with results of previous recordings, comparison with limit values, (see section 16.6), specified by standards, etc.

More recent systems of recording track defects have, in addition to previously mentioned material, equipment with the capacity to also measure rail characteristics, such as corrugation severity, rail head damage, side wear, etc.

16.6. Limit values of track defects

16.6.1. Limit values for high-, rapid- and medium-speed tracks

For each speed, two limit values are specified, (301), (304):

- *alert values* of track defects, which, when reached, require scheduling the intervention of track maintenance teams. These values will be designated as L_{inf},
- *upper* values of track defects, which should not be reached, otherwise the deterioration in track quality may become irreversible. Upper values will be designated as L_{sup}.

The decision to realize maintenance works should be taken between the limits L_{inf} and L_{sup}.

Tracks are usually classified in four categories, depending on train speed, as follows, (297):

I. high-speed tracks (V > 200 km/h),
II. rapid-speed tracks (140 km/h < V ≤ 200 km/h),
III. medium-speed tracks (100 km/h < V ≤ 140 km/h),
IV. low-speed tracks (V < 100 km/h).

According to the French railways, the standard deviation for longitudinal, transverse, and horizontal defects and for track categories I, II, III is given in Table 16.1.

Table 16.1.

Standard deviation (mm) of longitudinal, transverse, and horizontal defects for a length of 300m for various categories of tracks, according to the French railways, (304)

Track category	I V>200km/h		II 140km/h<V≤200km/h		III 100km/h<V≤140km/h	
	L_{inf}	L_{sup}	L_{inf}	L_{sup}	L_{inf}	L_{sup}
Longitudinal defect LD	0.6	0.8	0.7	1.0	0.8	1.2
Transverse defect TD	0.4	0.6	0.5	0.7	0.6	0.8
Horizontal defect HD	0.9	1.4	1.0	1.6	1.2	2.0

16.6.2. Limit values for medium- and low-speed tracks

As discussed in section 16.5, decisions about track maintenance for medium- and low-speed tracks are taken in relation to the maximum values of defects as recorded by the recording vehicle. Values of track defects that necessitate, when reached, intervention and track maintenance are specified as *intervention limits*.

Table 16.2 illustrates absolute values of track defects for intervention of maintenance teams according to German, French and Dutch railways, and the UIC.

Table 16.2.
Intervention limit values of the various track defects for maintenance works according to some railways[*]
(3m basis for track twist, 10m basis for all other track defects)

Track speed (km/h)	Longitudinal defect LD (mm)				Transverse defect TD (mm)		Horizontal defect HD (mm)			
	DB	SNCF	NS	UIC	DB	NS	DB	SNCF	NS	UIC
140÷160	9	12	10	10	6	7	9	9	5	8
120÷140	11	13	11	10	7	7	11	10	6	8
100÷120	11	15	11	12	7	7	11	12	6	10
80÷100	11	17	12	12	7	8	11	14	8	10
60÷80	14	19	12	16	11	8	14	17	8	14
<60	14	21	12	16	11	8	14	20	8	14

Track speed (km/h)	Track twist (mm/m)			Track gauge (mm)		
	SNCF	NS	BR	DB	NS	BR
140÷160	3.0	3.3	3.3	-3/+20	-4/+6	-2/+4
120÷140	3.0	3.3	3.3	-3/+20	-4/+9	-2/+4
100÷120	3.3	3.3	3.3	-3/+20	-4/+9	-2/+4
80÷100	4.0	3.3	3.3	-3/+20	-4/+9	-2/+4
60÷80	5.0	3.3	3.3	-3/+20	-4/+12	-2/+4
<60	6.0	3.3	3.3	-3/+20	-4/+12	-2/+4

16.6.3. Acceptance values

Acceptance values are values of track defects that can be left after the execution of track maintenance, since it is practically impossible (and very costly) to attain a geometrically perfect track.

Thus, after the execution of maintenance works, some low values of track defects can be left, as long as geometrical characteristics of the track comply with the values illustrated in Table 16.3.

---

[*] DB: German Railways, SNCF: French Railways, NS: Railways of Netherlands, UIC: International Union of Railways, BR: (former) British Railways (today Network Rail).

16.6.4. Emergency values

If track defects surpass locally *emergency* (or immediate action limit) values as illustrated in Table 16.4, an immediate reduction of permitted speed must be imposed, until maintenance teams intervene and reduce the values of track defects.

Table 16.3.
Acceptance values of track defects after execution of maintenance works

Track speed (km/h)	>250	200÷250	120÷200	>80	<80
Longitudinal defect (mm)	2	3	3	4	5
Horizontal defect (mm)	2	3	3	4	5
Track twist (mm/m)	1	1	1	1.5	1.5
Track gauge (mm)	±2	±2	±2	±3	±3

Table 16.4.
Emergency values of track defects and maximum permitted speed

Emergency values of track defects		Maximum permitted speed (km/h)
Longitudinal defect LD (mm)	Horizontal defect HD (mm)	
LD ≥ 18	HD ≥ 15	140
LD ≥ 20	HD ≥ 17	120
LD ≥ 23	HD ≥ 19	100
LD ≥ 25	HD ≥ 22	80
LD ≥ 27	HD ≥ 25	60
LD ≥ 30	HD ≥ 28	40

16.6.5. Limit values according to European specifications

The European technical specifications for interoperability (TSI) deal with only two track defects: variations of track gauge and track twist, (134).

The limit values of variations of track gauge for an immediate action (according to the TSI) are given in Table 16.5, (134).

The maximum value[*] of track twist according to the TSI is 7 mm/m for track speeds ≤ 200 km/h and 5 mm/m for track speeds > 200 km/h. As the base-length for recording track twist may vary from one technique to another, European specifications set the limit value of track twist in relation to the base-length of measurement, (Fig.16.7), (134).

Furthermore, the TSI prescribe that variations between nominal and real value of cant should be for standard gauge tracks (e=1,435 mm) at maximum ±20 mm, (134).

---

[*] Maximum values of TSI refer to immediate action values, whereas values of Table 16.2 refer to intervention limits for the scheduling of preventive maintenance.

Table 16.5.
Limit values of variations of track gauge according to
the European Technical Specifications for Interoperability, (134)

Speed V (km/h)	Limit values of variations of real track gauge from nominal track gauge (mm)	
	Minimum track gauge (mm)	Maximum track gauge (mm)
V≤120	- 9	+ 35
120<V≤160	- 8	+ 35
160<V≤230	- 7	+ 28
V>230	- 5	+ 28

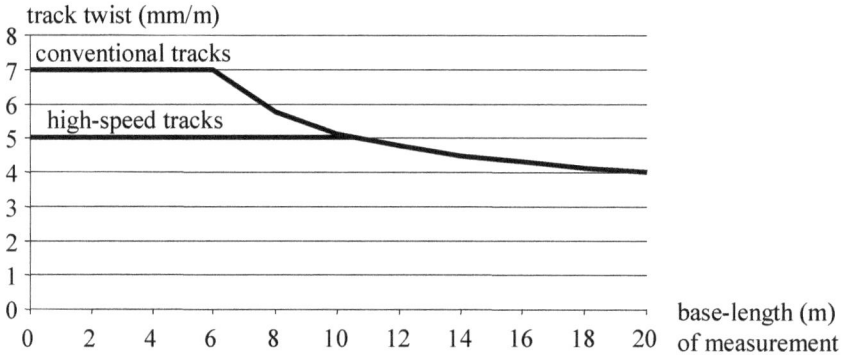

Fig. 16.7. Limit values of track twist in relation to the base-length of mea-surement, according to the European Technical Specifications for Interoperability for high-speed and conventional tracks, (134)

16.7. Progress of track defects

We will examine now how an initial track defect will evolve as a function of traffic load. Knowledge of the way that track defects evolve may help a timely scheduling of remedial action by track maintenance teams, before the limits previously given are exceeded, (292).

16.7.1. Longitudinal defect

A series of tests and statistical analyses has shown that a defect, present in a track after maintenance, progresses rapidly up to a critical traffic load on the order of 2 million tonnes, beyond which defect progress is much slower, (301), (305). This means that up to this traffic load the track has not been fully stabilized and shows signs of instability. Transverse resistance of a track after maintenance is only 50% of its value when fully stabilized.

16.7.1.1. Mean settlement of track

The evolution of the mean settlement of track is given by the following semi-empirical formula, (305):

$$m_e(T) = a_1 + a_0 \cdot \log \frac{T}{T_r} \qquad (16.3)$$

where:

T_r = $2 \cdot 10^6$ tonnes,

a_1 : mean settlement for a traffic load T_r,

a_0 : settlement increase rate (mm/decade), mainly depending on subgrade quality, with mean values $2 \div 6$ mm/decade.

The ratio $\dfrac{a_0}{a_1}$ illustrates the slow progress of the mean settlement after a traffic load of $2 \cdot 10^6$ tonnes is reached and was found to have values $0.25 \div 0.70$,

$$\frac{a_0}{a_1} = 0.25 \div 0.70 \qquad (16.4)$$

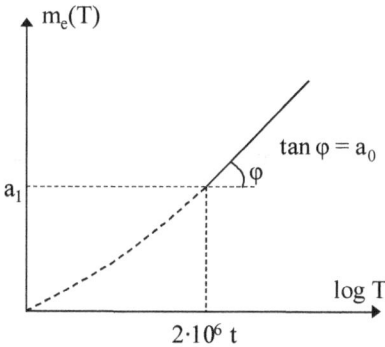

Fig.16.8. Progress of the mean value $m_e(T)$ of the track settlement as a function of traffic load

16.7.1.2. Standard deviation of longitudinal defects

For medium- and high-speed tracks, the standard deviation of longitudinal defects $sd_{LD}(T)$ is used as a measure of the quality of track. A series of statistical studies has yielded the following semi-empirical formula (of a form similar to formula 16.3), (305):

$$sd_{LD}(T) = c_1 + c_0 \cdot \log \frac{T}{T_r} \qquad (16.5)$$

where:

c_1 : standard deviation of longitudinal defects for a traffic load $T_r = 2 \cdot 10^6$ t,

c_0 : rate of increase of standard deviation of longitudinal defects as a function of traffic load, with mean values $0.1 \div 0.2$ mm/decade.

16.7.1.3. Interval between maintenance sessions

Let sd_{LD}^{lim} be the limit value of longitudinal defects, specified by the limits set in section 16.6. From equation (16.5) we deduce that the limit traffic load T_{lim} between two successive maintenance sessions will be:

$$T_{lim} = 2 \cdot 10^6 \cdot 10^{\left(\frac{sd_{LD}^{lim} - c_1}{c_0}\right)} \tag{16.6}$$

Since the parameter c_0 is almost constant, the determining factors for the time interval T_{lim} between two successive maintenance sessions are the terms sd_{LD}^{lim} and c_1, the latter amounting to the track condition after maintenance. An increase of time T_{lim} is therefore possible by improving the initial condition of the track after maintenance, i.e. by a better quality of track maintenance works.

In the case of medium- and low-speed tracks, mean values of longitudinal defects are used, instead of the standard deviation of the longitudinal defects, while the form of the above equations remains the same.

16.7.2. Transverse defect

Transverse defects have a pattern of progress similar to equation (16.5). The evolution of the standard deviation of transverse defects is given by the following formula:

$$sd_{TD}(T) = u_1 + u_0 \cdot \log\frac{T}{T_r} \tag{16.7}$$

where coefficients u_1 and u_0 (with mean values for u_0: $0.1 \div 0.4$ mm/decade) are defined similarly to coefficients c_1, c_0 of equation (16.5), (305).

16.7.3. Horizontal defect

Track loading on the horizontal plane differs from vertical loading in two main respects:
♦ traffic load effects are much more irregular and discontinuous,
♦ stresses developed should, for safety reasons, remain within elasticity limits.

Like other types of defects, horizontal defects progress relatively fast for an initial traffic load T_r on the order of $2 \cdot 10^6$ t and thereafter slow down considerably. Their evolution may also be approximated by a semi-logarithmic formula in relation to traffic load, which, however, in many instances has

442

shown deviations and a large dispersion. The following formula has been suggested for the evolution of the mean value of the horizontal defects, (305):

$$m_{HD}(T) = d_1 + d_0 \cdot \log \frac{T}{T_r} \tag{16.8}$$

where coefficients d_1, d_0 are defined as in equation (16.3), with mean values $d_1 = 0.6 \div 1.0$ mm and $d_0 = 0.15 \div 0.30$ mm/decade.

The ratio $\dfrac{d_0}{d_1} = 0.2 \div 0.3$, which illustrates how slow the progress of the horizontal defects is after a traffic load $T_r = 2 \cdot 10^6$ t.

16.7.4. Gauge deviations

Gauge deviations mainly depend on subgrade and rolling stock type and therefore their evolution is difficult to determine in terms of the various parameters of the track.

16.7.5. Track twist

Evolution of track twist is also of semi-logarithmic form:

$$sd_{twist}(T) = g_1 + g_0 \cdot \log \frac{T}{T_r} \tag{16.9}$$

where coefficients g_0 and g_1 (with mean values for g_0: $0.2 \div 1.0$ mm/m/decade) have a rather large dispersion and are defined as those of equation (16.5), (305).

16.8. Mechanical equipment for maintenance works

Modern railway technology has a panoply of maintenance machines, of which the following can be highlighted, (289), (298):

i) Heavy leveling and lining tamping machines, (Fig. 16.9), which should be used, to the extent feasible, only in scheduled maintenance operations, where leveling and lining operations are systematic. A necessary condition for their use is that the ballast be sound, free of soil contamination, of proper granulometric size, and of adequate mechanical strength. The performance of such equipment averages about $500 \div 800$ m of length of track per hour, although theoretical rates given by constructors have higher values, and even up 3,000 m for express tamping machines, (289).

Tamping is the operation whereby track defects are rectified and includes the following stages:

– A surveying team initially determines the elevation or horizontal correction, which should be given to the track. Alternatively, and without the need of a surveying team, modern recording vehicles can provide the geometrical rectification at the longitudinal and horizontal level that should be given during tamping.

–   The tamping machine makes a first tamping on the track. It moves the track left, right, or up, depending on the track defects which should be rectified. It lowers the tamping blades and compacts the ballast under the sleeper, (Fig. 16.9).

–   The recording vehicle passes and measures the remaining defects (acceptance tolerances), (see Table 16.3). However, modern tamping machines have equipment to record values of remaining defects after tamping, without the need of a recording vehicle.

In addition to tamping machines, track maintenance also requires other machinery, such as:

♦   Ballast compacting and stabilizing machines, which follow the tamping machines and contribute to the increase of the stability and transverse resistance of the track.

♦   Ballast profiling machines, which give the ballast the appropriate cross-section profile, (Fig. 16.10).

♦   Ballast cleaning machines, which are used when small size (<22 mm) ballast grains are more than 30% of the total ballast volume. The ballast cleaning machine removes to a depth of 25 cm beneath the sleeper

Fig. 16.9. Tamping machine in the course of track maintenance

Fig. 16.10. Ballast profiling machine

all ballast stones smaller than 35mm. The performance of cleaning machines is around 400 m of track per hour, (289).

♦ Formation rehabilitation machines. As discussed in section 9.1, a good design of the subgrade should result in no need to intervene during the renewal of ballast. However, improvement of the bearing capacity of the subgrade is necessary in some cases and is conducted with the formation rehabilitation machines, which insert in the top of the subgrade an additional formation layer consisting of a blend of sand and gravel.

ii) Light (portable) tamping machines, the use of which also requires a sound ballast material. Since this equipment is easily transported, it is highly flexible and should be employed in:

- limited operations on discontinuous sections of the track, of up to about 300 m in length, where the use of heavy machinery would be inappropriate and expensive,
- repetition of tamping on particular sections of the track,
- leveling of switches and crossings,
- (as an exception only) for the systematic maintenance of track sections, where heavy machinery is not available or is unable to be used in that particular track.

iii) Hand tools such as the fork, the pickaxe, etc., which are now considered virtually obsolete, but can still be used in the following cases:

♦ on track sections with ballast in a state of advanced disintegration, where mechanical tamping is not possible without adding new sound ballast,

♦ in the case of isolated, local, and urgent repeat tamping, where the extent of the operation does not justify the use of even light ballast tamping equipment.

Maintenance of track with the use of only manual means may result in ten times or more man-hours compared to fully-mechanized maintenance (289).

During scheduled maintenance sessions, the track equipment is inspected and any damage is rectified. For this purpose, the following equipment is used:

– bolt and screw (both screwing and unscrewing) machines,
– machines for drilling holes into timber sleepers,
– rail cutting machines,
– machines for drilling holes in rails, etc.,

– rail grinding machines, in order to smooth irregularities at the rail surface, which can be a result either of track defects (short-pitch corrugations, see section 10.9.4.4) or of train operation. The grinding of rail can be achieved with the help of either rotating stones or of stones oscillating longitudinally, (290). Grinding of rail may be necessary at any moment between successive maintenance sessions; average value of grinding of rails in Europe is once per 1÷3 years of operation, in relation to traffic load.

16.9. Scheduling of maintenance operations

Railways are complex systems and consist of discrete subsystems, the interactions of which are neither simple and obvious nor easy to anticipate. Figure 16.11 illustrates a block diagram of the entire maintenance procedure and the parameters involved. In this chart two processes are apparent, each opposing the other, (298), (304):

- The traffic process, which, by the track-rolling stock interaction, tends to increase track defects and to destabilize the system as a whole.
- The maintenance process, which strives to reduce defects and restore the track to its previous good condition.

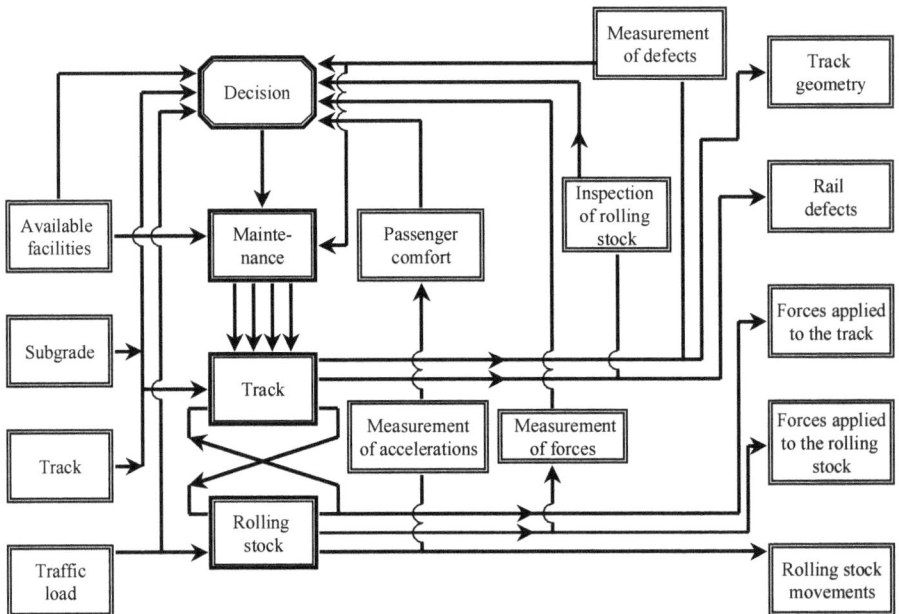

Fig. 16.11. Block diagram of the interactions between the various subsystems and parameters determining track maintenance operations, (304)

The two above mentioned processes should be in equilibrium, which incidentally is the basic purpose of maintenance works. This equilibrium can be achieved only by timely and rational scheduling, which:

♦ is based on systematically sorted information from past maintenance operations,

♦ optimizes the use of the mechanical equipment,

♦ assigns priorities correctly along the railway infrastructure, on both regional and local levels.

Figure 16.12 illustrates a flow chart of the successive stages of track maintenance and renewal within the frame of preventive maintenance. In order to make a more efficient use of both the human and mechanical resources, it is necessary to draw up such diagrams both at the strategic management level and during maintenance works. Telematics can also contribute to a rational scheduling of track maintenance, (294).

For main tracks, the tamping of ballast is conducted once per 3÷5 years of operation, while ballast replacement and renewal is conducted once per 15÷30 years, depending on the strength and mechanical properties of ballast, and the traffic of the line.

16.10. Technical considerations for track maintenance works

When performing maintenance operations, the following considerations should be kept in mind:

– Leveling adjustment is mandatory with any horizontal operation, no later than the next day and in any case before the track stabilizes.

– If the leveling adjustment is performed by heavy machinery, the machine should perform horizontal adjustment simultaneously.

– If the leveling adjustment is performed by heavy machinery, no additional elevation adjustments should be made before track stabilization, which is brought about by line traffic.

– In the event that, after the elapse of the stabilization period, defects not completely rectified are still found, supplementary adjustments should be performed by light equipment, without having to relift any elevated sections of the track.

As we have previously analyzed, a sensitive period follows after maintenance (until a traffic load on the order of 2 million tonnes), during which defects evolve rapidly. On lines with medium-traffic load (group UIC 4), this period corresponds to about 1÷4 months. On lines with high traffic (groups UIC 1, 2, 3) this period corresponds to 15÷40 days. During this time, the track should be under continuous and careful surveillance, so as to monitor the progress of the various defects and make timely local

447

interventions with light (or heavy, if necessary) machinery, whenever defect accumulation is unusual or excessive. The sensitive period after maintenance is therefore the key to track longevity and to the reduction of future maintenance expenses. If the measures mentioned above are not taken during this period, problems will frequently arise later and increased efforts will be required to restore track geometry, (297).

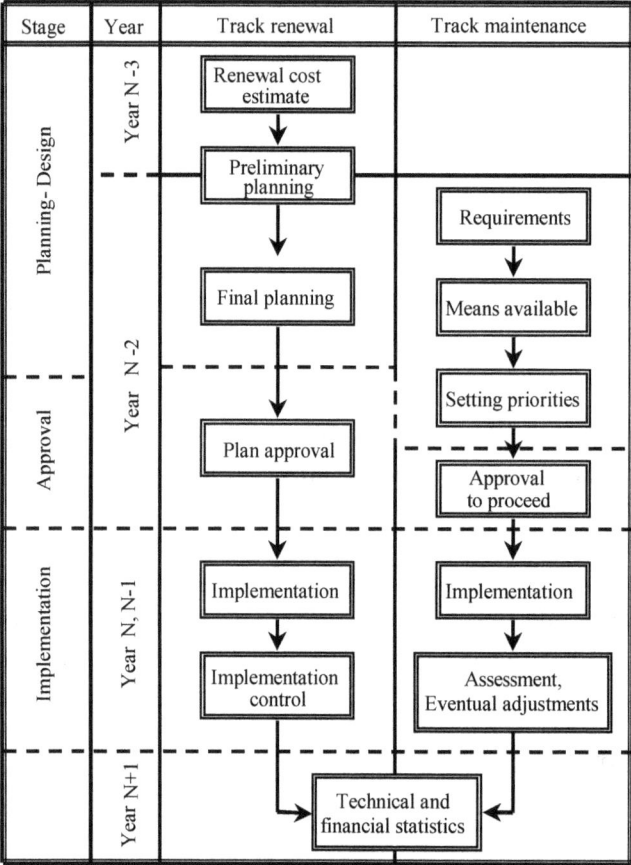

Fig. 16.12. **Flow chart of the various planning and implementation stages of track renewal and maintenance operations**

16.11. Optimization of track maintenance expenses – The RAMS analysis

16.11.1. Optimization of track maintenance expenses

The maintenance of track (but also of rolling stock and signaling systems) is essential for safe and comfortable rail transport, which should be realized at the lowest possible cost. As analyzed in sections 6.1 and 7.2, all problems of railways should be examined within a systems approach and during the life cycle of railways, which may be as long as 50 years for rails and concrete sleepers or shorter for other components. Rationalization and optimization of partial costs can be achieved if they are integrated within the following two concepts:

- the *life cycle cost* of a rail component, which is understood as the sum of all expenses required to support the specific component from its conception and fabrication, through its operation, to the end of its economic life, (291),
- the *life cycle costing*, which is a methodology of systematic economic evaluation of life cycle costs of all rail components and procedures over a period of analysis.

Track maintenance represents 25÷35% of operation costs of a unified railway company.

16.11.2. The RAMS analysis

During recent years, a global conception of the railway product has been developed and is known under the initials *RAMS*. The RAMS analysis consists of the following:

- *Reliability*, understood as the probability that a rail component can perform its required functions under given conditions and for a given time interval,
- *Availability*, understood as the ability of a rail component to perform its required functions under given conditions at a given time, provided that the required external resources (human, funding) are ensured,
- *Maintainability*, understood as the probability that a maintenance action of a rail component can be carried out under prescribed procedures and resources within a given time interval,
- *Safety*, understood as the freedom from unacceptable risks of harm.

16.11.3. Track maintenance by own resources or by outsourcing

The maintenance of track was conducted until some decades ago almost exclusively by railway personnel, with the use of equipment that belonged to the railway company. However, in recent years, some infrastructure managers prefer to outsource part or all the maintenance activity of track, (see also section 6.6.3). If such a procedure of outsourcing is decided upon, special attention should be paid to the following:

- ♦ critical terms of the contract of outsourcing (such as delivery date, unpredicted events, real total final cost, etc.) should be analyzed in detail by using sensitivity analysis,

♦ make sure that both total cost (over the whole life cycle) and quality of track are better for the infrastructure manager, when outsourcing the specific activity,

♦ quantify the probability that the service will not be delivered in accordance with the requirements and foresee alternative solutions in the case of delays, failure, etc.,

♦ identify all eventual risks, try to limit them, and provide alternative solutions.

16.12. Condition-based maintenance, Big data, and Artificial Intelligence

Condition-based maintenance is a maintenance strategy which is based on measuring (either continuously or at certain intervals) the condition of any component (e.g. ballast) of rail infrastructure; in relation to these measures it is possible to assess how close to failure are strength reserves of the material or the component under consideration and to take in time the appropriate measures and actions, so as to avoid failure.

Preventive maintenance is based on regular inspection and records of the track and on previous engineering experience, both of which lead railway engineers to provide a maintenance plan with details on maintenance and replacements that should be conducted. In the case of preventive maintenance, the failure of any material or component of the track is not a random event, but it can be predicted within a particular time span. In this way, however, it is not ensured that any item of the track has reached its ultimate resistance or operation capabilities.

Condition-based maintenance is not in conflict with preventive maintenance, which has been the strategy used by many railways around the world for decades. Condition-based maintenance can be combined with preventive maintenance and thus lead to *predictive* maintenance: the time schedule of maintenance is not fixed, it changes in relation to the actual situation of the track; all track components are replaced while working; no track component is replaced before it reaches its limits; some maintenance may be programmed out of schedule. Predictive maintenance proceeds in the following way:

• establish guidelines for the limit values of track defects and for the mechanical characteristics of track materials,

• install monitoring sensors or conduct systematic records of the condition of the track, with the use of recording vehicles but also with the help of either recording equipment placed on-board of commuter trains running on the track or wayside equipment,

• all these recorded data are part of what is known as big data,

- laser scanners and color cameras transform the recorded data into figures and cross-sections of the actual situation of the track,
- successive series of big data constitute what is known as cloud data: sets of data describing accurately at different time intervals the situation of any track component or material,
- cloud data are continuously comparing the situation of the track with limit values prescribed in guidelines,
- consecutive waves of cloud data are integrated as elements of a unified set and can constitute what is known as internet of things: the course and de-tailed characteristics of any item of the track are accurately described,
- recorded data or big data or cloud data or internet of things can be intro-duced (with the help of machine learning techniques) to artificial intelli-gence algorithms and more particularly to image analysis. The computer compares continuously what really exists with the limit situation that should not be overridden and in addition it provides a forecast for the like-ly evolution from the existing to the limit situation, (295),
- the time schedule of preventive maintenance is continuously adapted and the computer calculates the actualized time schedule and the required equipment and staff,
- in this way, the lifetime of every asset and component of the track is opti-mized, maintenance and replacement are conducted at the right moment, safety during maintenance works is increased, quality conditions (speed, low values of acceleration, low noise emissions, etc.) of rail transport are greatly improved.

Thus, predictive maintenance (i.e. a combination of preventive mainten-ance with condition-based maintenance) can become a basic component of a more competitive, less costly, and efficient railway.

16.13. Track maintenance, vegetation, and weed control

The issue of vegetation, appearing along the track, was analyzed in section 9.16. Weeds can cause serious detrimental effects on the ballast and the sub-grade by:

- contaminating the ballast with dirt and vegetation debris, which affect free drainage,
- accelerating the decay of components such as concrete sleepers, not only by chemical action but by the expansion of roots in cracks and crevices,
- by obscuring the track, and thus defects normally observed by the naked eye would not have been seen on routine visual inspections.

Arsenic-based chemicals were introduced for weed control in the 19th century and continued to be used in varying degrees in some countries until the 1930s, but are not used today. During the 1930s, sodium chlorate was introduced as a chemical weed control and with the addition of fire depressants (such as calcium chloride) was made reasonably safe without any toxic effect. Since 1950, herbicides have been extensively used, particularly hormone selective weedkillers.

Application rates are 1÷20 kg/hectare, the average being 4÷8 kg/hectare. Herbicides must be applied evenly over the area, at the lowest possible volume per hectare if in liquid form, at the greatest speed, with maximum safety, and taking all measures to prevent any environmental harm.

Spray trains can also be used. Their capacity is reported to reach 300 km/day; however, the daily average observed over a four month season in the UK was 130 km/day, (302).

Herbicides are harmful for the environment. Thus, railways reduced greatly the quantities of herbicides they use for weed control. More specifically in Europe, railways consume around 400 tonnes of herbicides per year, against 130,000 tonnes of total sales of herbicides per year in this continent, (293).

Application rates of herbicides per track kilometer are 0.50 kg and costs amount to 300÷400 € per kilometer of track, (293). Herbicides are spread over the track usually with spraying trains, but also with backpack spraying, road-rail vehicles, or with small equipment, (293).

Herbicides are a chemical product for weed control. Another chemical product for weed control is organic acids. Other more environmentally friendly ways to control weed are: hot water, wet steam, and electrical radiation (electro-weeding).

17 Slab Track

17.1. The dilemma between ballasted and non-ballasted track

17.1.1. Advantages and weaknesses of ballasted track

Until the early 1970s, railway tracks were laid on ballast, which is replaced every 15÷30 years, whereas maintenance of track takes place every 3÷5 years. Maintenance and renewal of ballast are conducted under extremely difficult conditions in the intervals between successive trains, usually during the night, and the available time for maintenance or renewal works is usually smaller than 3÷5h.

The increase of speed beyond 200 km/h resulted for ballasted tracks in a disproportionately greater increase of maintenance costs, which for high-speed tracks are almost double that of conventional tracks (V<200 km/h). Although ballasted track has sufficient mechanical characteristics (high transverse resistance, low stresses and settlements), life cycle and maintenance considerations oriented railway managers and engineers to the use of non-ballasted (slab) track instead of ballasted track, particularly for high-speed tracks.

However, the ballasted track has the advantage of ensuring high flexibility of the track, much lower construction cost, possibility to easily rectify track defects or differential settlements, absorption of dynamic effects, and emission of lower levels of noise (compared to slab track), (314).

17.1.2. The non-ballasted track

In non-ballasted tracks, a concrete slab (reinforced or prestressed) or an asphalt layer replaces the ballast; the rail can lie either directly on the slab or on sleepers, which in their turn lie on the concrete slab. Below the concrete slab, a ballast concrete layer and an anti-frost layer are interposed, (Fig. 17.1).

Thus, the non-ballasted track uses a series of successive layers (in the same way as ballasted track) in order to gradually reduce stresses from rail to subgrade, so that stress values at the subgrade are lower than its bearing capacity.

453

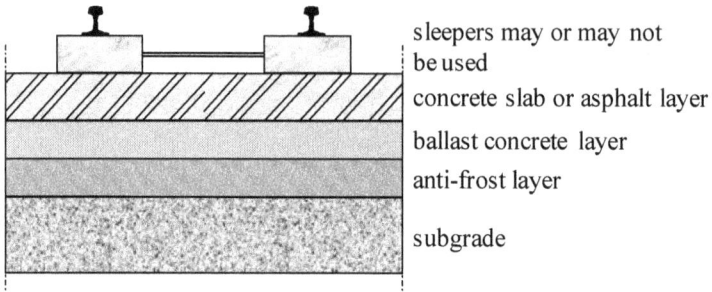

Fig. 17.1. Non-ballasted track

The basic advantage of non-ballasted track is its low maintenance cost and excellent and uninterrupted operation conditions, in comparison to ballasted track. In addition, as the concrete slab has a lower thickness compared to ballast, the non-ballasted track results in a reduction of the required cross-section for tunnels, something that reduces the overall tunnel construction cost. Slab track has a long lifetime (50÷60 years), more than double compared to ballast (15÷30 years). Increased transverse resistance and passengers' comfort are also among the advantages of slab track.

However, non-ballasted track is not free of disadvantages, the most important being its higher construction cost. Savings in maintenance, however, can recover this additional cost of the slab track within a number of years, depending on the economic conditions of each country, (see below section 17.8).

Once the slab track is installed, it is very difficult to overcome eventual differential settlements and, therefore, the use of slab track must be restricted to areas where a good and stable subgrade quality can be provided. However, when slab track is used, noise levels are higher compared to ballasted track.

A question, which has not been answered yet, is what will happen at the end of the life cycle of a slab track (50÷60 years) and how can a fresh concrete slab replace an old one without interrupting traffic.

17.1.3. First trials, tests, and evolution of slab track techniques

Among the first trials for slab track techniques, tests were run in the former West Germany in 1959 and in Japan in the early 1960s. Slab track was investigated by the Research Department of the UIC in a test track constructed in the United Kingdom in 1967 for this purpose. The first slab track was constructed at the railway station of the city of Rheda in the former West Germany in 1972. Since the 1980s, an increasing number of kilometers of slab track were constructed in many countries, such as Germany, Japan, the Netherlands, Italy, South Korea, China, etc.

Concerning applications of slab track in high-speed lines, these were realized in Japan after the mid-1970s and were followed by applications in the high-speed lines Paris-London (mid-1990s), Cologne-Frankfurt (2002), Brussels-Amsterdam, and extensively in China, where almost 29,000 kilometers of new high-speed lines were built on concrete slab until 2020. Figure 17.2 illustrates the evolution worldwide of kilometers of high-speed lines built on slab track, (307).

Kilometers of lines on slab track

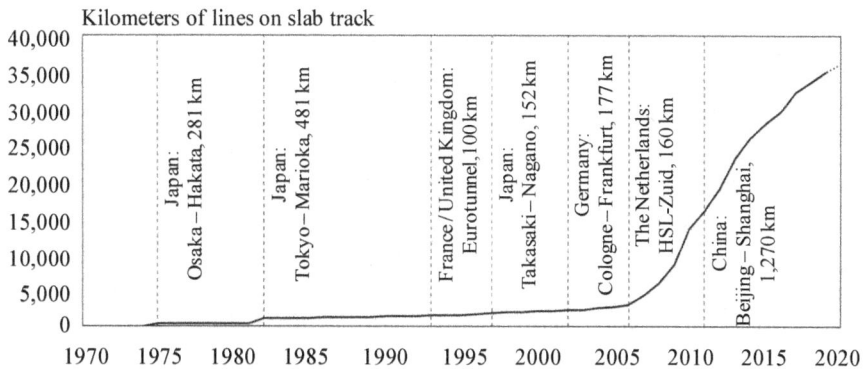

Fig.17.2. Evolution of the number of kilometers of slab track worldwide in high-speed railways, (307)

17.2. A variety of forms of non-ballasted track

The various forms of non-ballasted track can be classified as follows:

- slab track with sleepers, which are embedded in a reinforced concrete slab (Rheda technique) or in a monolithic concrete slab (Züblin technique) or directly on an already constructed track (Stedef technique). Most of these techniques are named after the places where they were applied for the first time,
- prefabricated slab track without sleepers (Shinkansen and Bögl techniques),
- embedded rails in a concrete slab,
- sleepers placed on an asphalt layer.

Slab track systems with sleepers embedded on concrete (Rheda, Züblin) are more noisy than systems where sleepers are positioned directly on the concrete slab (Stedef).

17.3. Slab track with sleepers

17.3.1. The Rheda technique

In the Rheda technique, (Fig. 17.3), an anti-frost layer of 30 cm is placed on top of the subgrade and above it a 30 cm thick ballast concrete layer, upon which a reinforced concrete trough of a thickness of 18 cm is placed. Mono-block or twin-block sleepers, spaced 65 cm apart, are embedded, with the use of filling concrete, in the concrete trough.

The concrete trough has a mechanical compressive stress of 300 kg/cm^2 for the cylinder test (equivalently 370 kg/cm^2 for the cube test). The ballast concrete layer has a mean compressive stress of 150 kg/cm^2. Its granulometric composition contains grains greater than 2 mm at 55÷85% of the total weight and grains smaller than 0.063 mm at less than 15% of the total weight.

In recent evolutions of the Rheda technique, the sleepers have a hole through which steel bars are placed longitudinally in order to avoid loosening. The Rheda technique has been used extensively in open track, tunnels, and bridges in Germany.

The recording of track defects on a slab track of the Rheda type gave smaller values of the various track defects and the track gauge almost unchanged, compared to the ballasted track.

The Rheda technique is suitable for speeds of 300 km/h and even more, has a rather simple assembly, but is noisy and any reparation or renewal has a high cost, (311).

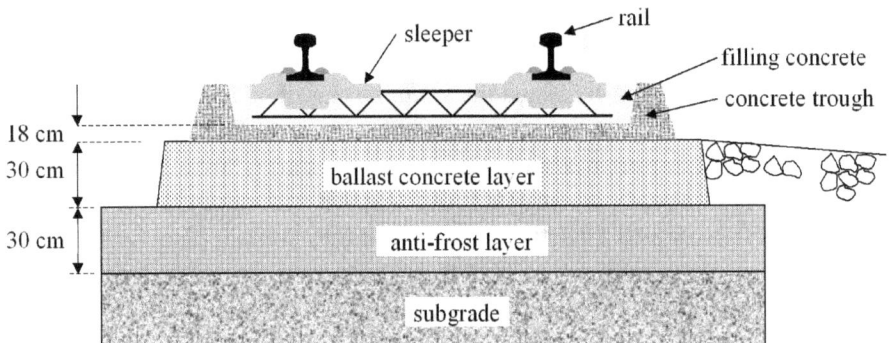

Fig. 17.3. The Rheda technique

17.3.2. The Züblin technique

The Züblin technique, (Fig. 17.4), differs from the Rheda technique in that the monoblock or twin-block sleepers are embedded directly in a monolithic 20 cm thick concrete slab with a mechanical compressive stress of 300 kg/cm^2 for the cylinder test (equivalently 370 kg/cm^2 for the cube test). In the Züblin technique, sleepers are positioned in the fresh concrete, whereas in the Rheda technique the concrete slab is already constructed and sleepers are positioned on the concrete slab with the use of filling concrete. The Züblin technique has been extensively used in Germany, Netherlands, and elsewhere.

The Züblin technique has similar technical performances with the Rheda technique (suitable for speeds of 300 km/h and even more, noisy, difficult reparation), however, it is more automated and permits a more accurate construction.

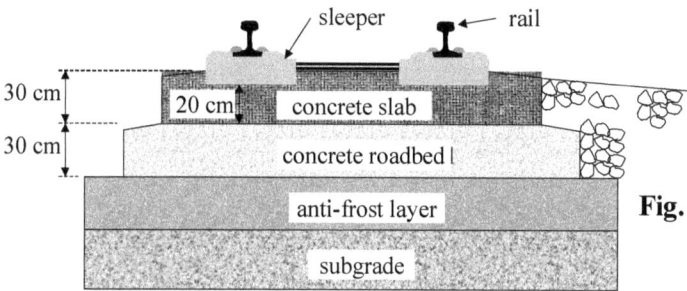

Fig. 17.4. The Züblin technique

17.3.3. The Stedef technique

In the Stedef technique, (Fig 17.5), sleepers are positioned on an already constructed concrete slab. A rubber layer 4.5mm thick is placed between the sleeper and the slab, (314). The Stedef technique has been used in tunnels in France and in the Channel Tunnel.

The Stedef technique is more costly than the Rheda and Züblin techniques, but is less noisy and any replacement of failed parts is easier.

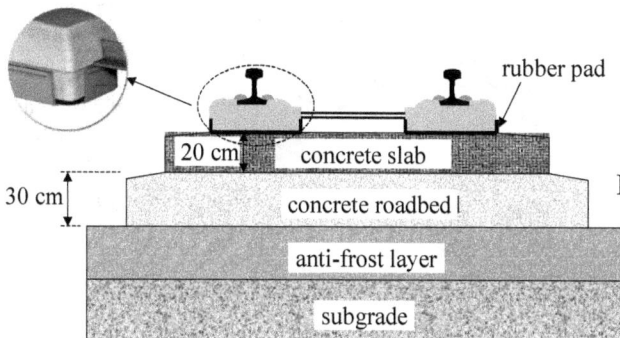

Fig. 17.5. The Stedef technique

457

17.4. Slab track without sleepers

In slab track without sleepers, the rails are positioned directly on prefabricated prestressed concrete slabs, (Fig 17.6). In order to absorb the increased dynamic effects, an asphalt roadbed of a thickness of 40 cm is interposed between the slab track and the roadbed. This technique has been extensively used in Japan and is known under the name Shinkansen technique. The horizontal dimensions of the concrete slab are 4.95 m × 2.34 m and its thickness is 16 cm in tunnels and 19 cm in open track. As illustrated in Figure 17.7, cylindrical stoppers are used in order to prevent lateral and longitudinal movements of the track. The Shinkansen technique provides a low construction time and an easy replacement of rails and fastenings.

A variation of the Japanese technique is the Bögl technique, (Fig. 17.7), with geometrical dimensions of the prefabricated slab of 6.45 m × (2.55÷2.80) m, a thickness of 20 cm and a compressive stress of 450 kg/cm^2 for the cylinder test (equivalently 550 kg/cm^2 for the cube test). In the Bögl technique, slabs are of reinforced concrete in the longitudinal direction and of prestressed concrete in the lateral one. The Bögl technique provides, similarly to the Shinkansen technique, low construction time.

Adjustment and rectification of track defects is easier in slab track systems without sleepers, compared to slab track systems with sleepers.

In the category of slab track without sleepers, we can also consider the embedded rail technique, (Fig. 17.8), in which between the rail and the concrete slab an elastic material is interposed, (313).

The embedded rail technique has low levels of emitted noise, low construction and maintenance costs, but any adjustment of track defects is costly.

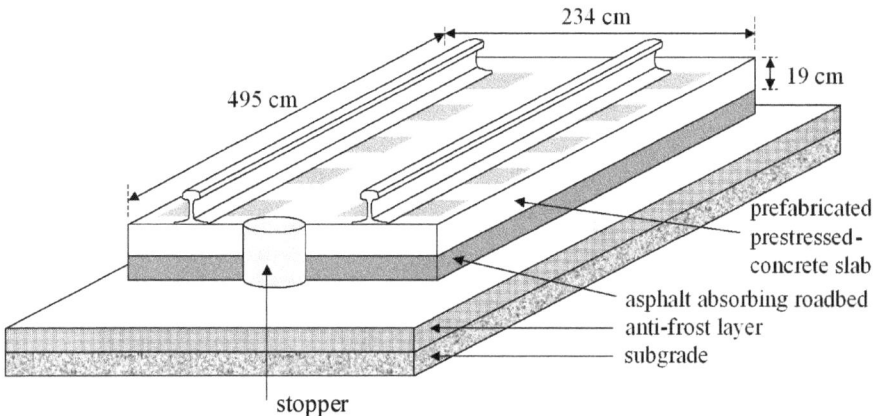

Fig. 17.6. The Shinkansen technique

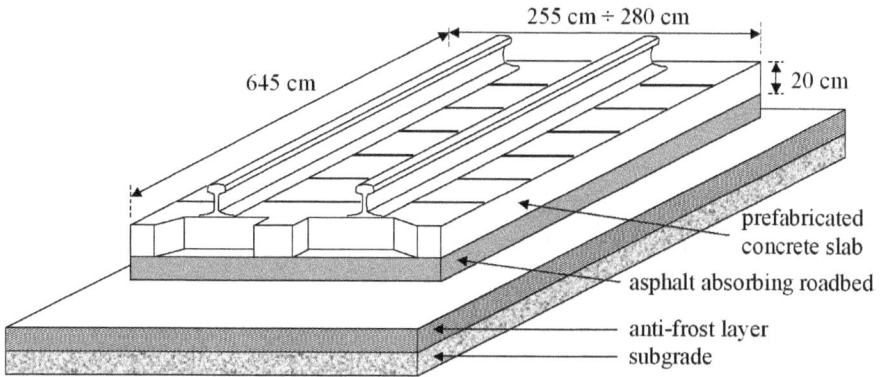

Fig. 17.7. The Bögl technique

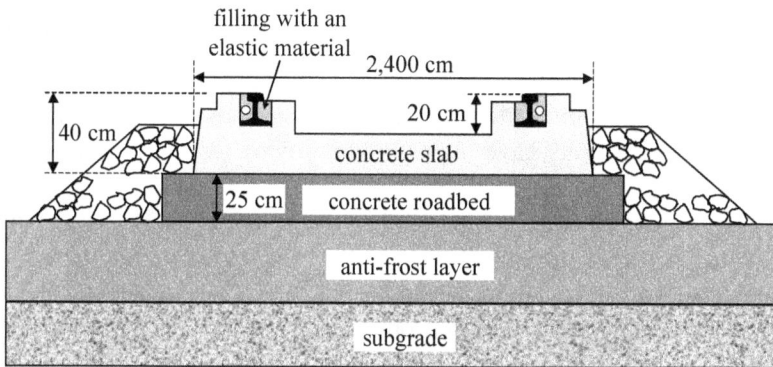

Fig. 17.8. The embedded rail technique

17.5. Non-ballasted track on an asphalt layer

Ballast may be replaced by an asphalt layer, (Fig. 17.9), of a thickness of $25 \div 30$ cm, on top of which sleepers are positioned. Asphalt has the same mechanical characteristics as in road engineering and is placed in-situ with the use of similar equipment.

A track on an asphalt layer is the most flexible among the track techniques without ballast. Thus, it easily permits rectification of any track defects or irregularities, is less noisy, and has low construction cost. The big handicap of a track on an asphalt layer is its low transverse and longitudinal resistance; due to the vulnerability of asphalt in high temperatures, a track on an asphalt layer is unsuitable for temperatures (in the open air) higher than 50°C.

459

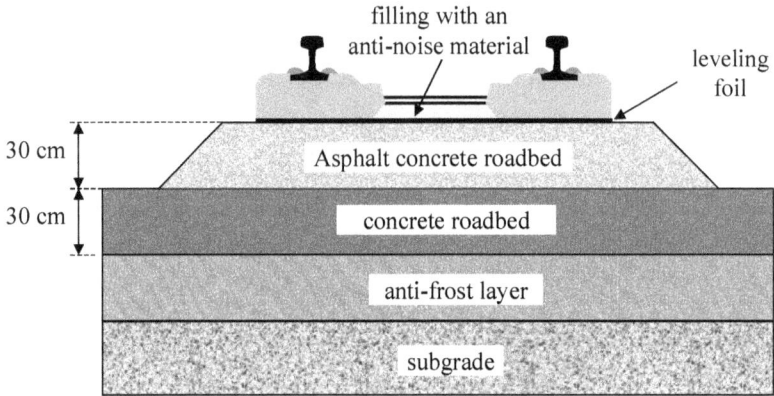

Fig. 17.9. Non-ballasted track on an asphalt layer

17.6. Mechanical behavior of slab track

17.6.1. Application of the finite element method for the modeling of slab track

Traditional methods simulated slab track as a multi-layer system. However, many experimental results have highlighted the need for a more accurate simulation.

The finite element method can be used for the accurate analysis and modeling of the mechanical behavior of slab track, (306), (312). Figure 17.2 illustrates the mesh of such a finite element model, (312). In order to take into account the dynamic effect of the problem, which may be critical for slab track but negligible for ballasted track, the equation of dynamics is used:

$$M \cdot \ddot{U}^{(i)}_{t+\Delta t} + C \cdot \dot{U}^{(i)}_{t+\Delta t} + K_{(t)} \cdot U^{(i)} = R_{t+\Delta t} - F^{(i-1)}_{t+\Delta t} \qquad (17.1)$$

in which:

M : the mass matrix,
C : the damping matrix,
K : the stiffness matrix,
U : the displacement vector,
R : the vector of external loads,
F : the vector of forces exerted on system's nodes,
i : number of iteration,
t : time.

Application of this model has been made in the case of a concrete slab with a maximum compressive strength of 300 kg/cm^2 and a subgrade of good quality (S$_3$).

17.6.2. Stresses and settlements in the case of slab track

The dynamic nonlinear finite element model presented in Figure 17.10 has given the following values for the vertical stresses under the axle load, (312):
- between sleeper and concrete slab: 1.96 kg/cm^2,
- top of the subgrade: 0.60 kg/cm^2.

As far as the vertical settlement under the axle load is concerned, the model has given the following values:
- ♦ top of concrete slab: 0.34 mm,
- ♦ top of the subgrade: 0.30 mm.

Figure 17.11 illustrates the elastic line of the concrete slab.

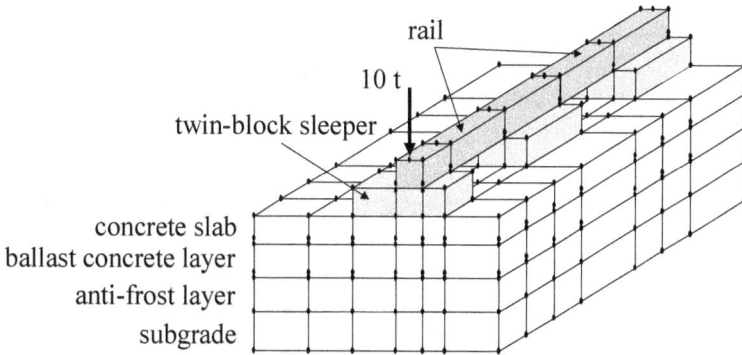

Fig. 17.10. Mesh of a model for slab track with the use of the finite element method, (312)

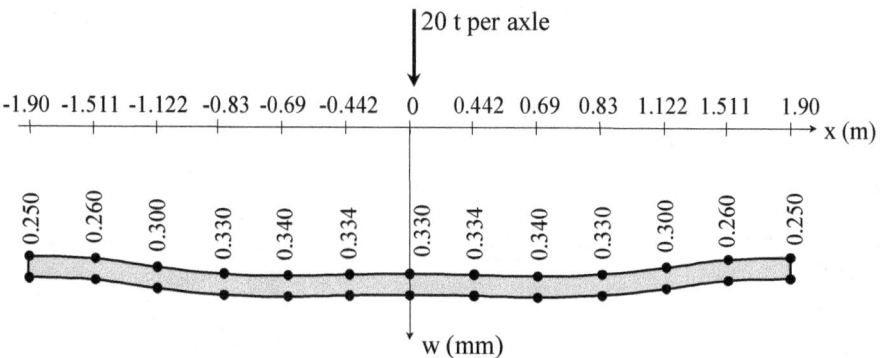

Fig. 17.11. Elastic line of the concrete slab, (312)

17.7. Transition between ballasted and slab track

Slab track has a higher stiffness and a lower flexibility, compared to ballasted track. However, overall track quality and passengers' comfort cannot change from one section of the track to the other. For this reason, a transition zone between ballasted and slab track should be designed as carefully as possible. Each slab track technique has its peculiarities for the transition zone. Figure 17.12 illustrates a transition zone design for the Rheda system, in which:

– the transition zone has a ballasted track section and a slab track section,
– in the transition slab track section, the ballast concrete layer under the slab track is extended from 30 cm to 50 cm,
– in the transition ballasted track section, ballast grains and stones are stuck to each other. In the part AB of this section, (Fig. 17.12), the ballast concrete layer is extended and partially replaces the subballast, whereas in the other part BC the subballast layer is extended,
– two auxiliary rails are placed along the transition zone in the inner part of each running rail,
– the anti-frost layer is also extended to a significant part of the transition zone.

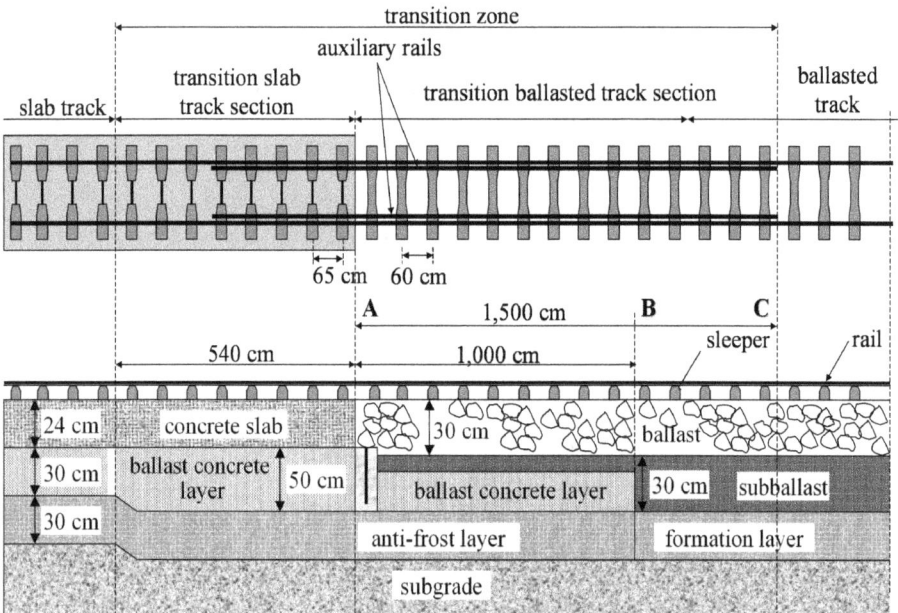

Fig. 17.12. Transition zone between ballasted and slab track

17.8. Costs and construction rates of slab track

Slab tracks are designed with a top speed of 250÷350 km/h for high-speed railway lines and 160 km/h for metro and suburban railway systems and an estimated life cycle of 50÷60 years. In Germany, construction cost of the slab track was 680 €/m for the Rheda technique and 575 €/m for the Züblin technique, against a cost of 365 €/m for ballast construction (all values of year 2008). These costs do not include the increased earthwork and subgrade costs for slab track. The cost of non-ballasted track on an asphalt layer is around 630 €/m. In France, construction costs for slab track are reported to be double that of ballasted track.

The consideration of a great number of slab track systems (principally the Rheda and Züblin techniques) in Germany revealed maintenance costs to be approximately 10% of that of ballasted track, whereas in Japan (Shinkansen technique) maintenance costs of slab track amount to 20÷30% of maintenance costs of ballasted track, (308), (310).

The Cologne-Frankfurt track, constructed in 2002 entirely on a concrete slab with the use of the Rheda technique, had an average construction cost of 26.7 million €/km of track, (see also section 5.2.2).

Figure 17.13 illustrates the comparative construction and maintenance costs for ballasted and slab track for Japan.

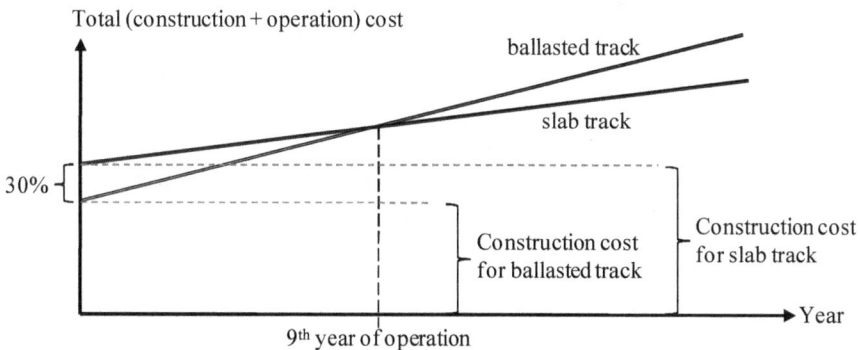

Fig.17.13. Comparative construction and maintenance costs of ballasted and slab track for Japan, (25).

The comparative construction cost of the various forms of slab track in relation to ballasted track (construction cost of which is given the value 1) is as follows, (307):
– Rheda 1.5÷3.0,
– Züblin 1.25÷1.5,

- Stedef 1.5÷3.0,
- Shinkansen 1.3÷1.7,
- Bögl 1.5÷1.8,
- Embedded rails 1.2÷1.5,
- Asphalt layer 1.4÷1.7.

Due to the difficulties during construction of each slab track technique, the various systems have different construction rates, as follows (in meters per day): Rheda 500, Züblin 200, Shinkansen 350, Bögl 350, Embedded rails 200, Asphalt layer 200÷270, (307).

17.9. Monitoring and repair of slab track

Any railway track undergoes extensive deterioration under traffic load. Ballasted track is a flexible system and restoration of any kind of defects or irregularities can be rather easily conducted. A non-ballasted track is more inflexible than a ballasted one; thus, cracks appearing on the concrete slab would require remedy works with injections of fibre-reinforced polymers.

Thus, monitoring of the situation of concrete slab at any moment is critical and can be conducted with the help of the following devices:

- *measuring devices* of strain, in order to keep track of the evolution of any crack in the concrete,
- *electric sensors*, placed in the concrete slab, that can collect in real time all data related to the slab track, transmit information to maintenance and safety centers, detect any coming problems, and schedule in time the necessary repairs,
- *fibre optic sensors*, which are embedded in the concrete slab and usually complement traditional electric sensors; they can control continuously the structural integrity of the concrete slab and locate in time any potential failure points at even an early stage.

18 Train Dynamics

18.1. Train traction

Any moving object must overcome the various resistances developed during its motion and, to achieve it, consumes some quantity of energy. A train moves thanks to the tractive force usually provided by the locomotive, which hauls a number of passenger or freight vehicles. The locomotive may be powered by either internal-combustion (diesel) engines, in which case there is diesel traction, (see section 20.4), or by electric motors, in which case there is electric traction, (see section 20.5). Since 2018, some locomotives are fueled from products derived from hydrogen, (see section 20.13).

Any hauled vehicle consists of the vehicle body, carrying passengers or freight, and the wheels. The body is supported by the wheels either directly on their axles, (see section 19.3), or on bogies, (see section 19.4). Wheels which provide traction are referred to as driving wheels, whereas wheels which do not provide traction are known as trailing wheels.

The distinction between tractive and hauled vehicles is less clear in diesel-electric powered vehicles, where only certain of the otherwise identical passenger vehicles have driving wheels.

In order to ensure train operation at a particular speed, adequate tractive force should be provided to overcome the various forces resisting train motion.

18.2. Resistances acting during train motion

During train motion, resistance forces arise, which must be overcome by the tractive force developed by the locomotive. These resistance forces are:
- running resistance R_L (mechanical and aerodynamic) in horizontal rectilinear motion,
- resistance R_c caused by track curves,
- grade resistance R_g caused by gravity on gradients, positive when moving uphill, negative when moving downhill,
- inertial (or acceleration) resistance R_{in} caused by inertia due to acceleration on starting and when speed is not constant.

465

Total resistance R is the sum of R_L, R_c, R_g, R_{in}. The resistance per unit weight of rolling stock is called *specific resistance r*.

Many of the formulas given below are *empirical* or *semi-empirical* and include coefficients with values found for a particular type of rolling stock (e.g. BR: (former) British railways (now Network Rail); DB: German railways; SNCF: French railways, etc.) and for specific operating conditions.

18.3. Running resistance R_L

18.3.1. General equation for the running resistance

Running resistance of a train is given by the general equation (18.1), (322), (324):

$$R_L = A + B \cdot V + C \cdot V^2 \tag{18.1}$$

In this equation:
♦ the terms A+B·V include the various mechanical resistances. The first term A (which does not depend on speed but only on rolling stock characteristics) represents the rolling resistances and those generated by friction between the wheel flange and the rail on curves. The second term B·V represents the various mechanical resistances, which are proportional to speed (rotation of axles and shafts, mechanical transmission, braking, etc.),
♦ the third term $C \cdot V^2$ represents the aerodynamic resistances.

The parameters A, B, and C can be expressed as functions of the rolling stock characteristics by the following formulas (R_L in kg, V in km/h), (324):

$$A(kg) = \lambda \cdot M \cdot \sqrt{\frac{10}{m}} \tag{18.2}$$

where:　M :　total train mass (tonnes),
　　　　　m :　mass per axle (tonnes),
　　　　　λ :　parameter with values depending on the rolling stock type, e.g. for SNCF vehicles $0.9 < \lambda < 1.5$,

B·V (kg)　=　0.01 M·V for good quality track and rolling stock on bogies, $\tag{18.3}$

$C \cdot V^2$(kg)　=　$k_1 \cdot S \cdot V^2 + k_2 \cdot p \cdot L \cdot V^2$. $\tag{18.4}$

In formula (18.4), the first term represents the aerodynamic resistances arising at the train front and rear and the second term represents the aerodynamic resistances generated longitudinally along the surface p·L of the train,

where: k_1 : a parameter depending on the shape of the train front and rear. For instance, in conventional medium- and low-speed SNCF rolling stock, $k_1 = 20 \cdot 10^{-4}$, while for TGV trains, $k_1 = 9 \cdot 10^{-4}$, (324),

S : front surface cross-sectional area (in m^2) (commonly around $10 \div 12$ m^2),

k_2 : parameter depending on the condition of the surface $p \cdot L$. As an example, in conventional SNCF rolling stock, $k_2 = 30 \cdot 10^{-6}$, while for TGV rolling stock, $k_2 = 20 \cdot 10^{-6}$,

p : partial perimeter (in meters) of the rolling stock down to the rail level, with common values around 10 m,

L : train length (m).

Figure 18.1 illustrates the increase of mechanical and aerodynamic resistances as a function of speed. We can remark that aerodynamic resistance is crucial at high speeds and trains are given a suitable aerodynamic shape in order to reduce it, (316).

Figure 18.2 illustrates the running resistance (as a function of speed) and the power required to overcome this resistance. We see that in order to increase speed by 50% from 200 km/h to 300 km/h, engine power has to be increased by more than 250%.

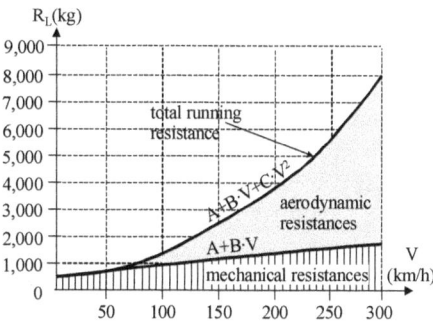

Fig. 18.1. Mechanical and aero-dynamic resistances as a function of speed, (48)

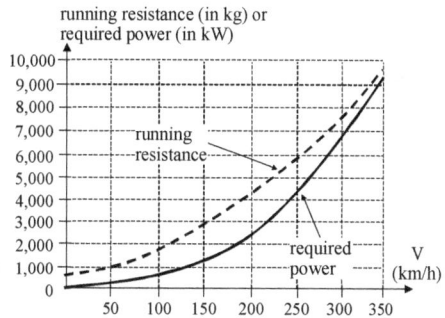

Fig. 18.2. Running resistance and required traction engine power (at zero gradient) as a function of speed (case of the French TGV), (327)

18.3.2. Empirical formulas of some railways for the running resistance

The values of parameters A, B, C of equation (18.1) depend on the characteristics and peculiarities of the rolling stock. The various rolling stock manufacturers and the various railways have developed empirical formulas for these parameters. Formulas in use by various railway authorities worldwide are given below.

18.3.2.1. Formulas of the French railways

18.3.2.1.1. Diesel or electric locomotives

The running resistance is given by the empirical formula, (325):

$$R_L(kg) = 0.65 \cdot L + 13 \cdot n + 0.01 \cdot L \cdot V + 0.03 \cdot V^2 \qquad (18.5)$$

where: L : locomotive mass (tonnes),
 n : number of axles,
 V : speed (km/h).

18.3.2.1.2. Hauled rolling stock

Due to the dissimilarity of the hauled rolling stock types, the various formulas present a large spread; they are simplified by merging the terms $B \cdot V$ and $C \cdot V^2$ of equation (18.1). The common practice is to calculate the specific running resistance r. Therefore, (325):

- For passenger rail vehicles on bogies:

$$r(kg/t) = 1.5 + \frac{V^2 (km/h)}{4,500} \qquad (18.6)$$

- For standardized UIC-type vehicles:

$$r(kg/t) = 1.25 + \frac{V^2 (km/h)}{6,300} \qquad (18.7)$$

- For passenger vehicles on axles and express freight vehicles:

$$r(kg/t) = 1.5 + \frac{V^2 (km/h)}{2,000 \div 2,400} \qquad (18.8)$$

- For block freight vehicles:

$$r(kg/t) = 1.2 + \frac{V^2 (km/h)}{4,000} \qquad (18.9)$$

18.3.2.1.3. Electric passenger vehicles

Electric passenger vehicles (including traction motors) are commonly used in high-speed trains and in suburban commuter services. The total running resistance R_L in the case of electric commuter trains can be calculated by the formula, (325):

$$R_L(kg) = \left(1.3 \cdot \sqrt{\frac{10}{m}} + 0.01 \cdot V\right) \cdot P + C \cdot V^2 \qquad (18.10)$$

where $C = 0.0035 \cdot S + 0.0041 \cdot \dfrac{p \cdot L}{100} + 0.002 \cdot N$ (18.11)

with:

P   : total mass of the electric passenger vehicle (in tonnes),
m   : mass per axle (tonnes),
V   : speed (km/h),
S, p, L : as in equation (18.4), (section 18.3.1),
N   : number of raised pantographs, (see section 20.8.5).

18.3.2.2. Formula of the American railways

American railways use the modified Davis formula for the specific running resistance, (320):

$$r(lb/t_s) = 0.6 + \frac{20}{M} + 0.01 \cdot V(mph) + \frac{k}{m \cdot n} \cdot V^2 \qquad (18.12)$$

where:

1 lb   : 0.454 kg,
1 t$_s$: short tonne = 2,000 lbs = 907.2 kg,
M   : train mass,
m   : mass per axle,
n   : number of axles in train,
mph : miles per hour (1 mph = 1.61 km/h),
k   : C·S,
C   : air resistance coefficient (from tables),
S   : vehicle cross-sectional area (in square feet).

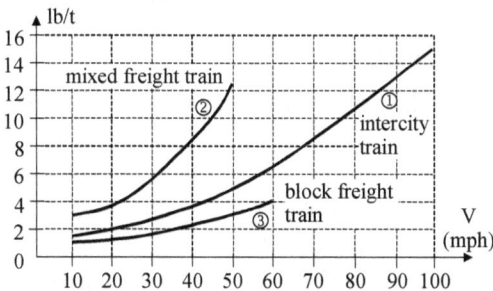

Fig. 18.3. Specific running resistance according to the American railways, (320)

Figure 18.3 illustrates the specific running resistances for various rolling stock types:

① intercity trains, V=80 mph, m=25 lb/axle, train composition of 16 vehicles with a total mass of 1600 t$_s$,

② mixed freight trains, V=50 mph, m=15 lb/axle, average vehicle mass: 45 t$_s$, total train mass: 3,000 t$_s$,

③ block freight trains, V=60 mph, m=60 lb/axle, train of 21 vehicles, each 240 t$_s$.

18.3.2.3. Formulas of the German railways

The German railways use the Strahl formula for freight trains, (Fig. 18.4), (320):

$$r(kg/t) = \frac{1}{10} \cdot \left[25 + k \cdot \left(\frac{V(km/h) + \Delta V}{10} \right)^2 \right] \tag{18.13}$$

and the Sauthoff formula for intercity trains:

$$r(kg/t) = a + 0.0025 \cdot V + 0.48 \cdot F_e \cdot \frac{n_w + 2.7}{M} \cdot \left(\frac{V + 15}{10} \right)^2 \tag{18.14}$$

where:

k   : 0.5 for mixed freight trains and 0.25 for block trains,

V   : train speed (km/h),

ΔV : head wind speed (it takes in usual situations the value 15 km/h),

a   : coefficient taking the value 1.0 for roller bearings and the value 1.9 for plain bearings (roller bearings do not have rolling elements, whereas plain bearings (called also shafts) do have),

F_e : coefficient related with train front area characteristics, taking usually the value 1.45,

n_w : number of wagons,

M : train mass (in tonnes).

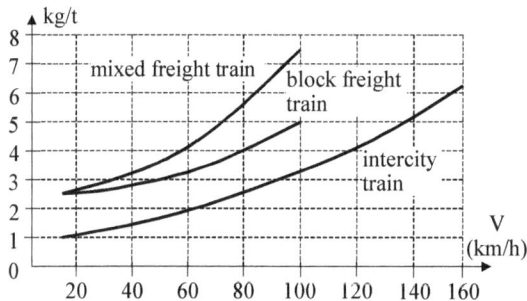

Fig. 18.4. Specific running resistance according to the German railways, (320)

Figure 18.4 illustrates the specific running resistance for various freight rolling stock types according to the German railways.

18.3.2.4. Formulas for broad and metric gauge railways

For *broad gauge* (e=1.676m) railways, the following formulas have been suggested, (Fig. 18.5), (320):

- *passenger trains*
    $$r(kg/t) = 0.6855 + 0.02112 \cdot V (km/h) + 0.000082 \cdot V^2 \tag{18.15}$$
- *freight trains*
    $$r(kg/t) = 0.87 + 0.0103 \cdot V + 0.000056 \cdot V^2 \tag{18.16}$$

For *metric gauge* (e=1.000m) railways, the following formulas have been suggested, (Fig. 18.6), (320):

- *passenger trains*
    $$r(kg/t) = 1.56 + 0.0075 \cdot V (km/h) + 0.0003 \cdot V^2 \tag{18.17}$$

- *freight trains*

$$r(kg/t) = 2.60 + 0.0003 \cdot V^2 \qquad (18.18)$$

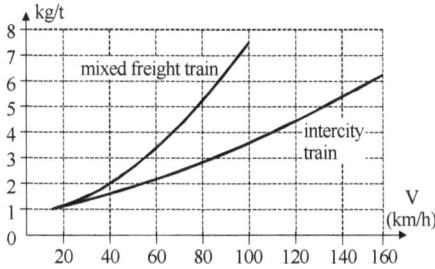

Fig. 18.5. Specific running resistance for broad gauge railways, (320)

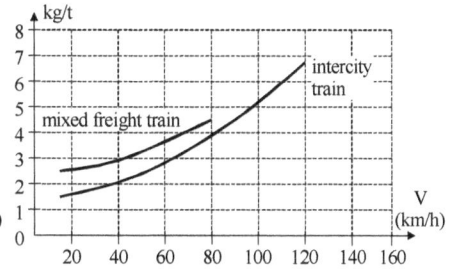

Fig. 18.6. Specific running resistance for metric gauge railways, (320)

18.3.3. Resistances developed when running in a tunnel

Compared to operation of a train in open air, operation in a tunnel has certain peculiarities caused by sudden increases in pressure (with an unfavorable influence on passenger comfort), increased aerodynamic resistances, problems arising when trains cross, and finally the need to ensure proper ventilation.

18.3.3.1. Pressure problems

When a train enters a tunnel, the front section (the head) of the train compresses the air at the entrance, giving rise to a compression wave, (Fig. 18.7), the amplitude of which increases as the train proceeds, reaching a maximum when the rear section (the tail) of the train enters the tunnel. At this moment the vacuum left behind the train creates a decompression wave. The compression wave at the train front, which propagates at the speed of sound along the tunnel, is reflected by tunnel walls and returns in the form of a decompression wave. With respect to the decompression wave generated by the tail of the train inside the tunnel, it undergoes corresponding changes and finally returns

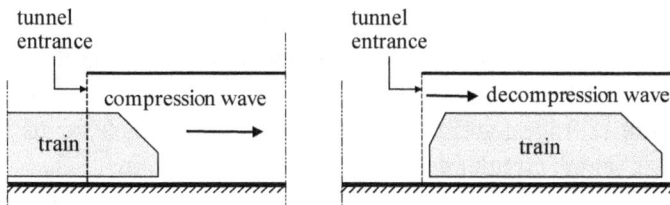

Fig. 18.7. Compression and decompression waves when a train enters a tunnel

471

in the form of a compression wave. When all these waves are superposed, they give rise to pressure fluctuations progressively diminishing in amplitude as a function of time, (48), (317).

At the exit of the train from the tunnel, a part of the compression wave is reflected back toward the entrance. The remaining part of the compression wave, called micro-pressure wave, radiates out and causes noise and vibration-related problems. The strength of the micro-pressure wave increases with train speed, (316).

It should be noted, however, that passenger discomfort is caused not so much by pressure variations as by the rate of pressure variation. During abrupt changes of weather, the pressure may change by up to 1,300 mm H_2O^*, whereas a 1,000 m increase in altitude causes a pressure drop of 1,100 mmH_2O, with no significant discomfort. In contrast, during train motion in a tunnel, pressure changes are much smaller but also much more annoying, due to high pressure change rate. The human body can adapt to significant changes in pressure, provided that they are not abrupt, (331).

The comfort limit for pressure variation of the human ear is around 900 Pa. The UIC and many railway authorities recommend as limit value for the pressure variation at the human ear 500 Pa and for the pressure variation rate 200÷300 mmH_2O/sec, (Fig. 18.8), (316).

Factors affecting passenger comfort, therefore, include both pressure variation Δp and pressure variation rate $\Delta p/\Delta t$. It has been shown that passenger comfort is not significantly affected as long as, (328):

$$\Delta p \cdot \frac{\Delta p}{\Delta t} < c \qquad (18.19)$$

where c is a constant, the value of which is different in the various railway authorities.

Figure 18.8 illustrates recorded values of pressure variations, which depend greatly on rolling stock characteristics. We see that until a speed of 220 km/h is reached, passenger comfort is not significantly affected. Beyond this value, however, pressure variations and their rates of change become significant.

In order to reduce pressure waves (and specifically micro-pressure waves at the exit of a tunnel), rolling stock constructors tried to reduce the cross-sectional area in the front of the train and experimented with many forms for the nose of the train. Figure 18.9 illustrates the effect on pressure variation for three forms (ellipsoid, paraboloid, circular cone) of nose shape of a train, (321).

---

[*] 1mm H_2O = 9.81 Pa.

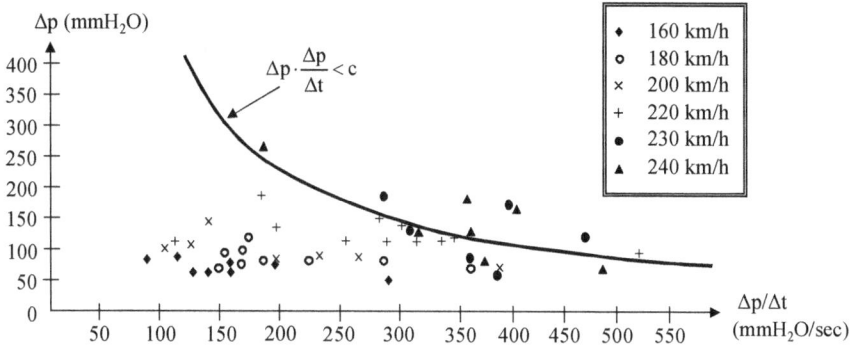

Fig. 18.8. **Pressure variation (Δp) and pressure variation rate ($\Delta p/\Delta t$) as a function of speed, (328)**

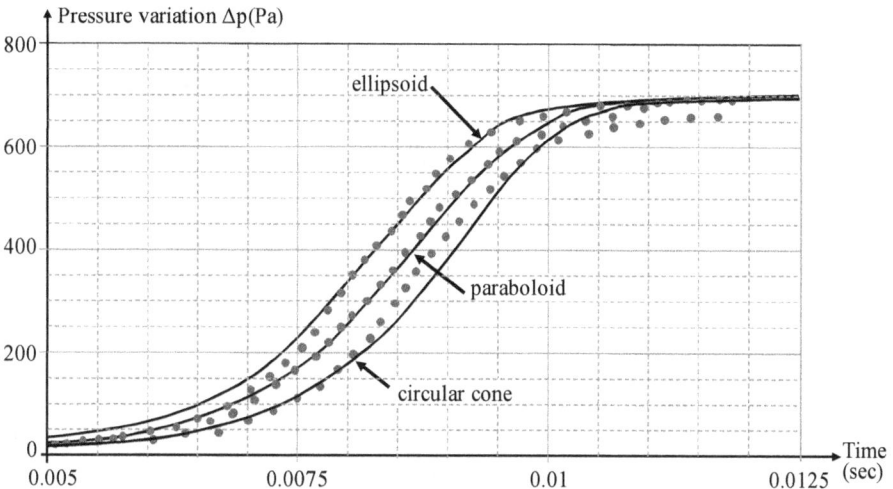

Fig.18.9. **Effect on pressure variation of the form (ellipsoid, paraboloid, circular cone) of nose shape of a train with a front surface cross-sectional area of $12.8\,\text{m}^2$, running at a speed of $200\,\text{km/h}$ in a tunnel of a section of $46\,\text{m}^2$, (321)**

18.3.3.2. Increased aerodynamic resistances in tunnels

In order to reduce increased aerodynamic resistances in tunnels, lateral openings are made along the tunnel with spectacular results, (Table 18.1).

In the Channel Tunnel, which is composed of two single track tunnels, lateral openings every 375m resulted in a reduction of the power required to overcome aerodynamic resistance from 13.5 MW to 5.8 MW at a speed of 140 km/h, (see also section 2.4), (48).

Table 18.1.

Comparative running resistance for a train weighing 705 t in tunnels with and without lateral openings, (325)

	In the open air	In tunnel		
		Openings every 250 m	Openings every 500 m	No opening, one train in tunnel
Total running resistance (kg)	6,480	8,170	8,830	14,930
Required power (kw)	2,820	3,550	3,840	6,500

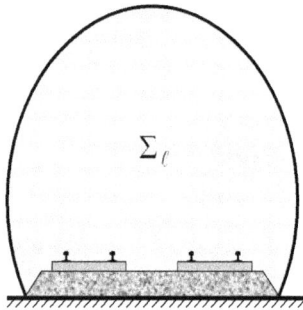

Fig. 18.10. Effective tunnel cross-section Σ_ℓ

In order to reduce the aerodynamic resistances in tunnels, efforts are made to reduce the S/Σ_ℓ ratio[*], where S is the front surface cross-sectional area of the train and Σ_ℓ the effective tunnel cross-sectional area, (Fig. 18.10). Thus:

– in single track tunnels, $\dfrac{S}{\Sigma_\ell} = 0.30 \div 0.50$,

– in double track tunnels, $\dfrac{S}{\Sigma_\ell} \sim 0.15$.

It is evident that an excessive reduction in the S/Σ_ℓ ratio would lead to an inordinate and expensive increase of tunnel cross-section.

18.3.3.3. Crossing of trains in tunnels

When a train crosses another in a tunnel, compression waves generated by the first train strike the other and conversely. As the faster train gives rise to stronger effects, the slower train is obviously subjected to greater stresses.

Tests conducted by the Italian railways have shown that aerodynamic effects when two trains cross in a tunnel do not affect significantly passenger comfort, mainly because of their short duration (a few tenths of a second), (328). Human hearing is disturbed by extraneous influences only if they last longer than half a second. With respect to damage of the rolling stock, (mainly eventual fracture of window glasses), the above tests have shown no significant risk at high speeds, (328).

---

[*] called also blockage ratio.

18.3.3.4. Tunnel cross-section requirements at high speeds

All the aforementioned reasons entail that the tunnel cross-section increases as speed increases. Table 18.2 gives the effective cross-sectional area Σ_ℓ for various speeds and for double track tunnels, whereas Figure 18.11 illustrates the dimensions of a tunnel with a running speed of 300 km/h. However, in the design of high-speed tunnels (V>200 km/h), emphasis should be put not only on the distance between tracks (usually 4.20÷4.70 m, see sections 7.10.3 and 12.6) and the cross-sectional area Σ_ℓ (55÷100 m²) but also on the performance and the mechanical resistances of the rolling stock (particularly the glass parts).

The required cross-sectional area of a tunnel (illustrated in Table 8.2) is higher for short-and medium-length tunnels (with lengths up to 5 km). For longer lengths of tunnels, the cross-section may be reduced. Thus, in the design of the high-speed railway in California, USA, tunnels with a length of 10 km had a cross-section by around 15÷20% lower, compared to tunnels with lengths up to 5km.

Table 18.2.
Required tunnel cross-sectional area Σ_ℓ for a double track tunnel at various speeds, (318)

V_{max} (km/h)	160	200	240	300
Σ_ℓ (m²)	40	55	71	100

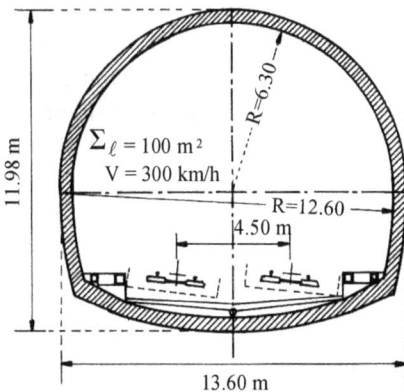

Fig. 18.11. Cross-section of a high-speed tunnel

18.3.4. Comparative running resistance between railways and road vehicles

The running resistance of a rail vehicle (passenger or freight) is far lower than that of a road vehicle, five times lower for passenger transport and four times lower for freight transport. The lower railway running resistance is firstly due to the lower coefficient of friction of the metal wheels on metal rails and secondly to the lower aerodynamic resistance per unit of length of a train, because of its great length (compared to road vehicles).

18.4. Resistance R_c due to track curves

Additional resistance on curves is caused by:
- friction between wheel flange and rail,
- wheel slippage on the rails, since the axles of a bogie or of a two-axle rail vehicle are always parallel.

The specific resistance r_c, occurring along curves, can be expressed by the following formula, (325):

$$r_c(kg/t) = \frac{k}{R} \qquad (18.20)$$

where: k : parameter with values between 500÷1,200, the average being 800,
        R : radius of curvature in the horizontal plane (m).

18.5. Resistance R_g caused by gravity

In a rail vehicle running along a straight level track, the force component perpendicular to the direction of gravity is zero. However, when the plane of the track is inclined (when the train is running uphill or downhill), a force component R_g develops parallel to the plane of the track, (Fig. 18.12), and in the case of an uphill gradient this component is an additional resistance to vehicle motion.

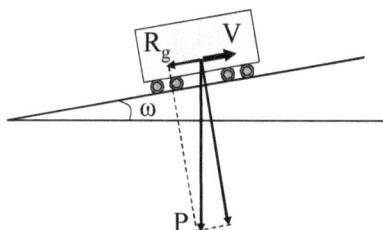

As the longitudinal gradient of railway tracks is small and seldom exceeds 20‰, the angle ω is very small and therefore it can be assumed that $\sin\omega = \tan\omega$. Consequently:

Fig. 18.12. Gravity resistance

$$R_g = P \cdot \sin\omega = P \cdot \tan\omega = P \cdot i \qquad (18.21)$$

where i is the longitudinal gradient.

Some railways unify resistances due to layout curves and to gravity in a common term.

18.6. Inertial (acceleration) resistance R_{in}

Resistance forces arising from the acceleration of a train are given by the equation of dynamics and depend on the geometry and the materials of rolling stock vehicles. Inertial resistance is proportional to train mass and acceleration.

If α is the acceleration, the specific inertial resistance r_{in} can be calculated from the formula:

$$r_{in}(kg/t) = \frac{\alpha}{g} \cdot q \qquad (18.22)$$

where:

q : a mass coefficient taking into account both the fixed and the rotating masses of the rolling stock, such as shafts, electric motors, etc.

If M_{rot} are the rotating masses and M the total train mass, then:

$$q = 1 + \frac{M_{rot}}{M} \qquad (18.23)$$

Measurements have shown that an acceleration of 1 cm/sec^2 results in an inertial resistance of 1 kg/t, which is approximately as much as that from an uphill gradient of 1‰.

18.7. Starting force and traction force of a train

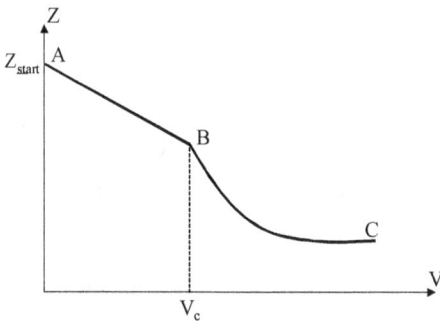

Fig. 18.13. Traction force Z – running speed V diagram of a diesel train

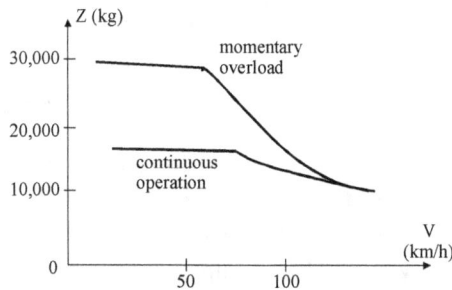

Fig. 18.14. Traction force Z – running speed V diagram of an electric train

Starting force is the force required to put a train into motion and is denoted as Z_{start}. The starting force should overcome the sum of all resistances generated during train motion. If all vehicles of a train departed simultaneously, the starting force would have to be very high. In practice, however, this is never the case, since the train does not start as a block, due to the gaps between the successive vehicles. Once the train departs, the force necessary to continue its motion is called *traction (or tractive) force Z* and is much smaller than the starting force. If Z refers to the force required per tonne of rolling stock, then it is called *specific traction force z*.

In diesel traction, (Fig. 18.13), the force developed by the traction engine decreases with increasing speed, while the maximum traction force Z is developed when the train starts to move. As speed increases,

477

the traction force decreases, at first linearly (segment AB) and as speed increases further, resistance plummets (segment BC) leveling off at a minimum, corresponding to the maximum speed of the tractive vehicle.

Electric trains, (Fig. 18.14), can sustain momentary overloads, in which case the traction force is greater than that in continuous operation and therefore higher speeds are attainable.

Figure 18.15 illustrates the diagrams of the specific traction force z as a function of longitudinal gradient in the cases of passenger and freight trains. Usual values of the specific traction force are 10÷20 kg/t for passenger trains and 10÷30 kg/t for freight trains, (326).

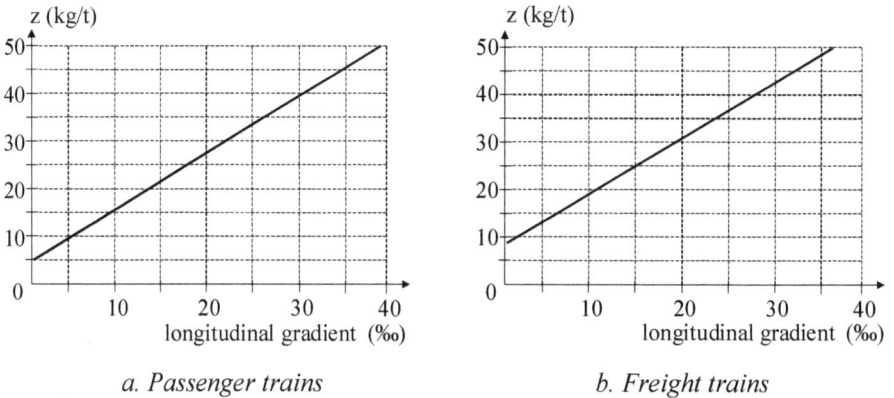

a. Passenger trains   *b. Freight trains*

Fig. 18.15. Specific traction force z of a train in relation to longitudinal gradient

18.8. Adhesion forces

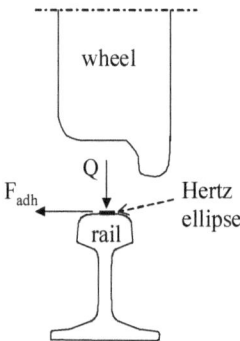

Fig. 18.16. The adhesion force F_{adh}

The contact between wheel and rail occurs along an elliptical surface known as the Hertz ellipse, (Fig. 18.16, see also sections 7.7 and 10.6.1). Along the Hertz ellipse, the adhesion forces F_{adh} that are developing are necessary to ensure continuous rotation of the wheel. This requires that the adhesion force F_{adh} be equal to or greater than the traction force Z, (Fig. 18.17).

478

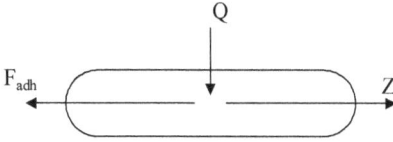

Fig. 18.17. Traction force Z and adhesion force F_{adh}

The adhesion coefficient μ is defined as the ratio of the horizontal adhesion force F_{adh} to the vertical wheel load Q:

$$\mu = \frac{F_{adh}}{Q} \qquad (18.24)$$

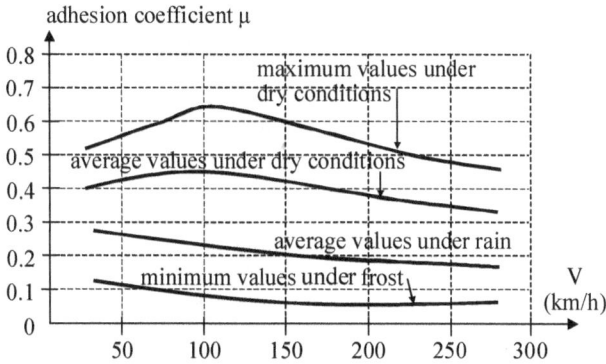

Fig. 18.18. The adhesion coefficient μ in relation to train speed and weather conditions, (324)

The adhesion coefficient μ depends mainly on weather conditions but also on train speed, (Fig. 18.18), (321). To satisfy the condition $F_{adh} \geq Z$, the minimum required values of μ have been surveyed and are given in Table 18.3, (324).

Table 18.3.
Minimum required values for the adhesion coefficient μ, (315), (324)

Traction mode		Braking mode (downhill motion, electric braking)	
V (km/h)	μ_{min}	V (km/h)	μ_{min}
160	0.30		
200	0.10	0÷200	0.095
300	0.07	200÷300	0.060

Concerning the impact of the various track and rolling stock characteristics on the adhesion coefficient, it was found that, (323):

- increasing wheel diameter from 700 mm to 920 mm caused little increase of adhesion,
- changing the inclination of the rails on the sleepers, (see section 7.9, Fig. 7.12), from 1/40 to 1/20 decreased adhesion by 17%,
- increasing wheel load from 8 t to 12 t, decreased adhesion by 12%.

The adhesion coefficient μ can be expressed as a relation of train speed by the following empirical formula, (320):

$$\mu = \frac{7.5}{V(km/h) + 44} + 0.161 \qquad (18.25)$$

It is known from kinematics that for a motor wheel to perform properly, the theoretical peripheral speed of the wheel should be greater than the actual translational speed, (Fig. 18.19):

$$V_{rot} > V_{trans}$$
$$V_{rot} = 2 \cdot \pi \cdot r_0 \cdot n \qquad (18.26)$$

where: r_0 : rolling radius
n : number of revolutions

Otherwise, there will be:
♦ braking, if $V_{rot} < V_{trans}$,
♦ wheel skid, if $V_{trans} < V_{rot} \neq 0$,
♦ wheel lock, if $V_{rot} = 0 > V_{trans} \neq 0$.

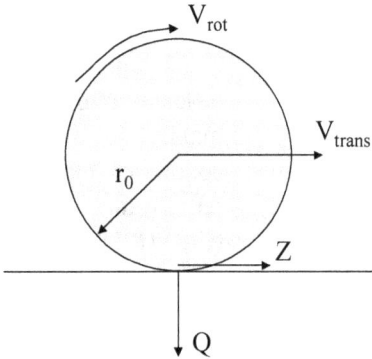

Fig. 18.19. Speeds and forces on a wheel

It is evident that keeping train speed constant or increasing it requires that the traction force be equal to or greater than the total resistance developed during train motion.

18.9. Required power of the engine of a train

The traction force necessary for train motion is ensured by the engine power, which can be distinguished into nominal power and effective power. Nominal power is the power as specified by the engine manufacturer. However, auxiliary devices of the engine absorb a part of the nominal power and another part is lost during transmission from the motor shaft to the wheels. The remaining power is termed effective power and is the part actually available to power the motor wheels and the train as a whole.

Power is measured in either horsepower (hp, ps, cv) or kilowatts (kW). Engine power N (in horsepower) can be calculated by the formula:

$$N = \frac{Z(kg/t) \cdot V(km/h) \cdot P(t)}{270} \qquad (18.27)$$

where Z is the traction force and P is the train weight.

Therefore, it is evident that train power depends on speed, and this should be specified every time that a power value is given. Table 18.4 gives the power required for the operation of various types of trains.

Power often refers to unit weight of rolling stock, in which case it is termed specific power N_e (kW/t or Ps/t). A parameter which determines the course of a train is the distance required to attain a final speed, (Fig. 18.20).

Table 18.4.
Power required by various types of trains, (324)

Type of train	Weight (tonnes)	Speed (km/h)	Power (kW) at a 5‰ gradient
Passenger	800	160	4,400
Freight	1,800	100	6,350
Suburban	190	140	1,050
High-speed (TGV)	418	300	6,850

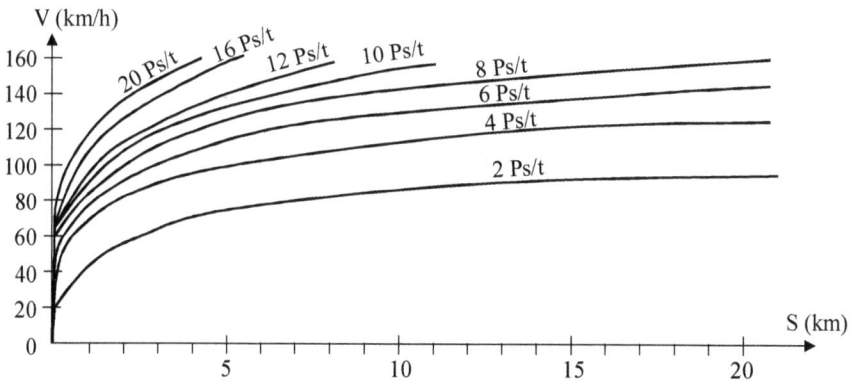

Fig. 18.20. Required distance S as a function of specific power to enable train speed starting from zero to reach a final value

18.10. Values of train acceleration and deceleration

In order to increase speed, a train develops longitudinally acceleration, whereas in order to reduce speed it develops deceleration. The values of acceleration and deceleration of a train depend on the type of rolling stock (passenger, freight) as well as on the distance within which the train must attain its maximum speed. The shorter this distance, the higher the values of acceleration and deceleration, as is the case with metropolitan and suburban railways. For reasons pertaining to human physiology, maximum acceleration should not exceed 1.0 m/sec^2.

Typical longitudinal *acceleration* values for various types of rolling stock are:
- freight trains: $0.2 \div 0.4$ m/sec^2,
- intercity trains: $0.4 \div 0.6$ m/sec^2,
- suburban trains: $0.6 \div 0.8$ m/sec^2,
- metros: $0.8 \div 1.0$ m/sec^2.
- trams: $0.8 \div 1.2$ m/sec^2.

Typical longitudinal *deceleration* values for various types of rolling stock are:
- conventional freight trains: 0.10 m/sec^2,
- express freight trains: 0.25 m/sec^2,
- passenger trains: $0.40 \div 0.50$ m/sec^2,
- suburban railways, metros: 0.60 m/sec^2.

A critical parameter for passenger comfort is the variation of acceleration per unit time, known as jerk. Jerk should not exceed the value of 1.5 m/sec^2/sec. The lower the value of jerk, the higher the value of longitudinal acceleration that can be accepted by the human body, (Fig. 18.21), (331).

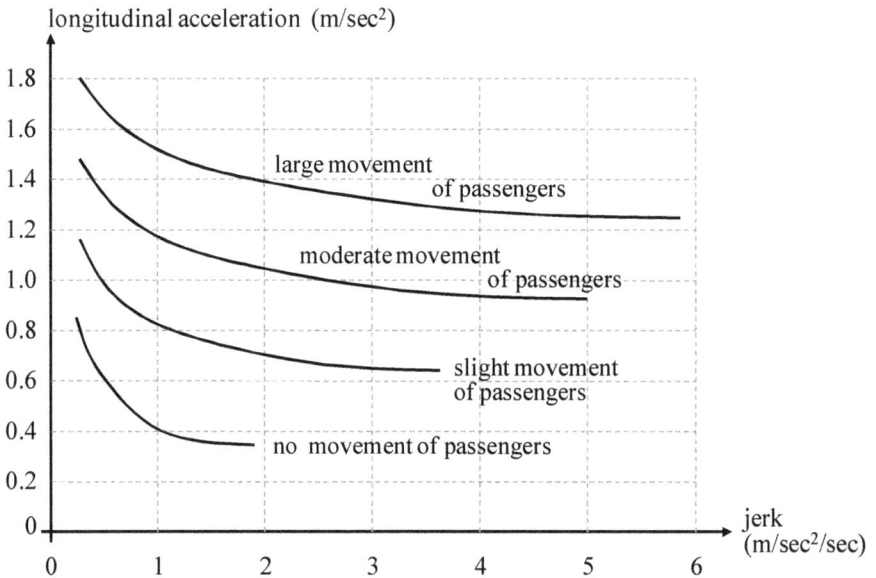

Fig. 18.21. Relation between jerk and longitudinal acceleration, (331)

18.11. Train braking

18.11.1. Braking systems

Train braking is achieved with the help of one of the following two brake types, (315), (319), (330):

– *Shoe* (or *block* or *tread*) brakes. They operate with the help of the friction developed on the wheels by the pressure of metal or synthetic shoes. Both wheels of an axle are provided with braking shoes.

– *Disc* brakes. The braking action is achieved by friction on steel discs or cast iron fixed to the axle. A basic disadvantage of disc brakes is the generation of high temperatures reaching 500°C and even higher (up to 680°C), (319), (330).

The following methods are used for transmission of the braking force:

• *Air braking*, using changes of air pressure in special conduits, initiated in the driver's cab by operating a valve. This system has the disadvantage that braking is not simultaneous on all train vehicles.

• *Electropneumatic braking*, developed in the 1960s, to reduce the transmission delay of the braking operation to the vehicles in a train. In this system, air pressure is modified simultaneously at all wheels by electrically actuated air valves at each brake. The system is operated by an electric signal transmitted on a line along the train.

• *Electromagnetic braking*, developed in recent decades so as to confront the great increase in train speeds. In this type, the braking action is applied directly to the rails. Special shoes with electromagnets, which carry a current during braking, achieve braking. Electromagnetic braking may function independently or in combination with other systems.

• *Electrodynamic braking*, doing away with brake shoes' wear, since deceleration is obtained by converting the electric traction motors into electric generators. The power generated by braking is used for auxiliary purposes. In the case of electric locomotives, the energy recovered may be returned to the power network, through the pantograph. The recovered energy is 3%÷6% in intercity trains, 20% in mass transit and freight trains, and 40% in trains on high-gradient tracks, (320).

Rail vehicles are provided with anti-skid devices, which monitor wheel rotation and modify the braking force whenever wheel locking is detected. In consideration of train braking, special concern needs to be given to poor adhesion conditions that can be created under certain weather conditions due to rain, ice, and leaf deposits.

Finite element analysis gives the possibility of studying both mechanic and thermodynamic behavior of braking systems, (Fig. 18.22).

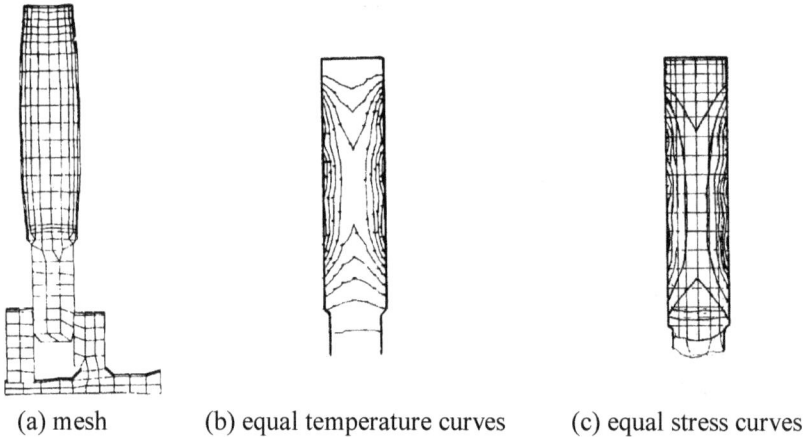

(a) mesh   (b) equal temperature curves   (c) equal stress curves

Fig. 18.22. Analysis of a disc brake with the use of the finite element method

18.11.2. Braking distance

Empirical formulas have been suggested to calculate the braking distance L for the various train categories, (325).

Freight trains (V<70 km/h)

The braking distance can be calculated by Maison's formula:

$$L(m) = \frac{4.24 \cdot V^2 \, (km/h)}{1{,}000 \cdot \varphi \cdot \lambda + 0.0006 \cdot V^2 + 3 - i} \tag{18.28}$$

where   i : track gradient (‰ or in mm/m). Track gradient is regarded positive downhill and negative uphill,

φ: friction coefficient depending on track gradient. Values of φ are:
   $\varphi = 0.10$, for i < 15‰
      $= 0.10 \div 0.00133$ (i–15), for i > 15‰

λ: braking percentage, defined as the ratio of the braking weight to the total vehicle weight and expressing the braking force required for braking one tonne.

Braking percentage λ is a critical factor for the braking distance. Table 18.5 gives values of λ for various types of rolling stock and brakes. In any case, formula (18.28) gives the possibility to calculate the braking percentage λ in relation to the braking distance L, the train speed V, the gradient i, and the friction coefficient φ.

Table 18.5.
Braking percentage λ for various types of rolling stock and brakes

	braking percentage λ
Normal brake:	
- Tractive vehicles with axle load P=15÷20t	80÷95%
- Hauled vehicles with axle load P=15÷20t	65÷90%
Emergency brake:	
- Tractive vehicles	160÷220%
- Hauled vehicles	130÷220%

Passenger trains (V=70÷140 km/h)

The braking distance is given by Pedeluck's formula:

$$L(m) = \frac{\varphi \cdot V^2 (km/h)}{1.09375 \cdot \lambda + 0.127 - 0.235 \cdot i \cdot \varphi} \qquad (18.29)$$

with the various parameters defined as in formula (18.28).

Diesel-electric passenger vehicles

The braking distance is given by the formula:

$$L(m) = \frac{0.0386 \cdot V^2 (km/h)}{\gamma - \dfrac{i}{100}} \qquad (18.30)$$

where γ is the deceleration (m/sec^2).

Other empirical formulas

Previous formulas, developed by the French railways, are also suggested by the UIC, (329). However, the German railways use Minden's formula for the braking distance, which is:

braking of passenger trains $\qquad L(m) = \dfrac{3.85 \cdot V^2 (km/h)}{[6.1 \cdot \psi \cdot (1 + \lambda/10)] + i} \qquad (18.31)$

braking of freight trains $\qquad L(m) = \dfrac{3.85 \cdot V^2 (km/h)}{5.1 \cdot \psi \cdot \sqrt{\lambda - 5} + i} \qquad (18.32)$

with the parameter ψ taking values between 0.5÷1.25 (in relation to the brake type characteristics) provided by nomographs, (326).

Figure 18.23 illustrates the braking distance for low and medium speeds and various rolling stock types.

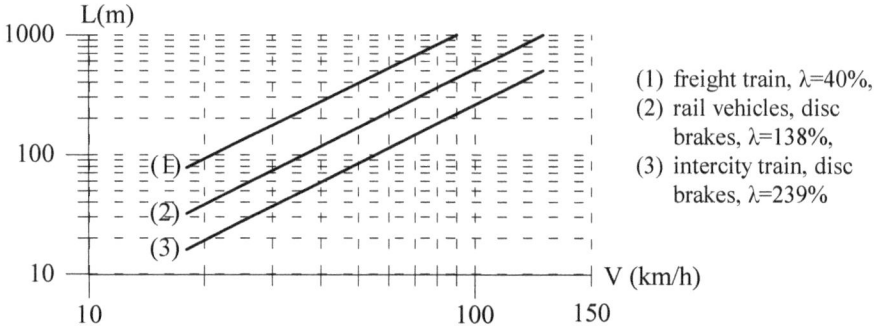

Fig. 18.23. Braking distance L in relation to speed (at zero gradient) for medium and low speeds

(1) freight train, λ=40%,
(2) rail vehicles, disc brakes, λ=138%,
(3) intercity train, disc brakes, λ=239%

The braking distance calculated by the above formulas is augmented by at least 10% as a safety margin (depending also on the signaling system). The greater the speed, the longer the braking distance, (Table 18.6). Figures 18.24 and 18.25 illustrate the braking distance at high speeds for the signaling systems TVM 300 and TVM 430 of the French high-speed train (TGV).

Table 18.6.
Braking distance
in relation to speed

V (km/h)	Braking distance (m)
160	1,300 ÷ 1,400
200	2,500 ÷ 3,000
270	6,000 ÷ 8,000
320	7,500 ÷ 9,000

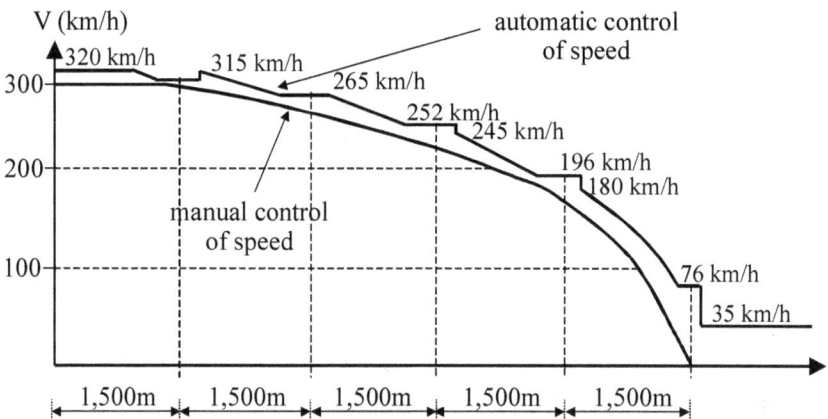

Fig. 18.24. Required braking distance at high speeds (track with zero gradient) – Signaling system TVM 300

486

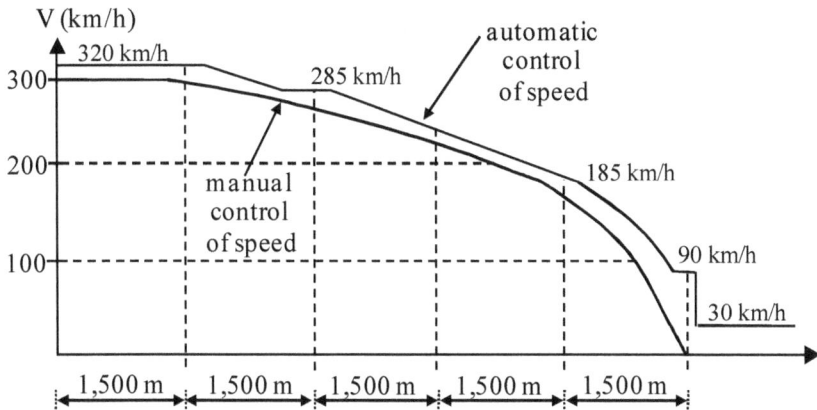

Fig. 18.25. Required braking distance at high speeds (track with zero gradient) – Signaling system TVM 430

18.11.3. European specifications concerning braking

According to the European technical specifications for interoperability relevant to rolling stock and concerning braking of trains, (333):

– the train braking system should ensure that the train's speed can be reduced or maintained on a gradient and that the train can be stopped within the maximum allowable braking distance,

– the primary factors that influence the braking performance are: the braking power, the train mass, the train running resistance, the speed, the available adhesion,

– the main functional and safety requirements of braking systems include:

• a main brake function used during operation for service and emergency braking,

• a parking brake function used when the train is parked,

• the main brake system may be continuous or automatic,

• when running, the driver should be able to check from the driving position the following: the status of the train brake control command line and the status of the train brake energy supply,

• in order to check emergency braking, tests should be carried out on dry rails at the following speeds: 30, 60, 80, 120, 140, 160, 200 km/h and at the maximum design speed of the specific rolling stock type.

19 Rolling Stock

19.1. Components of a rail vehicle

Every rail vehicle (passenger or freight) requires a set of parts and devices in order to move: wheels, axles, bogies, springs, couplings, and buffers.

19.2. Wheels

19.2.1. Geometrical characteristics and materials

On standard gauge tracks, the wheel diameter of the hauled rolling stock ranges from 0.84 m to 0.92 m. As the continuing tendency is to increase wheel load, one would expect wheel diameter to increase as well. This, however, is not feasible beyond current wheel diameters, because larger wheels would on the one hand increase weight and therefore manufacturing and operating costs, and on the other hand result in a greater vehicle floor height from track level. This would be detrimental to both the stability of the vehicle and the space available within it, since the loading gauge of the track (that is the free space around the rolling stock) is fixed and cannot be changed.

For standard gauge tracks, the wheel diameter of locomotives ranges between 0.85÷1.10 m, whereas for metric gauge tracks wheel diameter averages 0.75 m. The diameter of rail vehicles for standard gauge tracks has an average value of 0.90 m. Figure 19.1 illustrates the geometrical characteristics of a wheel according to the UIC.

Two main parts can be distinguished in a wheel, (Fig. 19.2):

– the tire, which is the external part of the wheel and comes in contact with the rail. Since it is subject to great wear, the tire is made of a material highly resistant to wear,

– the internal disc of the wheel. The external part of the disc inside the tire is the wheel rim.

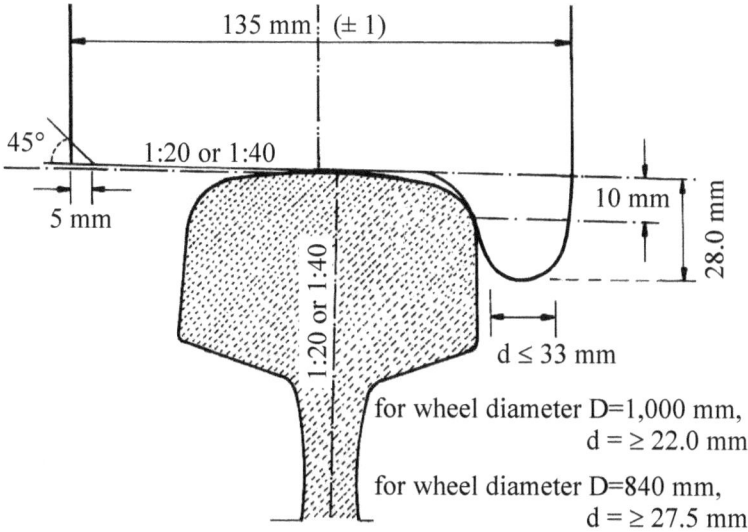

Fig. 19.1. Running surface of a wheel according to the UIC, (336)

Fig. 19.2. Tire and wheel rim of a rail vehicle, (320)

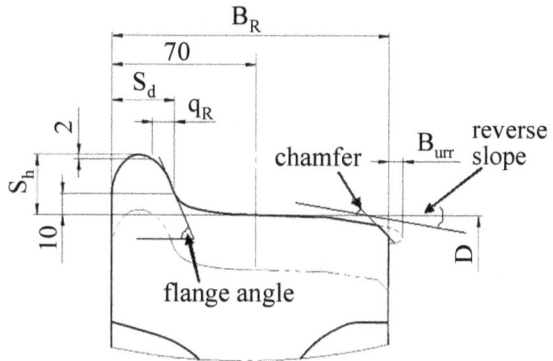

Fig. 19.3. Details of geometrical characteristics of a wheel, (333)

Tire thickness ranges between 65÷70 mm; a tire is considered worn out when its thickness is reduced to 30 mm. The first tire material was soft iron, but it wore out quickly and was difficult to weld properly. Accordingly, it was replaced by hard steel, which, however, should have low brittleness. In metro vehicles, elastic wheels are more and more commonly used.

According to the European Technical Specifications for Interoperability (TSI), maximum and minimum values for the geometrical characteristics of a wheel are given in Figure 19.3 and Table 19.1, (333).

Table 19.1.

Maximum and minimum values of the various geometrical characteristics of a wheel according to the TSI, (333)

Geometrical characteristic	Wheel diameter D (mm)	Minimum values (mm)	Maximum values (mm)
Width of the rim ($B_R + B_{urr}$)	$D \geq 330$	133	145
Thichness of the flange (S_d)	$D > 840$	22	33
	$760 < D \leq 840$	25	
	$330 \leq D \leq 760$	27.5	
Height of the flange (S_h)	$D > 760$	27.5	36
	$630 < D \leq 760$	29.5	
	$330 \leq D \leq 630$	31.5	
Face of flange (q_R)	≥ 330	6.5	

19.2.2. Wheel defects and reprofiling

Wheel rims suffer from a number of defects due to: thermal phenomena, the form of the wheel profile, fatigue of the wheel-rail contact, and shelling (i.e. loss of material), (337). The frequency with which these defects appear is different in relation to the predominance of passenger or freight traffic, as it can be observed when comparing American and European railways.

The analysis of stresses in the wheel has revealed maximum values in the range of 2 t/cm^2 ÷ 3 t/cm^2, (334). For this reason, many railways have set 4 t/cm^2 as the limit of wheel stress, (344).

Due to wheel wear, a reprofiling is necessary and is conducted in relation to wear and traffic, usually every 100,000÷250,000 km of traffic, (344).

19.2.3. Life cycle of a wheel

The life cycle of a wheel may be short in the case of a tramway (250,000 km of traffic) or high in the case of a high-speed train (2,000,000 km), (Fig. 19.4), and is strongly affected by the nature of traffic (passenger or freight) and the value of the axle load.

19.3. Axles

Wheels are connected in pairs on axles, for which each vehicle includes at least two. The increases in vehicle weight, combined with the need to keep the stresses of the track within reasonable limits, have led to the addition of a third and then a fourth axle.

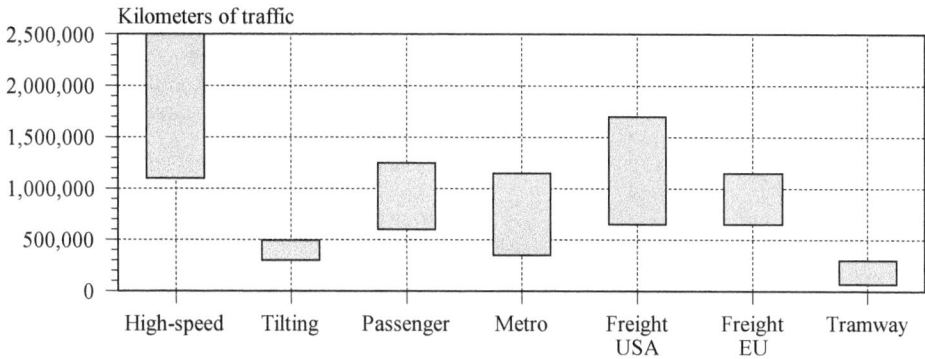

Fig 19.4. Life cycle of wheels in relation to the type of traffic, (337)

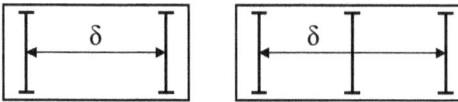

Fig. 19.5. Wheel-base of a vehicle

The distance δ between the two most distant fixed axles of a vehicle is termed the wheel-base δ of the vehicle, (Fig. 19.5). The greater the wheel-base of a vehicle, the more stable it is on straight track but the more difficult it will be to run on curved track. The maximum wheel-base length δ enabling a rail vehicle to operate in depots with a curve of radius R is given by the formula:

$$\delta_{max} = 0.30 \cdot \sqrt{R} \qquad\qquad (19.1)$$

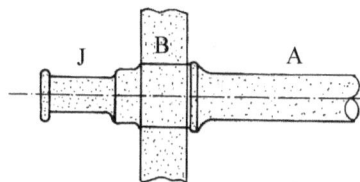

Fig. 19.6. Parts of an axle

An axle is composed of the following parts, (Fig. 19.6):
♦ the axle-journal J, which is supported by the bearing,
♦ the wedging region B, which is the part of the axle wedged into the wheel body,
♦ the main body A of the axle, located between the two wheels.

The vehicle load is applied to the bearings and thence transmitted to the journals and the wheels. There are two kinds of bearings, journal bearings and rolling-contact bearings, which are in turn distinguishable into ball bearings and roller bearings, (342).

Stresses in driving axles are both torsional and bending, while stresses in trailing axles are only bending.

491

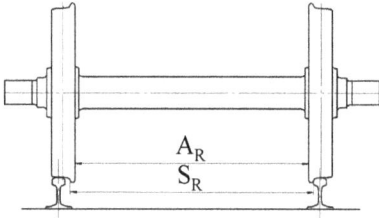

Fig. 19.7. Details of geometrical characteristics of an axle

According to the TSI, minimum and maximum values of the geometrical characteristics of an axle, in relation to wheel diameter, are given in Figure 19.7 and Table 19.2.

Table 19.2.

Maximum and minimum values of geometrical characteristics of an axle according to the TSI, (333)

Geometrical characteristic	Wheel diameter D (mm)	Minimum values (mm)	Maximum values (mm)
Front to front dimension (S_R)	D > 840	1,410	1,426
(distance between active faces)	760 < D ≤ 840	1,412	
	330 ≤ D ≤ 760	1,415	
Back to back distance (A_R)	D > 840	1,357	1,363
	760 < D ≤ 840	1,358	
	330 ≤ D ≤ 760	1,359	

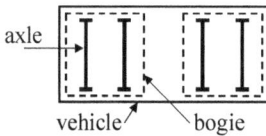

Fig. 19.8. Bogie

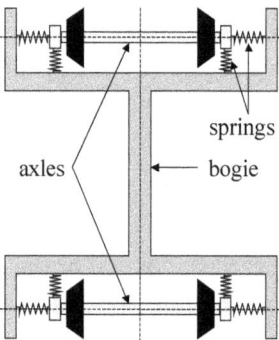

Fig. 19.9. Conventional type of bogies and location of springs

19.4. Bogies

19.4.1. Definition and functions of a bogie

The increase in the number of axles of railway vehicles gave rise to the need to segregate the axles into groups. This is achieved by means of bogies, where two or more axles are mounted on the same frame, (Fig. 19.8). In commonly used bogies, (Fig. 19.9), the axle body and the wheels are rigidly joined, and as a result they rotate at the same angular speed. The bogie frames are connected to the rail vehicle body and to the axles through springs and shock absorbers, providing the ve-

492

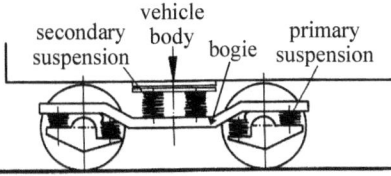

Fig. 19.10. Primary and secondary suspension of a rail vehicle

hicle with primary and secondary suspension, (Fig. 19.10).

Bogies perform the following functions, (341):

– support rail vehicle body firmly,
– run stably on both straight and curved track,
– ensure good ride comfort by absorbing vibrations generated by track defects and by minimizing the impact of centrifugal forces, when a train runs on curves.

19.4.2. Forms of bogies

Bogies are classified into two-axle, three-axle, etc., based on the number of axles. The two-axle bogie is the most common.

Another classification of bogies is into articulated and non-articulated types, (Fig. 19.11). Two non-articulated bogies usually support one rail vehicle body, whereas one articulated bogie supports the back end of the proceeding vehicle and the front end of the rear vehicle. Although the articulated bogie has some disadvantages, such as a complex structure, increased axle load (due to the support of one vehicle body by one bogie), and difficult maintenance, it offers many advantages, including a lower center of gravity, better ride comfort (since vehicle ends do not overhang bogies), and less effect of running noise on the passenger, as there are not seats over the bogies.

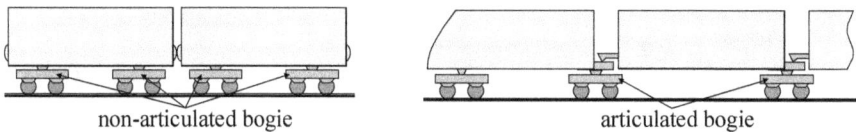

non-articulated bogie   articulated bogie

Fig. 19.11. Non-articulated and articulated bogie

19.4.3. Components of a bogie

The basic components of a bogie are:
– the rotating beam which allows the bogie to rotate around the vehicle body on curves, isolates the body from vibrations generated by the bogie, and transmits traction forces from the bogie to the body,
– the bogie frame, which accommodates various bogie equipment and is generally fabricated by welding together two side beams and two cross beams into

an H-shaped frame. The thickness of these beams was initially 6 mm, then increased to 9 mm and later on to 12 mm, but it was finalized at 8÷9 mm, in order to reduce weight, (341),

- suspension devices, which affect conditions of operation of the bogie and comfort of passengers,
- transmission devices, which consist of gear and flexible couplings to transmit motive power generated by the motor or the engine to the axle.

19.4.4. Self-steering bogie

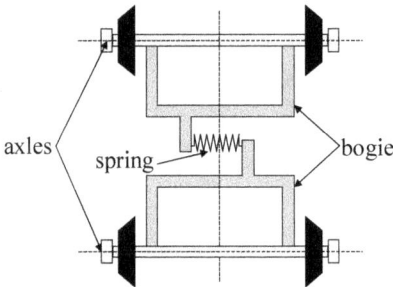

Fig. 19.12. Self-steering bogie

When a train runs on a curve at high speeds, the wheels exert high lateral forces on the rails and cause wear and tear of wheel flanges. These lateral forces can be reduced to one half or one third with the use of self-steering bogies, (Fig. 19.12), which allow the wheels and axles to move more freely, and thus the axle centerline is aligned on the radius of curvature, (343).

19.5. Springs

Springs are used between parts of the same rail vehicle as well as between successive vehicles.

If P is the load applied on a spring and Δl is its length variation, the work energy stored in the spring is:

$$W = \frac{1}{2} \cdot \Delta l \cdot P^2 \tag{19.2}$$

Depending on the type of rail vehicle, constraints are set to the maximum value Δl of springs as follows:

- locomotives, Δl: 10÷15 mm,
- passenger vehicles, Δl: 50÷70 mm,
- freight vehicles, Δl: 15÷25 mm.

19.6. Couplings and buffers

Couplings and buffers are devices used to interconnect rail vehicles to form trains. Their main purpose is to transfer from one vehicle to the other various forces (such as braking, etc.), electric current, etc.

For passenger comfort reasons, springs in passenger vehicle couplings have a low value of length variation Δl ranging between $12\div20$ mm. Contrariwise, in freight vehicle couplings, Δl is ranging between $30\div50$ mm.

As opposed to coupling springs, buffer springs in passenger vehicles should have a high value of Δl (ranging between $50\div70$ mm), so as to absorb the various shocks and vibrations thoroughly and quickly. A similar requirement is not necessary for freight vehicles, for which Δl ranges between $30\div50$ mm.

Hook couplings were used in the past to connect the vehicles in a train. However, automatic couplings are implemented today, ensuring automatically the coupling of successive rail vehicles, in particular the connection of brake air pipes and electric circuits, (340).

Buffers are employed to keep constant spacing between rail vehicles and to absorb shocks. In standard gauge tracks, buffer height above rail level is $0.90\div1.25$ m.

According to the European Technical Specifications for Interoperability, (333):

- the level of the centerline of buffers should be $0.98\div1.065$ m above rail level in all loading and wear conditions,
- the standard screw coupling system between vehicles should be non-continuous and should comprise a screw coupling permanently attached to the hook, a draw hook, and a bar with an elastic system,
- the height of the centerline of the draw hook should be $0.95\div1.045$ m above rail level.

19.7. Design of rolling stock

Design of rolling stock is a field of cooperation between the humanities and engineering sciences. The time spent by a passenger in a train should be considered as a moment of his/her life by respecting the individuality and personality of each passenger and with the concern to provide the possibility of enjoying travel time by relaxing, working, or doing anything else.

The design of each type of rolling stock should be analyzed in relation to the specific characteristics of traffic (intercity, regional, suburban), travel time, level of technology, cultural attitudes (habits) of clients, and cost of purchase and maintenance. A good design should ensure the following: safety, security, space and comfort, modularity of the space, calmness, a low noise level, lighting in relation to needs, easy access (particularly for people with disabilities and the elderly), and a view to the external physical environment.

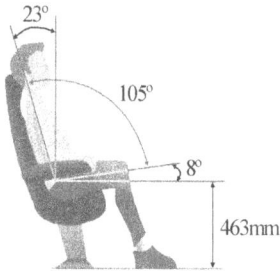

Fig. 19.13. Ergonomy of seats of second class of the high-speed Paris-Lyon train

Ergonomy, the required space, and adaptation of technology to human needs are key factors for a good design. Figure 19.13 illustrates the geometrical characteristics of a seat of the Paris-Lyon high-speed train. The whole analysis can be conducted with the use of the finite element method.

Until some years ago, the only priorities in rolling stock design were technology and economy. This has changed, as aesthetics, decoration, and a more human environment make part of a good design, which must bring out a new conception of human values, of innovation, and of optimizing travel time.

There is a tremendous variety of types of rolling stock for passenger and freight traffic. Table 19.3 (next page) illustrates the characteristics of some high-speed trains.

The reader can look for technical details in the web sites of constructors, which are https://www.mobility.siemens.com/global/en.html for Siemens, www.alstom.com/our-solutions/rolling-stock for Alstom (which absorbed Fiat Ferroviaria), https://rail.bombardier.com/en.html for Bombardier (which absorbed ABB).

19.8. Localization of the position of a rail vehicle with the use of GPS or other satellite systems

As explained in section 1.14, localization of the accurate position (coordinates x, y) at any moment (t) of a rail vehicle and of its speed (V) may be done with the use of GPS (Global Positioning System) and GSM (Global System for Mobile Communications) applications. Information on the position and speed of the train is transmitted from the satellites to the train and next to the operation control center, (see section 20.10.4), and is compared automatically to what has been planned. Any difference detected is transmitted automatically and the course of the train is rescheduled. Such systems are in use not only for high-speed trains (for instance the system named Localys for Thalys trains), but also for low-traffic lines, such as the European system named Locoprol (Low cost satellite signaling and train protection for low-density traffic railway lines).

GPS is a satellite system which was developed in the USA and is the first satellite system available for use by the general public. The last two decades, other satellite systems have been developed by other countries: Galileo in Europe,

Table 19.3.
Technical and operating characteristics of some types
of high-speed rolling stock, (compiled from data of constructors)

Type of rolling stock (year in service)	Thalys (1996)	Eurostar (2015)	ICE 3 Multiple-system (2000)	ICE 3 Single-system (2013)	China high-speed (2017)	Japan high-speed (2011)	Korean high-speed (2004)	Velaro E (2007)
Line operated	Paris - Brussels - Amsterdam - Cologne	London - Paris - Brussels	High-speed lines of Germany		Beijing - Shanghai	Various lines	Seoul - Busan	Madrid - Barcelona
V_max (km/h)	320	320	330 (220 for DC)	320	400	320	300	350
Capacity (seats)	377	894	413	444	556	731	955	404
Length (m)	200	400	200	200	209	253	388	200
Width (m)	2.90	2.95	2.95	2.95	3.36	3.35	2.90	2.95
Wheel diameter (m)	0.90	0.90	n.a.	n.a.	n.a.	n.a.	n.a.	n.a.
Axle load (tn)	17	17	16	14	17	13	17	17
Weight (empty/loaded)	n.a. / 385 tn	n.a. / 815 tn	436 tn / n.a.	454 tn / n.a.	n.a.	453.5 tn / n.a.	701 tn / 771 tn	434 tn / n.a.
Power supply / Traction power	25 kV 50 Hz / 8,800 kW 15 kV 16.66 Hz / 3,689 kW 1.5 kV DC / 3,680 kW 3.0 kV DC / 5,120 kW	25 kV 50 Hz / 12,200 kW 0.75 kV DC / 3,400 kW 3.0 kV DC / 5,700 kW	15 kV 16.7 Hz / 8,000 kW 25 kV 50 Hz / 8,000 kW 1.5 kV / 4,300 kW 3.0 kV / 4,300 kW	15 kV 16.7 Hz / 8,000 kW	25 kV 50 Hz / 9,750 kW	25 kV 50 Hz / 6,000 kW	25 kV 60 Hz / 13,560 kW	25 kV 50 Hz / 8,800 kW
Specific characteristics	Signaling equipment: TBL 1, TBL 2, TVM, ATB, Induzi, PZB/LZB, KVB, ERTMS Level 2		Starting force: 30 tn	Starting force: 30 tn		Signaling: ATC, Acceleration From 0 to 320 km/h: 380 sec	Acceleration: 0.45 m/sec²	Starting force: 28.3 tn, Acceleration From 0 to 320 km/h: 380 sec

Glonass in Russia, BeiDou in China, IRNSS in India, GZSS in Japan, etc. Thus, GPS has become just one of the many different sets of satellites that can provide data for the localization of the position of any object anywhere on or near the earth's surface.

The existence of many satellite systems and the need to exploit the advantages of all these led to the creation of the Global Navigation Satellite System (GNSS), which is a group of satellites that provide data for the localization of the position of an object. GNSS has access to multiple satellite systems at every moment and has equipment compatible for a continuous use for all navigational satellites placed under its umbrella. This compatibility results from the fact that GNSS uses geosynchronous (as opposed to geostationary) orbiting satellites with a unique orbit radius, (332).

Older satellite systems (such as GPS) were working in frequencies around 1,500 MHz and were providing an accuracy of around 3m, whereas new ones are working in frequencies around 1,100÷1,200 MHz and are providing an accuracy 10 times higher, on the order of 30 cm, (332).

GPS or GNSS provide the absolute localization of a train. The relative localization can be achieved with the use of other systems (such as the Doppler radar, wheel sensors, balises on the track, etc.) and needs a number of sensors which can be placed either on the track (balises, track circuits (see section 21.3.2)) or on-board (wheel sensors, inertial sensors). It is noteworthy that GPS and GNSS, which provide an absolute localization, are also based on on-board equipment.

19.9. Tilting trains

19.9.1. Needs which gave rise to the tilting technology

Many of today's railway lines were built almost one century and a half ago, at a time when technology and transport requirements recommended speeds which, by today's standards, are considered low. As a result, the track layout of many railway lines features curves with a small radius, particularly in mountainous areas.

During the past six decades, railways have tried to adapt to market requirements by improving the basic track components (rails, sleepers, and ballast), in the majority of cases, however, without addressing the problem of small-radius curvatures. There were some exceptions in this respect though. For instance, in some parts of the world new dedicated high-speed lines have been constructed, and improvements were made to the track layout of some existing lines. Despite these efforts, however, the majority of railway lines still feature almost the same layout as when they were first constructed. Thus, in most cases, railways must reduce travel times without necessitating, for cost reasons, the construction of new lines or the improvement of the layout of the existing ones.

Tilting trains could offer a low-cost solution in this respect, as they operate on existing tracks and can attain higher speeds (as compared to conventional trains), thanks to a mechanism which tilts the vehicle body of the train when negotiating curves, thus giving it an additional superelevation. Tilting train technology, under the right circumstances, offers an adequate alternative to high-cost layout improvement.

Nevertheless, the tilting train solution should always be thoroughly studied in each case; it should be examined whether, (338):

♦ the reductions in travel times are sufficient, taking into account what other transport modes (such as airplane, private car, and bus) can offer,

♦ any improvements to the track, the signaling, and power supply systems will be required,

♦ the return of investment will be satisfactory,

♦ the cost of operation will be competitive, as compared to that of other transport modes.

19.9.2. Tilting technology

Tilting trains try (and often fully succeed) to reduce cant deficiency in curves, (see section 14.2.2), by tilting the vehicle body in relation to the wheel-base, (see section 14.2.3, Fig. 14.4). There are two different tilting technologies, (339):

– *the passive method*: whereby the vehicle suspension increases when nego-tiating curves, so that the turning point of the vehicle remains above its center of mass. This method, which is applied by the Spanish Talgo, per-mits a tilting angle of 3° to 5° between vehicle body and axles,

– *the active method*: whereby a larger tilting angle, up to 8°, is achieved, which is calculated as a function of the non-compensated centrifugal acce-leration. When the train enters a transition curve, the transverse accelera-tions developed at the bogie are detected by accelerometers. Instructions to begin rotation of the vehicle body around its axis are transmitted by an elec-tronic device located at the front of the train. This technique is applied by the Italian Pendolino and ETR, the Swedish X2000, and the German VT610.

Two methods of detection of curves to be run by tilting trains have been developed, (339):

♦ *an on-board system for curve detection*: whereby bogie-mounted accele-rometers detect the transverse accelerations of the bogie. A curve detection system, called a gyroscope, which is placed at the front of the train, detects the moment that a train starts moving on a transition curve. At that moment, it transmits immediately an electronic signal, which initiates the tilting of the vehicle body (in relation to the acceleration detected). The technique is used by the tilting systems in continental Europe,

♦ *an electromagnetic system for curve detection*: whereby in-track devices transmit data concerning the track layout characteristics to a computer located on the train, so that, at the appropriate time, tilting of the vehicle body is initiated. This technique, which is applied in Japan and previously on the APT trains in the UK, could be regarded as more efficient than the previous one, but has the disadvantage that it requires in-track detection devices.

19.9.3. Technical and operating characteristics of tilting trains

The main technical characteristics of tilting trains are as follows, (339):

Angle of tilting
The trains featuring a passive tilting system (Talgo) achieve an angle of tilting of 3° to 5°, whereas the trains featuring an active tilting system achieve angles of tilting up to 8°.

Maximum speed
All electric tilting trains have high performance with respect to speed, ranging from 200 to 250 km/h. The diesel tilting trains feature a maximum speed of 160 km/h and are primarily used for suburban services.

Relation of speed V_{max} to the radius of curvature R
The relation of speed V_{max} to the radius of curvature R depends on the values of cant and cant deficiency. For *conventional* rolling stock, the relation between V_{max} and R(m) is generally:

$$V_{convent.\ train}^{max}(km/h) \cong 4.7 \div 5.0 \cdot \sqrt{R(m)} \qquad (19.3)$$

For *tilting* trains, this relation becomes:

$$V_{tilting\ train}^{max}(km/h) \cong 6.0 \cdot \sqrt{R(m)} \qquad (19.4)$$

Thus, an increase in speed up to 20%÷25% is achieved by tilting trains in curves, as compared to conventional trains.

Additional superelevation
The increase in speed results from the additional superelevation, induced by the tilting system, which can reach up to 150÷200 mm.

Mechanism of tilting

Three different kinds of tilting mechanism have been developed: pneumatic, hydraulic, and electric. In order to reduce the forces exerted on the rail, the technique of self-steering radial bogies is applied, (339).

Axle load

All tilting trains have a low axle load, ranging between 13÷17 tonnes, as is the case with high-speed trains.

Track gauge and geometrical characteristics of vehicles

Tilting trains feature a high adaptability to the different track gauges (1.435 m, 1.524 m, 1.067 m, 1.000 m) and geometrical requirements of the rolling stock.

Signaling

Generally, the use of tilting technology is accompanied by an increase of the maximum speed of tilting rolling stock in straight track (compared to conventional trains). This results in an increase of braking distances, thus requiring certain changes with respect to signaling.

Power supply

The power supply system also requires certain adaptations, the extent of which depends on the specific requirements of the railway and country concerned.

Loading gauge

Tilting technology rolling stock has been designed so that the loading gauge is sufficient to allow for the additional superelevation, which is achieved in curves. Thus, no problems arise when tilting trains run in tunnels.

Track characteristics and defects

A high quality of track (rail UIC 60, concrete sleepers, ballast with a minimum thickness of 35cm and a high hardness) is required for the operation of tilting trains. If the maximum speed does not exceed 160 km/h, timber sleepered track is also adequate.

Track defects and frequency of track maintenance are almost similar to that of tracks run by conventional trains.

19.9.4. Reductions in travel times by tilting trains

Tilting trains have achieved a reduction in travel times between 12% and 33%, (338).

However, when the application of tilting trains is not accompanied by an increase in speed in straight track (compared to conventional rolling stock), the reduction in travel times (only as a result of higher speeds in curves) ranges from 12% to 20%, with a mean value of 15%. This must be considered as being the direct effect of tilting, since conventional trains can also achieve an increase in speed on straight track.

19.9.5. Cost of tilting trains

The cost of tilting rolling stock constructed in Italy is reported to be $3\div5\%$ higher compared to conventional rolling stock. However, French plans for a tilting TGV reported higher cost differences, on the order of $10\div20\%$.

According to studies conducted by the French railways, a reduction in travel time of 1 minute achieved by tilting trains costs between $1.5\div4.5$ million € for speeds up to 160 km/h and between $9\div18$ million € for speeds higher than 160 km/h, as compared to an average cost of $35\div50$ million € for non-tilting TGV trains, (see also section 5.2.2, Table 5.1). The additional costs for running tilting trains result from the higher costs of tilting rolling stock and from additional costs in track and signaling.

19.10. Maintenance of rolling stock

19.10.1. Objectives and scheduling

Maintenance of both the infrastructure (track, signaling, electrification or diesel power supply) and the rolling stock (pulling (locomotives), pulled (rail vehicles)) is critical for a high-quality and safe rail transport, which operates at the lowest possible cost. As infrastructure is a rigid system, the most vulnerable parts of the railway system are signaling and rolling stock. Maintenance of locomotives is analyzed in chapter 20. The general objectives of maintenance of rolling stock are to ensure safety and proper function of the rail system, increase lifetime of all rail components, and provide a high level of comfort and well-being for passengers.

Scheduling of maintenance of rolling stock can be driven by one of the following parameters: time (from the previous maintenance), kilometers run (from the previous maintenance), alarming condition of some component or material.

19.10.2. Levels and works of maintenance

We usually distinguish four (and sometimes five) levels of maintenance of rolling stock:
- level 1 (daily inspection). It consists in inspecting externally the state of breaks, electric systems, pantographs, and other essential parts of rolling

stock. The maximum time between two successive inspections of level 1 is 48 hours,

- level 2 (monthly inspection). In this case, the in-situ inspection concerns the operation and functions of breaks, electric equipment, pantographs, etc. Two successive inspections of level 2 depend on the type of rolling stock and are conducted for high-speed trains every 30 days or 30,000 km of operation, (38),
- level 3 (bogie overhaul). At level 3 the following are inspected: breaks, wheels, gears and other parts. Maintenance of level 3 is conducted for high-speed trains every 18 months or after 600,000 km of operation,
- level 4 (general overhaul). It consists in disassembling railway vehicles and checking and repairing every component or even small part of a vehicle. Maintenance of level 4 is conducted for high-speed trains every 36 months or after 1,200,000 km of operation.

19.10.3. Equipment and staff

Maintenance of rolling stock requires specialized equipment and staff with a high level of expertise. It can be realized by the railway company in specialized workshops, by the manufacturer in its workshops, or in workshops belonging both to the manufacturer and the railway company.

19.11. Preventive, corrective, and condition-based maintenance of rolling stock

As with infrastructure, maintenance of rolling stock can be conducted according to one of the following methods: preventive, corrective, condition-based. Preventive maintenance is in principle a scheduled maintenance based on predefined intervals, whereas corrective maintenance consists in unscheduled repairs in case of a coming (or already realized) failure.

Preventive maintenance is based on regular inspection and takes into account data and experience from the past. It aims at minimizing the probability of failure of the rail vehicle component under consideration and usually suggests specific time intervals or hours of operation of rolling stock between successive maintenance sessions. Preventive maintenance does not take into account the remaining resistance reserves of any component or material, does not optimize lifetime, and may become expensive if it is based only on data of the past, which disregard recent technical evolutions and achievements.

Corrective maintenance of rolling stock aims at maximizing the lifetime of any component, minimizing any unnecessary repairs, and replacing or maintaining any component just before a coming failure or fault. If it is not based

on a continuous and ongoing measurement of the health and mechanical reserves of a material, it may prove dangerous. Service levels under corrective
maintenance are generally below acceptable comfort levels.

Condition-based maintenance stands on a continuous and regular monitoring of a part or a material of the rolling stock, so as to replace it just before
failure. It can be combined with preventive maintenance and, thus, lead to an
optimization of the lifetime of any part of the rolling stock. Condition-based
maintenance requires a number of sensors positioned at the more sensible
points of the component under consideration and can make use of artificial
intelligence methods (particularly artificial neural networks).

On the basis of monitoring the situation of a specific material or item and
taking into account a series of maintenance data of the past (both of the company owning the train but of other companies also) railway authorities can
decide the following: the accurate time to conduct rolling stock maintenance
and achieve longer intervals between successive maintenance, lower rates of
unpredicted failures, corrective maintenance practically reduced to zero, lower
maintenance costs, higher availability of vehicles, lower operation costs, and
finally a higher competitivity of the rail sector.

20 Diesel and Electric Traction, Hydrogen Trains

20.1. The various traction systems

The railway vehicle which provides the necessary traction power for the movement of a train is often referred to as a locomotive. Traction power may use steam, diesel, or electric power. Very recently, trains fueled from hydrogen have also been developed.

The first power generation system used for traction was steam. Indeed, the spread of the railways was primarily due to the industrial exploitation of the steam invention, which coincided with the first industrial revolution (steel, coal, steam, (see Fig.1.3)). The first steam engine appeared in 1804 and was used for railway traction in the 1830s. Thereafter, it remained for more than 120 years the principal traction mode for railways.

20.2. Steam traction

20.2.1. Operating principle of the steam engine

Steam engine operation is based on the following principle, (Fig. 20.1). The wheel T is connected to the rod MD by the crank TM. The rod MD is connected to the piston rod DE of the steam cylinder, thereby converting the reciprocating motion of the rod DE, generated by steam power, into wheel rotation. The wheels of the motor axle, connected to the rods (one on each side), are known as the main driving wheels. A single driving axle is not sufficient to provide the required traction force and therefore other driving axles are also provided. However, since the latter cannot be directly connected to the rod, their wheels are coupled to the main driving wheels by rods called coupling rods which, in turn, are successively connected to the wheel cranks TM, T_1M_1, etc. These axles are known as coupled axles. The coupled axles and the axles of the main driving wheels are the driving axles. The total number of coupled axles seldom exceeds five or at most six, but is never less than two.

505

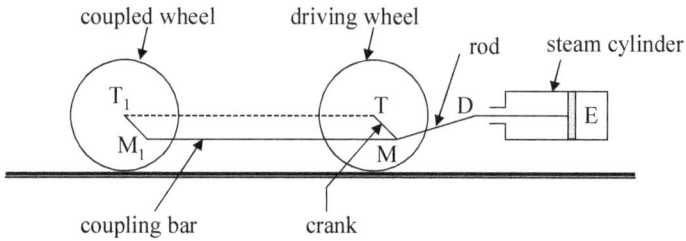

Fig. 20.1. Operating principle of the steam engine

A steam locomotive may use coal or petroleum as fuel. Biomass has also been suggested as a source of fuel. According to the aforementioned operating principle, the thermal energy liberated by either coal or petroleum is stored as dynamic energy resulting from steam pressure and is converted, when necessary, to train kinetic energy.

20.2.2. Main parts of a steam locomotive

The main parts of a steam locomotive are:
- the vehicle, mainly comprising the rolling devices, the frame, the coupling devices, buffers, suspension, etc., as well as the driver's cab, in which all equipment and instruments for locomotive operation and control and for running the train as a whole are located,
- steam generation equipment, i.e. the boiler and associated parts such as water pump, etc.,
- the engine, i.e. steam cylinders, pistons, slide valves, distribution devices,
- various auxiliary systems, e.g. air compressors for braking, central heating, sand boxes to increase adhesion between driving wheels and rails, lubrication and braking devices, several safety systems, etc.

20.2.3. Disadvantages and obsolescence of the steam locomotive

Presently, steam traction is commercially employed only on a few railway lines in Africa and Asia, whereas in Europe and Northern America it has been a museum item for quite some decades now and is used only for the operation of touristic trains in some routes. There are many reasons that steam locomotives are no longer used and were replaced by diesel or electric locomotives, (324):
- low fuel efficiency. Only about 6% of the energy liberated by coal combustion is used in steam locomotives for train traction,
- poor technical performance. Most steam locomotives cannot exceed a power of 3,000 Hp (only a few of them have a power up to 6,000 Hp) and a maximum operating speed of 120÷140 km/h,

- the need to maintain a large number of water supply facilities,
- high maintenance cost,
- time-consuming fuel replenishment procedure. A steam locomotive can operate autonomously only 12÷14 hours,
- increased fire hazard,
- harm to the environment (atmospheric pollution, noise).

20.3. From steam traction to diesel traction and electric traction

20.3.1. From steam traction to diesel traction

Diesel traction of trains was introduced during the 1930s but was systematically developed during the 1940s and 1950s. Diesel locomotives are driven by a diesel internal-combustion engine. In comparison to steam traction, diesel traction offers far higher efficiency, a lower operating cost (by at least 50%), much better performance (power, speed), cleaner operation, improved passenger comfort, convenience, and less strenuous work for the train driver.

20.3.2. From steam traction to electric traction

The emergence of electric railway traction dates back to 1879. The first implementation of electric power in railway vehicles was restricted to urban areas, with the development of the electric tramways between 1880÷1914 and with the first metro lines at the end of the 19th century. Thus, electric traction was first introduced in railways in 1900, when it was adopted in the London and Paris metro lines and in the mountainous railway lines of Switzerland.

Since 1920, electric traction has been extensively used, especially after 1950. Its operating cost is almost half compared to the cost of diesel traction, but it necessitates a higher initial expense due to the fixed installations that are required (contact wires, pantographs, substations, etc.), (see sections 20.5.1 and 20.8.1). Electric traction is accordingly used only on high-traffic lines.

20.3.3. Gas turbine locomotives

Gas turbine locomotives were developed in the 1960s and early 1970s. The basic element of a gas turbine locomotive is the gas turbine engine, operating by the expansion of overheated and compressed gaseous combustion products. After the energy crisis of 1973 and the abrupt increase of oil prices, gas turbines were no longer cost-effective, due to their high energy consumption, and they have been abandoned.

20.4. Diesel traction

20.4.1. Operating principle of the diesel engine

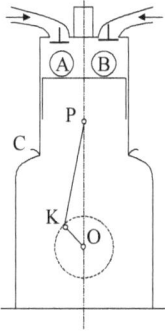

Fig. 20.2. Schematic representation of diesel engine

The basic element of the diesel engine, (Fig. 20.2), is the cylinder C, inside which the piston P moves by reciprocating motion. This reciprocating motion is transmitted as rotary motion by the rod PK and the crank OK to the main driving crankshaft O. On the cylinder cover are located the valves A and B, permitting functions in the following order:
- suction,
- compression and injection of gaseous fuel,
- combustion and expansion of gaseous combustion products,
- exhaust.

A third valve is provided near the other two, letting in compressed air to start the engine. All three valves are spring loaded and are opened and closed at appropriate times by levers and a camshaft.

The aforementioned description refers to a single action engine in which suction, injection, and exhaust occur in the cylinder on the same side of the piston, specifically on the upper side.

The motor function of a diesel engine is carried out in four cycles, as follows:

1. *Suction stroke.* This involves the time period in which the piston, starting from the upper dead point (UDP), with valve A open and valve B closed, descends to the lower dead point (LDP), while cylinder C is filled with fresh air.

2. *Compression stroke.* This corresponds to the ascent of the piston from the LDP, when valve A and valve B close and remain closed, to the UDP. Thus, the air pressure in the cylinder increases from 1 atmosphere to 30÷40 atmospheres, causing the temperature to reach 400÷500°C. Shortly before the piston reaches the UDP, fuel is injected under pressure into the cylinder. The fuel droplets subsequently ignite at the moment when the piston reaches the UDP. The pressure within the cylinder C then reaches 50÷80 atmospheres and the temperature reaches 1,800÷3,000°C.

3. *Expansion stroke*. This corresponds to closed valves and to a piston travel from UDP to LDP under the action of expanding gases. Valve B opens at the appropriate moment.

4. *Exhaust stroke*. This covers the piston travel from LDP to UDP with valve B open. Combustion gases are forced out, while a large part of the gases is released from the cylinder during the first moments of this stroke. The piston pushes the remainder gases during its ascent to the UDP, when valve A opens.

Upon the completion of the above cycle, the piston returns to its initial position and the process is repeated. It is possible, however, to complete the entire cycle in two strokes, in which case we have a two-stroke engine. With respect to the number of cylinders, there are diesel motors with 4, 5, or 8 in-line cylinders or 8÷12 cylinders in a V arrangement. Motor speeds range from low (750 rounds per minute, rpm) to high (1,200÷1,600 rpm) and can provide power up to 4,000 Hp. Diesel locomotives in the USA operate at 1,000 rpm, whereas in Europe they have higher motor speeds, which in some cases can reach 2,000 rpm. Low-speed motors are heavier for the same power. In order to withstand high temperatures, cylinders have double walls for water circulation in between.

20.4.2. Transmission systems

In diesel locomotives, drive power transmission from the motor to the wheels is achieved by the following methods:

- by hydrodynamic transmission and hydrodynamic speed shifting (e.g. of the Voith type),
- by hydrodynamic transmission and mechanical speed shifting (e.g. of the Mekydro type),
- by electrical transmission: the diesel engine drives an electric generator which feeds in turn electric motors joined with the driving wheels. In the case of an electrical transmission, no gearboxes are employed and operating conditions are identical to those in electric locomotives. Diesel locomotives of this type, termed diesel-electric locomotives, are essentially direct current generator systems supplying the motors of the driving axles. If traction requirements are high, several diesel locomotives in series may be used in the same train,
- by other means, e.g. by hydrostatic transmission, or by purely mechanical transmission.

20.4.3. Requirements of diesel locomotives

A diesel locomotive should meet the following requirements:

- pulling capability of medium and heavy loads on a level track, uphill, or downhill, with a high transmission box efficiency at medium and increased speeds,
- overload capability, on the one hand in the low-speed range, and on the other hand uphill at full load,
- capability to brake with no slippage at increased speeds, as well as to keep within speed limits downhill without using mechanical brakes,
- motor operation within the favorable operating region,
- high reliability and low maintenance cost.

20.4.4. Advantages and disadvantages of diesel traction

Diesel traction, in comparison to electric traction, requires no additional costs for track equipment and provides autonomy.

However, diesel traction has the following disadvantages compared to electric traction:

- lower performance (power, force, speed),
- higher energy consumption,
- more air pollution and noise,
- higher maintenance costs.

20.5. Electric traction and its subsystems

In contrast to diesel traction, where the energy required for train operation is generated within the diesel locomotive itself, the energy needed for electric traction is transmitted to the electric locomotive by an external subsystem, the power supply subsystem.

20.5.1. Power supply subsystem

The power supply subsystem includes:

- *substations*, where the voltage is stepped down and (in certain electric traction systems) the alternating current (AC) frequency is converted or the AC is rectified into direct current (DC),
- *overhead contact wires* or *conductor rails* to convey the electric energy from the substations to the electric locomotive.

The electric substations may obtain electric power:

- either from the national high-voltage power network at a frequency of 50 Hz in Europe or 60 Hz in the USA,
- or from a separate high-voltage distribution network, at a frequency of 16⅔ Hz, considerably lower than that of the national network. This sepa-

rate network may be connected to the national network or may be independent, i.e. it may have its own power generating plants.

Therefore, when planning the electrification of a railway line (existing or under construction), the proximity of the national power network to the railway line as well as the energy available from the power network should be considered.

In substations, the characteristics of the electric energy obtained from the power network are changed (voltage reduction and/or frequency conversion and/or rectification from AC to DC) and the converted energy is channeled through the transmission line to the rail vehicles. Substation spacing ranges from 15÷70 km and mainly depends on the electric traction system but also on the line traffic load.

As a rule, the transmission line from substations to vehicles is in single-phase configuration. Electric traction engines obtain electric power from a conductor, which may be:

– either an overhead contact wire, used in intercity and suburban railways and (sometimes) in metros,
– or a conductor rail (one or two), used in metros and in some suburban railways.

When only one overhead contact wire or conductor rail is provided, grounding of current is done through the rails. Either one or both rails may be used.

20.5.2. Traction subsystem

The traction subsystem includes the electric traction engine with all its equipment and devices. In this subsystem, electric energy is converted into mechanical energy, which is used to operate the train.

In the case of an overhead contact wire, electric power is transferred to the vehicle through a pantograph. In the case of third or fourth rail conductors, collector shoes on the vehicles pick up the power, (see section 20.8.6).

20.5.3. Requirements and priorities

The two aforementioned subsystems, power supply and traction, have different requirements and, depending on the priority assigned to energy transmission (power supply subsystem) or energy use (traction subsystem), various electric traction systems have been developed.

20.6. Electric traction systems

20.6.1. Direct current traction

Direct current has a better performance compared to alternating current as regards the traction subsystem. For a long time, therefore, from the beginning of the 20th century until about 1950, priority was given to good motor operation. As series-excited DC motors offered the best operating conditions for railway traction until some decades ago, railway engineers sought an electric traction system using direct current. Early electric transmission systems, therefore, operated at the same voltage as the traction motors. The main voltages employed were:

– *750 V*, mainly used on third and fourth rail systems,
– *1.5 kV*,
– *3 kV*.

The above voltages are far lower than those employed on national power networks (150 kV, 220 kV, 280 kV) and too low for efficient power transmission. DC railway traction therefore necessitates large cross-sections of the contact wire (400÷900 mm^2) and closely spaced substations. Spacing of substations is 15÷20 km in the case of 1.5 kV and 35÷40 km in the case of 3 kV, (351).

Thus, direct current railway traction, though more efficient as regards the traction subsystem, proves less efficient when it comes to the power supply subsystem. DC traction corresponds in 2018 to 35% of electric railway lines worldwide and has mainly been used in France, Spain, Italy, Japan, certain parts of the UK, Russia, and India.

According to the UIC, direct current traction systems for speeds up to 250 km/h must comply with the following requirements; standard height of contact wire: 5.0÷5.5 m (minimum 4.9 m, maximum 6.2 m), maximum permissible average contact force for 220 km/h<V<250 km/h: 26 kg, for 200 km/h<V<220 km/h: 22 kg, for 160 km/h<V<200 km/h: 18 kg, maximum span length 65 m, maximum lateral deflection of contact wire at support ≤30 cm, (351).

20.6.2. Alternating current traction

Alternating current has a better performance compared to direct current as regards the power supply subsystem, but encounters problems in the traction subsystem. AC motors meeting the requirements of traction engines are series-excited AC motors with a collector, which, however, face problems related to the AC frequency. The need therefore initially arose to use AC at a frequency lower than the 50 Hz or 60 Hz used at the national power network.

20.6.2.1. Alternating current traction at 15 kV, 16⅔ Hz

In this system, electric substations may obtain power from either of two sources:

– from the national power network (at a frequency of 50 Hz or 60 Hz), in which case there is voltage reduction and frequency conversion in the substations,

– from a separate network carrying low-frequency AC, in which case there is only voltage reduction in the substations.

AC traction at 15 kV, 16⅔ Hz is used in Central Europe (Germany, Austria, Switzerland) where substations are supplied from special low-frequency AC power plants, and in Northern Europe (Sweden, Norway) where substations are supplied from the 50 Hz national power network. However, since 1996, the frequency of 16⅔ Hz was increased by 0.2% to 16.7 Hz in tracks of Austria, Switzerland and the former West Germany. AC traction 15 kV, 16⅔ Hz corresponds in 2018 to 11.7 % of electric railway lines worldwide, (Fig. 20.4), substations are spaced 20÷50 km apart and overhead contact wires have considerably smaller cross-sections (on average 100÷120 mm^2, from 80 mm^2 to 150 mm^2) than in DC traction. This system, however, has disadvantages concerning the operation of motors, mainly their great susceptibility.

20.6.2.2. *Alternating current traction at 25 kV, 50 Hz*

To overcome the disadvantages of the two previously described systems, it was necessary to seek a traction system that combines the advantages of both systems without presenting any of their disadvantages. This was achieved after 1950 with the development of efficient lightweight ignitron rectifiers, later superseded by thyristors, which in their turn have been superseded during the 1980s by the 'gate turn off' technology, (see section 20.10.3). In this electrification system, substations are supplied from the national electric network and simply step the voltage down to 25 kV, 50 Hz, which is transmitted to the locomotive through the contact wire and the pantograph, (see Fig. 20.12). In the locomotive, the voltage is again stepped down, rectified, and applied to the series-excited DC traction motors, (358).

The 25 kV, 50 Hz system represents in 2018 53.9% of electric railway lines worldwide and is almost exclusively used in new electric railway traction facilities. Substations are spaced at distances of 50÷70 km and contact wires have cross-sections 3÷5 times smaller than in DC systems. Thus, French railways employ, in their electrified lines at AC 25 kV, 50 Hz, typical cross-sections of the contact wire of 107 mm^2 in non-high-speed lines and 150 mm^2 in high-speed lines. In 2018, out of 285,249 kilometers of electrified lines *worldwide*, 25.4% used DC 3 kV, 5.4% used DC 1.5 kV, 11.7% used AC 15 KV, 16⅔ Hz, 53.9% used AC 25 kV, 50 Hz, while 4.2% used other electrification systems, (1).

A comparison of the construction cost for traction systems using 1.5 kV DC and systems using 25 kV, 50 Hz AC yields figures lower by 30% for the latter than for the former, (Fig. 20.3), (356).

Figure 20.4 illustrates the traction systems for the various European countries and Figure 20.5 the basic components and characteristics of each system. In 2018, out of 175,926 kilometers of electrified lines in *Europe* (Russian Federation included), 35.5% used DC 3 kV, 5.1% used DC 1.5 kV, 19.6% used AC 15 kV, 16⅔ Hz, 37.4% used AC 25 kV, 50 Hz, while 2.4% used other electrification systems, (1).

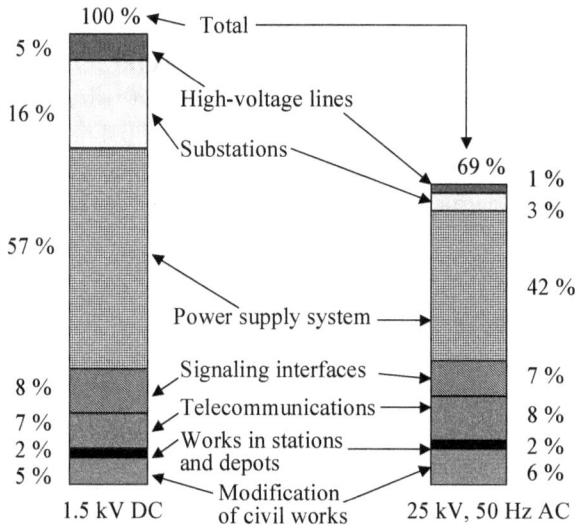

Fig. 20.3. Construction cost of DC and AC traction systems (economic data of Western Europe), (356)

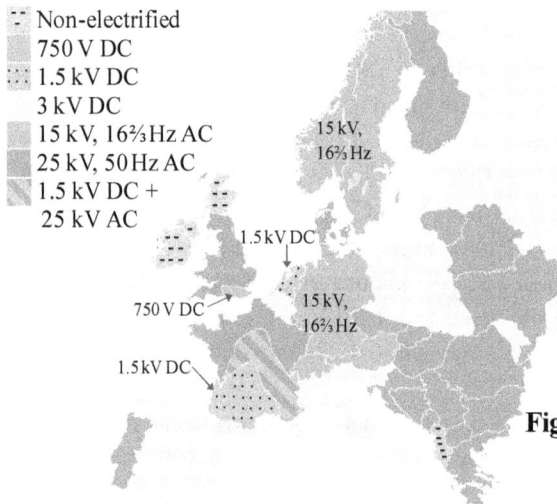

Fig. 20.4. Electric traction systems in various European countries

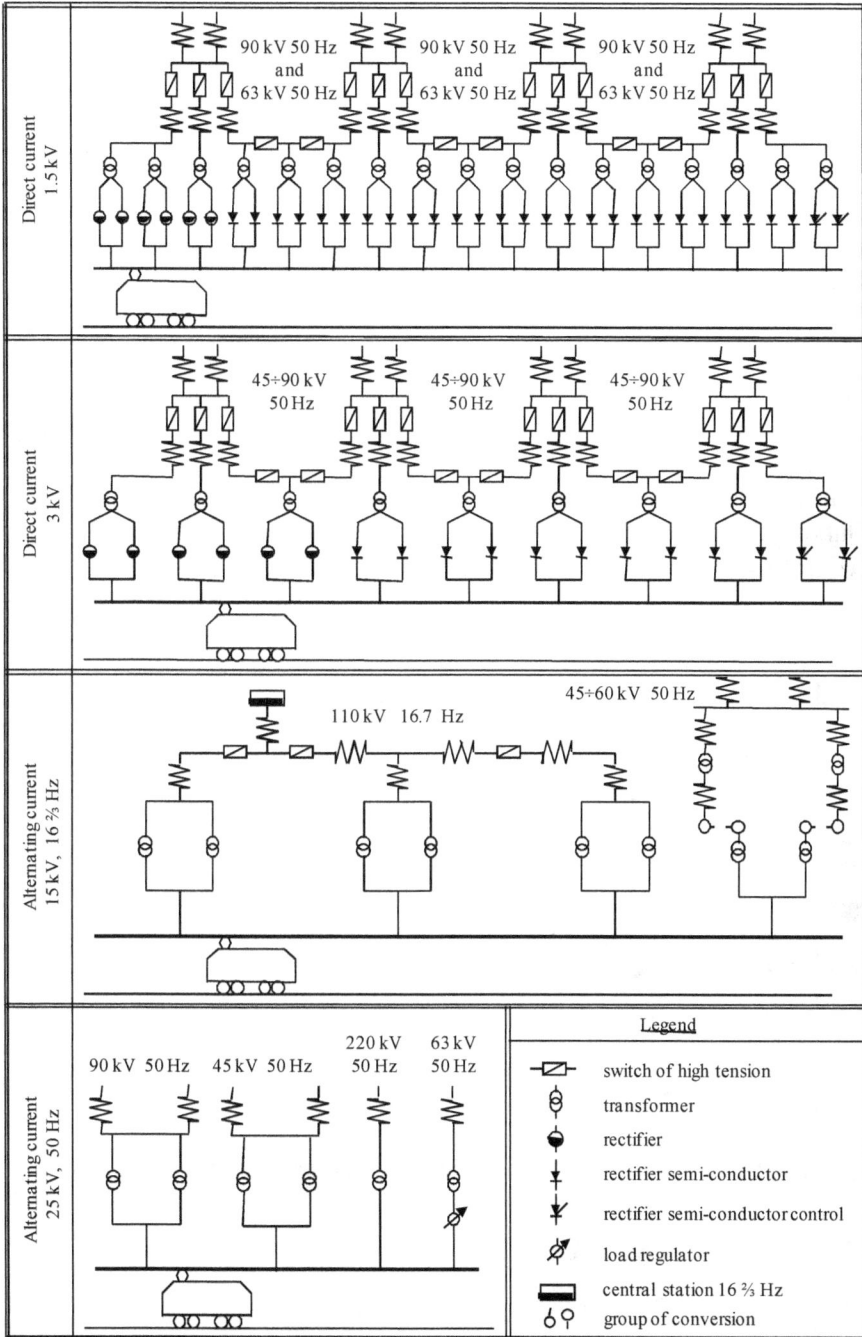

Fig. 20.5. Basic components and characteristics of the various electric traction systems, (360)

20.6.3. Advantages and disadvantages of electric traction compared to diesel traction

A basic advantage of the electric locomotive is its specific power (50÷55 kW/t), more than double the specific power of the diesel locomotive (20÷25 kW/t), (Fig. 20.6), (347).

Electric locomotives can sustain momentary overloads (when starting, on steep gradients, etc.), in contrast to diesel ones, as long as acceptable lifetime and maintenance cost constraints are considered.

Furthermore, along lines crossing high-altitude areas, no power drop is observed with electric traction engines. This is not the case with diesel engines, as the density of the air entering the engine is significantly reduced.

In the case of long tunnels, electric traction is mandatory due to the limited air supply.

Finally, electric engines cause during operation little, if any, atmospheric pollution, while maintenance is far simpler and easier than with diesel engines. Nevertheless, it should be noted that even diesel trains pollute much less than private cars and trucks, (see section 23.2).

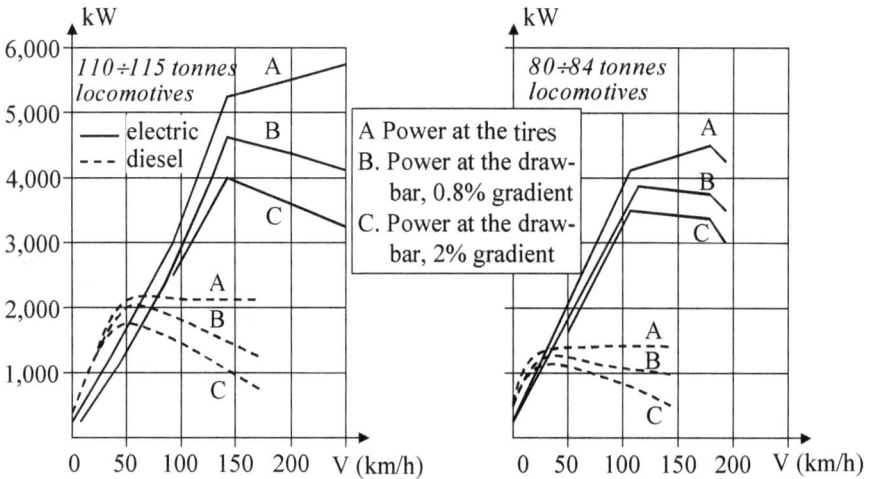

Fig. 20.6. Comparative power of electric and diesel locomotives, (347)

20.7. Feasibility analysis before electrification

20.7.1. Feasibility analysis parameters and procedure

When conducting a feasibility analysis to justify electrification of a railway line, two cost factors should be taken into consideration:

516

- fixed installations costs, including overhead contact systems and substations, which do not depend on traffic volume,
- operating and maintenance costs, which depend on traffic volume.

The quantity commonly taken into account is the total annual costs as a function of the line's traffic, an index of the latter being the energy consumed annually per kilometer of line. Figure 20.7 illustrates a comparative presentation of the annual costs of diesel and electric traction.

We can remark that electric traction is not cost-effective for low-traffic lines. However, as the point beyond which electric traction becomes cost-effective is approached, a more detailed investigation of the problem is required.

The period of feasibility analysis usually covers 20÷25 years; costs (initial construction cost and annual operating costs) are converted for each traction system to constant prices by the present value method. A feasibility analysis of electric traction involves many uncertainties, particularly with respect to the price of liquid fuel in the next 20÷25 years, the discount (actualization) rate whereby the various costs are converted to constant prices, the length of the feasibility analysis period, etc. It is accordingly advisable to also perform a sensitivity analysis (which aims to examine the impact of the variation of one parameter of the problem (e.g. construction cost) to the result of the feasibility analysis), (16).

annual costs

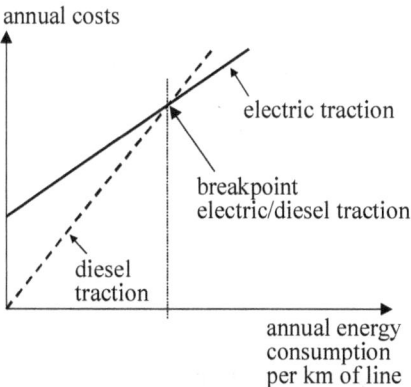

Fig. 20.7. Annual costs as a function of energy consumption per kilometer of line for diesel and electric traction

20.7.2. Criterion for selection of the lines to be electrified

In most cases, the need arises to reach a conclusion easily and quickly as to whether or not electrification of a particular line is advisable and then conduct a detailed feasibility study. The various railway networks and authorities have established simple criteria to this effect; the most widely used among them is the number of trains on a line or (more precisely) the energy consumption per kilometer of line.

The criteria in question vary from one railway to the other, since the particularities of each country as regards cost of labor, cost of energy, cost of borrowing, etc., are involved.

A criterion, which, however, can only be used to make a first approximate estimation, is the number of trains running on the line. For example, until 1973 (when the cost of energy was low), a line had to be run daily by at least 30 trains per direction to qualify for consideration of electrification. After the energy crises of 1973 and 1979, the criterion became around 15 trains per direction daily, and it has changed ever since in relation to the increase of the cost of energy.

However, given that a train may transport passenger or freight with a varying number of vehicles, a more representative criterion may be the annual energy consumption per kilometer of line, which in the case of a double track amounts at $1.0 \div 1.3$ MW/km for a high-speed line and $1.7 \div 2.5$ MW/km for a heavy-freight line. For instance, the French railways consider in principle an annual consumption of 70 MWh/km of line as the electrification cost-effectiveness threshold, while the German railways estimate this limit at 150 MWh/km, (355). Criteria, therefore, may differ significantly from one railway to the other. However, many governments and institutions (such as the EU) are considering electrification as just one of the measures to combat the greenhouse effect and create a green economy and growth. Based on this line of thinking, the above criteria may be further lowered in the future.

When the traffic load or the energy consumption on a particular line exceeds the above limits, a detailed feasibility study should be conducted, as described in section 20.7.1, before any decision to electrify the line is made.

20.8. Overhead contact system

20.8.1. Parts and components of the overhead contact system

The overhead contact system includes, (354):

♦ Feeder conductors, contact conductors (touching the pantograph), suspension wire ropes, guy wires.

♦ Conductor support structures, (Fig. 20.8), which may consist of poles or frames.

♦ Insulators, post brackets, (Fig. 20.9), tensioning devices (usually every 1,200 m), counterweights, various mounting hardware, wires connecting the poles to the contact wire and to the ground, and conductors for connection to the substations.

As illustrated in Figure 20.9 the overhead contact system is suspended from the post brackets, which in turn are mounted by insulators on supporting poles, erected 3.25 m÷3.80 m from the track axis (this distance is increased in curves by 40 cm maximum). The post brackets are usually zinc-plated steel pipes.

a. pole-supported *b. frame-supported*

Fig. 20.8. Support of the overhead contact system

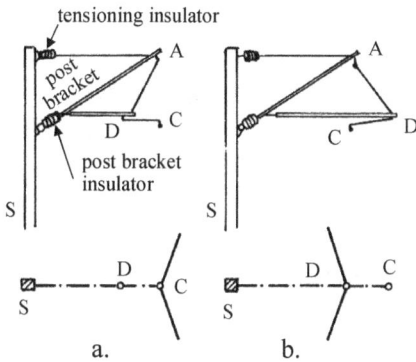

Fig. 20.9. Insulators and post brackets (C is the supply point)

a. catenary in traction
b. catenary in compression

20.8.2. Calculation of the characteristics of the contact wire with the use of physical models

Calculation of the cross-section and other characteristics of the contact wire is performed on the basis of the permissible voltage drop from the substations to the locomotive switchboards, allowing a fluctuation of no more than 10% from the nominal value.

Theoretical calculation of the voltage drop is based on the assumption that the passing load is constant, which, however, is not the case, as the number of trains, their positions on the track, etc. may vary. Calculation of the characteristics of the contact wire can be conducted either with the use of the finite element method, (see below section 20.8.3), or with the help of a small-scale physical model, where:

- substations are simulated by constant-voltage sources, complemented by suitable resistors simulating the internal circuits of the stations,
- current feeder and return wires are simulated by suitable resistors,
- trains are simulated by variable resistors, which can be connected to various points of the line,
- suitable measuring instruments in the physical model give a direct reading (as a function of substation distance and transmission line cross-section) of substation output voltage, total current at each substation, voltage at engine switch-boards, etc.

This physical model was used extensively until the 1980s and enabled the testing and verification of various combinations of transmission conductors, substation distances, etc. and the selection of the optimum solution.

20.8.3. Calculation of the contact wire with the use of the finite element method

The finite element method can be used for the accurate and dynamic analysis of both the mechanical and electrical behavior of the contact wire and it can take into account the following: geometrical parameters of the overhead contact system, conductor characteristics (cross-sections, materials and their appropriate constitutive law, etc.), pantograph's mass, the spring and damper characteristics of the pantograph, number and spacing of pantographs, and aerodynamic effects, (350). Such a calculation has been conducted on the high-speed line Paris-Marseille and permitted calculation of the following, (Fig. 20.10):

- contact force between contact wire and pantograph,
- oscillations of the pantograph,
- accurate position of the transmission line at any moment.

Finite element analysis has provided for the contact wire (for AC 25 kV, 50 Hz and speeds up to 350 km/h) the following results, (350):

- ♦ a cross-section of 150 mm^2,
- ♦ a required minimum mechanical resistance of 43 kg/mm^2,
- ♦ a linear maximum resistance of 0.148 Ohm/km in 20°C,
- ♦ a conductivity of 80%,
- ♦ a medium deflection of the contact wire of 6cm at the speed of 300km/h and 9 cm at 350 km/h.

During the analysis, the following European standards (or other, if any) on overhead contact systems and power supply should be taken into account:

- EN 50119, (2020), Overhead contact system,
- EN 50149, (2012), Copper and copper-alloy contact wires,
- EN 50163, (2020), Voltage systems on railway power-supply networks.

In any case, according to UIC regulations, in the case of 25 kV, 50 Hz traction, the contact wire voltage, in order to ensure a normal traction engine power supply, should have a maximum value of 27.5 kV, a normal value of 25 kV, a minimum value of 19 kV, and only a momentary fall to 17 kV, (348), (349).

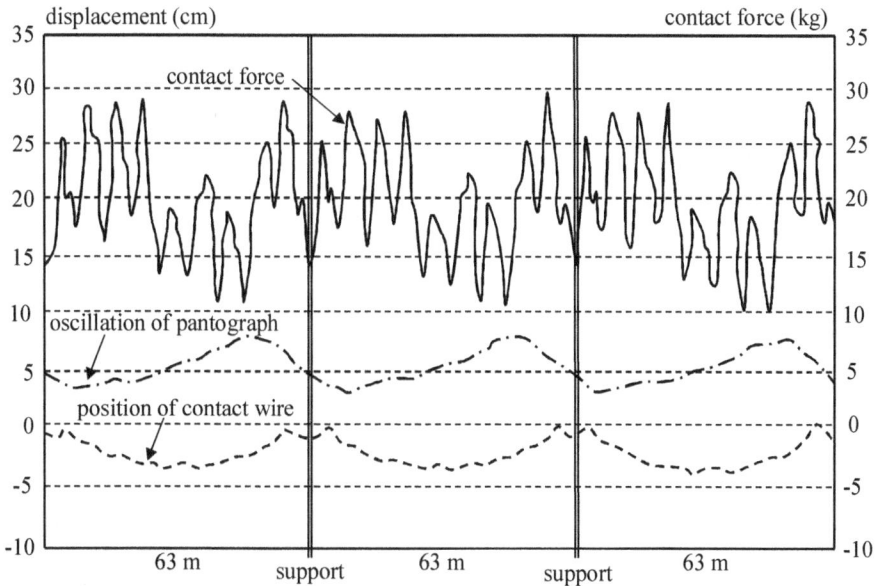

Fig. 20.10. Results of application of the finite element method for the calculation of the overhead power supply system of a high-speed track, (350)

20.8.4. Suspension of overhead contact systems

Various suspension methods of overhead contact systems are being used, (Fig. 20.11), depending mainly on train speed, but also on weather conditions (wind speed and direction) and on pole spacing. With low speeds (up to 120 km/h), simple suspension is adequate, whereas with medium and high speeds catenary-type suspension is mandatory, (358).

However, the contact wire oscillates at the transverse level with a maximum displacement at the order of 20 cm. This transverse oscillation may cause a quick wear of the parts of the pantograph, in touch with the contact wire.

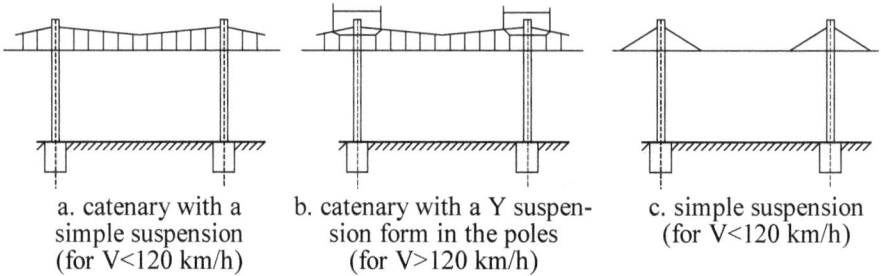

a. catenary with a simple suspension (for V<120 km/h)

b. catenary with a Y suspension form in the poles (for V>120 km/h)

c. simple suspension (for V<120 km/h)

Fig. 20.11. Suspension methods of overhead contact systems

Whenever several tracks are laid parallel (stations, tunnels entrance-exit, bridges, etc.), it is advisable to reconfigure and eliminate certain tracks in order to reduce the total number of tracks to be electrified, (359).

20.8.5. The pantograph

The pantograph transfers electric power from the overhead contact wire to the railway vehicle, (Fig. 20.12).

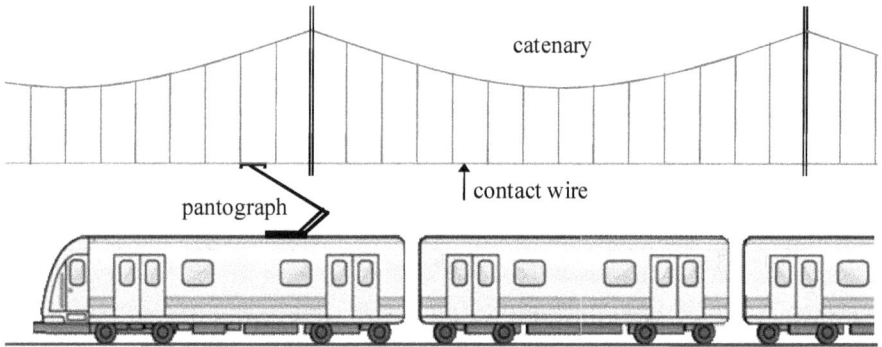

catenary

pantograph

contact wire

Fig. 20.12. Pantograph, catenary, and locomotive

According to the European technical specifications for interoperability, (361):
- the working range of a pantograph should be at least 2,000 mm,
- the contact point of pantograph to the contact wire should be at a height 4,500÷6,500 mm above rail level,
- the static vertical force exerted by the pantograph head on the contact wire should be at the range 60÷90 Nt for AC systems, 90÷120 Nt for DC 3 kV systems, 70÷140 Nt for DC 1.5 kV systems.

The most widespread pantographs have lengths of 1,600 mm (Europe, with a height of 300 mm) or 1,950 mm (Germany, with a height of 250 mm) or

1,880 mm (Japan, with a height of 260 mm), a trapezoidal (mostly) or rectangular cross-section, and should be designed according to a relevant technical specification (e.g. EN 50206-1, 50405, 50367 for Europe).

20.8.6. Power transmission by conductor rail

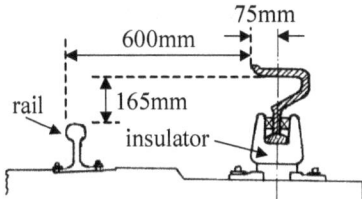

Fig. 20.13. Power supply by conductor rail

As mentioned in section 20.5.1, electric power may be supplied to locomotives by using either an overhead contact system or conductor rails (one or two). Conductor rails are placed aside to the track and are mainly used in metros and some suburban railways. Systems with one conductor rail are often called third rail systems, whereas systems with two conductor rails are called fourth rail systems.

The conductor rail solution, (Fig. 20.13), is preferable in the case of increased traffic loads, for which very large overhead line cross-sections would be otherwise necessary. The conductor rail is equivalent to an overhead contact system with a 900 mm^2 cross-section and in the case of tunnels it allows a smaller loading gauge, and therefore leads to savings in construction costs.

In the vicinity of level crossings or turnouts, the third rail is interrupted and special insulated cables ensure power supply continuity. Special attention should be paid to safety, possibly by covering the conductor rail with insulating plates at level crossings, passages, and personnel working areas. Conductor rails are more sensitive to snow and frost than overhead systems. In some metros (for instance, the London underground) two conductor rails are used (fourth rail system) to avoid grounding of current on the running rails. Indeed, return of the current in third rail and overhead systems is provided by the running rails of the track, whereas in fourth rail systems (two conductor rails) it is provided by the fourth rail; thus, neither running rail carries any current in fourth rail systems.

A conductor rail must be treated as an extremely dangerous component and nobody should be permitted to approach it in less than 30 cm, unless special protective measures are taken (such as coverage of third rail by some kind of shield constituted of insulating material). Standing water in contact with the conductor rail or any liquid spilled over it should be also treated with extreme care.

Until the early 1950s, steel conductor rails were extensively used, iron conductors later on, and recently aluminum-steel composite rails. Permissible intensity is 2,800 A for an iron conductor rail and 4,700 A for an aluminum composite rail for a maximum temperature of 85°C, a critical temperature of the environment of 40°C and a conductor cross-section of 5,100 mm² (specification of the metro of Berlin).

Because of the great mass of the conductor rail, length variation for extreme temperature differences (-30 °C ÷ +80° C) may become high, and for this reason joints are placed every 45÷60 m.

Conductor rails are placed at the rail level and beyond the track gauge, (Fig. 20.13).

20.8.7. Electrical and power characteristics of some high-speed tracks

Table 20.1 recapitulates the principal electrical and power characteristics of some high-speed tracks.

Table 20.1.
Characteristics of electrification of some high-speed tracks in Europe, (346)

	Paris-Lyon	Paris-Marseille	Madrid-Barcelona	Berlin-Hannover	Cologne-Frankfurt
Length of new line (km)	394	206	450	189	135
Maximum gradient (‰)	35	20	25	12.5	40
Maximum total power per train (kW)	14	18 (duplex train)	12	19	19
Traction system	25 kV, 50 Hz	25 kV, 50 Hz	25 kV, 50 Hz	15 kV, 16.7 Hz	15 kV, 16.7 Hz
Average distance between substations (km)	52	51	50	38	19.6
Average power of substations (MW)	40	60	30	30	30
Power per km of line (MW/km)	0.8	1.2	0.6	0.8	1.5
Maximum current intensity per track (A)	~1,000	1,000	1,000	1,000÷1,500	1,500

20.9. Poles supporting overhead contact line

20.9.1. Pole material

The poles supporting the overhead contact line may consist of cast steel, or zinc plated steel, or prestressed concrete, or reinforced concrete.

20.9.2. Pole spacing

The spacing between supporting poles ranges between $50 \div 75$m depending on the following factors: pantograph oscillations, locomotive transverse motions, weather conditions (intensity of snow, wind speed).

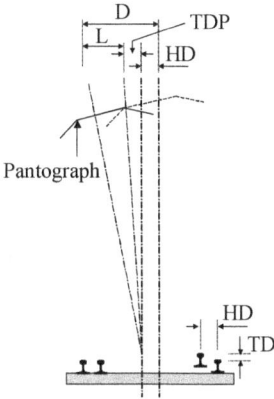

Fig. 20.14. Pantograph oscillations

Figure 20.14 illustrates the transverse displacement D of the pantograph, resulting from the addition of, (348), (359):
— the horizontal defect HD,
— the transverse defect TD, which is reflected on pantograph displacement multiplied by the ratio μ:

$$\mu = \frac{\text{height of contact wire}}{\text{track gauge}}$$

$$\text{and } TDP = TD \cdot \mu$$

— the transverse displacement L of the locomotive, depending on the speed of the train, the height of the overhead wire, the locomotive suspension springs, etc.

Both longitudinal and transverse pantograph motion have to be calculated in detail. It should be stressed that the primary constraint on maximum train speed (574.8 km/h in test runs in 2007) is the maximum permissible pantograph oscillations and to a lesser degree the metal-to-metal (wheel-rail) contact, (352).

20.9.3. Pole foundation

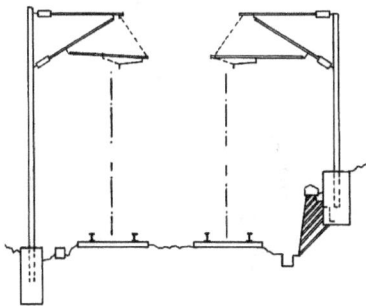

Fig. 20.15. Erection of electric traction poles

When erecting poles for electric traction, special care is required both at the excavation and at the filling-up stages, (Fig. 20.15), so as to minimize eventual settlement of the ground.

When the subgrade is of good quality, foundation of the poles has the form illustrated in Figure 20.16 and calculation of moments M is conducted according to the equation: $M = c \cdot B \cdot L^3$

with coefficient c depending on soil characteristics, (359).

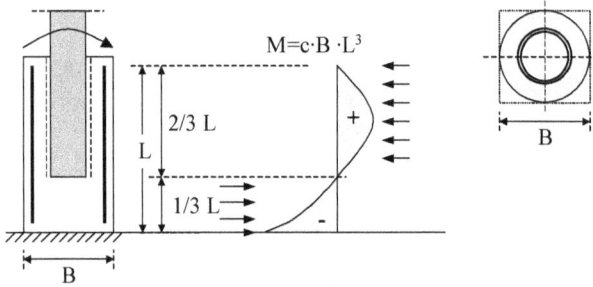

$$M = c \cdot B \cdot L^3$$

2/3 L

L

1/3 L

B

B

Fig. 20.16. Calculation of pole foundation on good-quality subgrade

In the case of poor subgrade, poles are erected on a concrete slab with usual dimensions 2.0 m×3.5 m, at a depth of 1.1÷1.2 m.

20.10. Substations

20.10.1. Substations feeding direct current systems

Substations feeding DC systems, in addition to stepping the three-phase voltage down, also rectify the AC into DC.

Rectification was initially performed by AC motor – DC generator couples, later superseded by mercury-pool rectifiers and more recently by silicon rectifiers.

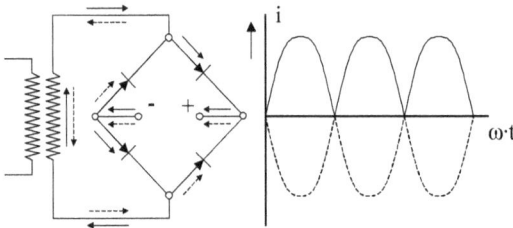

Fig. 20.17. Function of a DC substation

A modern substation includes a voltage transformer with one or two output voltages, and a rectifier assembly, (Fig. 20.17). Silicon diodes or thyristors have been used as rectifiers, but since the mid-1980s, they have been replaced by 'gate turn off' technology, (see below section 20.10.3).

20.10.2. Substations feeding alternating current systems

In AC substations, (Fig. 20.18), only the voltage is being stepped down, and therefore substations in this case are simpler than DC substations. AC substa-

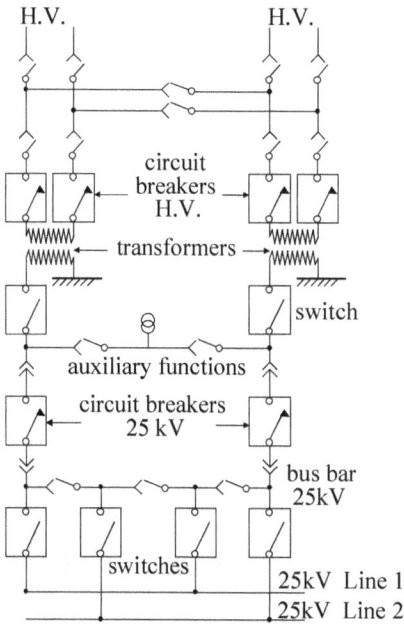

Fig. 20.18. AC substation 25 kV, 50 Hz

tion design should take into particular consideration the risk of short-circuiting, which can be prevented by the addition of appropriate devices.

20.10.3. From thyristors to 'gate turn off' technology

Thyristors were extensively used until the mid-1980s. The introduction at that time of the 'gate turn off' technology, (Fig. 20.19), permitted omission of the commutating circuits, thus enabling a distinct reduction of current losses, (Fig. 20.20, next page).

Strictly speaking, 'gate turn off' is nothing but a special type of thyristor; it is a high-power semi-conductor device, which has the capability not only to turn on the main current with a gate drive circuit, but also to turn it off, thanks to a third lead, the gate lead.

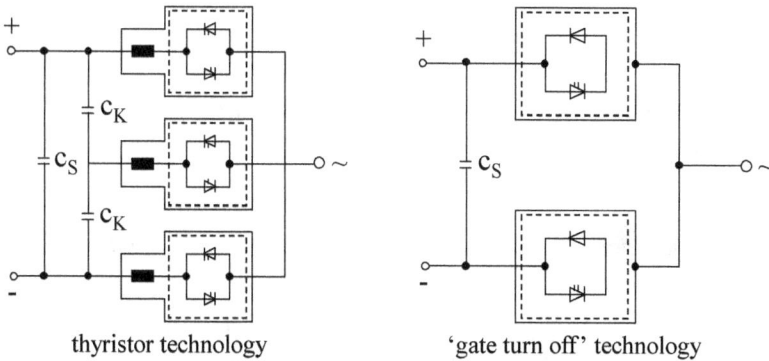

thyristor technology   'gate turn off' technology

Fig. 20.19. Thyristor and 'gate turn off' technologies

20.10.4. Operation control center

Nowadays, substations and the systems supplied by them are remote-controlled and monitored from an operation control center, which is equipped with a visual panel showing the tracks, substations, and the sections supplied (and

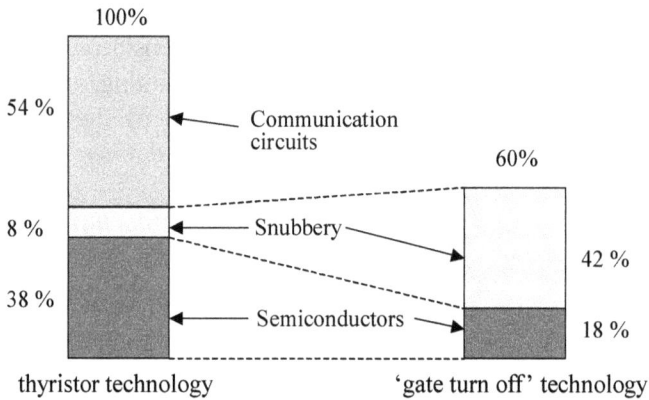

Fig. 20.20. Current losses for thyristor and 'gate turn off' technologies, (320)

therefore controlled) by each substation. Remote control is achieved by using a signal code composed of different frequencies. Electronic control circuits in recent years have made possible execution times on the order of 0.3 sec and more recently even 0.1 sec.

The operation control center is the heart not only of controlling energy power of trains but also of the safe, reliable, and on time operation of trains. It can monitor in real time the path of any train and thus control, supervise, and modify any train schedule, detect early enough any conflicting schedules (e.g. trains operating on the same track on opposite directions or very close to each other), adapt their routes, and optimize capacity of the track, traffic conditions, and quality of service. Advanced technology and qualified staff are among the prerequisites for an efficient operation control center. Any movement of a train in conflict or non-compatible with schedule can be immediately stopped from the operation control center by breaking off the current to the train.

20.10.5. Interference of electric traction with telecommunication and signaling systems

In addition to power transmission lines (in the case of electric traction), telecommunication and signaling cables are also running (usually underground) alongside railway tracks. In order to prevent interference between the power cables on the one side and telecommunication and signaling cables on the other, voltages induced in the telecommunication and signaling network should be calculated precisely. Installations composed of steal and situated near the track may also be affected. The magnetic field created by current used by the train equipment may be strong and may affect neighboring television, personal computers, and hospital equipment. In such cases, a risk analysis should be conducted.

Problems may also arise in areas where traction power cables intersect with lines of the public power network.

20.11. Synchronous and asynchronous motors

Electric motors may be classified into the following three general categories, (Fig. 20.21):

♦ *Direct-current motors.* The inductor is fixed (stator) and carries DC. Induction takes place between the stator and the moving part or rotor, which is supplied with DC through brushes, so that the rotor windings carry alternating current. Motor speed is adjusted by varying the DC voltage applied to the motor as well as by varying the induced magnetic field. The direction of rotation is reversed by inverting the inductor connections (polarity reversal).

♦ *Synchronous motors.* The inductor is rotating (rotor) and carries DC. Induction takes place between the rotor and the fixed part (the stator), which carries three-phase AC. Rotation speed is adjusted by varying the frequency of the three-phase alternating current. Reversing the AC phase sequence reverses the sense of rotation.

♦ *Asynchronous motors.* The inductor is fixed (stator) and carries three-phase AC. Induction takes place between the stator and the rotating part (rotor) which carries three-phase AC. Speed is adjusted by varying the three-phase AC frequency. Reversing the inductor phase sequence reverses the sense of rotation.

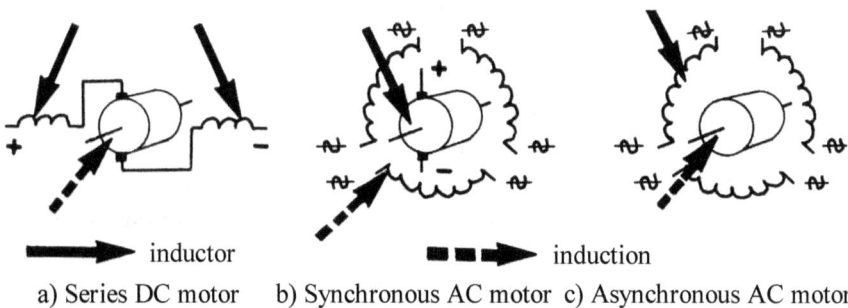

inductor   induction

a) Series DC motor   b) Synchronous AC motor   c) Asynchronous AC motor

Fig. 20.21. The three categories of electric traction motors

Asynchronous motors offer the following advantages:
– lighter weight, about half compared to synchronous motors of the same power,

- higher efficiency and torque and less track loading,
- simple construction, reliability, and low maintenance.

Most electric locomotives use direct-current motors. The French railways, for instance, employ electric locomotives of the BB series, manufactured by Alstom, weighing 90 tonnes, with a power of 4,400 kW and a speed of 160 km/h; the Swedish Railways use R/C-series locomotives manufactured by the former ABB; the same is with British locomotives of Class 91; the German railways use E 181.2-series locomotives, manufactured by Krupp, with a power of 3,300 kW and a speed of 160 km/h, but in their locomotives of series E120 they use asynchronous motors. Examples of asynchronous motors are the German high-speed ICE, the high-speed Eurostar London – Paris, while the French high-speed trains in the lines Paris – Lyon and Paris – Bordeaux have synchronous motors. Asynchronous motors were also used in the Japanese Shinkansen (until 1990, but later were replaced by synchronous motors) and the Chinese high-speed trains.

Synchronous and asynchronous motors are practically equivalent concerning power. They are more efficient than direct-current motors because of their greater speed of rotation. Asynchronous technology is expanding rapidly, in spite of complicated electronic command systems. The choice between synchronous and asynchronous motors must be based on an analysis of the purchase, operation, and maintenance cost of each one of them.

20.12. Maintenance of locomotives – Depot

A critical factor for the good operation of the rolling stock is efficiency and in time maintenance. Maintenance of locomotives was in the past mostly preventive but nowadays it is combined with condition-based maintenance, (see section 19.11), in order to optimize the lifetime of all components of a locomotive, and is based on the following principles:

- specialization of staff and equipment,
- timely scheduling of maintenance sessions,
- appropriate mechanical and computer equipment for the accurate monitoring of any deficiencies,
- continuous control and evaluation of results,
- reduction of cost.

For electric traction locomotives, various routine inspections and maintenance must be performed: two-day inspection, weekly inspection, monthly technical inspection, two-month maintenance, four-month maintenance, yearly maintenance, general overhaul every 10 years, general overhaul every 20 years. Similar levels of maintenance are encountered with diesel locomotives. In order to optimize the use of rolling stock, railways conduct the so-called RAMS (Reliability, Availability, Maintainability, Safety) analysis, (see also section 16.11).

Maintenance up to the four-month level can be performed in the local de-pot. Beyond this level, repairs are conducted at a maintenance facility.

Maintenance hours required for diesel locomotives are more than double compared to electric ones. Maintenance costs of diesel locomotives used for freight trains are per kilometer run 2÷3 times higher compared to electric lo-comotives, whereas for passenger trains are 3÷4 times higher. Overall, an electric locomotive, in relation to a diesel one, has a purchase cost by 20% lower on average, a maintenance cost by at least 25%÷35% lower, and an operating cost by 50% lower, (345).

20.13. Hydrogen trains

20.13.1. Hydrogen as a source of energy

Hydrogen trains are trains that use hydrogen as fuel. The idea to use hydrogen as a source of energy is not a new one, since hydrogen is abundant in nature (but not readily available), light, rich in energy per unit mass, and in addition it can be easily produced from water and stored.

Use of hydrogen as energy source implies production, storage, and energy conversion processes. Production of hydrogen can be achieved by means of one of the following techniques: thermal, electrolytic, photolytic. The thermal technique is the most common; high temperatures and partial oxidation libe-rate hydrogen from the chemical compounds that comprise it as one of their constituents. Storage of hydrogen implies to take into account a number of parameters such as weight, volume, efficiency, durability, and cost. Conver-sion of hydrogen into energy is a function of its expected use, (345).

Use of hydrogen as a source of energy does not cause any CO_2 emissions during operation but causes CO_2 emissions for its production.

20.13.2. Emergence and first applications of hydrogen trains

Hydrogen trains can be a good alternative solution to electric trains, in the case of regional lines with a low or medium traffic, which do not justify the electrification costs of the line. The first hydrogen trains (constructed by Als-tom) are in operation in Germany since 2018. Hydrogen trains convert the chemical energy of hydrogen to mechanical energy either by burning hydrogen in an internal combustion engine or by reacting hydrogen with oxygen in a fuel cell to run electric motors.

Hydrogen trains can be supplied by fuel tanks stored on the roof of the train; this, however, runs the risk to make trains too heavy. The range of autonomy of hydrogen trains was in 2021 around 600÷800 kilometers, (357).

Hydrogen trains can run on existing tracks, without any wayside additional infrastructure.

If use of hydrogen trains is generalized in the future, they could be fueled with the use of overhead wires or side conductors, a solution however which will require additional investments.

20.13.3. Advantages and disadvantages of hydrogen trains

Hydrogen trains can easily use existing tracks, they do not need any kind of electrification, they cause neither CO_2 emissions nor noise during operation, and they have cheap maintenance. However, they have less autonomy during operation (compared to diesel trains; electric trains have no limit in autonomy), are flammable (thus, causing safety concerns in case of accident), more expensive (for the production of hydrogen), and heavier.

20.13.4. Costs of hydrogen

Cost of production of hydrogen is estimated in Japan at $10÷$12/kg and in the USA at $7/kg. Cost of operation of a hydrogen train is estimated at the 1/3 compared to a diesel train, (357).

21 Signaling – Automations – Interoperability

21.1. Functions of signaling

21.1.1. Evolution of signaling

When the first trains made their appearance, it became clear that traffic regulation and safety rules were necessary.

As long as railway lines were few, with a small number of switches and crossings, the main concern (also valid to this day) was to ensure, before the departure of a train, that the line ahead to the next stop was clear. A series of accidents made necessary the posting of guards, who, by hand or flag signals, tried to notify the train driver on whether he should stop or proceed along his course.

Unavoidable human errors by the guards, however, led to the installation of signals visible day and night (semaphore signals), which had a clearer meaning to the driver than the flag signals used by guards. For a long time, semaphore signals were the basis of the regulation of railway traffic and are still employed in a number of countries to some extent. Such signals are usually illuminated during the night.

Advances in electric technology led, around the end of the 19th century, to the emergence of light signals, which have largely supplanted semaphore signaling. Light signals are still the principal tool of regulation and safety for railway traffic.

Until the 1970s the regulation of railway traffic was done with the use of fixed light signals along the track. As speed increases, however, the risk for the driver to overlook a signal also increases. At high speeds, additional signals within the driver cab (cab signals) are accordingly employed, providing continuous information on traffic and safety, (368), (372). Nowadays, cab signaling has been extended to railway traffic other than high-speed.

During the last decades, concern for safety resulted in technologies which assure a continuous transmission of information to the driver and an automatic control whether the permitted speed and signals compliance have been properly followed. Cellular telephones, GSM (Global System for Mobile Communications) techniques, and satellite technologies for the positioning of trains have contributed to such achievements.

Parallel to advances in electrical engineering, a number of achievements in technologies of automations permitted to replace many functions of the train driver by machine operations and provide since 1990 automatic trains, that is trains running on the network without driver.

21.1.2. Braking distance and signaling requirements

Due to metal-to-metal contact, rail transport has a low running resistance and thus a locomotive is capable of hauling much greater loads at higher speeds than a road vehicle with the same traction power.

On the other hand, adhesion forces between wheel and rail are lower than in a rubber-tired road vehicle and are at the origin of a serious disadvantage: the difficulty of stopping a moving train.

The braking distance of trains is 1,300÷1,400 m at a speed of 160 km/h, 2,500÷3,000 m at 200 km/h, and 7,500÷9,000 m at 320 km/h, (372). Therefore, due to the long braking distances, the protection of the train from obstacles on the track cannot be left to the vigilance and quick reaction of the train driver. Early warning of the driver is obligatory and is achieved by suitable signals and alarms.

21.1.3. Traffic safety and regularity

During train movement, three safety problems are arising (368), (372):

- Protection from another train, moving on the same track and in the same direction either in front of or after it. Due to the long braking distances, successive trains must be separated by large safety margins, which can be no shorter than the braking distance.

- In the case of single tracks, protection from trains moving in the opposite direction and prevention of a head-on collision. Accordingly, the movement of a train on any particular length of track is allowed only after ascertaining that the track is and will remain clear.

- Protection from trains moving on another track converging (by crossing or turnout) to the particular track.

The primary purpose of signaling is traffic safety. At the same time, however, it ensures traffic regularity, i.e. the presence of a train at a particular point at a specific moment and at a given priority. Thus, the degradation of traffic regularity may indirectly cause a reduction of safety.

On lines with a high traffic load, which approach track capacity (with the risk of saturation), signaling also aims at increasing traffic capacity, i.e. the maximum number of trains running on the specific track per unit time at a particular speed.

21.1.4. The regulatory framework

Operation of trains is governed by detailed rules specified for each railway network in the schedule service manual and in the general traffic regulation, with which the driver and the train schedule are under the obligation to continuously comply. In addition to regulation provisions, the course followed by the train should also conform to the instructions given by station dispatchers.

On lines with no (or out of order) light signaling, traffic regulations advise the driver about the action to be taken in each case and are usually materialized by semaphore signaling. Even on lines with light signaling, however, traffic regulations are important to traffic safety and are usually incorporated in computer software, which can regulate more or less automatically all technical and operating details of a train schedule.

21.1.5. Basic functions of signaling

Signaling must fulfill the following functions:
♦ separation of trains moving in the same direction,
♦ protection of trains passing through crossings or switches, by preventing the passage of another train on the same track,
♦ protection from a train moving in the opposite direction,
♦ train protection on level crossings,
♦ ensuring compliance of the train driver with speed limits, to prevent derailment,
♦ assisting both traffic safety and regularity.

21.2. Semaphore signaling

21.2.1. Visual and audible signals

Semaphore signaling is mainly visual. Audible signals are also used, however, mainly in the event of the driver ignoring a signal or a speed limit. Semaphore signaling is permanent or temporary (in the case of works or accident sites) and consists of devices activated mechanically. For this reason, semaphore signaling is often termed as mechanical.

21.2.2. Colors used in signals

Railway signaling uses the same colors as road signaling:
- red means that the train should stop immediately,
- green means that the line is clear and the train can move safely,
- yellow is warning that speed should be reduced because of an imminent prohibitory signal (red).

21.2.3. Types of signals

The various signals may be classified as follows:
- main signals,
  - home or entry signals,
  - exit signals,
  - intermediate signals,
  - block signals,
  - protection signals covering dangerous areas,
- advance signals,
- subsidiary signals,
- signaling boards, such as speed indicators, direction indicators, etc.

Semaphore signaling is used in a few cases today as the principal tool for regulating rail traffic. However, in networks where light signaling replaced semaphore signaling, the latter remains still in place as an alternative auxiliary mechanism in case of non-operation of light signaling.

21.3. Operating principles of light signaling – The track circuit

21.3.1. Definition of light signaling

Semaphore signaling cannot provide maximum safety to train traffic. The communication procedure between successive stations by exchange of cable messages is more reliable, but also time-consuming and largely limits the track capacity. Accordingly, on main railway lines, traffic is controlled by using light signaling.

Light signaling constitutes the electrical expression of the operating regulation of the particular line, taking into account the various imposed restrictions. Compared to traffic control by exchange of cable messages, light signaling carries out automatically, and therefore with very high reliability and speed, all specific functions and orders required for the safe running of trains, conditional, of course, on strict compliance by train drivers with the various signals.

21.3.2. The track circuit

21.3.2.1. Definition and components

A prerequisite condition before a train runs on a track is to determine whether any other train is present at some point of the track. This monitoring is continuously and automatically performed by track circuits, which are the basis of light signaling.

The track circuit is a railway subsystem, which in a simplified approach consists of, (Fig. 21.1), (372):
- the two rails of a track section AB,
- a relay at the entrance A and a power source at the exit B,
- the required insulating joints i for the two rails of track section AB in the longitudinal sense, i.e. each rail section AB is electrically insulated from the preceding and the following rail section,
- the necessary materials providing insulation of each rail from the sleepers (and therefore from the other rail of the track).

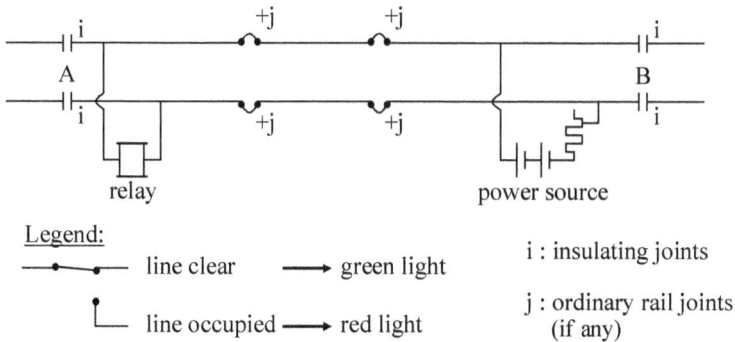

Legend:
—•— line clear   → green light
L— line occupied → red light

i : insulating joints
j : ordinary rail joints (if any)

Fig. 21.1. Parts of a track circuit

21.3.2.2. Operating principle of the track circuit

The track circuit is an electrical circuit which uses the two rails as transmission lines and is fed with a low current by a power source. When no train is present on track section AB, (Fig. 21.2.a), the current passes through the relay, which is activated and closes the electrical circuit, and thus the signaling equipment preceding section AB is displayed to a 'line clear' signal.

As soon as a wheel pair enters track section AB, (Fig. 21.2.b), the two rails are short-circuited through the wheels-axles interaction, the relay is no longer activated and the signaling circuit opens, causing the signal preceding track section AB to revert to the 'line occupied' signal.

In order to reliably detect the presence of a rail vehicle, however, at least two axles should enter track section AB.

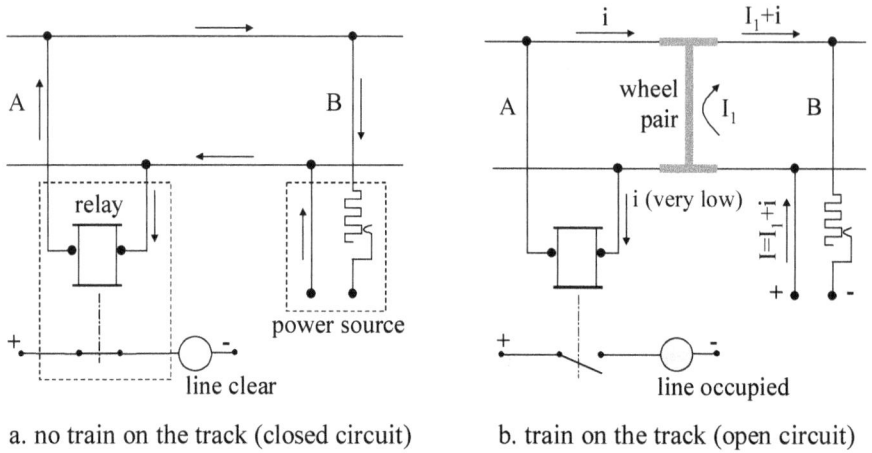

a. no train on the track (closed circuit)   b. train on the track (open circuit)

Fig. 21.2. Track circuit with no train (a) and with a train (b) on track section AB

21.3.2.3. The block section

When successive trains are moving in the same direction, they should be separated by a distance, termed a block section, at least equal to the braking distance d at the particular speed and usually equal to 1.5·d. In light signaling systems, a line is divided into successive track circuits AB, each constituting a block section.

At least one free track circuit should be interposed between successive trains. Let us consider the train positions shown in Figure 21.3. The light signal at the end of circuit 3 is green, at the end of circuit 2 (preceding circuit 3) is red (no entry), while the light signal at the end of circuit 1 (preceding circuit 2) is yellow (warning the driver to slow down because a red light will follow). If two track circuits were free between train A and train B, then the light signal in front of train A would be green.

21.3.2.4. Types of track circuits

The distance between successive stations may be divided into one or several track circuits. We will examine the case of one track circuit between two successive stations.

Figure 21.4 illustrates the signaling equipment in a station area. The track circuits of a station are distinguished into:
- Track circuit at the entrances of a station (01, 04).
- Switch track circuit. This is the designation given to the track circuit following the track circuit at the entrance of a station. It includes all electrically controlled switches on either side of the station entrance (02, 03).

– Stop track circuit. This is the track circuit in the parking or waiting tracks of a station (II), (see also section 15.13.3, Fig. 15.23).

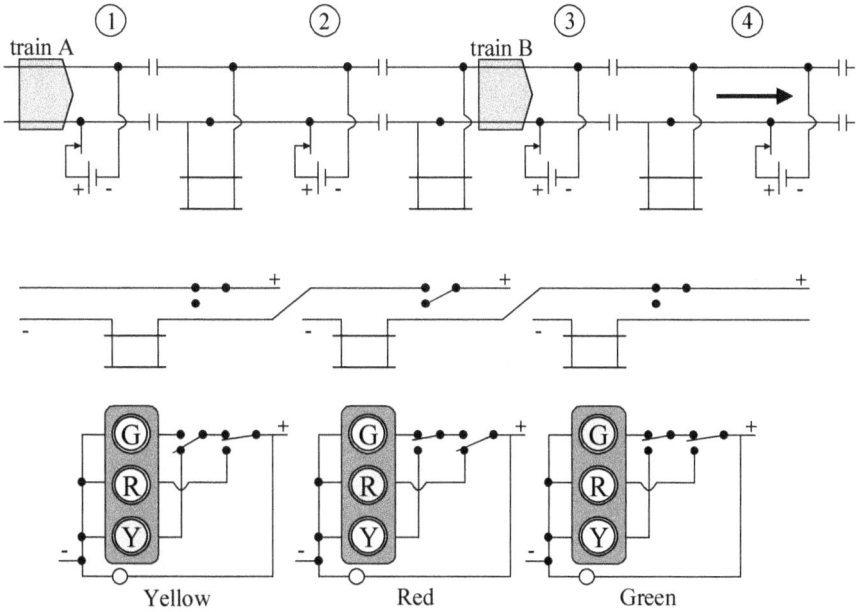

Fig. 21.3. Light signals in the case of successive trains

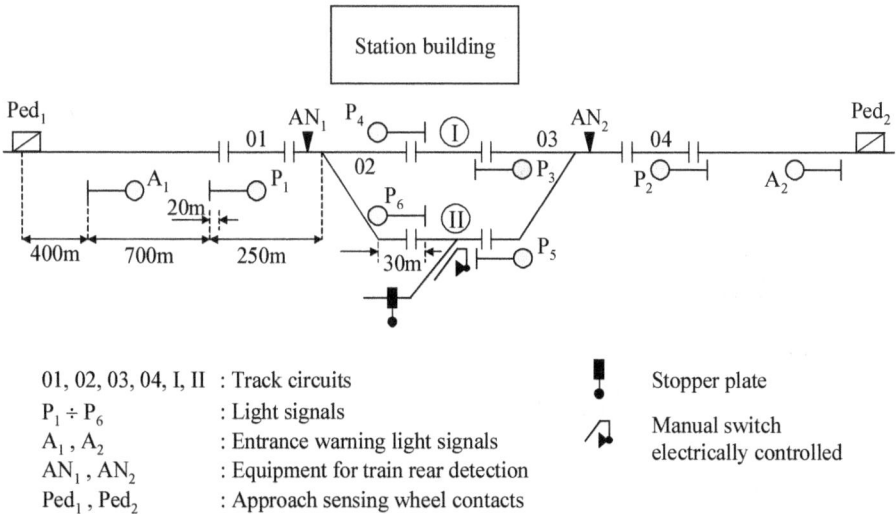

01, 02, 03, 04, I, II : Track circuits
$P_1 \div P_6$: Light signals
A_1, A_2 : Entrance warning light signals
AN_1, AN_2 : Equipment for train rear detection
Ped_1, Ped_2 : Approach sensing wheel contacts

: Stopper plate

: Manual switch electrically controlled

Fig. 21.4. Configuration of a signaling system in the area of a station

21.3.2.5. Track circuit relay

A relay is composed of four parts: the actuator, (Fig. 21.5.a), the armature, (Fig. 21.5.b), the base, and the cover.

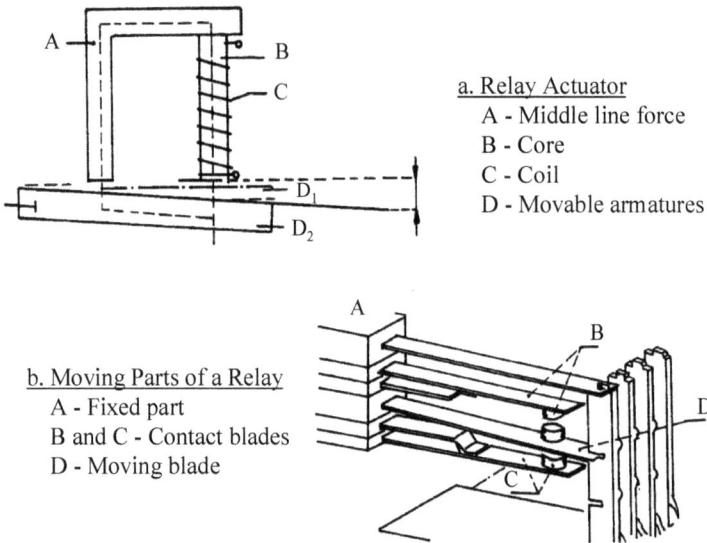

a. Relay Actuator
A - Middle line force
B - Core
C - Coil
D - Movable armatures

b. Moving Parts of a Relay
A - Fixed part
B and C - Contact blades
D - Moving blade

Fig. 21.5. Parts of a relay (a: Actuator. b: Moving parts)

21.4. Equipment and parts of a light signaling system

A light signaling system is composed of the following parts:
— train detection equipment (also including eventual treadles),
— light signals,
— point throwing machines (including point detectors, derailers, stopping blocks),
— interlocking equipment,
— electrical supply and feeding equipment.

21.4.1. Light signals

A light signal is composed of:
♦ the signal mast,
♦ the lights,
♦ the identification plate,

540

♦ the telephone sets, which enable the train driver to call the station dispatcher or traffic controller, provided that the station includes remote control.

The light signals are placed at the station entrance and exit. Advance signals are used to warn the train driver of the upcoming light signals.

21.4.2. Switch control devices

In a signaling system, the switches employed are usually electrically actuated but also (though not often) hydraulically or pneumatically, and their position is automatically monitored. Certain switches (normally of secondary importance) may be manually operated, but, as a mandatory requirement, their position is again electrically monitored.

21.4.3. Train integrity detectors

Entry of the first axles of a train into a track circuit does not guarantee that the entire train has entered the circuit, as part of the train may have been cut off. The integrity of the train as a whole is verified by the following procedure. A permanent magnet is mounted at the rear end of each train. At the entrance to each station, a so-called tail detector is located. This is an electromagnetic device mounted on the track and activated when the permanent magnet at the rear of the train is passing above it. Use of this, rather obsolete, equipment permits checking the integrity of the train.

21.4.4. Approach locking detectors

Traffic safety is ensured when successive trains cannot get closer than the braking distance. The relevant check is made by the so-called approach locking technique.

21.4.5. Local operation and display board

Each railway station, depending on its track configuration and the estimated traffic, is provided with a suitable local operation and display board.

On this board, the track and switch layout are displayed in clear schematic form and, by suitable luminous indications, the state of the light signals and the free or occupied condition of the tracks or track circuits are indicated. Defects or failures, if any, of the signaling system are shown by luminous indications on this local board.

The various operations of the local board are carried out by operating specific keys, whereby the station operator specifies a route, assigns a track, locks an exit light signal, etc. The local board includes certain controls, which are sealed under normal conditions. In a malfunction emergency, however, it is possible to restore normal system operation by unsealing and operating these controls.

21.4.6. Remote monitoring and control

21.4.6.1. Principles of operation

The remote monitoring and control system, enabling supervision of rail traffic from a central point, is used for better coordination and monitoring of a track section or of several successive trains. It is thus possible for a few operators to regulate a great number of tracks and trains.

All information in a remote controlled station is transferred by suitable devices and displayed on the central control board. Thus, the central operator has a complete picture of the situation at all stations in his area as well as of the situation in the various tracks (train locations, light signal status, occupied track circuits, switch positions, etc.). The board is updated automatically and continuously by special high-reliability coded signals.

The central operator sends the necessary instructions to the stations under his supervision in the form of high-reliability coded signals with the help of a control panel.

21.4.6.2. Equipment

The operation control center (or remote monitoring and control center) consists of:
- all monitoring devices of the position and speed of trains, which perform through color indications on a mosaic-type central control panel or, more recently, on a computer display, eventually projected on a large screen,
- all devices for the control and transmission of the instructions to satellite facilities. The latter are understood to be the stations, track switches, block posts, crossovers at double-track sections, etc. Each unmanned satellite position is provided with a type of remote control uninterruptible power supply.

21.4.6.3. Remote monitoring – Control of traffic safety

It should be stressed that the remote monitoring and control devices available to the central operator are not traffic safety equipment but simple means for transmission of instructions and reception of corresponding information. Traffic safety is at all times ensured:
♦ at satellite facilities, by the local safety installation, which permits light signals to function following all safety conditions,

♦ at the open line, by the automatic block system, which regulates the succession of trains.

21.4.7. Power supply equipment

The electric power necessary for operation of the signaling system is supplied by the national power network and is distributed to the various satellite facilities through transformers, rectifiers, and other power devices.

In the event of a power failure, a power generating couple is automatically activated at each station. Finally, in the case of malfunction of the motor generator couple, the supply of power to the signaling system is ensured by automatic switchover to a rechargeable battery.

21.5. Running procedure of trains in a light signaling system

The use of track circuits makes it easy to locate a train at any point on a track. Before scheduling a route from one station to another, the automatic signaling system checks by suitably coded signals that the track between the specific stations or the particular block sections is and will remain free of any traffic. The schedule is then carried out, with the simultaneous exclusion of any possibility to attempt another incompatible schedule.

The operator of the signaling system ensures the prerequisites necessary for the safe running of a train by means of automatic electric devices. These prerequisites are also known as safety interlocks and the main ones are given below.

21.5.1. Route interlock

Upon the arrival of a train at its destination (or upon departure from its origin), the track switches are locked at the position set by the scheduled route and any modification of their position before the scheduled train run is prohibited.

21.5.2. Single track interlock

When a single track circuit is provided between two stations, then the running of a train between the two stations rules out the movement of any other train on that particular track.

21.5.3. Approach interlock

This issue has been discussed previously, (section 21.4.4).

21.5.4. Interlocking of opposite schedules

Any scheduling of trains in opposite directions in stations is strictly prohibited.

21.5.5. Free way interlocking

In successive departure and arrival schedules, the arrival schedule should, as a mandatory requirement, precede the departure schedule.

21.5.6. Light signal interlocking

The order of succession of the various indications of light signals is ensured by the following interlock functions:
- a light signal may be opened only after the route interlock function is activated,
- a light signal is automatically closed upon finalization of a schedule,
- upon activation of a light signal, the indication corresponding to the track switch positions is selected,
- succession of light signal indications should be done in conformity to the traffic regulation,
- automatic switchover of a failing indication (e.g. due to a lamp failure) to an indication of a higher order of safety. For instance, if a yellow entrance signal fails, the red light signal is automatically turned on with simultaneous switchover of the preceding green to a yellow light.

21.5.7. Compatible and incompatible schedules

On the basis of the above, mutually compatible and incompatible schedules are laid out for each case.

21.6. Speed control

21.6.1. The various speed control systems

21.6.1.1. Automatic control and driver functions

For many decades, trains were equipped with the so-called dead man's handle (or emergency braking switch). This is an obsolete safety device, which immobilizes the train in the event that the driver loses consciousness. Although this is not a speed control system, it is a safety device which has been superseded by more advanced automations; these automations can substitute automatic train operation systems for the functions of the train driver.

The dilemma faced in recent years is whether the railways should opt for automatic train operation (with a marginal driver role) or the active role of the driver should be maintained, with the assistance of advanced automation systems. The first scheme could be implemented on metro lines, which have uniform traffic and are adequately protected, as they run in tunnels. On conventional railway lines, however, with a multitude of switches, non-uniform traffic, and a frequent need to insert unscheduled trains, completely automated train operation would lead to inflexibility (as regards dealing with unforeseen occurrences) and to a marginal role for the driver. The latter, with no apparent task, would not maintain the vigilance necessary to deal with unforeseen circumstances.

For the above reasons, fully automatic train operation, (see below section 21.7), is used principally in metro systems. In all other cases, the role of the driver remains in early 2020s essential, with continuous assistance and control by the indispensable automation systems. However, German, and French railways have plans to implement fully automatic trains in their networks in 2023.

Both train speed data collection by automatic control systems and speed control itself may be performed either at discrete intervals or continuously.

21.6.1.2. Intermittent speed control

The intermittent speed control operates prior to signals referring to speed limits, at track switches, at station entrances and exits, etc. The relevant data may be recorded either continuously or at discrete intervals.

The methods implemented may be either electromechanical (e.g. the so-called 'crocodile'[*], a technique employed by the French and other railways) or continuous electrical communication between a control panel and the train. The various methods include the automatic warning system, used in the United Kingdom, the Indusi[**], employed in the past by the German railways, and the 'balise[***] system', used by the French railways. Their differences notwithstanding, all systems rely on the same operating principle, (371): if at the beginning of a particular speed limit, train speed exceeds the speed limit by 5 km/h, the driver is notified by distinctive audible and visual signals. If the speed limit is exceeded by 10 km/h, the automatic braking mechanism is activated and the train is immobilized.

21.6.1.3. Continuous speed control

Continuous speed control depends on continuous communication between the track and the train. This is achieved by suitable equipment both in the track and in the driver's cab.

Continuous speed control also informs the driver about the specified speed (at each point of the route) and the actual speed at each moment.

---

[*] Though 'crocodile' is a means to control train speed, it aims primarily to provide evidence of the alertness of train diver.

[**] Indusi: (Induktive Zugsicherung) has been replaced by the Punktförmige Zugbeeinflussung (Post-shaped train control).

[***] Balise refers to a beacon or transponder installed between or beside the rails, so as to transmit information to trains at some appropriate places.

Continuous speed control is the first step to automatic train operation. The relevant technology, (see section 21.6.2.2 below), was developed in the early 1980s by the Norwegian and Swedish railways and was later adopted by many other railway networks, (369).

21.6.1.4. Speed control and interoperability

Speed control is an essential function of interoperability systems and it is analyzed below, (section 21.9).

21.6.2. Technical characteristics of train speed control systems

21.6.2.1. Electromechanical control

In the case of electromechanical control, a metal blade assembly, also known as a 'crocodile', is mounted in the middle of the track. A metal brush, mounted under the locomotive, contacts the blades.

Exceeding the speed limit or running a stop signal causes a weak 8,500 Hz AC voltage to be applied to the blade assembly. This frequency is sensed by a special receiver on the locomotive and triggers an audible light signal warning the driver. Should the latter fail to react within 5 seconds, the braking mechanism is automatically activated and the train stops, (371).

21.6.2.2. Track-locomotive continuous communication system

The relevant equipment is distinguished into units mounted on the track and units mounted on the locomotive*, (371).

On the track are mounted:

- on the one hand, devices transmitting coded information (concerning gradients, permissible speeds, red light signals if any, etc.),
- on the other hand, recorders (of the speed and other operating parameters) directly connected to the coded information transmission devices.

On the locomotive are mounted:
- ♦ A receiver, receiving the various data, transmitted in an electromagnetic induction mode, by the equipment mounted on the track. Advances in electronic technology make possible the transmission of a large amount of data. For instance, in the first generations of the French TGV, 2^{21} data could be transmitted on a continuous basis and 2^{28} on an intermittent basis. Advances in electronics permit to constantly increase this high number of data to transmit.

---

* The interaction track-locomotive operates as follows: running train and interlocking systems → trackside equipment → transmission of information → on-board equipment → driver and operation control center.

♦ A computer, which determines, on the basis of data detected by the receiver, the maximum permissible speed, the actual speed, and various other route parameters at each moment.

♦ Luminous panels on which the results of computer analysis are displayed.

21.7. The various degrees of automations in light signaling and operation of trains

The previous analysis illustrates clearly a gradual transition of signaling and operation systems of trains from purely electromechanical to more or less automated ones. This has been possible thanks to a progressive introduction of application of automations in railway problems. Thus, we can distinguish the following *five grades of automations* in train operation, (371):

– *automations of grade 0*: This grade does not include any form of automations and is based on manual operation by the train driver, without any kind of automatic train control (ATC) or protection (ATP). Route locking and maximum speeds are ensured by wayside signals, permanent operation rules, and voice radio communications. The role of train driver is central and critical.

– *automations of grade 1*: This grade of automations is also based on the manual operation of trains by the train driver, but with the simultaneous presence of automatic train operation (ATO) and protection. More particularly, ATP ensures route interlocking, the appropriate train spacing by taking braking distances into account, and train integrity. The driver of the train is responsible for setting the train in motion and stopping it, commanding train acceleration-deceleration, monitoring the track condition ahead, and making any required operation in the event of disruption.

– *automations of grade 2*: Grade 2 represents a semi-automatic operation of trains, as setting the train in motion or stopping it is done automatically. The driver of the train is responsible for closure and opening of doors, departure of the train, operation in the event of disruption. Trains with automations of grade 2 are fully equipped and operated with ATP, ATO.

– *automations of grade 3*: Grade 3 represents the driverless operation of trains, that is operation of trains without the presence of driver. Setting the train in motion or stopping it is done automatically. Closure of doors or operation in the event of disruption is the responsibility of train attendants. Trains with automations of grade 3 are fully equipped with ATP, ATO and no driver is required.

– *automations of grade 4*: In grade 4 automations, setting the train in motion or stopping it, closure or opening of doors, and operation in the event of disruption are all done automatically. Neither driver nor attendants are required on-board for the operation of trains.

Automations of grade 3 and 4 result in transfer of responsibilities for safety and operational functions from humans to complex electronic systems and are mostly employed in metro systems; however, they are not very frequent in railway systems (apart from metro), for which the level of automations does not exceed usually grade 2. However, automatic metro systems have recorded operation costs by 40% lower compared to metro systems with drivers, (378).

Automations in railway operation present a number of advantages such as: reduced costs for staff, non-dependence of safety on the vigilance of driver, efficient protection of staff and of anybody moving on the track, optimization of capacity, reduction of delays, in-time information of passengers on-board or in stations. However, automations require a number of adaptations related to signaling, rolling stock, detection of the accurate position of a train, and communication channels.

21.8. Train scheduling

The planning of a train schedule necessitates that the values of the following should be known:
- approved maximum loads,
- planned stop locations,
- running resistances and gradients,
- inertial coefficients of rotating masses,
- speed limits due to the track,
- speed limits due to the rolling stock,
- acceleration on starting,
- deceleration on braking,
- braking distances.

Characteristics of technical operation of trains should be also taken into account, when designing a train schedule, and more particularly:
- required travel times,
- train crossing in stations,
- best use of rolling stock, etc.

Optimization of track capacity requires the grouping of trains into two categories: fast (passenger) and slow (freight) ones. Within each category, the spacing of trains is related to the specific speed and to distances of braking; as previously analyzed, the higher the speed, the greater the braking distance.

Many computer software have been developed and are in use by the railways for the accurate calculation of train scheduling. Figure 21.6 illustrates such a scheduling for a double track, (370).

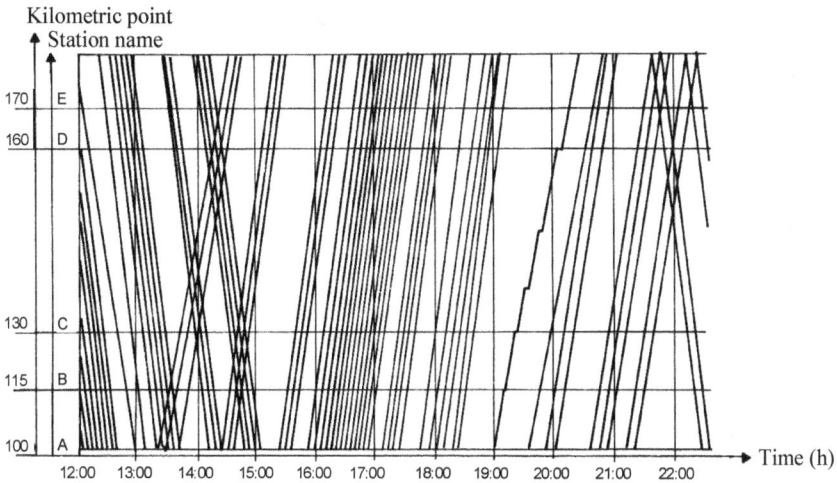

Fig. 21.6. Extract of a scheduling on a double track

21.9. Capacity of track

21.9.1. Definition of track capacity

The ability of the railway system to transport the maximum number of passengers and volume of freight under specific conditions of quality of service (speed, travel time, delivery time, comfort, reliability) is in the heart of operation services of rail (but also of any other transport) systems. As most railway tracks were constructed many decades ago, a critical issue is whether existing rail infrastructure is in position (with small-scale improvements, concerning in particular signaling) to absorb new additional traffic in the future, without a significant impact on punctuality and reliability of rail timetables.

Railways are a more rigid system than road transport, with one degree of freedom (against two for roads) and breaking distances far higher than those of road vehicles. Overtaking is possible for railways only in stations or in specific (though few) overpassing points of the rail network. The problem is more crucial for single tracks, where trains are moving in both directions. But even for double tracks, it is not easy, beyond a level of traffic, to add an additional train on an already existing schedule.

Depending on existing and estimated future demand, (see chapter 4), operation services of railways transform this demand into requirements for rolling stock and crew, and then in time schedules and specific timetables. Not all categories of demand are allocated the same priority, as freight and some intercity routes are considered of lower priority compared to high-speed and suburban routes. Any system, (such as the railways, which are a complex system), is an integrated set that can be broken down into subsystems, each one of which accomplishes a defined objective within a clear hierarchy of priorities.

The *capacity* of a track can be defined as the maximum number of trains that can be operated over a section of track in a given period of time (e.g. 1 hour or 24 hours) under specific levels of quality of service (delays, queuing, safety). Capacity is not a constant magnitude, but a varying quantity depending on the following: track characteristics (layout, overpasses, stations, turnouts, speed limits), speeds of various trains, types of rolling stock (affecting breaking distances), signaling and traffic control systems (particularly length of block sections), heterogeneity of traffic (coexistence of fast passenger trains (intercity), slow passenger trains (interurban), freight trains), access charges (higher for slots in congested hours), reliability and level of acceptable delays, priorities for the categories of traffic and trains considered as the most important, (375).

21.9.2. Theoretical, practical, used, and available capacity

Railway capacity depends on the way it is used. It is conditioned by *technical* parameters (track, signaling), *operational* model (travel times, succession of trains, level of service), and *priority* rules, and can be distinguished as follows, (376):

- *theoretical* capacity: it is derived from some kind of mathematical formula, based on the minimum time interval required between successive trains. Theoretical capacity represents an upper limit that can hardly be reached,
- *practical* capacity: it takes into account the real conditions of traffic; thus, its calculation is accomplished by taking into account timetables and schedules of trains and the intended level of service and delays. Practical capacity usually amounts at 60÷75% of the theoretical one,
- *used* capacity: it refers to the actual traffic volume for specific track sections over a period of time; in order to avoid congestion, used capacity must be lower than practical capacity,
- *available* capacity: it is the difference between used and practical capacity; it quantifies the capacity reserves and whether additional trains can run within specific time intervals (slots) and track sections.

It has been argued that capacity as such does not exist, (373). It is a trade between the need to run additional trains on a track and the level of delays of trains already running.

Figure 21.7 illustrates the relation between capacity and reliability. For low levels of reliability, capacity may become extremely high, however with harmful effects on the level of service (eventual extensive delays) and satisfaction of customers.

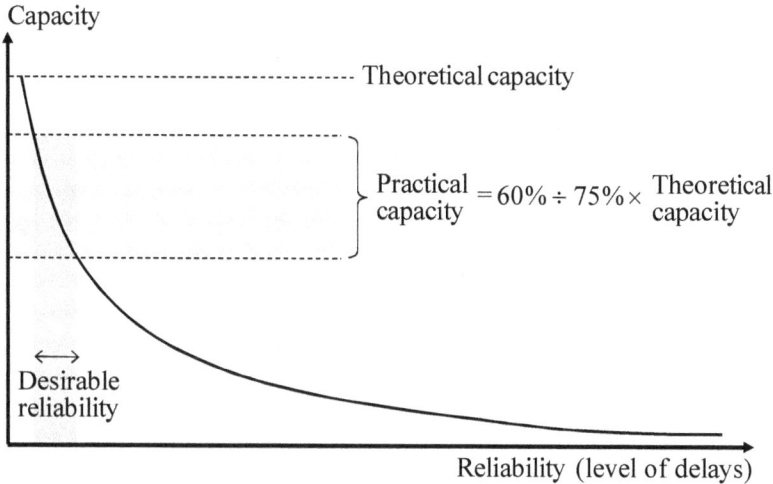

Fig. 21.7. Relation between capacity of track and reliability of train schedules

21.9.3. Models for the calculation of track capacity

21.9.3.1. Homogeneous traffic under ideal conditions

For tracks equipped with lateral light signals, block sections may have a length of 2 km or even lower and trains can follow each other every 3÷4 minutes. With the assumption that traffic is homogeneous, (i.e. it is composed of trains with the same speed, same length, with no stop, and succeeding each other at constant intervals (e.g. all 4 minutes)), the capacity of this track per hour will be 60 min/4 min=15 trains. Unfortunately, this situation occurs only on metros. Therefore, if a delay appears in a train, it affects all the following trains.

21.9.3.2. Delays and their effects

As railway traffic is composed of both fast and slow trains, making many stops and not regularly spaced, the target is to attain a practical capacity, which can absorb short delays and be as close as possible to the maximum theoretical one. Two approaches can be distinguished:
– increasing the time interval between successive trains (for instance in the previous example from 4 to 5 minutes),
– for every five schedules, having one schedule void (i.e. without traffic).

In this way, some short delays can be absorbed. Choice of the most suitable approach is made in relation to local conditions. In many cases, practical capacity is in the range of 60÷75% of the theoretical capacity, whereas for tracks with dense and homogeneous traffic it can reach 90% of the theoretical capacity.

Capacity may be increased if tracks are designed in some stations so that fast trains can overtake slow ones. This method is more efficient in the case of single tracks run by trains on both directions.

21.9.3.3. Homogenous traffic under real conditions

The capacity C of a single track run by homogenous traffic on the same direction and under real conditions (that is by taking into account some unavoidable delays) can be calculated by the following equation:

$$C = \frac{T}{t_a + t_b}$$
(21.1)

where: C: number of trains per track for a specific reference time (1h or 24h),

T: reference time, 60 min (1h) or 1440 min (24h),

t_a: average minimum headway between successive trains. Table 21.1 illustrates values of minimum headways in relation to the type of traffic and for various combinations of types of traffic. Values of Table 21.1 take into account a maximum length of 400 m for express passenger trains and 700 m for freight trains, (364),

t_b: buffer time; it is added to decrease the risks of delay propagation from one train to another.

Table 21.1.
Minimum headway (in minutes) between successive trains
of different types of traffic, (363), (364)

Combination of different types of rail traffic	Passenger, slow	Passenger, fast	Freight, slow	Freight, fast
Passenger, slow	1.64	2.58	2.53	2.64
Passenger, fast	9.25	1.49	2.39	2.49
Freight, slow	5.10	4.09	2.24	3.14
Freight, fast	5.96	3.88	2.57	1.68

21.9.3.4. Practical capacity for single and double tracks

If we apply eq. (21.1) for single and double tracks, by taking into account the characteristics of the signaling system and the length of block sections, we can calculate the values of practical capacity of tracks in relation to the degree of heterogeneity of traffic. Table 21.2 illustrates values of practical capacity for single and double tracks. On average, on a single track we can have 40÷80 trains/day on both directions, whereas for double tracks we can have 100÷200 trains/day on both directions.

Table 21.2.

Practical capacity (number of trains per day in both directions) for single and double tracks in relation to the degree of heterogeneity of track, the characteristics of the signaling system, and the length of block section, (363), (364)

Single or double track	Slightly hete-rogeneous traffic	Highly hete-rogeneous traffic
Single track, distance between overtaking stations:		
5 km	80 ÷ 120	60 ÷ 80
10 km	60 ÷ 90	40 ÷ 60
20 km	30 ÷ 40	20 ÷ 50
Double track, block system with relatively long block sections (>2 km), warning signals and block signals with 2 indications, possibilities for sidings at 20 km intervals	200 ÷ 300	100 ÷ 200
Double track, automatic block system with short block sections (1.5 km), block signals with 3 or 4 indications, possibilities for sidings at 20 km intervals	250 ÷ 400	100 ÷ 200

21.9.3.5. Some computer models for the calculation of track capacity

Every railway network or Infrastructure Manager has some form of computer model permitting calculation of track capacity by taking into account track and traffic characteristics. These models can be categorized as follows:

i. *analytical*; they are simplistic models based on eq.(21.1) Some representative models of this category are the model of UIC (known under Code 405R) and of Transportation Research Board (TRB),

ii. *synchronous;* they use mathematical modeling to optimize timetables, by taking into account discontinuous events, such as the addition of a slow (freight or suburban) train between fast trains. As most representative in this category is the model of UIC (known, under Code 406, as compression model), (373).

iii. *asynchronous;* they use time-step simulations based on train motion equations. Some asynchronous models are RailSys, RTC, among others.

In practice, the most frequently used models for the calculation of track capacity are the following:

a. *analytical* model of UIC (known under Code 405R); it is based on eq. (21.1) and on grouping the succession of trains by classes as follows: fast-fast, slow-fast, fast-slow, slow-slow,

b. models based on the *capacity utilization index,* which is the time taken to operate a squeezed or minimum technically possible timetable compared to the time taken to operate the actual timetable,

c. compression model of UIC (known under Code 406): it consists in compressing the existing train paths within a line section in a predefined time window.

21.9.4. Capacity optimization with the use of satellite technologies

Satellite technologies provide today enormous possibilities for the actual localization of any object moving on earth. The degree of consumption of capacity possibilities of a track can be easily and clearly visualized with the help of satellite technologies (GIS, GNSS), which can provide a colored image of the sections of a track approaching its limits of capacity, of bottlenecks on the network, and of available slots, (374).

21.10. Interoperability

21.10.1. Definition

Almost all railways have been designed following national needs and priorities. As a result, significant differences exist from one railway to another concerning track gauge, electrification, and signaling systems. International (and in many cases even national) rail services require changes of locomotives in the frontiers (or elsewhere) and in some cases transshipment of freight and transfer of passengers from one train to another. This situation causes delays, reduces quality of transport, increases costs, and must be tackled.

Interoperability can be defined as the ability of a rail system to allow the safe and continuous operation of trains, while achieving a specific level of performance. Thus, interoperability can refer to technical, operational, commercial, and legal issues and more particularly to the following subsystems of the rail system: infrastructure, energy, maintenance, signaling and control-command, rolling stock, traffic operation and management, tariffs and liability, and telematics. Among them, and in order to assure a safe and uninterrupted rail service, the most critical issues concern track gauge, electrification, and signaling. European Union Directives 48/1996, 16/2001, 50/2004, 57/2008, 797/2016 cover the various issues of interoperability and are detailed by relevant technical specifications, (134), (333), (361).

21.10.2. Interoperability of track gauges

When a vehicle runs on tracks of different gauges, the most efficient way to assure interoperability is to be equipped with axles of variable gauge, which at the frontier between two countries (or where different gauges exist) can be easily adjusted from one gauge to another, (366).

21.10.3. Interoperability of power systems

An electric locomotive necessitates special design and construction, so as to allow multi-current or multi-system operation, in order to run on more than one power systems. Currently, locomotives (like the Thalys high-speed train) equipped with systems permitting operation under *three* different power systems (25 kV 50 Hz, 1.5 kV, 3 kV) are in operation, as well as locomotives with the possibility to operate under *four* different power systems, (see section 19.7, Table 19.3), (367).

21.10.4. The European Rail Traffic Management System (ERTMS)

Table 21.3 illustrates the diversity of signaling systems in Europe, with nearly thirty different signaling and traffic regulation systems. The European Rail Traffic Management System (ERTMS) is a spectacular achievement for tackling this problem. ERTMS is composed of two components: the European Train Control and Command System (ETCS) and the Radio Communication System (GSM-R) (which sends information to the train driver). We can distinguish three levels of application in ERTMS:
- *ERTMS Level 1,* (Fig. 21.8). Track-based equipment, usually track circuits or axle counters, perform the detection of a train. The information is communicated to the driver from either the side signaling or using cab signaling.

  Transmission of data along the track is realized either in an intermittent way, with the use of the Eurobalise system, or in a semi-continuous way (Euroloop or radio in-fill).

  Eurobalise consists of the following components, (Fig. 21.8):
  - the Line-side Electronic Unit (LEU), which is a coder in interface between balise and usual signaling systems,
  - a balise situated on the track, which assures the exchange of information between soil and train, and between balise and the LEU,

555

Table 21.3.

Various signaling and train control systems in operation in Europe

Country	Train control system	Country	Train control system
Austria	PZB/LZB	Hungary	EVM
Belgium	BRS/TBL/TVM	Italy	BACC/RSDD
Czech Rep.	LS+AVV	Poland	SHP/KHP
Denmark	ZUB 123	Portugal	Ebicab 700
Finland	Ebicab 900	Spain	ASFA/LZB
France	TVM/KVB	Sweden	Ebicab 700/L10000
Germany	PZB/LZB/FZB	Switzerland	Signum/ZUB 121
Great Britain	AWS/TPWS	The Netherlands	ATB

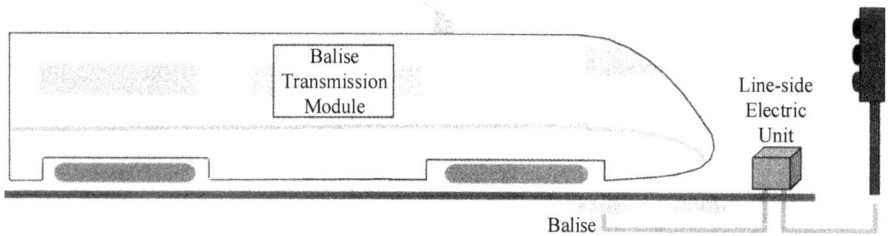

Fig. 21.8. European Rail Traffic Management System (ERTMS) Level 1

- an antenna and a reception system, known as Balise Transmission Module (BTM), which ensures the exchange of information between soil and board. Signals sent from the Eurobalise system to the balise use the frequency of 27.095 MHz (very close to the frequency of 27.115 MHz of KVB and EBICAB systems), whereas signals from the balise to Eurobalise antenna are sent at a frequency of 4.234 MHz,
- an on-board computer (Eurocab), in constant interface with the driver, for the continuous calculation of the position of a train, the correlation between permitted and actual speed, eventual emergency braking, etc.

If we want to work ERTMS Level 1 in a semi-continuous way, then it is necessary to install the Euroloop system, which consists of a cable running along the track and receiving messages which have been sent at frequencies between 1.8÷7.2 MHz.

ERTMS Level 1 can be used by itself or in superposition with a usual signaling system.

– *ERTMS Level 2*. In addition to the functions of ERTMS Level 1, in ERTMS Level 2, the transmission of data along the track is done by radio (GSM-R), (Fig. 21.9). The detection of trains is achieved by track-based equipment, usually track circuits or axle counters. Information is communicated to the train driver by cab signaling. In ERTMS Level 2, lateral signaling is no more necessary, but may continue to coexist with cab signaling. Coexistence, however, of the two modes of signaling may cause confusion or contradiction to the drivers. Authorization for the movement of a train is made continuously with the help of the radio through the soil to the train. In addition to ensuring interoperability, ERTMS Level 2 implemented in tracks with a dense traffic may augment track capacity by 10÷15%. Development of ERTMS Level 2 on a track ensures technical interoperability in both train control and signaling.

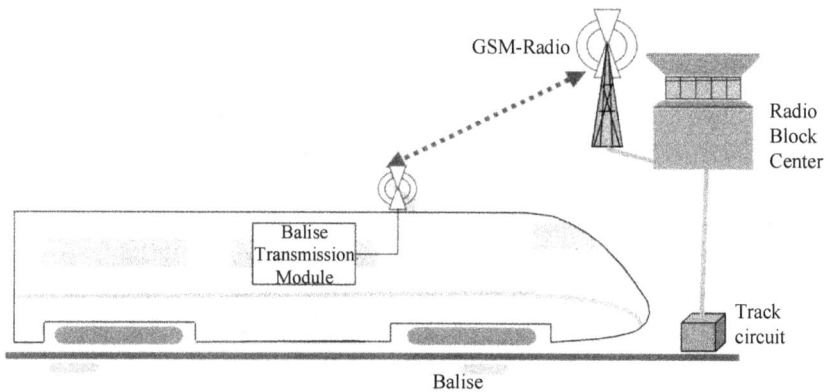

Fig. 21.9. European Rail Traffic Management System (ERTMS) Level 2

– *ERTMS Level 3.* Transmission of data along the track is done by radio (GSM-R). The detection of trains is achieved by train-based equipment reporting to the operation control center and the information is transmitted to the driver's cabin. In ERTMS Level 3, there is no more need for track circuit, (Fig. 21.10), which is replaced by a system of detection of the position of the train and of its integrity.

21.10.5. Costs and degree of deployment of ERTMS

ERTMS is a technological conception which emerged in Europe, but soon it expanded around the world, as many rail managers realized that through ERTMS it is possible to tackle a number of technical incompatibilities among railway technologies. Thus, ERTMS is deployed in European countries but also in a number of non-European countries, such as China, Turkey, Taiwan, India, South Korea, Saudi Arabia, Australia, Malaysia, Kazakhstan, Brazil, Mexico, etc.

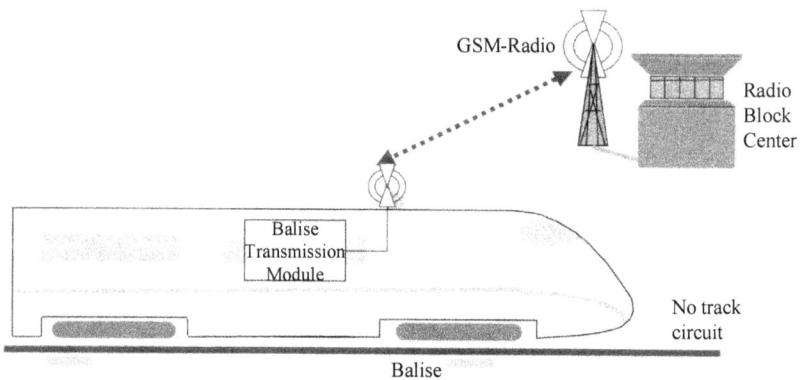

Fig. 21.10. European Rail Traffic Management System (ERTMS) Level 3

ERTMS requires equipment both on the track and on the rolling stock. Thus, newly constructed tracks can be equipped from the beginning with ERTMS and old tracks can be adapted by deploying wayside equipment which will permit the application of ERTMS. The same is with rolling stock: ERTMS equipment can be deployed in new vehicles from the beginning and in already existing vehicles as a supplementary equipment.

Worldwide, 105,185 km of tracks were equipped in 2019 with ERTMS, of which 51% in Europe, 32% in Asia, 12% in Africa-Middle East, 4% in Oceania, and 1% in South America. In 2010, 40,966 km of tracks were equipped with ERTMS worldwide and in 2014 83,125 km, (379). Average unit costs to install wayside equipment on the track for ERTMS Level 2 were per km of track 137,700 € in 2018, against 98,600 € in 2017, 88,800 € in 2016, and 140,000 € in 2015. It is clear that the ERTMS market presents high cost variations, due principally to the level of competition and the volume of orders. It is estimated that the medium-term costs to install ERTMS Level 2 on the track will be below (and perhaps well below) the threshold of 100,000 €/km of track.

The number of rail vehicles worldwide equipped with ERTMS (of level 1 or/and 2) were 13,219 in 2019, of which 69% in Europe, 19% in Asia, 6% in Africa-Middle East, 5% in Oceania, and 1% in South America. In 2010, 6,613 vehicles were equipped worldwide with ERTMS and in 2014 there were 11,155 vehicles.

Unit costs to install ERTMS equipment on a rail vehicle were on average 326,600 € in 2015, 293,900 € in 2016, 208,500 € in 2017, and they are expected to drop in mid-term below the level of 85,000 €/vehicle. However, even these low (compared to previous) values of costs are considered as discouraging for a massive installation of ERTMS equipment on rolling stock, (379). It is noteworthy that in 2020 only 20% of approved tractive rail vehicles in Europe were equipped with on-board ERTMS equipment, a percentage far lower than the desired one.

22 Safety – Level Crossings

22.1. Safety and railway accidents

22.1.1. Definition of safety and accidents

Safety is a main concern for all transport modes. Indeed, transport is a provisional situation aiming to move safely a person or freight from one place to another. Rail transport is based on technology and human actions; thus, it is inevitable for a moving vehicle that in a very few cases technical or human errors may cause undesired effects, such as damages to rolling stock, rail infrastructure or road vehicles (crossing the tracks), injuries and fatalities to individuals.

Safety is a concept that ought to include all measures and practices in order to preserve the life and health of individuals and the good quality of any equipment and material used during transport. According to the International Organization for Standardization (ISO), safety can be defined as the release from unacceptable risks, a risk being a combination of harm probability and gravity of harm. In rail transport, a risk can be measured in relation to undesired effects on materials (damages), humans (injuries, fatalities), and the transport process (delays).

The purpose of promoting safety measures is to reduce as much as possible any kind or form of *accidents*, understood as unfortunate events that happen unpredictably and unintentionally and result in harmful effects on human health and in damages on property. Accidents should be distinguished from incidents, understood as events that affect or could affect the safety of operation. However, some analyses consider accidents of minor importance as incidents.

Railway safety deals with the protection of life and health (of passengers, railway staff, other people crossing tracks or moving in the railway area) on the one side and property (damages to the rolling stock, the rail infrastructure, or the road vehicles) on the other.

22.1.2. Railway accidents as spectacular but also ordinary events

A few numbers of railway accidents (like train collisions or derailments) attract the attention of the media and the public, as they are spectacular events (the same being true for aircraft crashes). This leads to the misperception that the majority of railway accidents are due to collisions or derailments, whereas the truth is that most railway accidents leading to casualties take place in ordinary small-scale accidents.

It should be noted that the majority of rail fatalities do not concern neither passengers and railway staff nor level crossing accidents but at a percentage of more than 60% unauthorized persons moving without permission within the railway area and being hit by rolling stock, (see below Table 22.1).

22.1.3. Railway safety: a transverse and composite subject – Rail safety authorities

Railway safety is one of the main assets of railways, compared to other modes, and could and should be further strengthened, (377). This, however, proves to be a difficult task since railway safety is an extremely complex subject that crosses transversely almost all components of the railway system, but also of the road system (for level crossings). Safety depends on: i) technical, organizational, and managerial aspects, ii) professional qualifications and performances of employees, iii) the will of state authorities to introduce stricter regulations, iv) the communication channels towards the society and targeted groups, and v) psychological considerations for more self-control and discipline on the part of citizens. Thus, safety must be examined not as an isolated problem, but within a systems multi-disciplinary approach and with the involvement of many partners and authorities. The ultimate purpose is to create a safety culture, in which citizens are free to act as they wish but fully conscious of risks and of their effects when traveling, working, or dealing with railways.

The complexity of rail safety issues has led many governments to create independent bodies or authorities in charge of rail safety, which have brought about a knowhow of the actions to undertake. An overall presentation of such recommended actions can be found in the portal and numerous publications of such authorities, e.g. the Agency for Railways of the European Union (www.era.europa.eu), the Office of Rail and Road in the UK (www.orr.gov.uk), the Office of Railroad Safety in the USA (www.railroads.dot.gov/railroad-safety.com), and other national rail safety authorities.

21.2. Types and causes of railway accidents

22.2.1. Types of railway accidents

The various *types* of railway accidents can be classified in relation to the railway or human component involved in the accident and are as follows:

♦ *collisions* between two trains or between a train and obstacles within the structure gauge of the track,

♦ *derailments,* when at least one wheel of a train leaves the track and moves out of the rail,

♦ *level crossing* accidents, involving a railway vehicle and a road vehicle or a pedestrian in the area where a track crosses a road or a footpath on the same level,

♦ accidents to *persons,* caused by rolling stock in motion; surprisingly, it is the most common type of railway accident occurring when an individual (passenger, employee, unauthorized person) is hit by a train or by an object detached from a train, when a passenger or an employee falls from a train or is hit on-board while traveling. This category comprises also suicides occurring in the railway area, which however in the majority of safety analyses are not considered as railway accidents,

♦ *fires* (or explosions) in rolling stock, occurring when the latter is either in motion or stopped in the departure, arrival, or intermediate station,

♦ accidents occurring in the transport of *hazardous* chemicals and other dangerous materials,

♦ fall (intentionally or unintentionally) of individuals from *platforms* on the track.

22.2.2. Causes of railway accidents

Causes of railway accidents may be identified as follows:

– *errors* made by *train drivers*, such as ignoring or not noticing or passing light signals or signs at danger, excessive speed, failure to stop, driving under the influence of drugs or alcohol,

– *rolling stock deficiencies* and failures, due often to poor maintenance,

– *signaling* errors giving false information to the train driver, defective signals, signals out of operation,

– excessive values of *track defects*, a consequence of inadequate maintenance of track, which can cause derailment,

– *mechanical failures* in any of the constituents of track (rails, sleepers, fastenings), in the wheels, in bridges, or in tunnels,

– defects in the equipment or the operation of *train control* systems,

561

- defects or false operation of equipment in *level crossings*,
- *false switching*, resulting in guiding a train on tracks where other trains are moving or are stopped,
- mechanical and operational deficiencies of *buses, cars, trucks*, moving on the track in the area of a level crossing,
- *errors* of railway *personnel*, while executing *maintenance* or other works on the track, in depots, etc.
- *errors* of railway *personnel* during the *operation* of a train,
- *natural* causes, like landslides, earthquakes, floods, etc.
- acts of *suicide*, vandalism, sabotage,
- *improper loading* or *unloading* of cargo,
- *lack* of *attention* and *vigilance* of *humans*: while traveling on a train or waiting on the platform, while working near or on the tracks, while driving a car in level crossings, while working (employees), etc.

22.3. Categorization and indices of railway accidents

22.3.1. Categorization of railway accidents

There is no universal consensus among experts of what should be accounted for and included as a railway accident. For instance, most authorities do not consider suicides or accidents to trespassers in the railway area among the cases of railway accidents. As a result, data collected from various countries are not based on the same definition or basis of what constitutes a railway accident and consequently some comparisons may lack consistency.

At this point, a first issue of eventual confusion is the distinction between accidents and incidents. Both are reportable events of abnormality in the railway area, with differences, however, concerning the effects of this abnormality and their gravity. An accident is an unwanted or unintended sudden event or a chain of such events, leading to specific harmful consequences, such as fatalities, injuries, material damages. An incident is any occurrence associated with the operation of trains and affecting the safety of operation and eventually leading to property damage. There are regulations (like those of the EU) that consider some minor severity accidents as incidents, (Fig. 22.1), whereas other regulations clearly and fully distinguish accidents from incidents, (379), (380).

Railway accidents can be distinguished into significant and serious ones. According to the European Commission (Regulation 1192/2003 amended by Regulation 643/2018), a *significant* accident is any accident involving at least one rail vehicle in motion and resulting in at least one killed or one seriously injured person or in significant (>150,000 €) damages caused to rolling stock, track, other installations, or the environment, (379). Accidents in workshops, warehouses, and depots are excluded. A *serious* accident is considered any collision or derailment resulting in the death of at least one person or in serious injuries of five or more persons or in extensive damages (>2 million €) to rolling stock, track, and the environment. Under this definition, serious accidents are clearly a distinct small subset of significant accidents, (379).

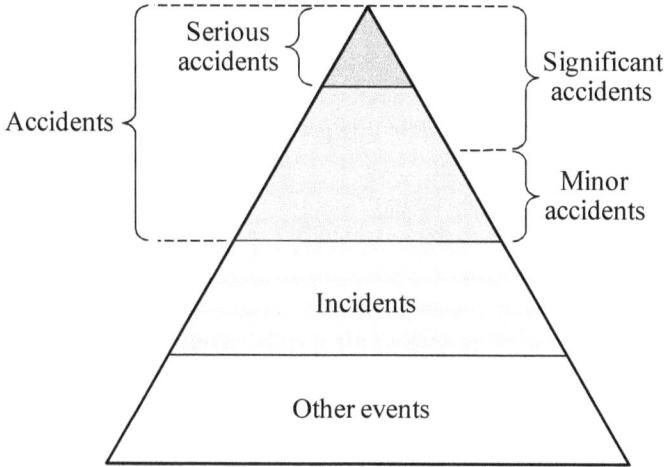

Fig. 22.1. Serious, significant, and minor accidents, incidents, and other events related to rail safety, according to the European Commission, (379)

22.3.2. Indicators for assessing rail safety

Though all societies consider human life and health as primary values, they allocate varying weight factors to deaths, injuries, and material damages. Thus, a number of indicators have been suggested for assessing rail safety and accidents: number of fatalities or of seriously injured, number of fatal, significant, or serious accidents, monetary value of material damages, total monetary value of deaths and injuries, (387).

If a global analysis or a specific component (e.g. level crossings) of rail safety is evaluated, then evolution over time of one or more of the above indicators can be used. Furthermore, if traffic is taken into account, the previous indicators should be reported per passenger-kilometer, or tonne-kilometer, or train-kilometer, or track-kilometer, in relation to the railway component considered.

Over the years, various composite indices have been suggested. To name one, the Fatality and Weighted Serious Injury index (FWSI), which takes into account fatalities (with a weight of 1) and serious injuries (with a weight of 1/10), is used mainly in the EU for accidents in level crossings, (379).

22.4. Evolution and statistical analysis of railway accidents

22.4.1. Relativity and inconsistency of statistical data

As analyzed previously, there is no general consensus worldwide about the definition of what can be reported as a railway accident and what should be included among its effects. Most authorities or countries do not consider suicides and accidents to trespassers in the railway area as railway accidents. Some countries focus on significant accidents in an effort to reduce the effects and severity of the most harmful accidents; however significant accidents are just a subset of total accidents. For this reason, any analysis of accidents data should carefully consider what these data include, in relation to what is defined as a railway accident, so as to check the consistency of data collected.

Another issue of concern is the relativity of considering some data (particularly old ones), since railway accidents are affected by technological achievements, regulations adopted, and safety measures. If a longer period is taken into account, so as to provide statistical validity, there might be a risk of not considering recent technical advances. For this reason, the following analysis will focus principally on available data of the period 2010÷2020.

22.4.2. Gravity and effects of the various types of accidents

The first step in the analysis of railway safety is to study the effects of accidents and their severity, which can be assessed by considering fatalities and serious injuries. The persons involved in railway accidents may concern passengers, employees, level crossing users, unauthorized persons, or other people. Table 22.1 illustrates fatalities and serious injuries in relation to the type of rail accident for the EU-28 countries for the year 2019. Table 22.1 testifies that in the total number of fatalities caused from railway accidents in the EU (suicides and trespasser excluded), collisions and derailments represent less than 2%, accidents in level crossings 32.5%, accidents to unauthorized and other persons hit by rolling stock in motion 62.9%, accidents to passengers and employees hit by rolling stock in motion 2.4%, and other types of accidents 0.2%

22.4.3. Statistical evolution of effects of railway accidents

Any reliable assessment for the advances in railway safety must stem from statistical analysis over recent years, preferably in relation to measures already taken and the conditions under which the most severe accidents occurred. Figure 22.2 illustrates the total number of fatalities and serious injuries in railway accidents for the EU-28 countries and the period 2010÷2019.

Table 22.1.

Fatalities and serious injuries of passengers, employees, level crossing users, and unauthorized persons in relation to the type of railway accident occurred in the EU-28 countries for the year 2019, (378)

	Fatalities						Serious injuries					
	passengers	employees	level crossing users	unauthorized persons	other	total	passengers	employees	level crossing users	unauthorized persons	other	total
Collisions	8	1	0	6	1	16	37	6	0	1	1	45
Derailments	0	0	0	0	0	0	0	1	0	0	0	1
Accidents involving level crossings	0	2	266	0	0	268	16	11	229	1	0	257
Accidents to persons caused by rolling stock in motion	8	12	0	496	22	538	33	17	3	211	33	297
Fires in rolling stock	0	0	0	0	0	0	0	0	0	0	0	0
Other	0	0	0	1	1	2	6	6	1	5	0	18
Total	16	15	266	503	24	824	92	41	233	218	34	618

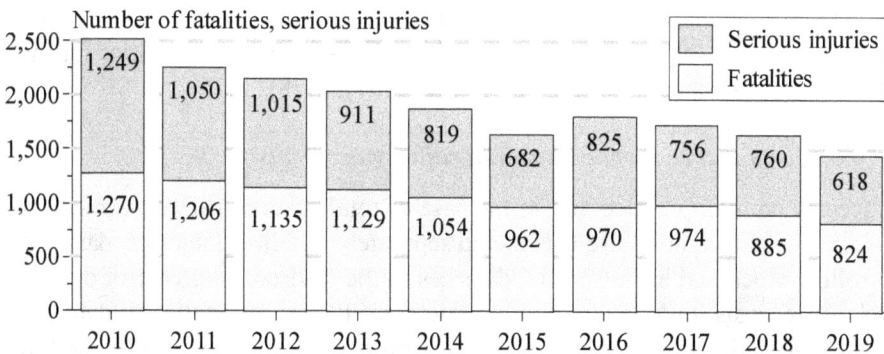

Number of fatalities, serious injuries

Serious injuries
Fatalities

	2010	2011	2012	2013	2014	2015	2016	2017	2018	2019
Serious injuries	1,249	1,050	1,015	911	819	682	825	756	760	618
Fatalities	1,270	1,206	1,135	1,129	1,054	962	970	974	885	824

Fig. 22.2. Evolution of the total number of fatalities and serious injuries from railway accidents in the EU-28 countries, (378), (379)

As accidents concerning passengers and employees usually have very distinctive causes, we can distinguish the effects of accidents on railway passengers from those on employees by using the passenger fatality rate (fatalities of passengers/billion p-km) and the employee fatality rate (fatalities of employees/billion train-km), (Fig. 22.2). From Figure 22.3 we can remark that for the EU-28 countries the decrease of the employee fatality rate (between 2006 and 2018) was slower than the decrease of the passenger fatality rate, a sign of the inefficiency of measures taken by the European railways for employee safety.

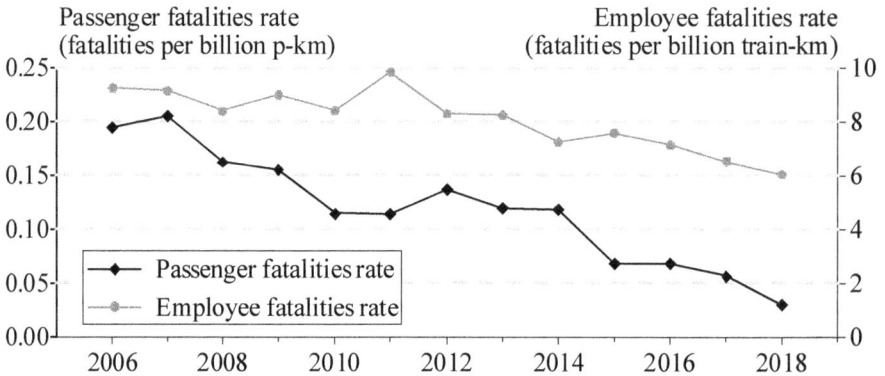

Fig. 22.3. Evolution of the fatality rate, resulting from railway accidents, for passengers and employees in the EU-28 countries, (379)

22.4.4. Suicides and trespasser accidents in the railway area

Though suicides and trespasser accidents in the railway area are not considered as railway accidents, a thorough rail safety analysis cannot ignore them, since they affect the life of individuals and the operation of trains. Thus, in the EU-28 in 2018, 2,637 fatalities from suicides were registered within the railway area (against 2,982 in 2012 and 2,614 in 2007) and 569 trespasser fatalities for the same year (against 655 in 2012 and 855 in 2007), (378).

22.4.5. Costs and economic impact of railway accidents

It seems common practice that in the case of total costs of railway accidents, the various partial costs are considered separately: fatalities, injuries, damages to rolling stock and infrastructure, damages to the environment, cost of delays, and other costs (administrative, productivity losses, reputational damage, modal shift, air pollution). Total costs of railway accidents are estimated at 3,811 billion € for the EU-28 countries* for the year 2018 and are allocated as follows: fatalities 76%, serious injuries 10%, material damages-delays-environment 10%, and other 4%.

---

* 555 million € for Germany, 520 million € for Poland, 315 million € for Italy, 285 million € for Hungary, 205 million € for France, 195 million € for Romania, 98 million € for Austria, 95 million € for the UK, 60 million € for Spain.

22.4.6. Railway safety in comparison with other transport modes and among various countries

The level of safety of a railway trip is far higher compared to other transport modes and is assessed by the fatality risk for a train passenger per billion passenger-kilometers. Taking into account the fatalities for the various transport modes for the EU-28 countries for the period 2014÷2018, the passenger fatality risk/billion p-km is 0.05 for a railway passenger, 0.5 for a coach/bus passenger, 2.45 for a private car passenger, 0.025 for an air passenger, (378), (379).

The railway passenger fatality index is for the period 2014÷2018 quite different in the various parts of the world: 0.05 in the EU-28, around 0.10 in South Korea, 0.15 in Australia, 0.175 in the USA, approaching 0.00 in Japan, (378).

22.5. Measures to improve railway safety

Railway accidents are the outcome of failures, the nature of which could be either technical or human. Railway safety can be increased by means of a compendium of measures, aiming to tackle and eventually eliminate one or more causes of railway accidents, as they have been analyzed in section 22.2.2. These measures are efficiently exploiting advances in the quality, strength, and reliability of materials, automations and light signaling, recording and monitoring of the state of each railway component, use of cameras for the surveillance of any behavior contrary to regulations, immediate transfer of information and orders (e.g. stopping or decelerating the movement of a train), organization and management methods, communication and information channels. The scientific and technological background of rail safety measures refers to the various rail components and has been analyzed in the relevant chapters of this book. Thus, measures to increase safety and reduce as much as possible rail accidents can include the following:

- advanced *signaling* systems and *ERTMS*, (see sections 21.3, 21.10.4), in order to ensure that every train moves on the right track and the right direction and that among successive trains there is appropriate distance, which is greater than the braking distance, (see sections 18.11, 21.1.2),
- *Automatic Train Protection* (ATP). With the help of ATP, (see sections 21.6, 19.8), at any moment the speed of a train is accurately recorded; thus, and if the running speed is higher than the permitted speed at a specific point, automatic braking is activated,
- automatic operation of *turnouts,* (see section 15.10), which is a combined result of achievements in signaling and automatic train control. Most accidents

in railway stations are the consequence of errors (either technical or human) in the operation of switches and crossings, which can lead a train to a different track from the one that was scheduled. Such an automatic operation has various hierarchical levels of control in order to reduce any risks of deficiencies in the signaling equipment,

- *train driver alertness and vigilance*, which can be assured by technical equipment such as the dead man mechanism, (see section 21.6). Any consumption of alcohol and drugs or use of mobile telephone (for private account) by the train driver should be strictly prohibited. Medical and psychological testing of drivers must be regularly conducted. However, the work conditions of the drivers should be carefully regulated and improved in order to reduce the number of working hours, stress, and workload, and in addition exclude any conditions of sleeping at the wheel (particularly in early morning or late night shifts). An ergonomic environment of work into driver's cabin is also essential,

- continuous *monitoring* of the condition of the *track* (values of track *defects* and track gauge, mechanical failures of rails, sleepers, fastenings). If the recorded values are at any point and moment higher than the limit accepted values, automatic braking of the train is activated, (see sections 16.5, 16.9),

- continuous monitoring of the condition of *rolling stock*, (see sections 19.10, 20.12), so as to take the appropriate measures for any deficiency recorded,

- for railways in which a *separation* of infrastructure from operation exists, there is the risk of lack of cooperation or of misunderstandings between services of the Infrastructure Manager (in charge of track safety) and services of Operation (in charge of rolling stock and operation safety). It has been reported that separation of the former unified railway activity is at the origin of a number of railway accidents that have occurred in recent years, (386). Railway *fragmentation* requires more formal safety processes, particularly regarding new entrants. In addition, if some sectors of the railway activity are privatized, *private* operators may take higher risks (e.g. by asking drivers to run at a higher speed than scheduled) and prioritize profit over safety,

- railway is the preferred mode for the transport of *dangerous* goods, as it provides higher safety compared to road transport. However, railways often cross city centers or are near inhabited areas, whereas highways are outside of urban activities. Regulations and control equipment should be strictly followed when loading or unloading a freight wagon, together with monitoring devices (alerting in time for any plausible danger), continuous staff training, control of working conditions and of protective and security equipment,

- installing *monitoring* cameras (to prevent, control, warn) at all perilous points such as level crossings, overpasses, platforms, vulnerable points for sabotage or suicide acts,
- *reducing* the number of *level crossings* at the minimum possible extent by removing, regrouping, or replacing them with overpasses; installing the appropriable *warning* system (with visual and sonor alert) and *protective* equipment (light signaling, light signaling with half or full barriers). *Cameras* and *sensors* (detecting anybody entering illegally) can warn, prevent further dangerous actions, transfer any relevant information and image, and stop any attempt of passengers, employees, trespassers, individuals, and drivers of road vehicles and pedestrians to perform illegal actions. The issue of level crossings is analyzed in detail in section 22.6 and thereafter.
- strict measures, equipment, and control for *track maintenance* works, (see section 16.8), and complete separation from the operation of trains. In the case of subcontractors, additional training may be necessary,
- information made available to the driver or the operation control center about imminent *natural disasters* (landslides, earthquakes, floods, heavy snow fall), so as either to stop the train or continue its path at a slow speed,
- physical *barriers* and obstacles in overpasses, bridges, extreme points of platforms, so as to discourage any kind of attempt to commit suicide, a criminal act, or vandalism,
- use of *drones* for aerial inspection, in conjunction with track monitoring and surveillance from the operation control center,
- reducing *communication errors*, such as misunderstandings of oral messages, misinterpretation of written instructions, difficulties in international communications. In such cases, it is deemed necessary to verify the specific information either with the use of mobile telephone or with a repetition of the sent information,
- appropriate slopes and widths in *platforms* of railway stations, monitoring and warning systems for any passenger not respecting the rules and safety distances,
- efficient communication to passengers, so as to stay alert and obey the warning signs, to follow instructions when on-board, to avoid staying on the edge of platforms, to avoid attempts to make use of shortcuts and cross tracks illegally, unless strict signs permit it. Any opening of train doors during movement should be blocked with the use of *central door locking* and *automatic doors*,

- regular *campaigns,* communication, publicity, and information conveyed to the public, so that everyone respects rules and regulations of railway safety and a new safety culture emerges in the society,
- continuous *education* of personnel and of all railway staff in charge of both infrastructure and operation,
- in spite of the most intensive efforts, railway accidents will not cease to occur. Railway authorities must make available and update on a regular basis concrete safety *crisis management* plans, providing immediate information towards health and police services, setting priority actions to be taken, allocating responsibilities in case of an accident ("who is doing what"), ensuring readiness of staff and equipment,
- *certification* for safety. Initiatives to undertake the appropriate measures for safety cannot be left only to railway undertakings. Thus, authorities of various countries regulate a series of measures and testifications before a train or driver is authorized to operate. In the European Union, a railway undertaking must hold a Safety Certificate in order to be granted access to the railway infrastructure. Up until June 2019 each member-state of the EU was responsible for issuing Safety Certificates; however ever since then, the jurisdiction of national authorities for the issuing of Safety Certificates has been transferred to the EU Agency for Railways, (see section 3.6), (386). An essential aspect of railway safety is the training and certification of staff, particularly of train drivers. This training covers operating rules, the signaling system, the knowledge of routes on which to drive, and emergency procedures. The railway undertaking should in addition prove that its rolling stock has been properly checked and approved.

22.6. Level crossings: Definition, classification, and indicators

22.6.1. Definition of level crossings

A *level crossing* (*LC*) (also referred to as railway level crossing or grade crossing) is defined as a location where a railway track crosses at grade a private roadway, footpath, or both. A LC is a level intersection of a railway track with a road (of any kind) or with a passage of pedestrians. Passages between platforms within stations or passages over tracks are not considered as LC.

Level crossings are the most vulnerable areas of a railway network. A number of accidents and incidents take place in LC and are defined as any collision of a train with a road vehicle, a person, a LC safety equipment, or any other occurrence on a LC that endangers or has the potential to endanger railway safety, (386).

Costs associated with railway safety in LC include:

♦ *humans* (loss of life[*], treatment of injured persons, quality of life of injured persons, lost labor hours, pain and suffering of relatives),

♦ *property damages* to rail rolling stock, rail infrastructure, road vehicles, buildings, and equipment,

♦ costs of *delays* (both to trains and road vehicles),

♦ costs of health emergency services, police, administration, legal services, etc.

22.6.2. Classification of level crossings

In any level crossing there is a minimum of signs (placed usually vertically, in some cases in addition horizontally) aiming to warn the driver of an approaching vehicle or a pedestrian about risks and dangers when crossing the track. However, in a number of LC with a denser rail and road traffic, safety cannot be left to attention, alertness, and vigilance of drivers of road vehicles and pedestrians; thus, some kind of protection to safeguard safety is necessary. In relation to whether a LC offers protection (in addition to road signs), level crossings are classified in *passive (*or *unprotected* or *open*) and *active* (or *protected*) ones. A passive LC does not have any form of prevention or protection which is activated when it is unsafe for use. Users of a passive LC are simply warned, with the help of signs and eventual alarming yellow lights on the road, that they are approaching a LC. A passive LC does not provide to the users any information when to cross it safely; it is the users' individual responsibility. In EU-28 countries, 53% of total LC are active (protected) and 47% passive (unprotected), (378). In the USA, active LC account for 54.7%, passive for 43.4%, and a remaining 1.98% do not have neither signals nor signs, (382). As protection techniques in LC, barriers (half or full) or gates are used, (see section 22.9.2). Any form of protection in LC requires to warn the road vehicle driver or pedestrian about a coming train; this is materialized with the use of visible devices (lights) placed at an appropriate height and audible devices (bells and other). However, lights and bells may constitute the only protection form in LC without barriers. Operation of barriers or gates of a LC can be done either *manually* or *automatically*.

Almost exclusively (at a percentage of 98% ÷ 99.5%), accidents in LC involve violation of regulations by drivers (of road vehicles) or pedestrians (no

---

[*] Accepted considerations for the monetary value of human life vary from some tens of thousands or few hundred thousand € for low-income countries to 1÷4 million € for high-income countries

attention of road driver or pedestrian, high vehicle speed, violation of the red light and (or) barriers, immobilized vehicle on the track). Passive LC (without protection) are 5 times more dangerous compared to active LC equipped only with light and sound warning, (384).

22.6.3. Average distance between level crossings

Any railway track separates the space it crosses. Communication and conduct of social and economic activities on the two sides of a track is ensured by level crossings. Depending on urban development, density of regional and local roads, policy of reducing or keeping the number of existing LC, we can observe great differences among countries regarding the number of LC per 100 km of track or (equivalently) the average distance between LC, which is: 2 km on average in the EU, 1 km in the USA, 7.5 km in Russia, 2.1 km in India, 3.1 km in Canada, 2.3 km in France, 2.4 km in Germany, 2.6 km in the UK, 3.1 km in Italy, (378), (379).

22.6.4. Maximum train speed for installing level crossings

The likelihood of an accident occurring at a level crossing increases with speed, because of high train braking distances, (see section 18.11.2). The various countries set a maximum train speed beyond which it is not allowed to install or keep a LC in a track, in relation to safety considerations in a particular country, the maximum speed of road vehicles, the average distance between LC, the severity of regulations, sanctions, and penalties. Thus, it is strictly excluded to install a LC on a track with train speeds higher than 200 km/h in Russia and many other countries, 160 km/h in Japan, Germany, the UK, 140 km/h in Finland, 120 km/h in China, (381).

22.6.5. Indicators for assessing safety performance in level crossings

Safety performance in level crossings can be assessed by means of a number of indicators, such as:
- number of fatalities, serious injuries/100 active LC,
- number of fatalities, serious injuries/100 passive LC,
- total costs of accidents in LC (both active and passive),
- number of accidents in LC per train-km or track-km.

In order to take into account both the traffic of the road (number of road vehicles) and the traffic of the track (number of trains), we can use as index the traffic moment, also known as circulation moment (in France, Spain, Italy, and elsewhere). The *traffic moment* is defined for a specific period of time (usually 24h) as the product of the number of road vehicles crossing a LC multiplied by the number of trains passing over the LC. The traffic moment is used by many authorities as the critical index before deciding whether a LC will be removed or kept, whether it will be equipped with some form of protection (and which one), and eventually whether it will be replaced by an overpass or a road flyover. The traffic moment must be representative of the situation of both road and rail traffic in the LC, by taking into account variations over time. Nowadays, it is possible to conduct road traffic counts on a continuous basis rather easily with the help of cameras, recording continuously the road traffic. If this is considered exaggerated, traffic moment can be based on the average of two different traffic count campaigns, each one including four successive days, one of which is not a working day.

22.7. Causes, statistical evolution, and effects of accidents in level crossings

22.7.1. Causes of accidents in level crossings

The causes of accidents in LC depend on whether a LC is passive (unprotected or open) or active (protected).

The maximum permitted train speed in *passive* LC is usually 120 km/h. LC can be left unprotected as long as the traffic moment is lower than a value in a range of 2,000÷10,000 trains×road vehicles/day, depending on the visibility conditions of the LC and the safety conditions of the country.

The only measure preventing a road vehicle driver or a pedestrian when approaching a passive LC is the existence of road signs and in some cases yellow-color flashing alarm lights. However, as there is no indication or suggestion whether somebody can safely cross a passive LC, this becomes an individual's responsibility; thus, safe crossing of a passing LC depends on the quality of road signs, the consciousness of risk and danger, vigilance, alertness, readiness and promptness of the road vehicle driver or pedestrian, good visibility over the track, and continuous alarming sounds from any coming train.

Active LC can be equipped with: i) automatically operated flashing lights and alarming sound signs (48% of total active LC in the EU-28), ii) manually operated flashing lights and alarming sound signs (18% of total active LC in the EU-28), and iii) automatically operated flashing lights, alarming sound signs, and barriers or gates (34% of total active LC in the EU-28).

Common causes of accidents in LC are as follows:
– in a few cases (0.5÷2%), the system of protection in the LC does not function properly or sends a false message and indication; this can be considered as the percentage related to the safety equipment of the LC. In the

remaining cases (98% ÷ 99.5%), the responsibility rests with the road driver or the pedestrian,

- lack of attention or vigilance on the part of the road driver or pedestrian, a consequence in many cases of not understanding an imminent danger, together with a high self-confidence over risks. This is the most frequent cause of accidents in passive LC,
- high road vehicle speed (often accelerating), in order to cross the LC before the train, and inability to stop under such circumstances. This happens principally in passive LC, but it may also occur in active LC,
- lack of composure and calmness of a road vehicle driver as he/she realizes, while violating regulations, that a train approaches,
- a road vehicle immobilized on the track,
- violation of red lights and barriers by road drivers and eventually at the same time maneuvering among barriers,
- violation of red lights and barriers by pedestrians, crossing often the track under the barriers or among them, and sometimes even pushing them.

22.7.2. Statistical evolution of accidents in level crossings

Accidents in LC are usually studied separately from railway accidents in general. The reason is that causes of accidents in LC are specific and rather limited in number and are systematically repeated over the course of time. Thus, evolution over time of the number of fatalities and serious injuries in LC can be a trustworthy index to assess the impact of measures already taken.

Figure 22.4 illustrates the evolution of fatalities and serious injuries in LC in the EU-28 countries and Figure 22.5 the evolution of fatalities and injuries in LC in the USA. Though a general decreasing tendency is clear in the EU for both fatalities and serious injuries, the rates are different and the curves are not steadily declining, and after a period of drop there are years with increased values. In the USA, the curve of fatalities has slightly increasing tendencies after 2012. Clearly, the phenomenon has erratic fluctuations. For the EU-28 countries, the coefficient of determination R^2 of the linear regression on the available data of Figure 22.4 is 0.68 for fatalities and 0.60 for serious injuries, whereas for the USA, (Figure 22.5), the coefficient of determination R^2 of the linear regression is 0.53 for fatalities and 0.71 for injuries. Therefore, any forecast based on statistical evolution risks to conclude inaccurate estimations, (see section 4.4). A number of studies and models, already developed, acknowledge this, (385).

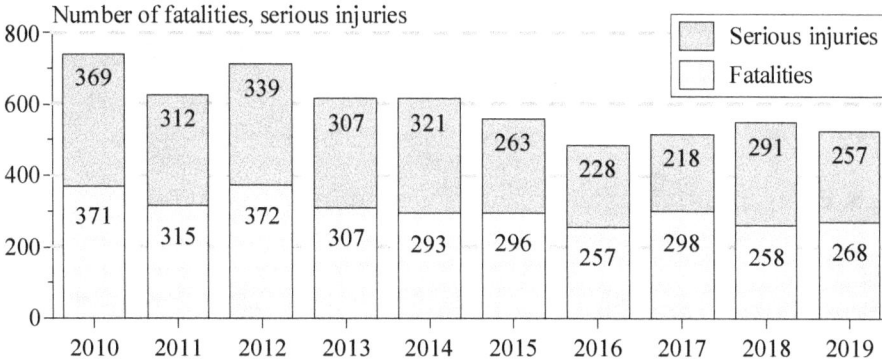

Fig. 22.4. **Evolution of the total number of fatalities and serious injuries in accidents at railway level crossings in the EU-28 countries (378), (379)**

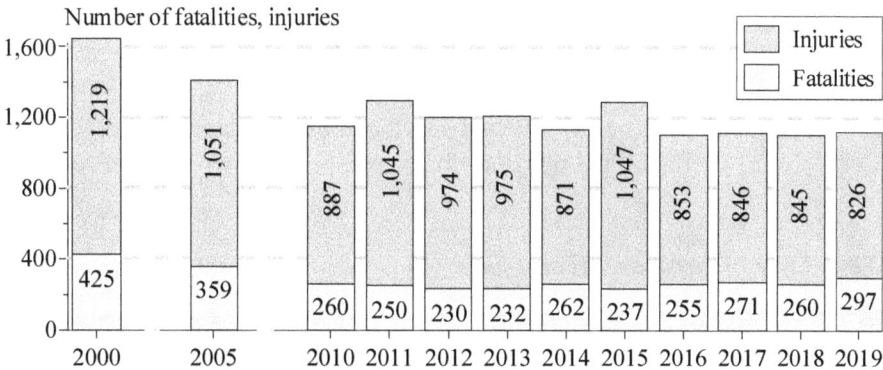

Fig.22.5. **Evolution of the total number of fatalities and injuries in accidents at railway level crossings in the USA, (382)**

22.7.3. Economic impact of accidents in level crossings

Accidents in LC have serious economic effects, which may refer to humans (fatalities, injuries), property damages (rolling stock, road vehicles, equipment of rolling stock, track), other (delays, police, emergency health services, insurance, administration). For the EU-28 countries, cost of accidents in LC is estimated to be 192 million € in 2016, and is allocated as follows: 81% fatalities, 10% serious injuries, 7% property damages, 2% delays, (379).

The cost for all actions, including infrastructure, to prevent a fatality to occur in LC is estimated for the countries of western Europe to 1÷2 million € and for the countries of eastern Europe to 0.6÷1 million €, (387). Costs to prevent a serious injury to occur are estimated for western Europe to 200,000÷300,000 € and for eastern Europe to 100,000÷150,000 €. The above costs are far lower than the costs of human life, and thus actions to increase safety in LC are fully justified from an economic point of view.

22.8. Policy, principles, and management strategy for level crossings

22.8.1. General policy and management strategy

Level crossings are a field of conflicting interests and of lack of understanding among railways, local authorities, state officials, road network responsibles. People often think that any LC exists to serve their needs and object to any thought of removing a LC. Local authorities stand often by citizens and put pressure on the railways to keep intact the existing situation, characterized usually by very short distances between LC. All the above risk to turn into a vicious circle, if state authorities do not impose a clear policy prescribing specific criteria regarding: when to keep an existing LC or remove and regroup it with some neighboring ones, when to install protective equipment (and of what kind), when to replace a LC by an overpass (even if it is costly), the level of continuous monitoring of any attempt to violate the regulations in protected LC, the severity of fines and penalties for non-respect and violation of regulations.

22.8.2. Case of passive level crossings

Passive LC are the most dangerous ones and some can be easily removed or regrouped every 2÷3 ones and replaced by one LC. This is particularly true for LC in agricultural or small-density areas, LC for pedestrians, LC used by heavy and slow-moving road traffic or heavy road vehicles passing with a periodic frequency. Any LC removed, replaced, or regrouped must be the object of a thorough analysis regarding adaptations in the road network, so as to continue providing road and pedestrian accessibility. If this is not properly done, removed LC risk to be replaced by inappropriate passages for road vehicles and pedestrians over the railway tracks, which is far riskier and more dangerous than a passive LC. In a removed LC, the road network must be at different level from the track, so as to render impossible for road drivers and pedestrians to make any attempt to cross the tracks illegally. For the LC that will continue to operate, the following must be applied:

- speed of an approaching train must not exceed 120 km/h (preferably 100 km/h),
- clear signs in the road network, warning the approaching road drivers and pedestrians,

- just before the track, flashing yellow-colored lights,
- good illumination of the LC during the night,
- video surveillance, which in addition can emit sound information signals when a train is approaching,
- good visibility.

A LC can be left unprotected as long as the traffic moment is lower than a value in a range of 2,000÷10,000 trains×road vehicles/day, depending on visibility conditions. Some authorities opt for the lower value (2,000), other ones for the higher (10,000).

22.8.3. Case of active level crossings

The first decision to take in an active LC is whether to install only road light signaling or equip the active LC with both road light signaling and barriers.

Protection of a LC with only road light signaling and without barriers should be permitted solely in exceptional cases, under very restrictive conditions, and for train speeds up to 140 km/h, (383), (384).

The solution of installing half barriers, shutting off a part of the road (the driving direction) can be used in combination with road light signaling for train speeds up to 160 km/h.

Full barriers, shutting off the whole width of the road, combined with road light signaling, are recommended for speeds above 160 km/h.

The general policy is to opt for an active LC as long as the traffic moment of the LC does not exceed 100,000 trains×road vehicles/day. Beyond this threshold, road network and railway tracks must intersect at different levels and thus a flyover or overpass must replace the LC.

In addition to the kind of protection described above, other measures are necessary to safeguard high levels of safety and are as follows:
– install cameras in all active LC in order to supervise any unauthorized use,
– inform in time road drivers and pedestrians approaching the LC about risks, dangers, fines, and penalties for unauthorized use,
– try to avoid long waiting times and long queues on the roads intersecting the LC,
– install good illumination of the LC and improve conditions of visibility and legibility of signs by appropriately improving the geometry of accessing roads (slopes, physical obstacles, trees),
– adapt appropriately the geometry of roads accessing a LC at a distance of at least 100 m,

– reduce the cases of failure and false signs of electronic systems protecting a LC, as much as possible,
– oblige drivers who are approaching a LC to reduce speed, but not lower than 30km/h, by installing radar systems to monitor the speed of road vehicles,
– adapt the levels of fines and penalties, so as to dissuade any attempt of violation of a red light or a barrier in a LC,
– special attention should be paid in LC crossed by trucks transporting dangerous goods, school buses, trucks carrying agricultural products, and particularly in LC crossed by tramway systems.

22.8.4. Replacement of level crossings with flyovers or overpasses

Replacing a LC with a flyover or an overpass is evidently the ideal solution, as in this case road vehicles and trains will move on different levels (preferably the tracks under the roads). However, such a solution is costly, in the order of 10÷20 million € per LC replaced by a flyover or overpass in France, and is applied in LC with high values of traffic moment, usually greater than 100,000 trains×road vehicles/day. Some countries apply a lower limit value for the traffic moment (at the range of 50,000÷100.000 trains×road vehicles/day) as a criterion to replace a LC with a flyover, (383), (384).

22.8.5. A strategy with clear priorities

In most countries, a small percentage of LC (10% ÷ 20%) are the most accident-prone ones and any strategic plans should get started from these most dangerous LC, as it is practically impossible to confront at the same time all safety problems in all LC.

22.9. Equipment of warning and protection in level crossings

22.9.1. Passive level crossings

As the crossing of a passive LC is a responsibility undertaken by any road driver or pedestrian, the equipment that could be installed consists of regulatory signs, warning signs, guide signs, and in some cases of pavement markings. Such equipment aims at providing clear messages and mandatory actions to any potential user of the LC, so as to minimize any risk and danger when crossing the LC. The general principle is to rationalize and organize the successive acts for any user of a passive LC: be informed and warned, understand the severity of a forthcoming danger, advance carefully, stop-look-listen, then look again, and lastly cross quickly, (382), (390).

The signs in a passive LC present a great variety and include, at a minimum, information and warning about the existence and the distance of the LC, the obligation of road drivers and pedestrians to yield priority to trains, and the stop sign. The lower line of such signs must be at a height of 1.5 m above the ground in agricultural roads and of 2.1 m in pedestrian passages, (382). As an excessive use of repeating signs could cause confusion to the user, it should be avoided.

In addition to vertical signs, and only as a supplement item, horizontal markings on the road pavement can be used. Lastly, yellow-colored flashing lights can be placed vertically just before the final sign of the LC.

22.9.2. Active level crossings

Crossing an active LC is not a responsibility of an individual who assesses (with the help of signs) safety conditions (as is the case in a passive LC), but a question of whether to conform with notifications of technical equipment, which make clear when to cross a LC and when to wait until it becomes safe to cross. Equipment in an active LC includes at least light signaling and can have in addition some form of physical protection (half barriers, full barriers, gates) deterring an individual from any attempt to violate regulations (as indicated by red light signals). Thus, the equipment of an active LC is composed of:

- signs (vertical, horizontal), similar to ones used in a passive LC,
- light signals, which can be posted either on vertical poles (placed on the shoulder of the road) or on horizontal cantilever beams (hang at a height of 2 m from the ground level). Vertical light signaling is mandatory in every active LC. The minimum visibility distance of a road driver approaching light signaling in a LC is 70 m for a road vehicle speed of 50 km/h, 90 m for 65 km/h, 150 m for 80 km/h, 220 m for 115 km/h, (389),
- audible bells (usually mounted on the light signaling poles), which supplement light signals and are more efficient in particular for warning pedestrians or bicyclists,
- some form of physical protection of the track, which can be:
  - half barriers at both sides of the track, blocking *half* of a two-way road in the direction of movement of road vehicles, (see Fig. 22.9, section 22.10.3).
  - full barriers at both sides of the track, blocking the *whole* width in both directions of movement of road vehicles, (see Fig. 22.9, section 22.10.3).
  - gates that block any movement towards the track and are used principally in private roads.

Half barriers are used for LC with train speeds up to 160 km/h and full barriers for speeds up to 200 km/h. Light signaling is mandatory to any LC with barriers or gates. Barriers can be operated either manually or automatically. Lifted barriers permit to cross the LC safely, while lowered ones strictly prohibit any crossing of the LC. Before the barriers begin descending from the lifted position, the lights in the LC turn from green to yellow (for 5 sec) and then to the red color. After operation in the red color for 6÷8 seconds, the barriers start to descend. Any failure or inconsistency in light signaling and barriers should be immediately transmitted to the operation control center. Until recently, LC were equipped with telephones, so as to make it easy for anyone to inform the authorities about faulty operation or malfunction.

Automatic operation of LC can be based on techniques of either track circuit or satellite communications. In the technique based on track circuit, (Fig. 22.6), when the locomotive enters the track circuit neighboring the LC under consideration, the information is transmitted to the operation control center, which activates the protection systems of the LC (light signals, barriers). In the case of the technique based on satellite communications, (Fig. 22.7), the locomotive is equipped with an on-board navigation unit employing multi-censor techniques. At a specific distance from the LC there is a monitoring point. Once the locomotive crosses it, the information is transmitted to the operation control center and the activation of the LC is done automatically. On-board navigation units can in addition detect any obstacle on the track, eventually a car immobilized in the area of the LC, and activate automatic braking of the train.

Operation of any active LC depends on electricity power supply; however, it is desirable that LC are equipped with an alternative auxiliary power supply system, providing operation up to 12 h in case of failure of operation of the principal power supply system, (381).

Barriers in front of an active LC must be placed at least 3 m (preferably 5 m) from the axis of the neighboring track, (see Fig. 22.9, section 22.10.3).

22.9.3. Illumination in level crossings

Illumination during the night is critical for both active and passive LC, but particularly for the passive ones. Luminaire poles must be placed 9 m from the external (to the road) right-of-way line of the track and should provide, at a distance of 30m from this external line, an illumination of 3 lux for concrete pavements and 9 lux for asphalt pavements, (382).

Whenever gates are used instead of barriers, red lights should be placed on the gates' arms in order to enhance visibility.

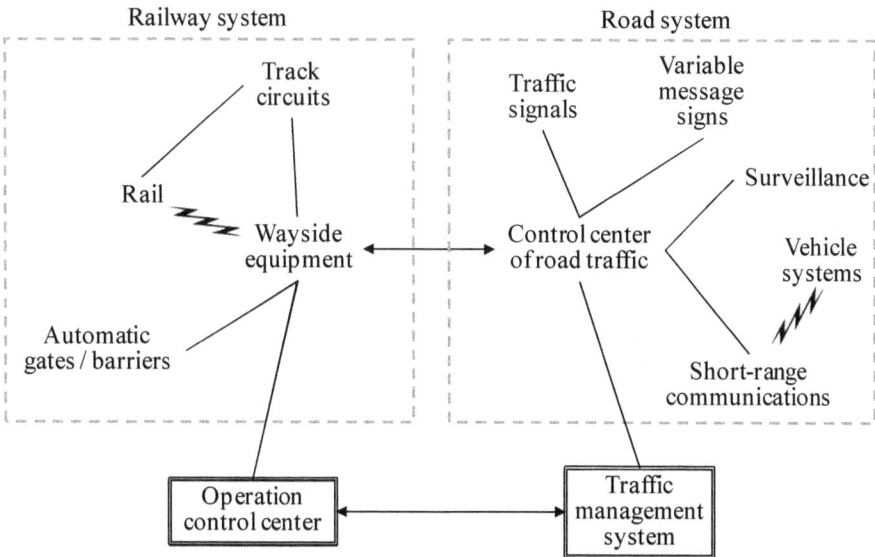

Fig. 22.6. Automatic operation of level crossings with the use of track circuit technique, (382)

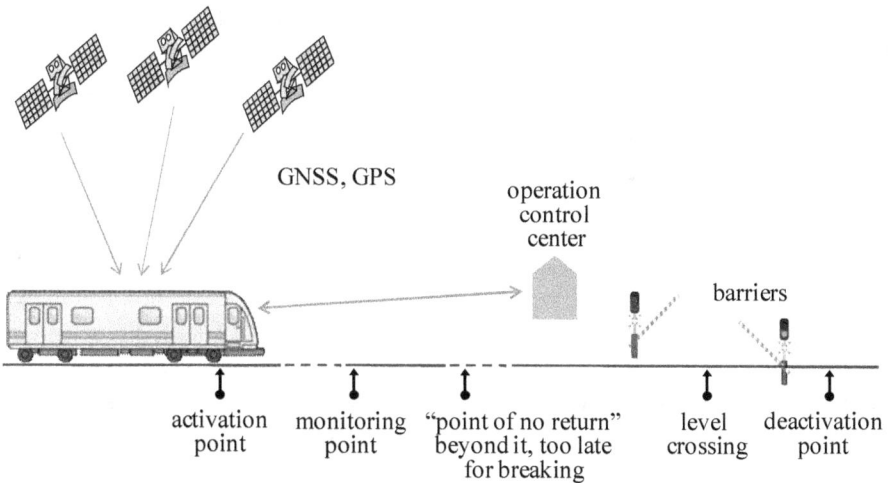

Fig. 22.7. Automatic operation of level crossings with the use of satellite techniques, (381)

22.9.4. Cost of equipment in level crossings

Cost of installing light signaling in a LC is on average $30,000, whilst installing light signaling with full barriers amounts to $150,000, (382).

581

22.10. Layout in the area of a level crossing

22.10.1. Design of road pavement and of cross-section in a level crossing

The running surface of the wheel of a train is the top part of the rail and is situated 15÷17 cm above the upper sleeper surface which constitutes, along with ballast, an almost plane surface. If the track layout is left in the area of a LC as it is in the open track, road vehicles would be obliged to run on the sleeper-ballast (almost plane) surface and climb over the rails, with whatever problems and risks this entails. Thus, it is necessary that both trains (on the one hand) and road vehicles and pedestrians (on the other) can move easily, safely, and without any major restrictions in the area of a LC. The evident solution is to put the running surface of cars at the same level with the top part of the rail, (Fig. 22.8), and create a small gap (of a width of 5 cm and a depth of 4÷5 cm) near the head of the rail, to allow the free movement of the flange of a rail wheel, (see Fig. 7.7, section 7.7). Such a plane running surface in the area of a LC can be achieved by filling the gap above the sleeper-ballast upper surface between the two rails by suitable material, which is for the surface of the filling area some kind of asphalt or rubber product (at a depth of 4÷5 cm) and just below it a roadbed layer (up to the sleeper-ballast upper surface), (Fig. 22.8).

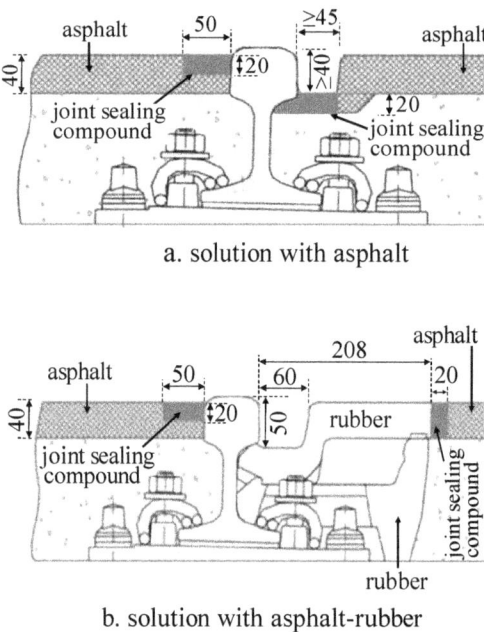

a. solution with asphalt

b. solution with asphalt-rubber

Fig.22.8. Cross-section of road pavement in the area of a level crossing (all dimensions in mm), (389)

The plane crossing surface in the area of a LC must extend transversely to the movement of road vehicles as much as the width of the traveled way and the shoulders of the road, plus 0.5 m on each side, according to British regulations, (389).

The road pavement in a LC, (Fig. 22.8), has a lower roadbed layer and an upper layer (of a depth of 4÷5 cm) for which various materials have been used: asphalt, asphalt panels, asphalt concrete, concrete, concrete panels, rubber panels, asphalt rubber panels, (389). Techniques based on concrete provide a rigid pavement, whereas techniques based on asphalt or rubber provide a flexible pavement. Whatever the material chosen, the most practical solution for its application is to use ready for application panels. Panels composed of asphalt, (Fig. 22.8.a), are simple to prepare but may crack rather easily and do not give a straight forward and easy access to the track components in cases of track maintenance works or replacement of some track components. Panels composed of rubber and asphalt, (Fig. 22.8.b), provide both flexibility and easy access to track, whenever this is necessary. Concrete panels are used in LC with a heavy freight traffic.

The pavement in the area of a LC must fulfill all requirements relevant to the mechanical behavior and the drainage of track, and in addition the electrical requirements related to signaling and electrification.

22.10.2. Sight distances in level crossings

The layout design of a LC must provide sufficient safety conditions by taking into account the required sight distances. The sight distance is a term used in highway engineering and defines the distance within which the driver of a vehicle running at a specific speed has a clear, unobstructed, and unambiguous view of the space in front of him. We usually distinguish the stopping and the crossing sight distance, (382), (390).

Sight distances (for both the road and train driver) in a LC are critical for the safe design of any (passive, active) LC. However, as there are many specifications of sight distances in the various regulations around the world, the reader ought to consult them carefully before tackling any relevant issue. In a LC, sight distances may refer either to the road driver (usually) and affect the signs to be placed on the road (and consequently the conditions of movement of the road vehicle) or the train driver and affect the signs on the track (and consequently the conditions of movement of the train). Thus, we distinguish the following sight distances in a LC:

- *stopping sight distance* d_{stop}, (Fig. 22.9): it is the sight distance along the road, measured from the nearest rail till the front surface of the road vehicle, which allows the vehicle to be safely stopped, if necessary, before the LC. It is clear that the stopping sight distance d_{stop} in a LC is a relation of vehicle speed V_v, deceleration, reaction time of the vehicle driver, and is found to be 66 m for V_v: 40 km/h, 112 m for V_v: 60 km/h, 170 m for V_v: 80 km/h, 243 m for V_v: 100 km/h, (390).
- *crossing sight distance* d_{cross}, (Fig. 22.9): it is the minimum sight distance of a road vehicle driver (before crossing the LC) towards an approaching

train, measured along the track from the intersection point track-road, which permits a road vehicle driver stopped before the stop sign (or the red light of the LC) to see an approaching train, cross safely the track(s), and leave free the track(s) from any obstacle. The crossing sight distance is a relation of train speed and of the type of road vehicle (car, truck), with the worst situation for heavy trucks. In such a situation, crossing sight distance is found to be 90 m for a train speed 20 km/h, 120 m for 40 km/h, 260 m for 60 km/h, 350 m for 80 km/h, 450 m for 100 km/h, 550 m for 120 km/h, (390).

- *pedestrian crossing distance* d_{ped}, (Fig. 22.9): it is the distance from the outer limit line of a LC (where a pedestrian can stand before crossing the LC) up to an approaching train on the track, which permits a pedestrian to see any coming train, decide and cross the LC safely, and leave the track free from any obstacles. Pedestrian crossing distance depends on whether the track is single or double and on train speed V_t and it is found to be for an average individual: 235 m for V_t: 100 km/h and a pedestrian crossing a single track, 350 m for V_t: 100 km/h and a pedestrian crossing a double track, 300 m for V_t: 140 km/h and a pedestrian crossing a single track, 485 m for V_t: 140 km/h and a pedestrian crossing a double track, (390).

All signs warning road drivers and pedestrians must be placed in such distances from the LC, so as to secure the minimum required sight distances. Before the principal warning signs placed in front of the LC, a number of advance warning signs must be placed along the road at a distance of 30 m for urban low-speed roads, 100 m for urban high-speed roads, 150 m for rural roads, 900 m for highways, (382), (390).

22.10.3. Horizontal alignment of roads and tracks in a level crossing

A railway track is a rigid system, with more demanding requirements concerning low values of the longitudinal gradient and the horizontal radius of curvature than any road system. This means that in the geometrical design of a LC, the road design must be adapted to the track and not inversely.

When designing the layout of a LC, it should be aimed that the road crosses the track at an angle of 90°, (Fig. 22.9), as this intersection angle provides better visibility at both sides of the LC. If an intersection angle rail-road of 90° is not feasible, the next preferable choice would be an intersection angle not smaller than 60°. It should be avoided to locate a LC on a horizontal curve of a track, as the cant to be given in a curve is not compatible with a plane level required for the road in the LC. If this cannot be secured, the cant of the track in curve must be zero.

Half or full barriers in front of the LC must be located at least 3 m (prefer-ably 5 m) from the axis of the closest to the barrier track, (Fig. 22.9). The stop sign before the barriers must be located at least 1.3 m before the barriers. The structure gauge (the space free from any obstacles) around the rolling stock in the area of a LC must be at least 2.5 m from the track axis. The stop-ping sight distance of road vehicles affects the location of the last stop sign before the barriers. The maximum length of road vehicles crossing a double track must not exceed 21 m.

Queuing in front of a LC is dangerous and should be avoided; thus, in cases of roads with dense traffic, the design of a LC should be extended to neigh-boring cross-roads, so as to avoid queuing before the LC.

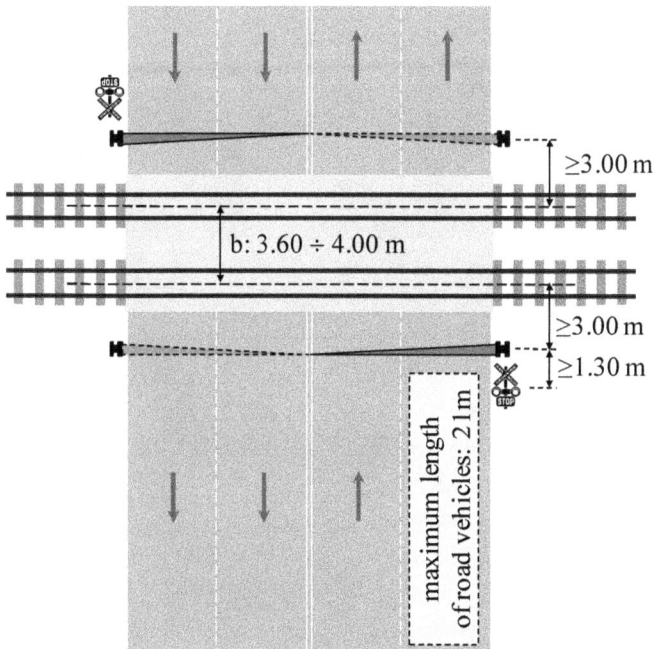

≥3.00 m

b: 3.60 ÷ 4.00 m

≥3.00 m
≥1.30 m

maximum length
of road vehicles: 21 m

Fig. 22.9. Positioning of barriers in the area of a level crossing, (389), (382)

As explained previously, in the area of a LC the upper surface of the road pavement must be at the same plane with the top of the rail and this must be ensured longitudinally at least 0.6 m from each rail end, according to the American specification, (Fig. 22.10). Beyond this distance, the road can be given the appropriate longitudinal gradient, with a limit value of 3% and a preferable value as in Figure 22.10.

It is quite crucial to include accessibility facilities for people with disabilities in the LC design.

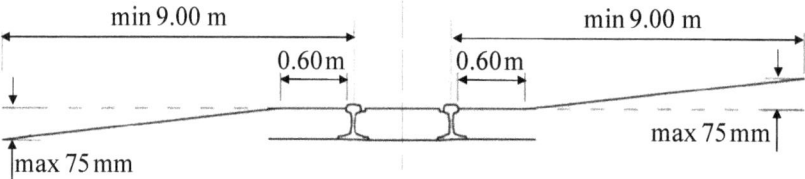

Fig. 22.10. Vertical alignment of a road in the area of a level crossing, (389)

23 Environmental Effects of Railways

23.1. Climate change, the transport sector, and sustainable development

23.1.1. Climate change

Every human activity has a minor or major effect on the environment. Up to a certain level of industrial production, the environment may absorb the effects of human activities through a natural procedure. However, beyond this level, climate change may appear and can be described as a significant and lasting change in the statistical distribution of weather patterns over periods from some decades to centuries or even thousands of years, (400). The origins of climate change can be traced to human activities but also to factors exogenous to human intervention, such as oceanic processes, solar radiation, plate tectonics, and volcanic activity. The question is whether at this point we have reached a level of human impact on the environment, beyond which climate change becomes irreversible.

The United Nations intergovernmental panel on climate change has concluded since the early 1990s that the balance of evidence suggests a discerned human influence on global climate. The analyses of authorities such as the NASA make clear that, (392), (400):

- the average global temperature has risen between 1900 and 2000 by 0.7 °C, between 2000 and 2010 by 0.20 °C, and between 2010 and 2020 by 0.24 °C, (392), (400). If no change occurs in the actual rates of global warming, average temperatures will rise by 2.6÷4.7 °C in 2100. Figure 23.1 illustrates annual changes of global temperature over the last 140 years,
- the global sea level has risen between 1900 and 2000 by around 20 cm and between 2000 and 2010 by 3 cm, with a rate of increase of 3.5 mm/year for the decade 2010÷2020. If no change occurs, a further rise at the global sea level of more than 30 cm should be expected by 2100, due principally to the melting of polar ice caps, (413),

587

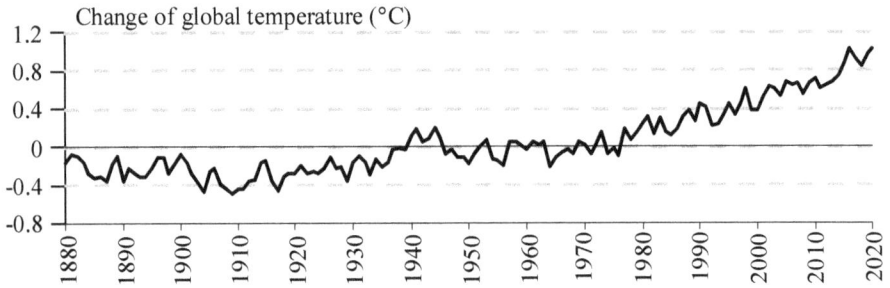

Fig. 23.1. Evolution of change in global temperature of the earth from 1880 to 2020, (392)

- the volume of the arctic sea ice was reduced from 7 million km^2 in 1980 to 6.25 million km^2 in 2000 and to 3.92 million km^2 in 2020, a reduction from 1980 to 2020 of 37%, (413),
- among 600 living beings tested, more than 75% present evidence compatible with an effort of adjustment to an increase in external temperature,
- known oil reserves will be exhausted by 2060,
- urban population is expected to double in the coming 30 years, from 3 billion in 2020 to 6 billion by 2050,
- there will be major shifts in the world's vegetation zones, deserts will become hotter, and desertification will increase,
- as storms are multiplying in numbers and intensity, wildfires rage well beyond their historic seasons, and oceans are rising and becoming warmer, there is no doubt that climate is changing at a level with no return.

Figure 23.2 illustrates the evolution of key factors of human activity and a rough forecasting of their plausible progress until 2100.

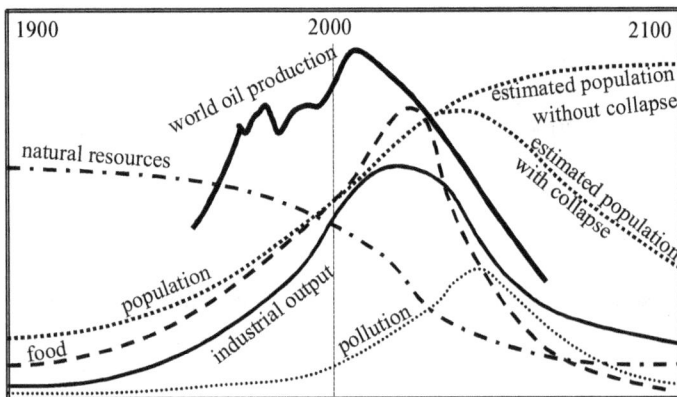

Fig. 23.2. Evolution of key factors of human activity between 1900 and 2100, (401)

23.1.2. The greenhouse effect and climate change

A greenhouse is a construction surrounded by glasses, which can collect and trap solar heat and keep warm inside. A similar phenomenon, named the greenhouse effect, is at the origin of convenient temperatures on earth, permitting life. The greenhouse effect is a natural process of regulating global climate, through retaining earth's heat. Indeed, when solar energy reaches the earth, a 30% is reflected back to space by the ozon layer and the clouds. The remaining 70% is absorbed by the air, the oceans, and the soil. As the earth is heating up, it subsequently radiates heat, the greatest part of which is trapped by the greenhouse gases, which are: carbon dioxide (CO_2), methane (CH_4), nitrous oxides (NO_x), and halocarbons. Carbon dioxide accounts for more than 3/4 of greenhouse gases. Without the greenhouse effect, the temperature on earth would be -18 °C and of course that would entail the lack of any trace of recognizable form of life.

However, due to human activities, most of which are based on the consumption of fossil fuels, since 1960 and particularly since 1980, (Fig. 23.3 and Fig. 23.4), we are producing more CO_2 and other greenhouse gases than in all the previous centuries. Thus, the greenhouse layer of the earth becomes thicker, causing increase in the global temperature of the earth. It is noteworthy that the lifetime until greenhouse gases are absorbed by a physical process is 5÷20 years for CO_2, 12 years for CH_4, at least 45 years for halocarbons, and 114 years for NO_x. The origins of greenhouse gases produced on the earth are in 2018 as follows: energy systems 34.4%, industry 24.1%, agriculture, forestry, and other land uses 21.5%, transport 14.3%, operation of buildings 5.7%. Principal emitters of CO_2 are in 2019 the following countries: China 30%, USA 15%, EU-28 9%, India 7%, Russia 5%, Japan 4%, other 30%, (392).

23.1.3. International initiatives and agreements

Confronting and combating climate change is a global problem that concerns all states and citizens, no matter where they live. However, conflicting interests of states and companies make it difficult to establish a global policy and adopt measures agreed by all parties involved, in spite of citizens' increasing sensibility.

The first coordinated and universally accepted effort to combat climate change was put forward in the United Nations convention on climate change, which was adopted in 1992 and set long-term objectives; it was extended by the Kyoto protocol, adopted in 1997 and entered in force in 2005; the principal objective of the Kyoto protocol was to reduce greenhouse gas concentrations in the atmosphere to a level that would prevent anthropogenic interference with climate. The Kyoto protocol ended in 2012 and was replaced by the Doha agreement of 2012, which in turn was replaced by the Paris agreement of 2015. In 2021, 190 countries ratified the Paris agreement, whose principal global target is to keep by 2050 the increase of global average temperature below 2 °C (preferably 1.5 °C) compared to pre-industrial levels. The Paris agreement allows each country to set its own emission reduction targets and adopt its own strategy for reducing them; it does not foresee any sanction to be imposed upon everyone who does not respect the undertaken commitments.

Figure 23.3 illustrates the evolution of world CO_2 emissions in various parts of the world during the last two centuries and Figure 23.4 the evolution of world CO_2 emissions in the various countries of the world during the last

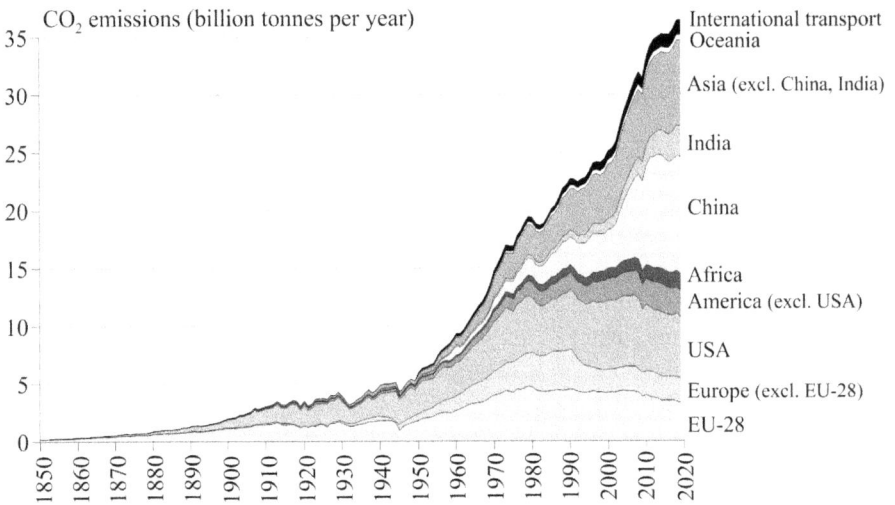

Fig. 23.3. Evolution of world CO_2 emissions since 1850, (412)

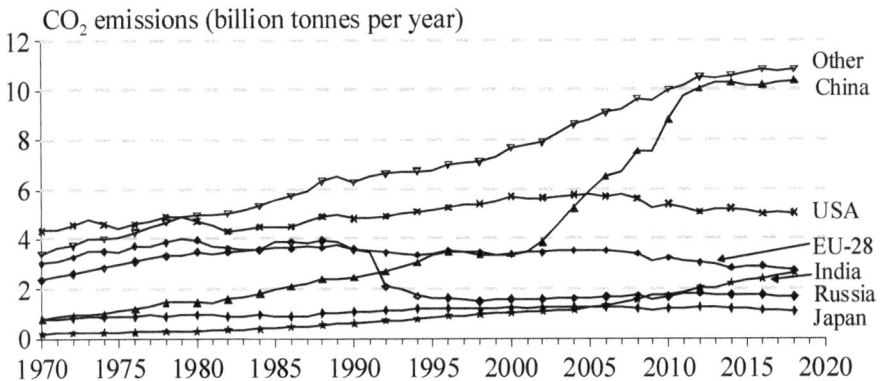

Fig. 23.4. Evolution of world CO_2 emissions in the various countries of the world during the last five decades, (412)

five decades, (412). It becomes clear that climate change is a global problem and thus calls for a collective action from all countries and citizens of the world.

The rates of growth of CO_2 emissions are correlated with the rates of growth of world GDP, as illustrated in Figure 23.5.

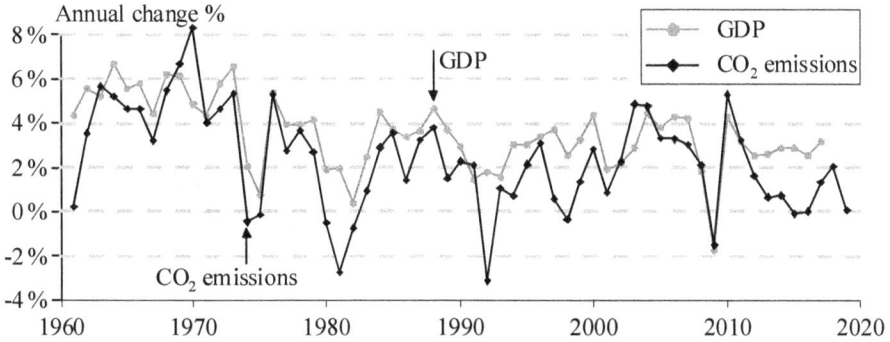

Fig. 23.5. Annual growth of global GDP and CO₂ emissions, (393)

23.1.4. Sustainable development

Awareness of the shortage of natural resources and of the effects of human activities on the environment has led world institutions and most governments to the adoption of the term 'sustainable development', which is understood as the kind of economic and social development in which resource use aims to meet human needs while preserving the environment, so that future generations can satisfy their needs and enjoy a level of prosperity not very different of that of generations between 1950 and 2020. Principal factors for the achievement of sustainable development are economic efficiency, environmental responsibility, and social equity, (401), (403), (408).

23.1.5. Transport and the environment

The transport sector has together with the industrial, tertiary, and household activities sectors a number of harmful effects on the environment, such as air and noise pollution, consumption of energy, accidents and safety, land occupancy, (404), (409), (410). Within the transport sector, however, railways are the least harmful to the environment mode of transport and this could prove a critical element for their development in the future.

The environmental effects of each transport mode (road, rail, air, sea) include passenger and freight traffic and may refer to the following:
- construction and maintenance of infrastructure,

- manufacture, maintenance, and disposal of rail and road vehicles, air-planes, ships,
- operation.

The consumption of transport by individuals is affected by their income and the GDP of the specific country, (see section 1.3.2, Figure 1.4). A causal relationship can be established between the individual consumption of transport C_{tr} and the GDP for various countries, as illustrated in Figure 23.6.

Conclusive evidence suggests that for many decades and all over the world the amount of time that people are willing to spend on travel has remained remarkably constant at approximately 1.1 hours per day, (see section 1.4.1). This means that as people have an increased income, they make use of faster modes of transport, a fact leading to more harm to the environment.

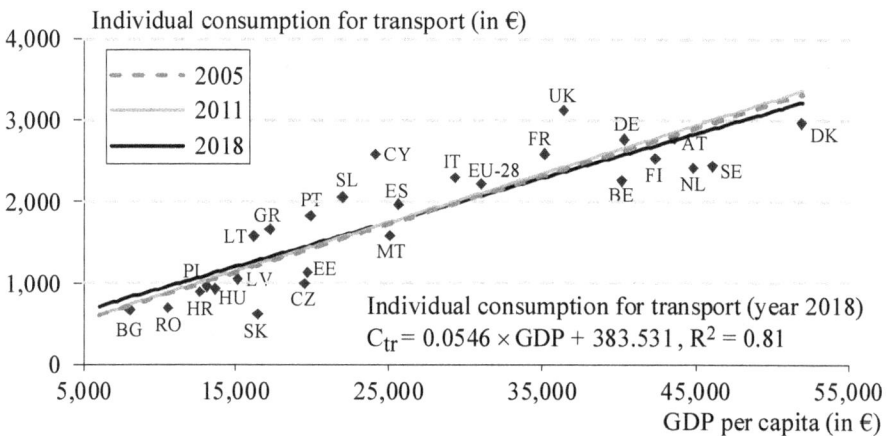

Legend: AT: Austria, BE: Belgium, BG: Bulgaria, CZ: Czech Republic, DK: Denmark, EE: Estonia, ES: Spain, FI: Finland, FR: France, DE: Germany, GR: Greece, HR: Croatia, HU: Hungary, IT: Italy, LT: Lithuania, LV: Latvia, PL: Poland, PT: Portugal, RO: Romania, SE: Sweden, SK: Slovakia, SL: Slovenia, MT: Malta, NL: The Netherlands, UK: United Kingdom, EU-28: European Union of 28 countries.

Fig. 23.6. A causal correlation between per capita GDP and individual consumption for transport, (393)

23.2. Air pollution and railways

23.2.1. Air pollutants from railways and other transport modes

Transport is an important air pollution emitter; though the problem is global, there is no consensus about the measures to be taken by the various countries around the world and the accurate measuring and recording of air pollution data. Thus, most analyses focus on specific parts of the world, e.g. the EU, USA, China, etc., (394). In the late 2010s, the transport sector was responsible in the EU countries for 81% of CO emissions, for 55.38% of nitrogen oxides (NO_x) emissions, for 12.08% of sulfur oxides (SO_x) emissions, and for 9.28% of non-methane volatile organic compounds, (394). The participation of each transport mode in the total amount of air pollutants for the EU-28 countries is illustrated in Table 23.1

Table 23.1.
Degree of participation (%) of the transport and the non-transport sectors and of the various transport modes in the emissions of CO, NO_x, SO_x, NMVOC for the EU-28 countries in the late 2010s, (394)

		CO emissions	NO_x emissions	SO_x emissions	NMVOC (non-methane volatile organic compounds emissions)
Non-transport sector		79.02%	44.62%	87.92%	90.71%
Transport sector	International aviation	0.70%	6.33%	0.81%	0.22%
	Domestic aviation	0.31%	0.71%	0.09%	0.11%
	International shipping	0.61%	14.74%	9.84%	0.62%
	Domestic shipping	1.28%	4.61%	1.21%	0.67%
	Rail transport	0.11%	0.87%	0.02%	0.10%
	Road transport	17.97%	28.12%	0.11%	7.57%

23.2.2. Specific emissions of air pollutants from railways and other transport modes

Figures 23.7 and 23.8 illustrate specific emissions (gr of pollutant (CO, NO_x, SO_x, NMVOC) per passenger-km or tonne-km) from railways and other transport modes.

23.2.3. The greenhouse effect and CO_2 emissions from railways and other transport modes

As analyzed previously, the greenhouse effect is at the origin of the existence of life on earth. However, human activities during the last 4÷5 decades, principally the burning of fossil fuels and deforestation, have led to the increase and accumulation of CO_2 concentrations around the earth, a fact that intensifies the natural greenhouse effect and causes additional global warming.

gr per p-km

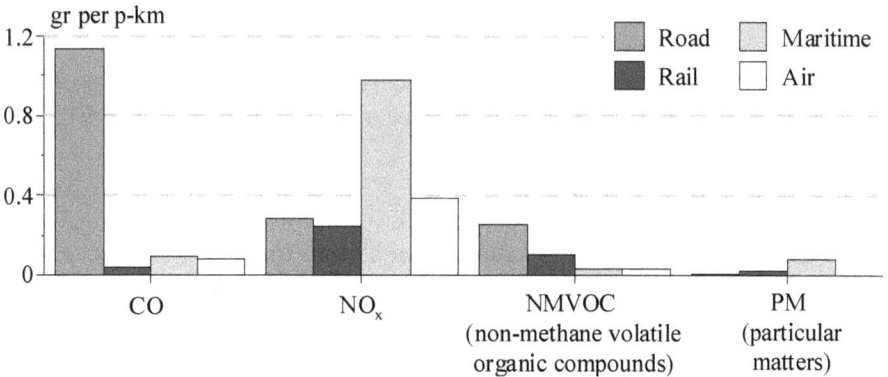

Fig. 23.7. Specific emissions (gr per p-km) of air pollutants from *passenger* transport for railways and other transport modes in the EU-28 countries, (395)

gr per t-km

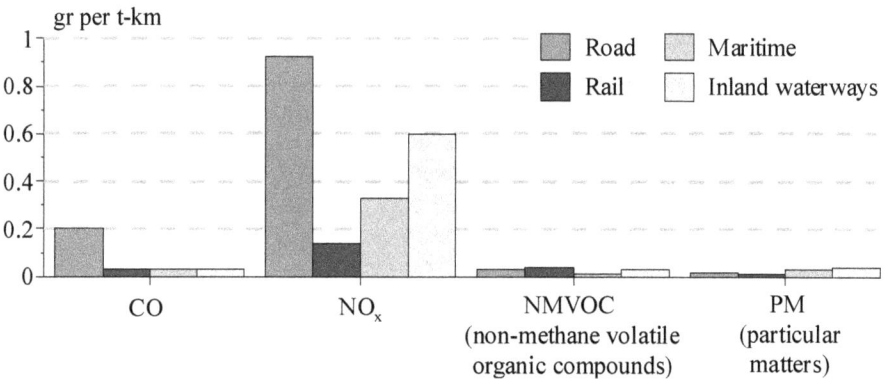

Fig. 23.8. Specific emissions (gr per t-km) of air pollutants from *freight* transport for railways and other transport modes in the EU-28 countries, (395)

Though CO_2 is the principal gas contributing to the greenhouse effect (by 76% worldwide), other gases (such as methane (by 16%), nitrous oxides (by 6%), halocarbons (by 2%)) also play a contributing role. The contribution of the transport sector to the emissions of greenhouse gases (ghg) is 15.9% worldwide, of which 3.7% from the USA, 1.8% from China, 1.7% from EU-28, 0.6% from India, and 0.5% from Russia.

Within the transport sector, the contribution of the various transport modes in ghg emissions was for the EU-28 countries for the year 2018 as follows: roads 71.0%, aviation 14.4%, navigation 13.6%, railways 0.5%, other 0.5%, (396). However, the situation varies at world level, where most railways are not electrified. Thus, ghg emissions from the various transport modes at world level for the year 2018 were as follows: private cars 44.4%, trucks 29.6%, aviation 11.3%, navigation 11.1%, railways 1.2%, and other 2.4%, (397), (398).

In spite of measures taken to control and reduce ghg emissions in the EU-28 countries for the past three decades (since 1990), Figures 23.9 and 23.10 illustrate that railways succeeded to reduce drastically their ghg emissions, domestic navigation to a lesser degree, whereas in the case of road transport these emissions increased and in the case of air transport ghg emissions saw a great increase, (398).

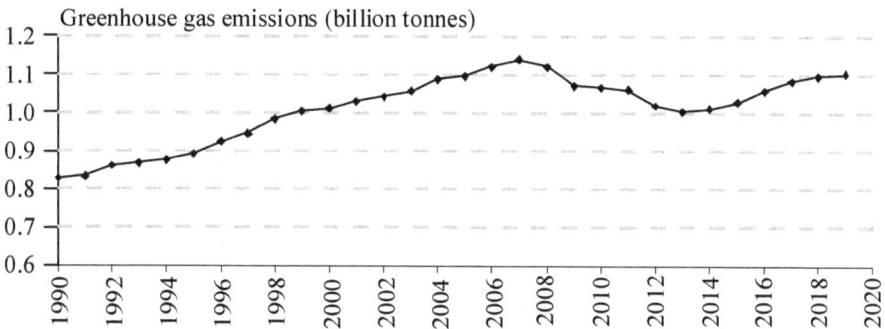

Fig. 23.9. Evolution of greenhouse gas emissions from various transport modes in the EU-28 countries between 1990 and 2019, (398)

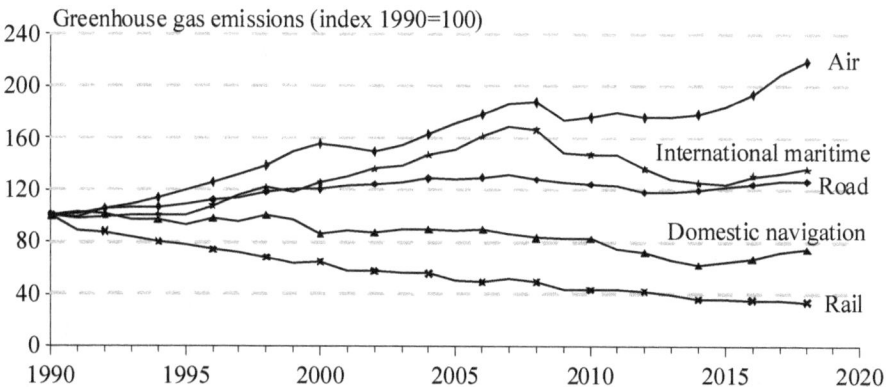

Fig. 23.10. Evolution of greenhouse gas emissions from various transport modes in the EU-28 countries between 1990 and 2019, (398)

23.2.4. Specific CO₂ emissions from railways and other transport modes

Specific CO_2 emissions from transport differ from one area of the world to the other, depending on the degree of electrification of railways, the load factor of the specific transport mode, the use of old or new technology to impart motion, etc.

Figure 23.11 illustrates average values of CO_2 emissions of railways and other transport modes for the EU-28 countries.

gr CO_2 per passenger-kilometer or tonne-kilometer

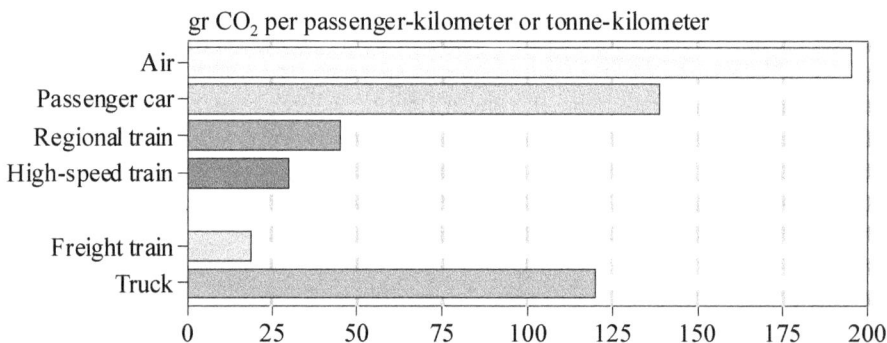

Fig. 23.11. Comparative CO_2 emissions of railways and other transport modes for the EU countries, (403)

Figure 23.12 illustrates the specific CO_2 emissions of high-speed trains, maglev systems, and aircrafts. Specific emissions do not vary *for railways* in relation to the distance travelled, whereas they vary greatly *for aircrafts*, (391).

It is noteworthy that specific emissions of gr CO_2/*passenger*-km were, on average for railways in Europe, 52 in 1990, 38 in 2010, and are expected to fall to 26 in 2030. Specific emissions for rail *freight* transport in Europe were (in gr CO_2/tonne-km) 31 in 1990, 18 in 2010, and are expected to fall to 15 in 2030, (397).

gr CO_2 per seat-kilometer (load factor = 100%, single departure)

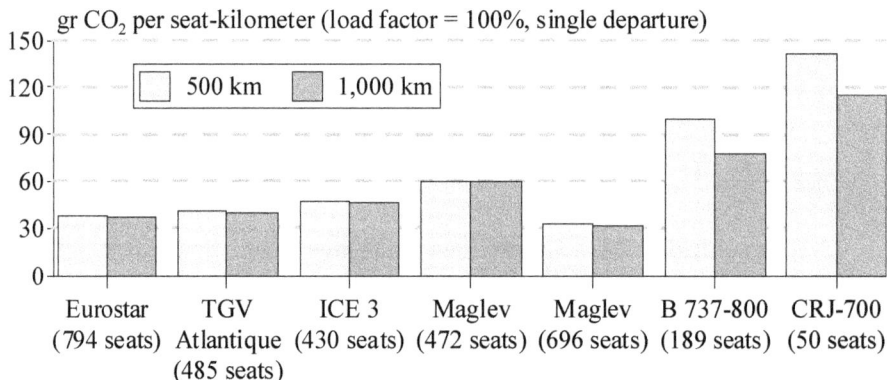

Fig. 23.12. Specific emissions of CO_2 per seat-km for high-speed trains, maglev systems, and aircrafts, (391)

596

23.2.5. CO_2 emissions for diesel and electric trains

Figure 23.13 gives for the EU countries specific CO_2 emissions for diesel and electric traction and Figure 23.14 CO_2 emissions by service type (high-speed, intercity, regional).

gr CO_2 per passenger-kilometer or tonne-kilometer

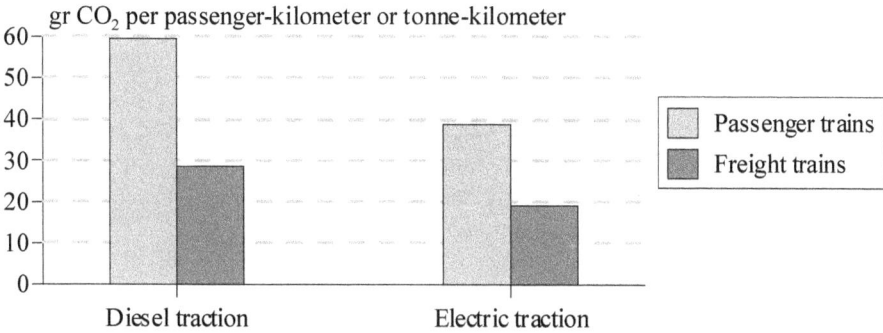

Fig. 23.13. Specific CO_2 emissions for diesel and electric traction for the EU countries, (403)

gr CO_2 per passenger-kilometer

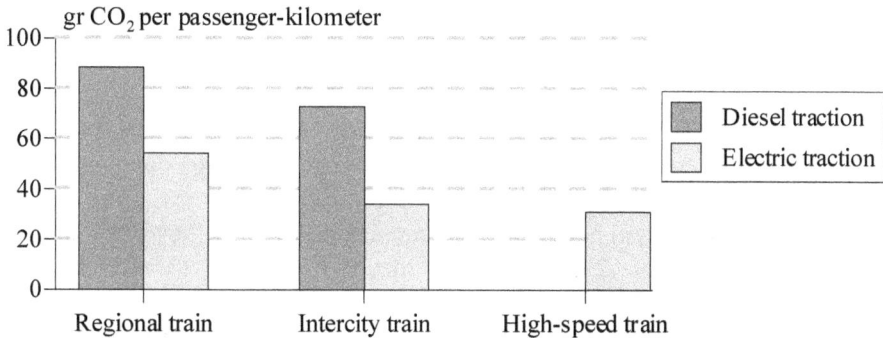

Fig. 23.14. Specific CO_2 emissions for high-speed, intercity, and regional trains for electric and diesel traction for the EU countries, (403)

23.2.6. Internalization of costs of CO_2 emissions

In section 5.7.2 it was analyzed that internalization of external costs in the real cost of transport, paid by the user, can become an efficient tool for tracking environmental problems. As a way to confront the greenhouse effect and CO_2 emissions, a carbon tax per tonne of carbon emitted has been suggested, with a value of 40÷50 US$/tonne of CO_2 in spring 2021 on the European market. If this internalization proceeds, something that is not very likely, a shift of traffic to the railways can be expected. Assessment of this shift of traffic may be approached as follows, (15).

597

First, a decision should be made about whether:
- internalization shall include only CO_2 emissions or all external costs,
- internalization shall be based on medium external cost or on marginal social cost.

A study on the internalization of all external costs for the EU countries was based on the increase of operation costs that would result and on cross-elasticities between rail and other transport modes. If internalization is conducted according to the average external costs, expected shift of traffic to the railways would be on the order of 12%÷15% for passenger traffic and up to 24% for freight traffic. If, however, internalization is conducted according to the marginal social cost, the expected shift of traffic for passenger and freight would be on the order of only 6%, (15).

23.3. Railway noise

23.3.1. Sources and damping of railway noise

Sources of noise from rail traffic have been analyzed in section 8.9.1 and are:
- noise from the engines of rolling stock,
- noise from wheel-rail interaction, plus (for electrified lines) noise from the contact between the pantograph and the contact wire, (see also section 20.8),
- aerodynamic noise.

Figure 23.15 illustrates noise levels for these three sources of railway noise in relation to speed, (405). Total noise is measured through the total pressure received by the human ear and is expressed on a logarithmic scale (dB(A)). Apparently, total noise is not the sum of the levels of the three sources of railway noise. Figure 23.15 illustrates that at low speeds (V<100 km/h) noise from the engines of rolling stock is dominant, at medium speeds (100 km/h<V<200 km/h) wheel-rail noise is dominant, whereas at high speeds (V>200 km/h) aerodynamic noise is dominant, (416). Concerning train type, however, the impact of the various sources of railway noise is different (Table 23.2), (405). Noise levels are attenuated by distance (though non-linearly) and are influenced more by distance than by changes in speed, (see section 8.9.3). Results of measurements of noise level in relation to distance, the type of train, and the speed were presented in section 8.9.5.

23.3.2. Noise indicators and maximum permitted level of rail noise

According to the European Directive 49/2002, related to the assessment and management of environmental noise, the so-called day-evening-night sound level L_{den} should be used as a basic noise indicator, which can be defined as follows, (405):

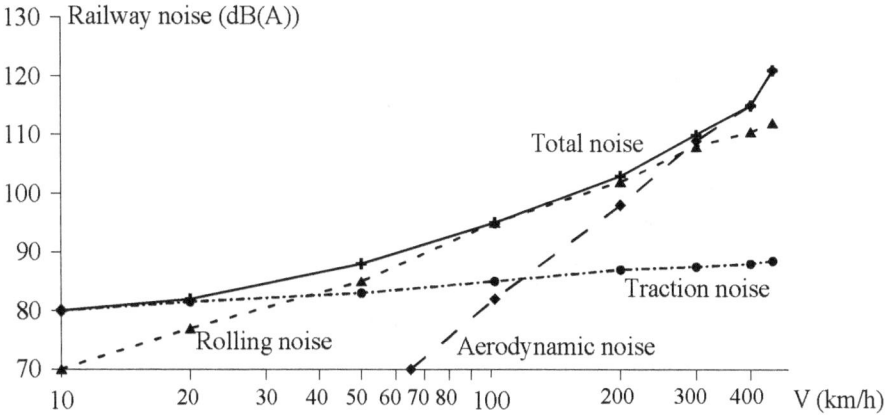

Fig. 23.15. The various sources of railway noise, (405)

Table 23.2.
Importance of sources of railway noise in relation to train type, (405)

Train type	Rolling noise	Traction noise	Aerodynamic noise
Freight trains	++	+	Not relevant
High-speed trains	++	+	++
Intercity or other long distance trains	++	+	+
Urban railways	++	+	Not relevant

$$L_{den} = 10 \cdot \log \frac{1}{24} \cdot \left(12 \cdot 10^{\frac{L_{day}}{10}} + 4 \cdot 10^{\frac{L_{evening}+5}{10}} + 8 \cdot 10^{\frac{L_{night}+10}{10}} \right) \qquad (23.1)$$

in which:

L_{day} is the A-weighted long-term average sound level determined over all the day periods of a year,

$L_{evening}$ is the A-weighted long-term average sound level determined over all the evening periods of a year,

L_{night} is the A-weighted long-term average sound level determined over all the night periods of a year.

The day is considered to have 12 hours, the evening 4 hours, and the night 8 hours. However, the evening period may be shortened by 1÷2 hours and the day or night period lengthened by 1÷2 hours accordingly, (405).

The European technical specifications for interoperability set the maximum noise emission for high-speed trains at the level of 87÷92 dB(A) as follows: 87 dB(A) for V=250 km/h, 91 dB(A) for V=300 km/h, 92 dB(A) for V=320 km/h, (134), (405). These levels are by 5%÷10% lower than the current emission levels in the more advanced European railways.

However, recommendations of the World Health Organization (WHO) for noise levels in living or working areas are far lower; for instance in order not to disturb people's sleep, noise level in sleeping rooms should in principle not exceed 32÷42 dB(A) at night. Thus, the need to attenuate and dampen emitted noise levels emerges. Acceptable noise levels in inhabitant areas vary across countries as follows: 55÷60 dB(A) during the day and 45÷55 dB(A) during the night.

23.3.3. Measures for the reduction of rail noise and related costs

The most efficient stage for the reduction of rail noise is during the decision making process concerning layout. In fact, layout design on embankment, viaduct, and bridge results in noise levels at the range of 75÷105 dB(A), whereas layout design in cut results in noise levels at the range of 50÷75 dB(A), (see section 8.9.4), (163), (406).

Other ways to reduce rail noise at the origin are, (405), (409), (417), (418):
- reduction of noise of the diesel engine (European Directive 26/2004 puts more strict terms),
- extensive use of rail dampers and resilient pads,
- appropriate grinding of rails, (see section 16.8),
- composite brake shoes, which in the case of freight trains could significantly reduce the emitted level of rail noise,
- use of new rolling stock, which if constructed after 2000 has emitted noise levels about 10 dB(A) lower compared to rolling stock constructed in the 1960s and 1970s.

If rail noise cannot be reduced at the origin, then the solution is passive methods of reduction, with most efficient among them the use of noise barriers, (406), (407), (409), (419). These should be placed as close as possible to the track and must have such a height that there is no direct visual contact between the receiver of the noise and the wheel of the rail vehicle. Implementation of noise barriers (of a non-absorbing material) at a height of 2 m and a distance of 3.50 m from the track results in a reduction of the perceived noise by approximately 10 dB(A). If, in addition, noise barriers have a noise absorbing material at the side of the track, noise reduction is further increased by 2÷5 dB(A). Noise reduction with the use of barriers is not affected by train speed, (170).

According to UIC, limit values of noise level at 7.5 m from the track axis and 1.2 m height should be for locomotives around 84÷85 dB(A) for a speed of 80 km/h and less than 99 dB(A) for a speed of 250 km/h, for passenger vehicles 79 dB(A) and for freight vehicles 83 dB(A) at a speed of 80 km/h (suburban traffic near inhabitant areas). Taking into account these limits, noise barriers can reduce rail noise levels just outside buildings in inhabitant areas at a level of 60÷70 dB(A), which can be further reduced with the use of insulated windows down to 50÷60 dB(A) in the interior of buildings.

Table 23.3 illustrates the sources of railway noise, suggested measures for the reduction of noise, expected results, and estimated costs.

Table 23.3.
Source of railway noise, suggested measure, level of impact,
expected reduction of noise, and estimated cost, (405)

Source of railway noise	Measure suggested	Level of impact (local, network, wide)	Expected reduction of railway noise	Estimated cost
rail	rail dampers	local	3÷7 dB(A)	300÷400 €/m
rail	slab track	local	5 dB(A)	n.a.
rail	rail pads	local	3÷4 dB(A)	n.a.
wheel	bogie shrouds together with low height barriers	local	8÷10 dB(A)	n.a.
wheel	wheel-tuned absorbers	network	2÷7 dB(A)	3,000÷4,000 € per wheel
rolling stock	K-blocks (type of brakes)	network	8÷10 dB(A)	4,000÷10,000 € per wagon
rolling stock	LL-blocks (type of brakes)	network	8÷10 dB(A)	500÷2,000 € per wagon
rolling stock	disk brakes	network	10 dB(A)	already established
wheel-rail contact	grinding of rails	local	10÷12 dB(A)	during maintenance of track
pantograph	shielding of pantograph	wide	5÷10 dB(A)	n.a.
propagation of vibrations	barriers 2m high	local	10 dB(A)	1,000 €/m
propagation of vibrations	barriers 3÷4m high	local	15 dB(A)	1,350÷1,700 €/m
propagation of vibrations	insulated windows	in house only	10÷30 dB(A)	3,000÷8,000 € per house with 4 windows

23.4. Energy consumption and railways

23.4.1. Energy consumption and the transport sector

With regard to the EU-28 countries in the year 2018, the transport sector consumed 33.9% of total energy, households 24.7%, industry 25.3%, services 13.4%, agriculture 2.4%, other activities 0.3%. Percentages of the consumption of energy at the world level for the year 2018 were as follows: transport 32.6%, industry 31.8%, households 19.1%, services 8.8%, other (agriculture, mining, etc.) activities 7.7%, (403), (414).

World energy demand came in 2018 from the following sources: oil 31.4%, coal 26.7 %, gas 22.9 %, renewable 9.7%, nuclear 5.0%, solid bio-mass 4.3%, (414). Under today's rates of consumption of fossil fuels, oil shortage is estimated likely around 2060, but known gas reserves will continue to serve the planet and satisfy world demand until 2100, (403). However, a massive shift from gasoline cars and airplanes to railways, electric cars, and cycling, in order to combat climate change, may change such an estimation.

23.4.2. Energy consumption within the transport sector

Within the transport sector for the EU countries in the year 2018, railways consumed 1.7% of total energy for transport activities, road transport 80.4%, domestic navigation 1.4%, air transport 0.8%, and other 0.7%, (414).

23.4.3. Energy consumption for diesel and electric traction

Figure 23.16 illustrates for the EU countries, what part of energy consumed by railways is used for diesel (28%) and electric (72%) traction. However, the situation may well be totally different in other parts of the world with fewer kilometers of electrified lines.

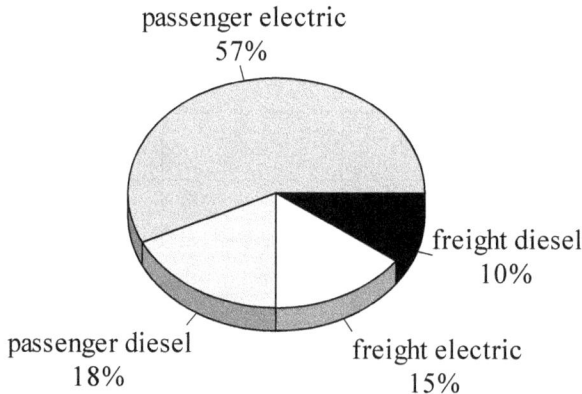

Figure 23.16. Consumption of energy for diesel and electric traction in the EU countries, (403)

23.4.4. Specific energy consumption of railways for passenger and freight traffic around the world

Figure 23.17 illustrates the specific energy consumption (energy consumption per unit of traffic) for passenger and freight traffic of railways at a world scale, for the USA, the EU-28 countries, China, Japan, and India, (403). Values of energy consumption differ from one area to the other, due to different degrees of electrification, the type (and age) of rolling stock, the axle load, etc.

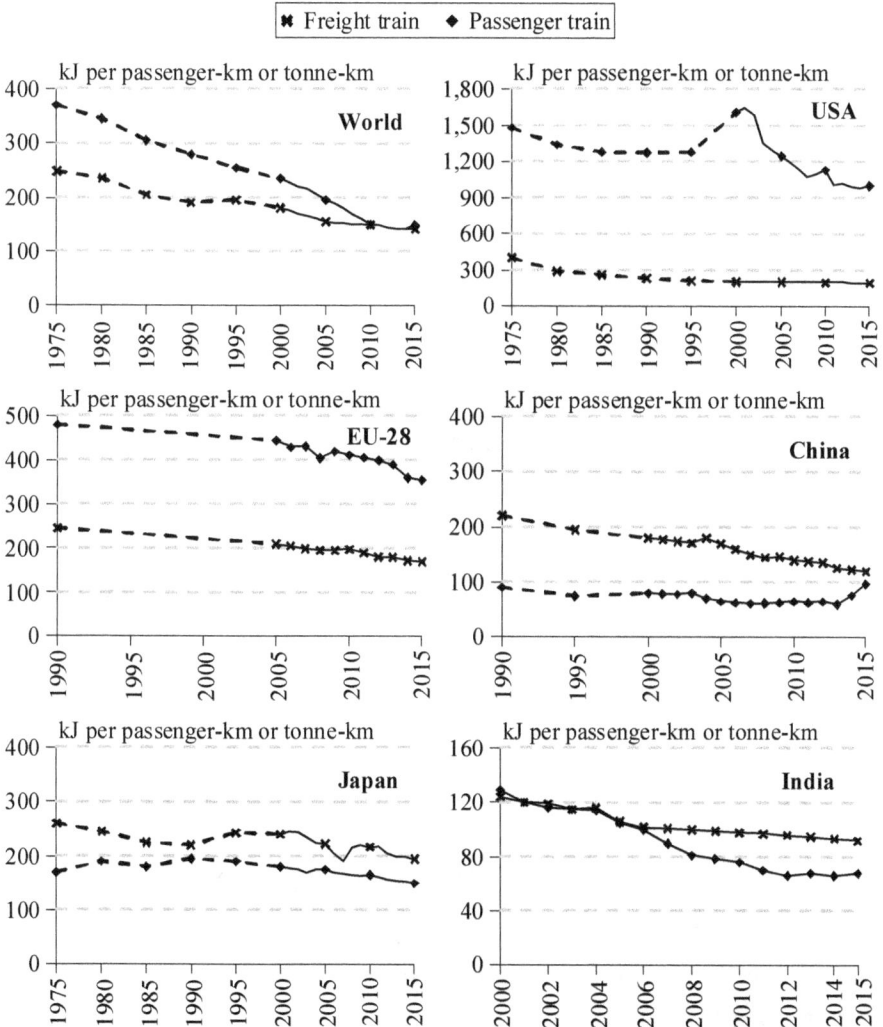

Fig. 23.17. Specific energy consumption of railways for passenger and freight transport, (403)

Specific energy consumption for both conventional and high-speed trains is in the range of 28÷39 Wh/seat-km and is not significantly affected by speed, (Fig. 23.18), but is strongly affected by longitudinal track gradient, (Fig. 23.19). Note that 1 kWh=3,600 kJ.

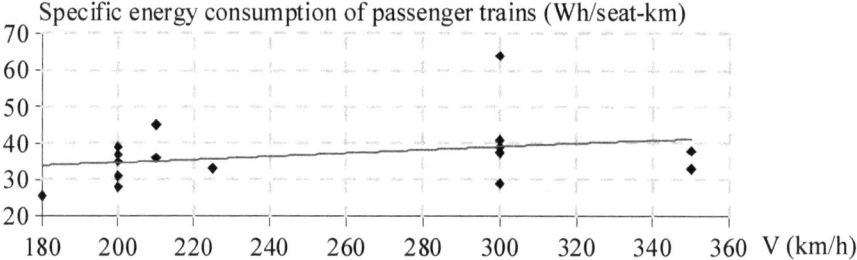

Fig. 23.18. Energy consumption of passenger trains in relation to speed, (403)

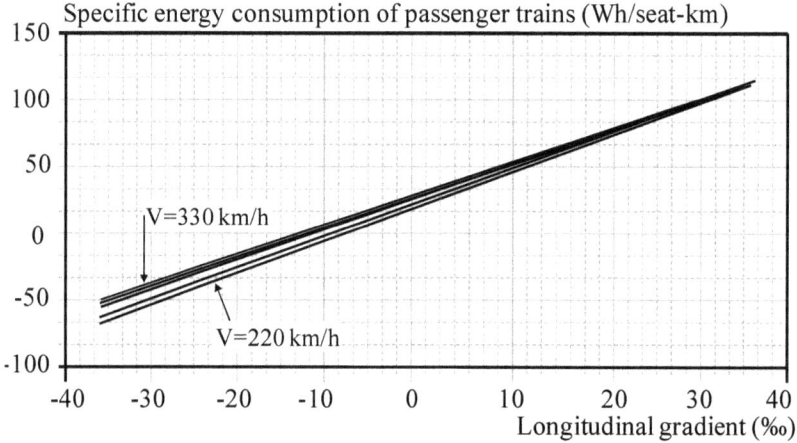

Fig. 23.19. Energy consumption of passenger trains in relation to speed and longitudinal gradient, (403)

23.4.5. Comparative specific energy consumption for railways and other transport modes

Figure 23.20 illustrates specific energy consumption[*] of some railway services in comparison with aircrafts.

The energy efficiency of rail transport in comparison with other transport modes is illustrated in Figure 23.21 for passenger transport and in Figure 23.22 for freight transport.

---

[*] Differences in the values of specific energy consumption for high-speed trains between Figure 23.18 and Figure 23.20 are the result of different train capacities and load factors, number of stops, level of technology, energy consumed for comfort reasons, etc.

Wh per seat-kilometer (load factor = 100%, single departure)

CRJ-700 (50 seats)

B 737-800 (189 seats)

Maglev (472 seats)

ICE 3 (430 seats)

TGV Atlantique (485 seats)

Eurostar (794 seats)

Maglev (696 seats)

Distance (km)

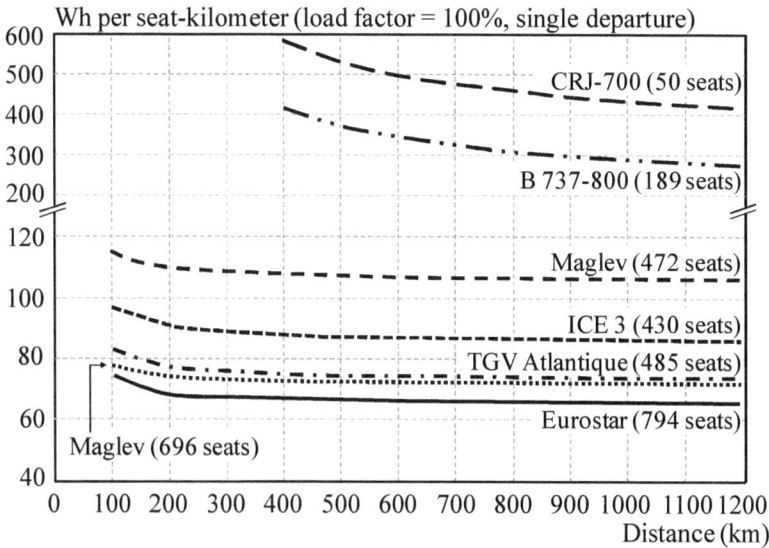

Fig. 23.20. Specific energy consumption for various high-speed railway services in comparison with aircrafts and maglev, (391)

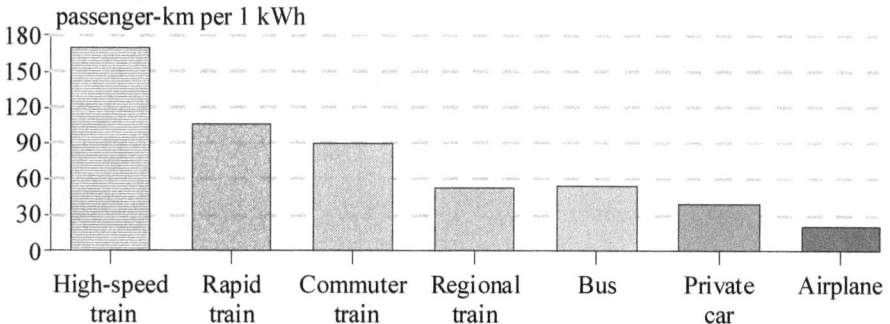

passenger-km per 1 kWh

| High-speed train | Rapid train | Commuter train | Regional train | Bus | Private car | Airplane |

Fig. 23.21. Number of passenger-km transported when consuming 1 kWh of energy for railways and other transport modes, (415)

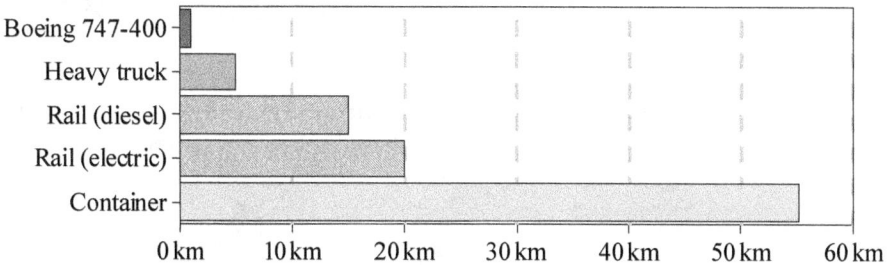

Boeing 747-400

Heavy truck

Rail (diesel)

Rail (electric)

Container

Fig. 23.22. Distance traveled by various transport modes for 1 tonne of freight when using 1 kWh of energy, (403)

23.5. Energy consumed in railways for comfort functions

Energy consumed by electric trains is easy to monitor analytically per kilometer of track and can be broken down in three categories:

- energy required to overcome the train's resistance to movement (rolling resistances, mechanical resistances, aerodynamic resistances), (see also section 18.3),
- energy required to provide comfort functions to passengers during traveling,
- energy lost between substations-pantograph and pantograph-wheel.

Figure 23.23 illustrates how energy is consumed in electric trains (conventional and high-speed). What becomes evident is the positive effect of using regeneration, that is regenerative braking, which feeds power back into the catenary; otherwise, this would be dissipated and lost, (420).

Oil prices present irregularities, (see Fig. 1.2), as a result of economic factors (recession-growth, needs of emerging economies), political factors (embargos, wars), the speculation of stock markets, psychological factors (fears of shortage). However, depending on the low or high values of oil, fuel costs as a percentage of total operation costs amount to $6\div10\%$ for rail passenger traffic, $10\div25\%$ for rail freight traffic and for trucking companies, and $15\div30\%$ for air traffic. Thus, fluctuations in oil prices do not critically affect the competitive position of the various transport modes.

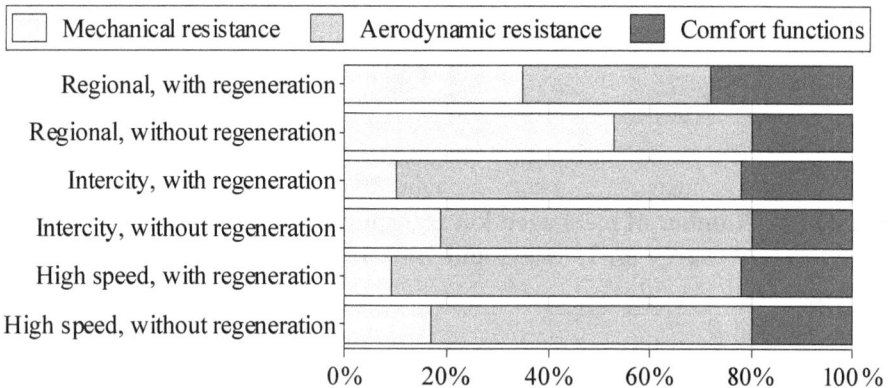

Fig. 23.23. Consumption of energy for various categories of electric trains to overcome resistances and assure comfort, (403)

List of References

CHAPTER 1

1. International Union of Railways (UIC), (2021), *Railway Statistics* 1970-2019, Paris.
2. European Union – Directorate General for Energy and Transport, (2021), *Energy and Transport in Figures*, Luxembourg.
3. International Transport Forum (ITF), (former European Conference of Ministers of Transport (ECMT)), (2021), *Transport Evolution 1970-2019*.
4. UIC, (2021), *Annual Reports*, Paris.
5. Profillidis V., (2021), 'Air – Rail Integration and Implications for Regional Airports', *International Conference*, University of Westminster, London, June 2021.
6. World Bank, (2021), *Railways Database*, Washington.
7. Chiambaretto P., Dumez H., Profillidis V., (2012), 'Air-Rail Intermodal Agreements as a Way to access New Markets for Non-Aligned Carriers: Lessons from France', *16th Air Transport Research Society World Conference*, Taiwan.
8. Thompson L., Tanaka Y., (2011), *High-Speed Rail Passenger Services: World Experience and U.S. Applications*, Thompson and Associates, Saratoga.
9. Profillidis V., Botzoris G., (2018), *Modeling of Transport Demand: Analyzing, Calculating, and Forecasting Transport Demand*, Elsevier.
10. Smith A., Benedetto J., Nash C., (2018), 'The Impact of Economic Regulation on the Efficiency of European Railway Systems', *Journal of Transport, Economics and Policy*, Vol. 53, No. 2.
11. Organization of the Petroleum Exporting Countries (OPEC), (2021), *World Oil Outlook*, Vienna.
12. Profillidis V., (2014), 'Quel Avenir pour le Ferroviaire Européen', *Transports*, No. 485.

13. Thompson L., (2010), *A Vision for Railways in 2050*, ITF, Transport and Innovation, Paris.
14. Leenen M., Wolf A., Peighambari A., (2014), *The Worldwide Market for Railway Industries,* SCI Verkehr, Hamburg.
15. UIC, (2008), *EURAIL 2025 – Strategic Planning of European Railways towards 2025*, Paris.
16. Profillidis V., (2016), *Transport Economics* – 5th Edition, Papasotiriou Ed., Athens.
17. UIC, (2015), *A Global Vision for Railway Development*, Paris.
18. Lin K., (2004), 'Making New Connections: Airport Rail Links in the United States', *Japan Railway and Transport Review*, No. 39.
19. Batisse E., (2003), 'Heavy Haul, a Challenge or an Opportunity for Europe's Railways', *Rail International*, September 2003.
20. Profillidis V., (2001), 'Separation of Infrastructure from Operation and the New Organization of the Railways', *Japan Railway and Transport Review*, No. 29.
21. Profillidis V., Botzoris G., Galanis A., (2018), 'Decoupling of Economic Activity from Transport-related Energy Consumption: an Analysis for European Union Member-Countries', *International Journal of Innovation and Sustainable Development*, Vol. 12, No. 3.
22. Profillidis V., Boilé M., (2001), 'Evolutions et Restructurations au Transport de Fret en Europe', *Transports*, No. 405.
23. Profillidis V., Botzoris G., (2001), 'Assessment of the Evolution of Environmental Effects of Transport', 1st International Conference on Ecological *Protection of the Planet Earth*, Xanthi, June 2001.
24. Profillidis V., (1998), *Theoretical and Practical Aspects concerning Land Access to Sea Ports*, ECMT, Round Table 113, Paris.
25. United Nations, (2018), *Enhancing Interoperability for Facilitation of International Railway Transport*, Bangkok.
26. ECMT (1995), *New Problems – New Solutions*, 13th International Symposium, Luxembourg.
27. EU – Agency for Railways, (2016), *Big Data in Railways*, Brussels.
28. Profillidis V., (1995), 'Light Rail Transit Systems: Present Trends and Future Prospects', *Journal of Light Rail Transit Association*, Jan. 1995.
29. Profillidis V. (editor), (1994), *Modernization of Railway and Airway Transport – The Impact of Liberalization*, International Conference, Democritus Thrace University, Xanthi, May 1994.
30. Central Japan Railway Company, (2018), *Annual Report*, Tokyo.
31. US Energy Information Administration, (2021), *Annual Energy Outlook*, Washington.

32. Agamez-Arias A., Moyano-Fuentes J, (2017), 'International Transport in Freight Distribution', *Transport Reviews*, Vol. 5, No. 4.
33. Profillidis V., (1991), 'Combined Transport between Greece, Europe and the Middle East-Present Trends and Future Prospects', *International Conference*, University of Trieste, September 1991.
34. Profillidis V., (1990), 'Light Rail Technologies in the 1990s', *International Conference on Electric Transport*, November 1990, Basel.
35. BP, (2021), *Crude Oil Prices from 1861*, http://www.quandl.com.
36. Estival J.-P., Profillidis V., (1985), 'For a New Strategy of the European Rail Networks', *Rail International*, July 1985.

CHAPTER 2

37. Profillidis V., Botzoris G., (2013), 'High-Speed Railways: Present Situation and Future Prospects', *Journal of Transportation Technologies*, Vol. 3, No. 2A.
38. UIC, (2019), '*High Speed Rail*', Paris.
39. Ranger St., (2018), *What is Hyperloop*, ZD Net.
40. Eurostar Group, (2005), 'Eurostar – A Seamless Journey to the Continent', *Japan Railway and Transport Review*, No. 40.
41. Andersen S., (2004), 'Überlegungen zur Anwendung der Magnetbahntechnik im spurgeführten Hochgeschwindigkeitsverkehr', *ZEVrail Glasers Annalen*, Vol. 128, No. 3.
42. Siemens, (2001), '*Transrapid: The New Dimension in Travel*', Erlangen, Germany.
43. Noulton J., (2001), 'The Channel Tunnel', *Japan Railway and Transport Review*, No. 26.
44. Najafi F.T., Nassar F.E., (1996), 'Comparison of High-Speed Rail and Maglev Systems', *ASCE, Journal of Transportation Engineering*, Vol. 122, No. 4.
45. Arduin J.-P., (1994), 'Development and Economic Evaluation of High Speed in France', *Japan Railway and Transport Review*, No. 9.
46. Taylor, C.L., Hyde, D.J., Barr, L.C. (2016), *Hyperloop Commercial Feasibility Analysis*, NASA – US Department of Transportation.
47. Brand M.M., Lucas M.M., (1989), 'Operating and Maintenance Costs of the TGV High-Speed Rail System', *ASCE, Journal of Transportation Engineering*, Vol. 115, No. 1.
48. Profillidis V., (1985), 'High-Speed Trains', *Technica Chronika* (Scientific Journal of Greek Engineers), Vol. 5, No. 3, Athens.

CHAPTER 3

49. Nikitinas V., Dailydka S., (2016), 'The Models of Management of Railway Companies in the European Union', *Procedia Engineering*, Vol. 134.
50. Johnson D., Nash C., (2012), 'Competition and the Provision of Rail Passenger Services: A Simulation Exercise', *Journal of Rail Transport Planning and Management*, Vol. 1, Issue 3.
51. RNE (Rail Net Europe), (2021), *'Network Statements of European Railways'*, Brussels.
52. United Nations, (2018), *Railway Reform in the ECE Region*, New York.
53. Association of American Railroads, (2011), *The Impact of Staggers Rail Act of 1980*, Washington.
54. Cantos P., Pastor J., Serano L., (2010), 'Vertical and Horizontal Separation in the European Railway Sector and its Effects on Productivity', *Journal of Transport, Economics and Policy*, Vol. 44, No. 2.
55. Cowele J., (2009), 'The British Passenger Rail Privatization: Conclusions on Subsidy and Efficiency from the First Round of Franchises', *Journal of Transport, Economics and Policy*, Vol. 43, No. 1.
56. CPCS, (2015), *Comparison of Canadian and United States Rail Economic Regulations*, Ottawa.
57. European Commission, (2006), *EU Policy and its Impact on the Rail System*, Brussels.
58. Profillidis V., (2006), 'La Législation Ferroviaire Européenne', *Transports,* No. 435.
59. Nash C., Nelsson J.-E., Link H., (2013), 'Comparing Three Models for Introduction of Competition into Railways', *Journal of Transport, Economics and Policy,* Vol. 47, No. 2.
60. Office of Rail and Road, (2020), *UK Rail Industry Financial Information*, London.
61. Ponti M., (2001), *What Role for the Railways in the East?*, ECMT, Round Table 120, Paris.
62. Intergovernmental Organization for International Carriage by Rail, (2011), *Convention concerning International Carriage by Rail (COTIF)*, Bern.
63. ITF, (2013), *Railway Efficiency*, OECD, Paris
64. Konno S., (1997), 'JNR Privatization – The First 10 Years and Future Perspectives', *Japan Railway and Transport Review*, No. 13.
65. International Railway Journal, (2018), *The Fourth Railway Package.*
66. Kurosaki F., (2018), 'A Study of Vertical Separation in Japanese Passenger Railways', *Case Studies in Transport Policy*, Vol. 6.

67. Profillidis V., (1990), 'Present Status and Future Prospects of Greek Railways – An Analysis of a Railway Network in a Difficult Situation', *Journal of Transportation Planning and Technology*, Vol. 14.
68. Profillidis V., (1987), 'A Methodology of Quantification of the Public Benefit that the Railways offer to the Society and a New Approach for the Appreciation of the Management of the Railway Undertaking', XVII *Panamerican Railway Congress Association*, Cuba.
69. World Bank, (1982), *The Railway Problem*, Washington.
70. Regulation 851/2006 of the European Community on *the Items to be included in the Definition and Scope of the Term Transport Infrastructure - A. Rail*, Brussels.

CHAPTER 4

71. Profillidis V., (2012), 'An Ex-Post Assessment of a Passenger Demand Forecast of an Airport', *Journal of Air Transport Management*, Vol. 25.
72. Department for Transport, (2012), *Rail Passenger Demand Forecasting Methodology*, London.
73. Börjesson M., (2014), 'Forecasting Demand for High Speed Rail', *Transportation Research Part A: Policy and Practice*, Vol. 70.
74. Profillidis V., Botzoris G., (2015), 'Air Passenger Transport and Economic Activity', *Journal of Air Transport Management*, Vol. 49.
75. Maddala G., (2010), *Introduction to Econometrics*, Wiley.
76. Profillidis V., Botzoris G., (2006), 'Econometric Models for the Forecast of Passenger Demand in Greece', *Journal of Statistics and Management Systems*, Vol. 9, No. 1.
77. Profillidis V., Botzoris G., (2005), 'A Comparative Analysis of the Forecasting Ability of Classic Econometric and Fuzzy Models', *Fuzzy Economic Review*, Vol. 10, No. 1.
78. Profillidis V., Botzoris G., (2004), 'A Time-series Model for the Forecast of Rail Passenger Demand with the use of the Least Median of Squares and the Singular Spectrum Analysis', *2nd International Conference on Research in Transportation*, Athens.
79. Profillidis V., Botzoris G., (2004), 'Econometric Models for the Forecast of Modal Split of Passenger Demand for Greece', *International Conference on Modelling and Simulation*, Valladolid, Spain.
80. Profillidis V., Botzoris G., (2003), 'The Market Survey: An Essential Tool for the Commercial and Tariff Policy of a Public Transport Undertaking', *2nd International Conference on Marketing*, UITP, Paris.

81. Tsai T.H., Lee C.K., Wei C.H., (2009), 'Neural Network Based Temporal Feature Models for Short-term Railway Passenger Demand Forecasting', *Expert Systems with Applications*, Vol. 36, No. 2.

82. Profillidis V., Botzoris G., Kolidakis S., (2018), 'Artificial Neural Networks: A Modern Tool for Empirical Modeling of Transport Demand', *5th Conference of Economics of Natural Resources and the Environment*, Volos, Greece.

83. Rail Freight Forward, '*Rail Freight Strategy to boost Modal Shift*', www.railfreigtforward.eu.

84. Profillidis V., Papadopoulos B., Botzoris G., (1999), 'Similarities in Fuzzy Regression Models and Application on Transportation', *Fuzzy Economic Review*, Vol. 4, No. 1.

85. Franses Ph., (1998), *Time-Series Models for Business and Economic Forecasting*, Cambridge University Press.

86. Elsner J., Tsonis A., (1996), *Singular Spectrum Analysis – A New Tool in Time-Series Analysis*, Plenum Press.

87. Cox E., (1995), *Fuzzy Logic for Business and Industry*, Charles River Media Inc.

88. Kolidakis S., Botzoris G., Profillidis V., Kokkalis A., (2020), Real-time Intraday Traffic Volume Forecasting - A Hybrid Application Using Singular Spectrum Analysis and Artificial Neural Networks', *Periodica Polytechnica Transportation Engineering*, Vol. 48.

89. Preston J., (1991), 'Demand Forecasting for New Local Rail Stations and Services, *Journal of Transport, Economics and Policy*, Vol. 25, No. 2.

90. Fowkes T., Nash C., (1991), *Analysing Demand for Rail Travel*, Avebury, London.

91. Moll S., Weidmann U., Nash A., (2012), 'Methodological Framework for Analyzing Ability of Freight Rail Customers to Forecast Short-term Volumes Accurately', *Transportation Research Record*, No. 2289.

92. Fowkes T., Nash C., Whiteing A., (1985), 'Understanding Trends in Intercity Rail Travel in Great Britain', *Transportation Planning and Technology*, Vol. 10, No. 1.

93. McGeehan H., (1984), 'Forecasting the Demand for Inter-Urban Railway *Travel in the Republic of Ireland*', *Journal of Transport, Economics and Policy*, Vol. 18, No. 3.

94. Jia, H.A.O., Lan, L.I., (2004), 'The Study of Comprehensive Forecast Model for Railway Freight Transport Volume', *Railway Transport and Economy*, Vol. 26, No. 11.

CHAPTER 5

95. European Commission, (2019), *Handbook of the External Costs of Transport*, Luxembourg.
96. Sanchez-Borràs M., López-Pita A., (2011), 'Rail Infrastructure Charging Systems for High-Speed Lines in Europe', *Transport Reviews*, Vol. 31, No. 1.
97. IRG (Independent Regulators' Group)-Rail, (2015), Updated Review of *Charging Practices for the Minimum Access Package in Europe*, Brussels.
98. International Transport Forum, (2019), *What is the Value of Saving Travel Time*, ITF, Round Table 176, OECD, Paris.
99. Thompson L., (2008), '*Railway Access Charges in the EU*', ITF, Paris.
100. European Commission, (2019), *Sixth Report on Monitoring Development of the Rail Market*, Brussels.
101. Nash C., (2005), 'Rail Infrastructure Charges in Europe', *Journal of Transport, Economics and Policy*, Vol. 39, No. 3.
102. Kopp A., (2005), '*Transport et Commerce International*', Transports, No. 431.
103. Chun-Hwan K., (2005), 'Transportation Revolution: The Korean High-Speed Railway', *Japan Railway and Transport Review*, No. 40.
104. European Commission, (2015), *Study of the Cost and Contribution of the Rail Sector*, Brussels.
105. INFRAS, IWW Universität Karlsruhe, (2004), *Facts of Competition in the European Transport Market*, Zurich.
106. Standard & Poor's, (2004), *Infrastructure Finance*, McGraw-Hill.
107. Crozet Y., (2003), *Time and Passenger Transport*, ECMT, Round Table 127, Paris.
108. Link H., (2004), 'Rail Infrastructure Charging and on-track Competition in Germany', *International Journal of Transport Management*, Vol. 2, No. 1.
109. Brambilla M., Erba S., Ponti M., (2003), 'Costs, Competition and the Role of the State in Freight Transport', *Trasporti Europei*, No. 23.
110. Baumgartner J.P., (2001), '*Prices and Costs in the Railway Sector*', Ecole Polytechnique Fédérale de Lausanne, Lausanne.
111. Almujibah H., Preston J., (2019), '*The Total Social Costs of* Constructing and Operating High-Speed Rail Lines', *Frontiers in Built Environment*, No. 5.
112. Button K., (2019), *Transport Economics*, Edward Elgar.
113. Profillidis V., (1996), 'The Logistic Chain and Railway Transport', *12th International Logistics Congress*, Athens.

114. ECMT, (1994), *Internalizing the Social Cost of Transport*, Paris.
115. Baumgartner J.P., (1991), *Economie des Transports*, Lausanne.
116. UIC, (1988), *Railway Statistical and Costs Terminology*, Paris.

CHAPTER 6

117. Marinov M., Viegas J., (2011), 'Tactical Management of Rail Freight Transportation Services', *Transportation Planning and Technology*, Vol. 34, No. 4.
118. Wee B.V., Reetveld P., (2008), 'The Myth of Travel Time Savings', *Transport Reviews*, Vol. 28, No. 6.
119. Friebel et al., (2007), 'Railroad Restructuring in Russia and Eastern Europe', *Transport Reviews*, Vol. 27, No. 7.
120. Wee B.V., (2007), 'Rail Infrastructure Challenges for Cost-Benefit Analysis and other ex ante Evaluations', *Transportation Planning and Technology*, Vol. 30, No. 1.
121. Profillidis V., Botzoris G., (2006), 'Public-Private Partnerships for Transport Infrastructure Projects and Impact on Planning and Operation', *3rd International Conference on Transport Research in Greece*, Thessaloniki, 2006.
122. Olsson N., (2006), 'Impact Analysis of Railway Projects in a Flexibility Perspective', *Transport Reviews*, Vol. 26, No. 5.
123. Profillidis V., Botzoris G., (2004), 'Recent Changes in Technology and Electronics and Impact on the Management of the Railway Undertaking', *20th European Conference on Operational Research*, Rhodes, July 2004.
124. Padilla-Angulo L., Friebel G., Mc. Cullough G., (2019), "Product Market Deregulation's Winners and Loosers: US Railroads between 1981 and 2001", *Journal of Transport, Economics and Policy*, Vol. 53, No. 3.
125. Jandova M., Poleta T., (2019), "Impact of on-track Competition on Public Finances", *Journal of Rail Transport Planning and Management*, Vol. 8, No. 2.
126. European Union (DG VII), UIC, and Community of European Railways, (1999), *Shaping the Future of Rail*, Paris.
127. Office of Rail and Road (ORR), (2020), *Government Support to the Rail Industry*, London.
128. Profillidis V., (1997), 'Possibilities of Financing Transportation Projects through Private Capitals', *International Conference «Present and Future of Greek Economy»*, Economic University of Athens, Greece.

129. Adler H., (1987), *Economic Appraisal of Transport Projects*, The World Bank, Washington.
130. Green T.J., (1983), *Long-term Planning*, Track Course, RIA, London.
131. McClintock A.G., Skinner R.N., (1983), *Project Management*, Track Course, RIA, London.

CHAPTER 7

132. Gesualdo A., Penta F., (2018), A Model for the Mechanical Behavior of the Railway Track in the Lateral Plane, *International Journal of Mechanical Sciences*, Vol. 146.
133. Zhang J., (2009), 'Analysis on Wheel-Rail Contact using Finite Element Method', *International Conference on Mechatronics Automation*, IEEE, Vol. 2, Hunan.
134. European Commission, (2014), *Regulation 1299/2014 on the Technical Specifications for Interoperability relating to the Infrastructure Subsystem*, Brussels.
135. Ubalde L., López-Pita A., Teixeira P., Saña A., (2005), 'Track Deterioration in High-Speed Railways: Influence of Stochastic Parameters', *8th International Conference on Railway Engineering*, University of Westminster, London.
136. UIC, Code 505, (2003), '*Railway Transport Stock – Rolling Stock Construction – Gauge*', Paris.
137. Liolios A., Profillidis V. et al., (2002), 'A Nonconvex Numerical Approach to the Dynamic Soil-Pipeline Interaction Induced by High-Speed Railway Traffic', *International Conference: Nonsmooth/Nonconvex Mechanics with Application in Engineering*, Thessaloniki.
138. Saito A., (2002), 'Why did Japan Choose the Narrow Gauge', *Japan Railway and Transport Review*, No. 31.
139. Kissel E., Missler M., (2001), 'The Use of Ballastless Track on the Lines of Deutsche Bahn AG: Interaction between Requirements, Operating Trials and further Development', *European Railway Review*, Vol. 7, No. 4.
140. UIC, (1999), *Common Recommended Practices for Metre Gauge Rolling Stock*, Paris.
141. The Permanent Way Institution (1993), *British Railway Track,* 6th Edition, Echo Press Ltd, Loughborough.
142. Fiedler J. (1998), *Grundlagen der Bahntechnik,* Werner–Verlag, Düsseldorf.
143. UIC, Code 714R, (2009) '*Classification of Lines for the Purpose of Track Maintenance*', Paris.

144. Semaly, (1988), *'Studies for the Metros of Lille and Strasbourg'*, Lyon.

145. ORE, D 161, RP 4, (1987), *The Dynamic Effects due to Increasing Axle Loads from 20 to 22.5 t*, Utrecht.

146. Profillidis, V. (1986), 'Applications of Finite Element Analysis in the Rational Design of Track Bed Structures', *Computers and Structures*, Vol. 22, No. 3.

147. Alias J., (1984), *La Voie Ferrée-Tome I: Techniques de Construction et d' Entretien*, Eyrolles, Paris.

148. Profillidis V., (1983), *La Voie Ferrée et sa Fondation-Modélisation Mathématique*, Ph.D. Thesis, Ecole Nationale des Ponts et Chaussées (ENPC), Paris.

149. ORE, C 116, RP 10, (1981), *Study of Optimum Rail Inclination and Gauge related to Wheel Profiles adapted to Wear*, Utrecht.

150. Peng D., Jones R., (2013), 'Finite Element Method Study on the Squats Growth Simulation', *Applied Mathematics*, Vol. 4, No. 5A.

151. Sauvage R., Richez G., (1978), 'Les Couches d'Assise de la Voie Ferrée', *Revue Générale des Chemins de Fer (RGCF)*, December 1978.

152. Prud' homme A., (1970), 'La Voie', *RGCF*, Paris, January 1970.

153. Sauvage R., Errieau J., (1970), 'Les Poses de Voie sans Ballast', *RGCF*, March 1970.

154. Kalker J., (1967), *On the Rolling Contact of Two Elastic Bodies in the Presence of Dry Friction*, Ph. D. Dissertation, Delft University.

CHAPTER 8

155. Zhang T.W., Lamas- López F., Cui Y.J., Calon N., D'Aguiar S.C., (2017), 'Development of a simple 2D model for Railway Track bed Mechanical Behavior based on Field Data', *Soil Dynamics and Earthquake Engineering*, Vol. 99.

156. Argilago A., (2016), *FEM and DEM double Scale Approach with second Gradient Regularization applied to Granular Materials*, Ph.D. Thesis, Univ. of Grenoble.

157. Sowmiya L., Shahu J., Gupta K., (2010), 'Three Dimensional Finite Element Analysis of Railway Track', *Indian Geotechnical Conference*, Bombay.

158. Lichtberger B., (2005), *Track Compendium*, Eurail Press, Hamburg.

159. Girardi L., (2003), 'Fabrication, Maintenance et Développement du Rail', *RGCF*, June 2003.

160. Suhairy S., (2000), *'Prediction of Ground Vibration from Railways'*, Swedish National Testing and Research Institute, Stockholm.

161. Panagiotopoulos P., (1993), *Hemivariational Inequalities, Applications in Mechanics and Engineering*, Springer, Berlin.

162. Wayson R.L., Bowlby W., (1989), 'Noise and Air Pollution of High-Speed Rail Systems', *ASCE, Journal of Transportation Engineering*, Vol. 115, No. 1.

163. Zicha J.H., (1989), 'High-Speed Rail Track Design', *ASCE, Journal of Transportation Engineering*, Vol. 115, No. 1.

164. Profillidis V., Humbert P., (1986), 'Etude en Elastoplasticité par la Méthode des Eléments Finis du Comportement de la Voie Ferrée et de sa Fondation', *Bulletin de Liaison des Laboratoires des Ponts et Chaussées*, Vol. 141.

165. Profillidis V. (1985), 'Three-Dimensional Elastoplastic Finite Element Analysis for Track Bed Structures', *Civil Engineering for Practicing and Design Engineers*, Vol. 4, No. 9.

166. ORE, D 117, RP 18, RP 25, RP 27, RP 28, RP 29, (1984), *Optimum Adaptation of the Conventional Track to Future Traffic*, Utrecht.

167. Salençon J., Halphen B., (1987), '*Elasto-plasticité*', ENPC, Paris.

168. Profillidis V., (1983), *La Méthode des Eléments Finis: Principes de Base et Techniques d' Application en Mécanique des Structures*, Textbook, French Railways, Paris.

169. Profillidis V., (1983), *Les Lois de Comportement Non-Linéaires en Mécanique-Traitement par la Méthode des Eléments Finis*, Textbook, French Railways, Paris.

170. ORE, C 137, RP 12, (1981), '*Railway Noise: Measurements of the Running Noise caused by Trains on Different Types of Bridges*', Utrecht.

171. Girardi L., (1981), 'Propagation des Vibrations dans les Sols Homogènes ou Stratifiés', *Inst. Techn. du Bat. et des Trav. Publ.*, No. 397.

172. Chang C., Adegoke C., Sellig F., (1980), 'Geotrack Model for Railroad Track Performance', *ASCE, Journal of Geotechnical Engineering*, Vol. 106, No. 11.

173. Zienkiewicz O., (1980), *The Finite Element Method in Engineering Science*, McGraw-Hill.

174. Ferreira P., López-Pita A., (2013), 'Numerical Modeling of High-Speed Train/Track System to assess Track Vibrations and Settlement Prediction', *ASCE, Journal of Transportation Engineering*, Vol. 139, No. 3.

175. ORE, D 71, RP 9, RP 10, (1978), *Stress in the Track, Ballast and the Subgrade under the Action of Repeated Loading*, Utrecht.

176. Eisenmann J., (1977), *Die Schiene als Träger und Fahrbahn*, Verlag Ernst, Berlin.

177. Zienkiewicz O., Valliapan S., King, I., (1969), 'Elastoplastic Solutions of Engineering Problems. Initial Stress-Finite Element Approach', *Journal for Numerical Methods in Engineering*, Vol. 1.

178. Zimmermann H., (1941), *Die Berechnung des Eisenbahnoberbaues*, Third Edition, Wilhelm Ernst und Sohn, Berlin.

CHAPTER 9

179. Morteza E., Hamidreza H.-N., (2013), 'Investigating Seismic Behavior of Ballasted Railway Track in Earthquake Excitation using FEM in Three Dimensional Space', *ASCE, Journal of Transportation Engineering*, Vol. 139, No. 1.

180. Budhima Indaratna, (2010), 'Field Assessment of the Performance of Ballasted Rail Track with and without Geosynthetics', *ASCE, Journal of Geotechnical and Geoenvironmental Engineering*, Vol. 136, No. 7.

181. Dhamge N., Atmapoojya S., Kadu M., (2012), 'Genetic Algorithm Driven ANN Model for Runoff Estimation', *Procedia Technology*, Vol. 6, 2012.

182. Quero D., Doan V.-T., (2002), 'Prise en Compte de l'Aléas Sismique de la Ligne du TGV Méditerranée', *RGCF*, February 2002.

183. Perlet J., (2002), 'Les Aménagements Hydrauliques de la Ligne du TGV Méditerranée', *RGCF*, February 2002.

184. Bowles J., (2001*), Foundation Analysis and Design* – 5th Edition, McGraw-Hill, New York.

185. Profillidis V., (2000), 'The Reinforcement Effect of Geotextiles in Railway Subgrades', *Rail International*, No. 7.

186. UIC, Fiche 719R (2008), *Ouvrages en Terre et Couches d' Assise Ferroviaires*, Paris.

187. UIC, Code 723R (1992), *Selection and Use of Weedkillers alongside Railway Tracks from the Standpoint of Environment Protection*, Paris.

188. Carter M., Bentley S., (2016), *Soil Properties and their Correlations*, Wiley.

189. UIC, Code 722R (1990), *Methods of Improving the Track Formation of Existing Lines*, Paris.

190. Profillidis V., (1985), *Geotextiles – Mechanical and Hydraulic Behavior – Applications*, Textbook, Thessaloniki.

191. Profillidis V., Kouparoussos A., (1984), 'Mechanical Behavior of the Railway Subgrade', KEDE, *Scient. Bulletin of the Ministry of Public Works of Greece*, Vol. 3-4, Athens.

192. Rowe K., (1984), 'Reinforced Embankments: Analysis and Design', *ASCE, Journal of Geotechnical Engineering*, Vol. 110, No. 2.

193. Société Nationale des Chemins de Fer Français, (1982), *Ouvrages en Terre Armée*, Paris.
194. Li X., Shi Y., (2019), 'Seismic Design of Bridges against Near-Fault Ground Motions Using Combined Seismic Isolation and Restraining Systems', *Shock and Vibration*, Vol. 2019.
195. ORE, D 117, RP 15, RP 16, (1981), *Filtration et Drainage*, Utrecht.
196. Sauvage, R., Langlade, J., (1981), 'L' Utilisation des Géotextiles dans les Plates-formes Ferroviaires de la SNCF', *RGCF*, July-August 1981.
197. Rankilor D., (1981), *Membranes in Ground Engineering,* John Wiley.
198. Hartmark H., (1979), 'Frost Protection of Railway Lines', *Engineering Geology*, Vol. 13, No. 1-4.
199. UIC, Question 714 (1978), *Adaptation de la Plate-forme dans l' Optique des Circulations à Grande Vitesse et de l' Augmentation de la Charge par Essieu*, Paris.
200. Tirant P., Sarda J., (1965), 'Chargements Répétés des Sols Fins Compactés et Non Saturés', *Bulletin de Liaison des Laboratoires des Ponts et Chaussées*, July-August 1965.
201. Ayres D., (1961), 'The Treatment of Unstable Slopes and Railway Track Formations', *Journal of the Society of Engineers,* Vol. 52, No. 4.

CHAPTER 10

202. WRIST, (2015), *Innovative Welding Process for New Rail Infrastructures*, European Commission, Brussels.
203. European Standard EN 13674-1, (2017), *Railway Applications, Track, Rail*, European Committee for Standardization, Brussels.
204. Betegon B. et al, (2009), 'Nonlinear Analysis of Residual Stresses in a Rail Manufacturing Process by FEM', *Applied Mathematical Modelling*, Vol. 33, No. 1.
205. Innotrack, (2008), '*Innovative Track Systems*', Brussels.
206. Thyssen, (2005), *Rail Sections*.
207. UIC, Leaflet 721, (2005), *Recommendations for the Use of Rail Steel Grades*, Paris.
208. ORE, D 185, RP 3, (1997), *Theoretical Modelling of Rail Corrugations and Validation by Measurement*, Utrecht.
209. Profillidis V., (1991), '*Mechanical Behavior of the Rail*', Professor G. Nitsiotas's Honorary Volume, University of Thessaloniki.
210. Edel K., Ortmann R., (1990), 'Fracture – Mechanical Characteristics of Rail Materials', *Rail International*, August-September 1990.

211. Tassily E., (1987), 'Propagation des Ondes de Flexion dans la Voie Ferrée considerée comme un Milieu Périodique', *RGCF*, March 1987.

212. Profillidis V., (1986), 'Continuous Welded Rail', *Bulletin of Greek Civil Engineers*, No. 172, Athens.

213. UIC, Code 860 (1979), *Technical Specification for the Supply of Rails*, Paris.

214. Sperring D., Squiers J., (1983), 'Rail Wear and Associated Problems', *British Railway Track Course*.

215. Mair R., Groenhout P., (1981), 'Croissance des Defectuosités Transversales dues à la Fatigue dans le Champignon des Rails de Chemin de Fer', *Rail International*, February 1981.

216. Tounend P., (1980), 'Analyse de la Probabilité et Coût des Défauts en Forme de Tache Ovale dus à la Fatigue des Voies en Alignement et en Courbe dans des Conditions de Fortes Charges par Essieu', *Rail International*, July-August 1980.

217. ORE, D 141, RP 1, (1979), *Statistical Study of the Evolution of Rail Defects in Relation to the Medium Axle Mass*, Utrecht.

218. UIC, (1979), *Catalogue of Rail Defects*, Paris.

219. Dang Van K., Gence P., (1978), 'Evolution des Critères de Fatigue-Application au cas des Rails', *RGCF*, December 1978.

220. Fowler G., (1976), *Fatigue Crack Initiation and Propagation in Pearlitic Rail Steels*, Ph. D. Thesis, Univ. of California.

221. ORE, D 117, RP 3, (1973), *Rail Behavior in Relation to Operation Conditions*, Utrecht.

222. Eisenmann J., (1970), 'Stress Distribution in the Permanent Way due to Heavy Axle Loads and High Speeds', *AREA*, Vol. 71.

223. ORE, D 71, RP 2, (1966), *Stress Distribution in the Rails*, Utrecht.

224. Yasojima Y., Machii K., (1965), 'Residual Stresses in the Rail', *Permanent Way*, Vol. 8, No. 26.

225. Timoshenko S., Langer B., (1932), 'Stress in Railroad Track', *ASME*, Vol. 54.

CHAPTER 11

226. *Technical Specifications of some Manufacturers of Fastenings* (e.g. Nabla, Vossloh, Pandrol), (2021).

227. UIC, (2013), *Sustainable Wooden Railway Sleepers*, Paris.

228. Profillidis V., (2001), 'The Mechanical Behavior of the Railway Sleeper', *Rail International*, No. 1.

229. European Standard, (2016), *'Railway Applications – Track – Concrete Sleepers and Bearers'*, European Committee for Standardization, Brussels.

230. Ferdous W., Manalo A., Van Erp G., Aravinthan T., Kaewunruen S., Remennikov A., (2015), 'Composite Railway Sleepers – Recent Developments, Challenges, and Future Prospects', *Composite Structures*, Vol. 134.

231. Bonewitz W., Fuhrer G., (1992), 'Einsatz von Elastomeren bei Schienenbefestigung bei Eisenbahnen und Nahverkehrsbahnen', *Die Bundesbahn*, No. 3.

232. SATEBA, (1992), *Twin-Block Railway Sleepers*, Paris.

233. FIP (Fédération Internationale de la Précontrainte), (1987), *Concrete Railway Sleepers*, Thomas Telfod Editions, London.

234. Profillidis V., Poniridis P., (1986), 'The Mechanical Behavior of the Sleeper-Ballast Interface', *Computers and Structures*, Vol. 24, No. 3.

235. Lindsey D., (1983), 'Rail Track Fastenings', *Track Course*, RIA, London.

236. Buekette J., (1983), 'Concrete Sleepers', *Track Course*, RIA, London.

237. Squires J.H., Sperring D.G., (1983), 'Theory and Development of Resilient Pads', *Track Course*, RIA, London.

238. Ticoalu A., Aravithan T., Karunasena W., (2008), 'An Investigation of the Stiffness of Timber Sleepers for the design of Fibre Composite Sleepers', *20th Australasian Conference on the Mechanics of Structures and Materials*, December 2008.

239. UIC, Leaflet 863V, (1981), *Technical Specification for the Supply of Non-treated Track Support (Wooden Sleepers for Standard and Broad Gauge Track and Crossing Timbers)*, Paris.

240. American Railway Engineering Association, (1982), *Concrete Ties*.

241. ORE, D 71, RP 8, (1973), *Load Distribution under the Sleeper,* Utrecht.

242. ORE, D 71, (1973), *Sollicitation de la Voie, du Ballast et de la Plateforme*, Utrecht.

CHAPTER 12

243. Trinh V.N. et al., (2012), 'Mechanical Characteristics of the Fouled Ballast in Ancient Railway Track Substructure by Large-Scale Triaxial Tests'*, Soils and Foundations*, Vol. 52, No. 3.

244. Indraratna, B., Thakur, P.K., Vinod, J.S., (2010), 'Experimental and Numerical Study of Railway Ballast Behavior under Cyclic Loading', *ASCE, Journal of Geomechanics*, Vol. 10, No. 4.

245. Suiker A., M., Selig E., Frenkel R., (2005), 'Static and Cyclic Triaxial Testing of Ballast and Subballast', *ASCE, Journal of Geotechnical and Geoenvironmental Engineering*, Vol. 131, No. 6.

246. European Standard EN 13450, (2002), *Aggregates for Railway Ballast*, European Committee for Standardization, Brussels.

247. Schmutz G., (2000), '*Ballast and Re-use of old Ballast*', Rail International, July-August 2000.

248. ORE, D 182, RP 4, (1995), *Standardized Technical Specifications and Description of the Quality Assurance System for Railway Ballast*, Utrecht.

249. Profillidis V., (1988), '*Mechanical Behavior of the Railroad Ballast*', 1st Congress of Geotechnical Mechanics, Athens.

250. Gray, P.S. (1983), '*Structural Requirements and Specifications of Ballast*', Track Course, RIA, London.

251. Stewart H., Selig E., (1982), '*Predictions of Track Settlement under Traffic Loading*', 2nd International Heavy Haul Conference, Colorado Springs, September 1982.

252. Société Nationale des Chemins de Fer Français (SNCF), (1979), *Constitution de la Voie Courante*, Paris.

253. Raymond G., Davies J., (1978), 'Triaxal Tests on Dolomite Railroad Ballast', *ASCE, Journal of Geotechnical Engineering*, Vol. 104, No. 6.

254. Brown, S. (1978), 'Repeated Load Testing of a Granular Material', *ASCE, Journal of Geotechnical Engineering*, Vol. 104, No. 6.

255. López-Pita A., (1977), 'Analyse de la Déformabilité du Ballast au moyen d' Essais en Laboratoire', *Associación de Investigation del Transporte*, Madrid.

256. ORE, D 117, RP 5, (1974), *Deformation of Track Ballast under Repeated Loading*, Utrecht.

257. Shenton M.J., (1974), *Deformation of Railway Ballast under Repeated Loading Triaxial Test*, Soil Mech. Sec., British Railways Res. Dept.

CHAPTER 13

258. Jing G., Aela P., (2020), 'Review of the Lateral Resistance of Ballasted Tracks', *Journal of Rail and Rapid Transit*, Vol. 234, No. 8.

259. Cléon L.-M., Parrot M., Tran-Ha S., (2002), 'Les Vents Traversiers sur la Ligne à Grande Vitesse Méditerranée', *RGCF*, February 2002.

260. Moreau A., (1987), 'La Vérification de la Sécurité contre le Déraillement', *RGCF*, April 1987.

261. Profillidis V., (1987), 'Parametric Analysis of Transverse Track Resistance and Application to the Design of the Ballast Section', *Scientific Bulletin of the Ministry of Public Works of Greece*, Vol. 1-2, Athens.

262. UIC, Leaflet 720R, (1986), *Laying and Maintenance of Track made up of Continuous Welded Rails*, Paris.

263. ORE, C 138, RP 8, (1984), *Permissible Maximum Values for the Y- and Q- Forces and Derailment Criteria*, Utrecht.

264. ORE, B 55, RP 8, (1983), *Prevention of Derailment of Goods Wagons on Distorted Tracks*, Utrecht.

265. ORE, C 138, RP 7, (1982), *Influence des Variations Oscillatoires de la Charge d' Essieu sur la Valeur Maximale Admissible de l' Effort Transversale du Point de Vue de Déripage de la Voie*, Utrecht.

266. Erchkov O.P., Kartzev V.J., (1980), 'Recherches Théoriques et Expérimentales sur les Mouvements des Véhicules Ferroviaires Circulant à une Vitesse de 200 km/h et Exigences Relatives à l'Entretien des Lignes à Grande Vitesse', *Rail International*, December 1980.

267. ORE, C 138, RP 5, (1980), *Effect of Train Speed on the Permissible Maximum Value of Load ΣY=S from the Point of View of Track Displacement*, Utrecht.

268. ORE, D 117, RP 8, (1976), *Influence of Various Measures at the Lateral Resistance of an Unloaded Track*, Utrecht.

269. Sauvage R., Amans F., (1969), 'Railway Track Stability in Relation to Transverse Stresses exerted by Rolling Stock – A Theoretical Study of Track Behavior', *Rail International*, November 1969.

CHAPTER 14

270. Koc W., (2012), 'Design of Rail-Track Geometric Systems by Satellite Measurement', *ASCE, Journal of Transportation Engineering*, Vol. 137, No. 1.

271. Pu H., Zhang H., (2019), 'Maximum Gradient Decision-making for Railways based on Convolutional Neural Network, *ASCE, Journal of Transportation Engineering*, Vol. 145, No. 11.

272. Claverie G., Crosaz Y., (2002), 'L'Insertion Paysagère de la Ligne Nouvelle Méditerranée', *RGCF*, February 2002.

273. Wilk A., et al., (2020), 'Evaluation of the Possibility of Identifying a Complex Polygonal Tram Track Layout Using Multiple Satellite Measurements', *Sensors*, Vol. 20, No. 16.

274. SNCF, (1998), *Voies Etroites-Particularités de Pose et d'Entretien*, Paris.

275. Taillé J.-Yv., (1990), 'Naissance d'une Ligne Nouvelle-Les Etudes de Tracé', *RGCF*, Paris.

276. UIC, Leaflet 703R, (1989), *Layout Characteristics for Lines Used by Fast Passenger Trains*, Paris.

277. Eliou N., Kaliabetsos G., (2014), 'A New Simple and Accurate Transition Curve Type for Use in Road and Railway Alignment Design', *European Transport Research Review*, Vol. 6, No. 2.

278. Busdy R.H., Drake D.G.H., (1983) 'Feasibility Studies and Outline Design', *Track Course*, RIA, London.

279. Hofer M., (1964), *Absteken von Kreisbogen*, Springer.

CHAPTER 15

280. Marinov M. et al., (2014), 'Analysis of Rail Yard and Terminal Performances', *Journal of Transport Literature*, Vol. 8, No. 2,

281. UIC, (2004), *Maximum Permissible Wear Profiles for Switches*, May 2004.

282. Butzbacher Weichenbau Gesellschaft (BWG), (2004), *Switches, Crossings and Slip Points*, Berlin.

283. ORE, C 184, RP 4, (1996), *Tests on Different Types of Crossings. Presentation of Results, Conclusions and Recommendations for Improving the Geometry of Crossings*, Utrecht.

284. UIC, Standard 711R, (1990), *Geometry of Points and Crossings with UIC Rails Permitting Speeds of 100 km/h or more on the Diverging Track*, Paris.

285. Bourda A., (1991), 'Un Système d' Information pour les Postes d' Aiguillage et de Circulation', *RGCF*, January 1991.

286. Deutsche Bundesbahn, (1988), *Merkblatt für den Entwurf von Gleisanschlüssen*, Frankfurt.

287. ORE, C 138, RP 8, (1984), *Permissible Maximum Values for the Y- and Q- Forces and Derailment Criteria*, Utrecht.

288. Lugg P., (1983), 'Crossings and Turnouts', *Track Course*, RIA, London.

CHAPTER 16

289. Plasser and Theurer, (2021), *Information Material for Laying, Tamping, and Maintenance Equipment of the Track*, Vienna.

290. Speno, (2021), *Technical Manuals for Grinding Machines*, Geneva.

291. Tzanakakis K., (2013), '*The Railway Track and its Long Term Behavior*', Springer, Heidelberg.

292. Andrade R., Teixeira F., (2011), 'Uncertainty in Rail-Track Geometry Degradation', *ASCE, Journal of Transportation Engineering*, Vol. 136, No. 12.

293. UIC, (2018), *Guidelines, State of the Art and Integrated Assessment of Weed Control and Management of Railways*, Paris.

294. Durazo-Cardenas I. et al., (2018), 'An Autonomous System for Maintenance Scheduling Data-rich Complex Infrastructure: Fusing the Railways Condition, Planning and Cost', *Transportation Research Part C: Emerging Technologies*, Vol. 89.

295. Takikawa M., (2016), 'Innovation in Railway Maintenance, Utilizing Information and Communication Technology', *Japan Railway and Transport Review*, No. 67

296. Dermenghem J.-P., Bimain A., Vilette Fr., (2002), 'La Maintenance de l'Infrastructure des Lignes à Grande Vitesse', *RGCF*, No. 108.

297. UIC, (1992), *Factors affecting Track Maintenance Costs and their Relative Importance*, Paris.

298. Profillidis V., (1986), 'Basic Principles for the Track Maintenance Works', *Technika Chronika (Scient. Bullet. of Greek Engineers)*, Vol. 6, No. 3, Athens.

299. UIC, Code 720, (2005), *Laying and Maintenance of Track made up of Continuous Welded Rails*, Paris.

300. Lewis R., (1983), 'Track Recording Machines', *Track Course*, RIA, London.

301. ORE, C 9, RP 9, (1983), *Tolérances en Service Admises dans la Superstructure de la Voie en Relation avec son Etat et la Marche des Véhicules*, Utrecht.

302. Waghorn D.W., (1983), 'Weed Control', *Track Course*, RIA, London.

303. Wilmott D. J., (1983), 'New Track Construction', *Track Course*, RIA, London.

304. Janin G., (1982), 'La Maintenance de la Géometrie de la Voie', *RGCF*, June 1982.

305. ORE, D 117, RP 2, RP 7, (1973), *Etude de l' Evolution du Nivellement en Fonction du Trafic et des Paramètres d' Armement*, Utrecht.

CHAPTER 17

306. Lei X., Zhang B., (2011), 'Analysis of Dynamic Behavior for Slab Track of High-Speed Railway Based on Vehicle and Track Elements', *ASCE, Journal of Transportation Engineering*, Vol. 137, No. 4.

307. Matias S.R., Ferreira P.A., (2020), 'Railway Slab Track Systems: Review and Research Potentials', *Structure and Infrastructure Engineering*, Vol. 16, No. 12.

308. Lay E., Ablinger P., (2002), 'Feste Fahrbahn Köln-Rhein/Main: Eine richtige Entscheidung', *Eisenbahningenieur*, Vol. 53, No. 12.

309. Xu L., Yu Z., Shi C., (2020), 'A Matrix Coupled Model for Vehicle-Slab Track-Subgrade Interactions at 3-D Space', *Soil Dynamics and Earthquake Engineering*, Vol. 123.

310. Darr E., Fierbig W., (1996), 'Stand der Entwicklung und des Einbahns der Festen Fahrbahn', *ZEV Glasers Annalen*, Vol 120, No. 4.

311. Henn W.D., (1992), 'System Comparison Ballasted Track-Slab Track', *Rail Engineering International*, No. 2.

312. Profillidis V., Poniridis P., (1990), 'Non-linear Analysis of a Slab Track run by Metro', *Scientific Journal of the Greek Laboratory of Public Works*, Vol. 105-106.

313. Brown J., (1983), 'Continuous Slab Track', *Track Course*, RIA, London.

314. Sauvage R., Errieau J., (1970), 'Les Poses de Voie sans Ballast', *RGCF*, March 1970.

CHAPTER 18

315. Günay M., Korkmaz M.E., Özmen R., (2020), 'An Investigation on Braking Systems used in Railway Vehicles', *Engineering Science and Technology*, Vol. 23, No. 2.

316. Niu J., Sui Y., Yu Q., Cao X., Yuan Y., (2020), 'Aerodynamics of Railway Train/Tunnel System: a Review of Recent Research', *Energy and Built Environment*, Vol. 1, No. 4.

317. André D., (2002), 'Aérodynamique dans les Tunnels du TGV Méditerranée', *RGCF*, February 2002.

318. ORE, C 218, RP 2, (1998), Draft UIC leaflet for Scaled Train Operations, *Determination of Railway Tunnel Cross-sectional Areas for Scaled Trains on the basis of Aerodynamics Considerations*, Utrecht.

319. Tsujimura T., Takao K., Sato K., (1993), 'Recent Trends of Brake Disc Materials', *Japanese Railway Engineering*, Vol. 32, No. 3.

320. ABB, (1992), *Traction Vehicle Technic for All Applications*, Mannheim.

321. Choi J., Kim K., (2014), 'Effects of Nose Shape and Tunnel Cross-Sectional Area on Aerodynamic Drag of Train Travelling in Tunnels', *Tunneling and Underground Space Technology,* Vol. 41.

322. ORE, C 179, RP 1, (1990), *Applicability of Computational Fluid Dynamics to Railway Aerodynamic Problems*, Utrecht.

323. Boiteux M., (1990), 'Influence de la Vitesse et des différents Paramètres Constructifs sur l'Adhérence en Freinage', *RGCF*, July-August 1990.

324. Metzler J.-M., (1989), *Généralités sur la Traction*, ENPC, Paris.

325. SNCF, (1988), *La Dynamique du Mouvement des Trains*, Paris.

326. Wende D., (2003), *Fahrdynamik des Schienenverkehrs*, Vieweg+Teubner Verlag, Wiesbaden.

327. Metzler J.-M., (1983), *Les Grandes Vitesses Ferroviaires*, ENPC, Paris.

328. Bianchi C., (1980), 'Fenomeni Aerodinamici della Marcia Veloce in Galleria', *Tecnica Professionale*, February 1980, Roma.

329. UIC, (2016), *Codes for Braking*: 540 V, 544-1, 543 VE, Paris.

330. Wasilewski P., (2020), 'Frictional Heating in Railway Brakes: a Review of Numerical Models', *Archives of Computational Methods in Engineering*, Vol. 27, No. 1.

331. Powell J.P, Palacin R., (2015), 'Passenger Stability within Moving Railway Vehicles: Limits of Maximum Longitudinal Acceleration', *Urban Rail Transit*, Vol. 1, No. 2.

CHAPTER 19

332. Otegui J., Bahillo A., Lopetegi I., Diez L.E., (2017), 'A Survey of Train Positioning Solutions', *IEEE Sensors Journal*, Vol. 17, No. 20.

333. European Commission, (2014 for locomotives and passenger rolling stock, 2013 for freight rolling stock), '*Technical Specification for Interoperability relating to the Rolling Stock Subsystem*', EU 1302/2014, EU 321/2013, Brussels.

334. Vionnet R., Pouillart Th., Viet J., (2005), 'Détermination des Contraintes Résiduelles par Ultrasons dans les Roues à la SNCF', *RGCF*, May 2005.

335. Mira L., Andrade A., Gomes M., (2020), 'Maintenance Scheduling within Rolling Stock Planning in Railway Operations, *Journal of Rail Transport Planning and Management*, Vol. 14.

336. UIC Code 510-2, (2004), *Trailing Stock – Conditions Concerning the Use of Wheels of various Diameters with Running Gear of Different Types*, Paris.

337. Stevenot G., Demilly F., (2002), 'Les Possibilités d'Amélioration de la Durée de Vie des Roues de Chemin de Fer', *RGCF*, May 2002.

338. Profillidis V., (2001), 'Tilting Trains – Operational Characteristics and Impact on Travel Time', *Public Transport International*, Vol. 1.

339. Profillidis V., (1998), 'A Survey of Operational, Technical, and Economic Characteristics of Tilting Trains', *Rail Engineering International*, Vol. 2.

340. Raison J., (1998), 'Les Equipements de Frein des Rames TGV', *RGCF*, March 1998.

341. Okamoto I., (1998), 'How Bogies Work', *Japan Railway and Transport Review*, Vol. 18, December 1998.

342. Yamanaka T., (1995), 'Vehicle Design Concept Towards 21st Century', *Japanese Railway Engineering*, Vol. 32, No. 4.

343. Joly R., (1988), 'Circulation d'un Véhicule Ferroviaire en Alignement et en Courbe-Bogie à Essieux Auto-Orientés', *Rail International*, April 1988.

344. UIC, Leaflet 510-5, (2007), *Technical Approval of Monoblock Wheels-Application Document for Standard EN 13979-1*, Paris.

CHAPTER 20

345. Logan K.G., Nelson J.D., McLellan B.C., Hastings A., (2020), 'Electric and Hydrogen Rail: Potential Contribution to Net Zero in the UK', *Transportation Research Part D: Transport and Environment*, Vol. 87.

346. Pagès M., Courtois Ch., (2005), 'L' Alimentation Electrique de la LGV Est Européenne', *RGCF*, July - August 2005.

347. RGCF, *Traction Electrique*, December 2004.

348. UIC, Code 600 OR, (2003), *Electric Traction with Aerial Contact Line*, Paris.

349. UIC, Leaflet 60608, (2019), *Conditions to be Compiled with the Pantographs of Tractive Units used in International Services*, Paris.

350. Courtois Ch., Viviant G., Augros D., (2002), 'Les Installations Fixes de Traction Electrique du TGV Méditerranée', *RGCF*, February 2002

351. UIC, Code 799-1, (2000), *Characteristics of Direct Current Overhead Contact Systems for Lines Worked at Speeds of over 160 km/h and up to 250 km/h*, Paris.

352. Lacôte F., (1998), 'Les Mutations du Matériel et de la Traction au XX Siècle', *RGCF*, July-August 1998.

353. Gigch V., Duin V., Heijsker V., (1996), 'Sizing the Traction Power Supply System with the Aid of Probability Theory', *Rail International*, Jan. 1996.

354. Kobayaski T., Ikeda, K., (1994), 'Development of New Types of Contact Wire for High-Speed Train on Shinkansen', *Japanese Railway Engineering*, Vol. 34, No. 1.

355. Köck F., (1990), 'Fahrzeugdiagnose der ICE – Triebkopfe und anderer Hochgeschwindigkeitsfahrzeuge', *ETR*, No. 6.

356. Metzler J.-M., (1990), *La Traction Electrique*, ENPC.
357. Kumar S., Verma Y.K., (2016), 'Hydrail – The Future of Railway Network', *International Journal of Modern Trends in Engineering and Science*, Vol. 3, No. 12.
358. UIC, Leaflet 606-2, (1986), *Installation of 25 kV and 50 Hz Overhead Contact Lines*, Paris.
359. Suddards A.D., (1983), 'Electrification, Construction and Installation', *Track Course*, RIA, London.
360. Oliveros Rives F., Rodriguez Mendez M., Megia Puente M., (1983), *Tratado de Explotación de Ferrocarriles*, Editorial Rueda, Madrid.

CHAPTER 21

361. European Commission, (2012), '*Technical Specification for Interoperability relating to the Control-Command and Signaling Subsystems*', EC 172/2012, Brussels.
362. Wung Jun-Feng, (2011), 'New Train Control Systems Suitable for Trains with Speeds up to 350 km/h', *ASCE, Journal of Transportation Engineering*, Vol. 137, No. 5.
363. European Commission, (2016), *Capacity Assessment of Railway Infrastructure*, Brussels.
364. Rothengatter W., (1996), *Bottlenecks in European Transport Infrastructure*, Brussels.
365. Baba Y., Hiratsuka A., Sasaki E., Yamamoto O., Miyamoto M., (2012), 'Radio-based Train Control System', *Hitachi Review*, Vol. 61, No. 2.
366. Giannakos K., Profillidis V., (2001), 'Un Projet d' Interopérabilité pour l' Europe du Sud-Est' *Rail International*, No. 6.
367. Giannakos K., Profillidis V., (2001), 'Technical Aspects of Railway Interoperability', *1st National Conference on Recent Advances in Mechanical Engineering*, American Society of Mechanical Engineers, Patras.
368. Alias J., (1993), *La Voie Ferrée – Tome 2: Signalisation*, ENPC, Paris.
369. Joing M., Cozzi Br., (1993), 'Gestion des Risques à la SNCF – Le Cas du Contrôle de Vitesse', *RGCF*, May 1993.
370. Alias J., (1993), *La Voie Ferrée – Tome 3: Exploitation Technique et Commerciale*, ENPC, Paris.
371. Blanc A., (1990), 'Le Contrôle de Vitesse', *RGCF*, December 1990.
372. ENPC, (1988), *Signalisation Ferroviaire*, Presses de l'Ecole Nationale des Ponts et Chaussées, Paris.
373. UIC, Code 406 R, (2013), *Capacity*, Paris.

374. Landex A., (2009), 'GIS Analyses of Railroad Capacity and Delays', *Proceedings of ESRI User Conference*, San Diego, California.

375. Sameni M.K., (2012), *Railway Track Capacity: Measuring and Managing*, Ph.D. Thesis, Univ. of Southampton.

376. Discembre A., Ricci S., (2011), 'Railway Traffic on High Density Urban Corridors: Capacity, Signaling and Timetable', *Journal of Rail Transport Planning and Management*, Vol. 1, No. 2.

CHAPTER 22

377. Singhal V. et al., (2020), 'Artificial Intelligence Enabled Road Vehicle-Train Collision Risk Assessment Framework for Unmanned Railway Level Crossings', *IEEE Access*, Vol. 8.

378. Eurostat, (2021), *Railway Safety Statistics*, Brussels.

379. European Commission, (2020), *Repost on Railway Safety and Interoperability*, Brussels.

380. UIC, (1999), *Priority Issues in Railway Safety*, Paris.

381. INRETS, (2009), *Safer European Level Crossing Appraisal and Technology*, Paris.

382. U.S. Department of Transportation, (2017), *Railroad-Highway Grade Crossings Handbook*, Washington.

383. French Parliament, (2019), *Proposition pour l'Amélioration de la Sécurisation des Passages à Niveau*, Paris.

384. Office of Rail and Road Regulation, (2021), *Managing Level Crossings*, www.orr.gov.uk/health-and-safety/level-crossings.

385. McCollister G., Pflaum C., (2007), 'A Model to Predict the Probability of Highway Rail Crossing Accidents', *Journal of Rail and Rapid Transit*, Vol. 221, No. 3.

386. EU Agency for Railways, (2018), *Safety Management System Requirements for Safety Certification or Safety Authorization*, Brussels.

387. EU Agency for Railways, (2018), *ERTMS: Making the Railway System Work better for Society*, Brussels.

388. UIC, Code 762, (2005), *Safety Measures to be taken at Level Crossings on Lines operated from 120 to 200 km/h*, Paris.

389. Vilotijevic M., Lazarevic L., Popovic Z., (2017), 'Railway/Road Level Crossing Design – Aspect of Safety and Environment', *International Conference on Energy Management of Municipal Transportation Facilities and Transport*, Springer.

390. Government of New South Wales, (2010), *Level Crossings – Engineering Manual TMC 521*, Sydney.

CHAPTER 23

391. Janic M., (2020), 'Estimation of Direct Energy Consumption and CO_2 Emission by High Speed Rail, Transrapid Maglev and Hyperloop Passenger Transport Systems', *International Journal of Sustainable Transportation*, Vol. 15, No. 9.

392. National Aeronautics and Space Administration (NASA), (2021), *Surface Temperature Analysis*, Goddard Institute for Space Studies, New York.

393. Profillidis V., Botzoris G., Galanis A., (2014), 'Environmental Effects and Externalities from the Transport Sector and Sustainable Transportation Planning - A Review', *International Journal of Energy Economics and Policy*, Vol. 4, No 4.

394. European Environment Agency, (2018), *Emission of Air Pollutants from Transport*, Brussels.

395. European Environment Agency, (2018), *Specific Air Pollutants Emissions*, Brussels.

396. European Commission, (2019), *Energy, Climate Change, Environment and Climate Action*, Brussels.

397. UIC, CER, (2015), *Transport and Environment-Facts and Figures*, Paris.

398. European Commission, (2018), *Greenhouse Gas Emissions from Transport in Europe*, Brussels.

399. Gwilliam K., Shalizi Z., (1995), *Sustainable Transport: Priorities for Policy Reform*, The World Bank, Washington.

400. NASA, (2013), *Global Climate Change – Vital Signs of the Planet*, California Institute of Technology, Pasadena.

401. Smith R., (1998), 'Global Environmental Challenges and Railway Transport', *Japan Railway and Transport Review*, No 18.

402. Rodrigue J-P, (2013), *The Environmental Impacts of Transportation*, Routledge, New York.

403. International Energy Agency – UIC, (2017), *Railway Handbook, Energy Consumption and CO_2 Emissions*, Paris.

404. Botzoris G., Galanis A., Profillidis V., Eliou N., (2015), 'Coupling and Decoupling Relationships between Energy Consumption and Air Pollution from the Transport Sector and the Economic Activity', *International Journal of Energy Economics and Policy*, Vol. 5, No 4.

405. European Parliament, (2012), *Reducing Railway Noise Pollution*, Brussels.

406. Pandya G.H., (2003), 'Assessment of Traffic Noise and its Impact on the Community', *Journal of Environmental Studies*, Vol. 60, No 6.

407. Galilea P., Ortúzar J., (2005), 'Valuing Noise Level Reductions in a Residential Location Context', *Transportation Research Part D: Transport and Environment*, Vol. 10, No 4.

408. Banister D., Button K., (1993), *Transport, the Environment and Sustainable Development*, E-FN SPON, London.

409. Profillidis V., Botzoris G., Galanis A., (2014), 'Environmental Impacts of Transport Modes and Transportation Planning', *2nd International Conference on Advanced Scientific Results*, University of Žilina, Slovakia, June 2014.

410. Botzoris G., (2005), 'Internalization of External Costs of Transport and Transportation Planning', *3rd International Conference on Ecological Protection of the Planet Earth*, Istanbul, Turkey.

411. Themelin L., (2003), 'Bruit au Freinage. Point de vue du Fabricant de Produit de Friction', *RGCF*, November 2003.

412. International Energy Agency, (2020), *Key World Energy Statistics 2020*, Paris.

413. NASA Goddard Space Flight Center, (2020), *Arctic Sea Ice Minimum at Second Lowest on Record*, Greenbelt.

414. European Commission, (2020), *Energy Consumption and Energy Efficiency Trends*, Brussels.

415. United Nations Economic Commission for Europe, (2017), *Trans-European Railway High-Speed Master Plan Study*, Geneva.

416. Westin J., Kågerson P., (2012), 'Can High-Speed Rail offset its Embedded Emissions?', *Transportation Research Part D: Transport and Environment*, Vol. 17, No. 1.

417. Raison J., Viet J.-J., (2003), 'Bruit et Matériel Roulant. Le Couple Roue-Semelles Composites. Réduction du Bruit de Roulement', *RGCF*, October 2003.

418. Dumitriu M., Cruceanu I.C., (2017), 'On the Rolling Noise Reduction by Using the Rail Damper', *Journal of Engineering Science and Technology Review*, Vol. 10, No. 6.

419. Li Y., Li Z., (2018), 'Application Effect of Chinese High-Speed Railway Noise Barriers', *Noise and Vibration Mitigation for Rail Transportation Systems*, Springer, Cham.

420. Wang J., Rakha H.A., (2017), 'Electric Train Energy Consumption Modeling', *Applied Energy*, Vol. 193.

Index

633

For Product Safety Concerns and Information please contact our EU
representative GPSR@taylorandfrancis.com
Taylor & Francis Verlag GmbH, Kaufingerstraße 24, 80331 München, Germany

www.ingramcontent.com/pod-product-compliance
Lightning Source LLC
Chambersburg PA
CBHW052115230326
41598CB00079B/3680